Contributions to Management Science

The series *Contributions to Management Science* contains research publications in all fields of business and management science. These publications are primarily monographs and multiple author works containing new research results, and also feature selected conference-based publications are also considered. The focus of the series lies in presenting the development of latest theoretical and empirical research across different viewpoints.

This book series is indexed in Scopus.

Faiz Hamid · Deep Mukherjee
Editors

Data Science for Modeling Managerial and Socioeconomic Problems

Concepts, Techniques, and Applications

Springer

Editors
Faiz Hamid
Department of Management Sciences
Indian Institute of Technology Kanpur
Kanpur, Uttar Pradesh, India

Deep Mukherjee
School of Economics and Public Policy
RV University
Bengaluru, Karnataka, India

Department of Economic Sciences
Indian Institute of Technology Kanpur
Kanpur, Uttar Pradesh, India

ISSN 1431-1941 ISSN 2197-716X (electronic)
Contributions to Management Science
ISBN 978-981-97-9059-3 ISBN 978-981-97-9060-9 (eBook)
https://doi.org/10.1007/978-981-97-9060-9

This Springer imprint is published by the registered company Springer Nature Singapore Pte Ltd.
The registered company address is: 152 Beach Road, #21-01/04 Gateway East, Singapore 189721, Singapore

Dedicated to our parents

Preface

Data Science is an interdisciplinary field with roots in statistics. Three separate and overlapping domains encompass data science: domain expertise, referring to traditional training in an academic field from where data originates (e.g., economics, management); computer science skills, enabling the development and application of algorithms for data storage, processing, and visualization; and statistical expertise, which focuses on modeling and summarizing data. The advent of data science has revolutionized the way we approach complex managerial and socioeconomic problems, offering sophisticated models to derive meaningful insights from data. This book, *Data Science for Modeling Managerial and Socioeconomic Problems: Concepts, Techniques, and Applications*, is designed to serve as a reference volume that bridges the gap between theoretical advancements and practical applications. By bringing together the expertise of contributors from various fields, this edited volume provides a user-friendly guide to understanding and applying data science methodologies to real-world problems.

Why Another Book?

One cannot become an expert in data science by reading a single book due to the field's vastness. This book is structured to progressively build the reader's understanding on select modeling techniques, starting with foundational concepts and moving toward interesting applications. Each chapter presents a blend of theoretical insights and practical case studies, ensuring that readers not only learn the techniques but also see their relevance and implementation in real-world scenarios. Unlike many existing texts that emphasize either abstract theories or simplistic examples, this book delves into advanced models and illustrates their use through detailed, real-world case studies. Each chapter offers a deep dive into specific methodologies, providing not only the theoretical underpinnings but also step-by-step guides for implementation using various software tools such as R and Python. Whether you are a practitioner seeking practical insights or a researcher exploring new methodologies, we hope this book inspires and informs your journey in the fascinating field of data science.

Plan and Organization of the Book

This reference book is designed to bridge the gap between textbooks and journal articles presenting advanced theoretical developments in data science. However, this volume does not aim to provide comprehensive coverage of state-of-the-art data science methods. The 15 chapters in this book comprise a representative sample, covering some cutting-edge methods and important problems in Management Science and Social Sciences. The first ten chapters cover methods from Statistics and Econometrics, and the next five chapters are on selected Machine Learning and Fuzzy techniques.

Following is an overview of the book's contents, summarizing each chapter's key themes, methodologies, and empirical applications. Chapter 1 presents methods for copula-based dependence modeling. It discusses the properties, estimation, and practical applications of bivariate copulas, highlighting their relevance in modeling complex dependencies in real-world data. Empirical examples demonstrate the utility of copula-based approaches in various domains, from finance to environmental science. Chapter 2 explores the use of Conway-Maxwell-Poisson (COM-Poisson) regression for modeling count data and discusses the properties of the COM-Poisson distribution, estimation methods, and model extensions. An empirical application to bike-sharing systems in the United States demonstrate the effectiveness of COM-Poisson regression in analyzing count data with overdispersion.

The next two chapters address discrete choice models from a statistical perspective. Chapter 3 discusses the Generalized Ordered Response (GOR) model and addresses the limitation of fixed thresholds in traditional ordered models by allowing variable thresholds. It applies the mixed GOR model to analyze quarterly passenger demand data from 300 US airports, illustrating model estimation using SAS statistical software. Chapter 4 delves into Conjoint Analysis for environmental evaluation, providing a thorough review of theoretical foundations and practical survey designs. Using a case study on the willingness to pay for organic rice in the Philippines, it explains the estimation of random parameter logit models using R, highlighting critical attributes like price and environmental quality.

Chapters 5 and 6 focus on Spatial Data Analysis. Chapter 5 covers the basics of the technique, including methods for geo-visualizing data, examining spatial heterogeneity with choropleth maps, and analyzing statistical distribution. The chapter also augments OLS regression model to incorporate spatial information and applies the methods discussed to examine the impact of child marriage on child malnutrition in India. Chapter 6 presents Geographically and Temporally Weighted Regression and applies the technique to study demographic changes in Poland, demonstrating the method's ability to analyze spatio-temporal patterns using R, and exploring factors influencing local fertility levels and economic impacts.

Chapter 7 explores econometric modeling techniques for assessing market risk in the Indian stock exchange. It discusses measures of market risk, estimation methods, and model evaluation techniques. Empirical results highlight the importance of accurately modeling market risk for informed decision-making in financial markets. Chapter 8 explores another popular econometrics topic—causal inference—through the Neyman-Rubin model and matching techniques. This chapter covers propensity score matching and genetic matching, demonstrating their use in evaluating training programs with observational data, and providing practical guidance using the R Matching package. Chapter 9 discusses how one can integrate Machine Learning with Econometrics. The author focuses on two techniques—LASSO regression and K-means clustering—for econometric analysis. The chapter explains the principles of supervised and unsupervised learning, applying R to identify skill claims and map skill clusters among informal workers in the job market in India.

The next two chapters discuss how to model discrete choice data from machine learning and multivariate statistics approaches. Chapter 10 presents two data dimensionality reduction techniques—principal component analysis and exploratory factor analysis and their application in assessing bank customer satisfaction in India. Through factor analysis, researchers gain insights into the underlying dimensions of customer satisfaction and identify actionable strategies for improving service quality. Chapter 11 provides a basic understanding of logistic regression and support vector machines for classification tasks. It applies these models to identify potential loan customers for a bank and assess their performance using appropriate measures of accuracy. Visualization tools for choosing between models and their implementation using R and Python are also presented.

Chapter 12 presents a novel forecasting approach, adaptive learning forecasting, for improving forecast accuracy by incorporating forecast error learning. The chapter explains the theoretical framework of the technique and demonstrates its practical application to agricultural markets. It exhibits how multivariate adaptive learning improves over univariate approaches, providing detailed methodological guidance. Chapter 13 explores the use of decision trees and random forests for predicting rare events using data on child health in India. The chapter addresses data imbalance challenges, introduces resampling techniques, and evaluates prediction accuracy of the models using Precision-Recall curves.

Chapter 14 discusses anomaly detection methods such as Mahalanobis distance and autoencoder neural networks and demonstrates their application in real-time automated vehicle health monitoring. Through empirical analysis, readers learn how anomaly detection techniques can improve system reliability and performance. The last chapter of the volume explores fuzzy models—a distinct modeling approach that differs from traditional statistical and econometric methods. Fuzzy models are mathematical tools for representing vagueness in relationships between variables and imprecise data. Chapter 15 focuses on leveraging Gaussian membership functions to develop a fuzzy mathematical model for linear regression, surpassing the limitations of conventional approaches. To demonstrate its application, the chapter presents a case study modeling the intricate relationship between employment and poverty in Middle East North African countries.

The code and datasets used in most of the chapters are available online as supplementary material on the book's webpage, allowing readers to replicate the analysis on their own.

Possible Use of the Book

This reference book is primarily aimed at Masters and Ph.D. students of Economics and Management Sciences. Graduate students studying economics, business administration, finance, public policy, and even management engineering should find this volume valuable. Graduate students can directly use it, with the only prerequisite being an intermediate course in probability and regression. This book is not intended as an introduction to R, Python, or programming in general. It is assumed that the reader is already familiar with these programming languages and can perform basic tasks such as writing functions, assigning variables, manipulating data, etc. For foundational knowledge in these languages, readers might refer to *Introduction to Data Science: A Python Approach to Concepts, Techniques and Applications* (Springer) by Laura Igual and Santi Seguí, and *Beginning Data Science with R* (Springer) by Manas A. Pathak.

The extensive coverage of the current volume provides an instructor with a lot of flexibility in selecting 8–10 chapters from this book for a one-semester course on Applied Data Science for graduate students. The most salient feature of the volume is the availability of data and code for most chapters, facilitating hands-on learning. This book can also be used as a complementary reference along with other advanced-level econometrics and data science textbooks. By meticulously integrating theoretical insights with practical applications, this book offers a unique perspective on data science, making it an invaluable resource for researchers, practitioners, and students alike.

Acknowledgment

Last but not least, we extend our heartfelt gratitude to the authors of this volume for the excellent chapters they produced. This book would not have been possible without the kind contributions of these scholars. We greatly appreciate their patience in dealing with the requests and comments from the editors. We would like to extend our heartfelt thanks to the publishing editor, Nupoor Singh, for her unwavering confidence in the book proposal, her exceptional patience and cooperation during the book's development, and her willingness to provide assistance whenever needed.

As editors, our aim has been to curate a collection that not only showcases the current state-of-the-art in data science but also inspires future research and innovation in the field. We hope this volume inspires innovative thinking and equips you with the tools to apply data science techniques effectively to tackle managerial and socioeconomic problems.

Enjoy your journey to the world of Data Science!

Kanpur, India Faiz Hamid
Bengaluru, India Deep Mukherjee
December 2025

Contents

1 **Copulas and Dependence Modeling with Examples** 1
Hans Manner

2 **COM-Poisson Regression Model: Application to Bike Sharing in the United States** 31
Suneel Babu Chatla and Galit Shmueli

3 **Generalized Ordered Logit Model: Application to Airline Demand Modeling in the United States** 69
Sudipta Dey Tirtha and Naveen Eluru

4 **Conjoint Analysis: Application to Consumer Preference for Organic Rice in the Philippines** 91
Shihomi Ara Aksoy

5 **Spatial Data Analysis: Application to Child Malnutrition in India** 129
Zakir Husain and Mousumi Dutta

6 **Geographically and Temporally Weighted Regression: Exploring Demographic Changes in Poland** 175
Maria Kubara, Kateryna Zabarina, and Katarzyna Kopczewska

7 **Econometric Modeling of Market Risk: Application to Indian Stock Exchange** 201
Suparna Biswas and Rituparna Sen

8 **Causal Inference with Matching: Evaluation of Programs and Policies** 229
Vikram Dayal and M. Rahul

9 **LASSO Regression and K-means Clustering: Application to Indian Labor Market** 249
Nandana Sengupta

10 Dimensionality Reduction Techniques: Application to Indian Bank Customer Survey Data 277
Maharnab Naha, Kanish Debnath, and Karan Singh Khati

11 Classification Using Logistic Regression and Support Vector Machine: Application to Identify Potential Loan Customers 311
Usha Ananthakumar

12 Adaptive Learning in Forecasting: Application to European Agricultural Markets 345
Foteini Kyriazi, Dimitrios D. Thomakos, and Efthymios Xylangouras

13 Decision Trees and Random Forest to Predict Rare Events: Case of Child Health in India 371
Dweepobotee Brahma and Debasri Mukherjee

14 Anomaly Detection Methods: Application to Automated Vehicle Health Monitoring 391
Khyati Thakkar, N. Hemachandra, and Tilak Raj Singh

15 Fuzzy Linear Regression: Analyzing Employment-Poverty Relationship in MENA Countries 427
Belhadj Besma and Kaabi Firas

Chapter 1
Copulas and Dependence Modeling with Examples

Hans Manner

Abstract This chapter introduces copula-based methods for dependence modeling. It presents empirically most useful bivariate copulas, explores their properties and dependence measures, and explains their estimation from a practical perspective. Additionally, the chapter covers model selection, goodness-of-fit testing, marginal distribution modeling, and data preparation techniques. Advanced topics are also discussed, including copula-based models for bivariate binary choice problems and extensions to higher dimensions, such as elliptical copulas, vine copulas, and factor copulas. To facilitate hands-on learning, we provide a small MATLAB toolbox for basic copula modeling, accompanied by two illustrative examples—one using simulated data and another based on real-world data. The chapter concludes with references to additional freely available software resources.

1.1 Introduction

Modeling dependence and multivariate distributions with copulas is a popular approach in Statistics and Econometrics. In the last two to three decades, copulas have become a standard tool in empirical and theoretical research. Numerous scientific papers and several books[1] have been written on methods involving and applications using these *dependence functions*. A good paper giving an overview of copula modeling for time-series data is Patton (2012). Fields of application are very diverse, but finance is the most popular one. In fact, an entire book is dedicated to copulas

[1] Popular textbooks dedicated entirely to copulas are Nelsen (2006) and Joe (2014).

Supplementary Information The online version contains supplementary material available at https://doi.org/10.1007/978-981-97-9060-9_1.

H. Manner (✉)
Department of Economics, University of Graz, Graz, Austria
e-mail: hans.manner@uni-graz.at

F. Hamid and D. Mukherjee (eds.), *Data Science for Modeling Managerial and Socioeconomic Problems*, Contributions to Management Science,
https://doi.org/10.1007/978-981-97-9060-9_1

for finance, and even that does not provide a complete treatment of this area of application; see Cherubini et al. (2004).

This chapter has two goals. First, it introduces copulas as a general modeling tool without reference to any specific applications. The applicability of the introduced methods is the central theme and the practical implementation of the methods is explained. In particular, as functionality for copula modeling is not included in standard econometrics software, a MATLAB toolbox is provided that helps the reader get some first hands-on experience of copula modeling. Its use is demonstrated in Section 1.4. Second, this chapter briefly presents some advanced copula-based models without going too deep into the technical details.

So, in simple words, what is a copula? Several answers are correct and, together, should give an intuitive understanding of the concept. A copula is a mapping from the unit cube $[0, 1]^d$ to the unit interval $[0, 1]$. It is a multivariate distribution function with $U(0, 1)$ marginals. Moreover, it can be seen as the component of a multivariate distribution that captures the dependence structure among the variables. Then how are copulas used? First, they can be used to construct multivariate distributions in a flexible and easy way. This is useful whenever a multivariate distribution is required as a component of a specific model. Second, estimating an appropriate copula model helps to understand the dependence between variables. For example, dependence may be stronger for large realizations of the variables than for small ones. Third, copulas are often used to simulate artificial data from multivariate distributions, for example to perform Monte Carlo simulations.

1.2 Basic Copula Methods

Simply put, a copula is a multivariate distribution function for random variables that follow a standard uniform distribution. In Section 1.2.1, some important preliminary concepts are reviewed and in Section 1.2.2 bivariate copulas[2] will be introduced. The most important copulas for empirical research are presented in Section 1.2.3. Estimation, model evaluation, and simulation are discussed in Section 1.2.4.[3]

1.2.1 Preliminaries

Consider a continuous random variable X and let x be a realization of this random variable. The cumulative distribution function (CDF) of X is given by

[2] Conceptually, the extension to multivariate models is straightforward, but the specific models can be much more complicated. Section 1.3 discusses this extension.

[3] This section draws extensively from the open access work Manner (2007).

$$F_X(x) = P[X \leq x] = \int_{-\infty}^{x} f_X(t)dt, \tag{1.1}$$

where $f_X(x)$ denotes the probability density function (PDF) of X. An important result, which will be crucial for our following discussion, is known as the **probability integral transformation** (PIT):

$$U = F_X(X) \sim U(0, 1). \tag{1.2}$$

In words, a continuous random variable transformed by its CDF will always follow a standard uniform distribution, i.e., $P(U \leq u) = u$. It can be interpreted as a standardization in the sense that a random variable X is mapped to its quantiles U, thus eliminating all specific properties of its initial distribution F_X. This result is not only important to understand copulas, but it is extremely useful for two purposes: goodness-of-fit testing and the simulation from specific distributions. Both of these tasks are also essential for copula modeling in practice.

Consider a situation where a specific distribution F is assumed for a given sample x_i with $i = 1, \ldots, n$, and we want to test whether this assumption holds. Then one computes the PIT for the sample,

$$u_i = F(x_i), \quad i = 1, \ldots, n,$$

and performs a specific goodness-of-fit test for the null hypothesis that the u_i come from a $U(0, 1)$ distribution. Various tests exist for this task, including, among others, the Kolmogorov-Smirnov test and the Neyman's Smooth test. Easy-to-use routines for these tests exist in software such as MATLAB or R. Both tests are applied in the applications in Section 1.4 and are included in the software that comes with this chapter.

Random number generation from a distribution of choice can be done in the following way, known as the inversion method. Generate a sample of n pseudo random numbers[4] $u_i^*, i = 1, \ldots, n$, from a $U(0, 1)$ distribution. Applying the inverse of the PIT, assuming invertibility,

$$x_i^* = F^{-1}(u_i^*),$$

we obtain pseudo random numbers x_i^* from the distribution F.

[4] The name *pseudo* is due to the fact that a computer cannot generate a truly random number. These numbers are generated by deterministic algorithms ensuring that, for all relevant purposes, these numbers behave as if they were truly random.

Example: Exponential distribution
Consider the exponential distribution, parametrized by its mean β, with CDF

$$F(x) = 1 - \exp(-x/\beta) \tag{1.3}$$

and quantile function

$$F^{-1}(u) = -\beta \ln(1-u). \tag{1.4}$$

Thus, using (1.3) a sample from an exponential distribution can be transformed into $U(0, 1)$ variables, and using (1.4) a sample from a standard uniform distribution can be transformed to be exponentially distributed.

Consider a pair of random variables X and Y with joint CDF $H(x, y)$

$$H(x, y) = P[X \leq x, Y \leq y] = \int_{-\infty}^{x} \int_{-\infty}^{y} h(s, t) ds dt, \tag{1.5}$$

where $h(x, y)$ is the joint PDF of X and Y. Recall that the marginal PDF of X is given by

$$f_X(x) = \int_{-\infty}^{\infty} f(x, y) dy$$

and that the conditional PDF of Y given $X = x$ is

$$f(y|x) = \frac{f(x, y)}{f_X(x)}.$$

X and Y are independent if $h(x, y) = f_X(x) f_Y(y)$. Conceptually, bivariate distribution functions are fairly straightforward. The difficulty lies in the fact that the construction of suitable models is not straightforward and the available standard models may not always match the properties of the empirical data. Copulas are a tool that can make the flexible construction of multivariate distributions an easy task.

1.2.2 Copulas and Related Dependence Measures

The following result, known as **Sklar's Theorem**, forms the basis for the copula approach to modeling multivariate distributions. It states that for any bivariate distribution function H, there exists a copula function C from $[0, 1]^2$ to $[0, 1]$ such that

$$H(x, y) = C(F_X(x), F_Y(y)). \tag{1.6}$$

In words, the bivariate CDF of X and Y is given by their copula evaluated at the marginal distributions. For continuous distributions, the copula is uniquely determined. We recall from the PIT in equation (1.2) that $U = F_X(X)$ and $V = F_Y(Y)$ are uniformly distributed, so the Copula $C(u, v)$ is the joint CDF of the random variables U and V and consequently it has $U(0, 1)$ margins. The copula can also be interpreted differently: For given marginal distributions F_X and F_Y, and any choice of copula C, the function H in (1.6) is a multivariate distribution. Hence, multivariate distributions can be constructed by coupling any marginal distributions with a copula.

It is important to note that the PIT can be seen as a standardization of random variables that removes their marginal behavior. Consequently, the copula represents the component of the bivariate distribution that fully describes the dependence between X and Y in a scale-invariant way. By using copulas, we can model and measure dependence in a way that remains unaffected by monotonically increasing transformations.

The following result is a straightforward consequence of Sklar's theorem and gives a first answer to the question how a copula function can be constructed:

$$C(u, v) = H(F_X^{-1}(u), F_Y^{-1}(v)). \tag{1.7}$$

In words, if a bivariate distribution is evaluated at $x = F_X^{-1}(u)$ and $y = F_Y^{-1}(v)$, thus considering it as a function of u and v, we obtain a copula $C(u, v)$. Why would you want to do that if you already have a bivariate distribution function in the first place? The reason is that the marginal distributions F_X and F_Y may not be appropriate for a specific application, but the dependence structure implied by the copula C defined in (1.7) may be. Then one could take that copula and use it to couple appropriate marginal distributions to obtain a desirable bivariate distribution.

Copulas define the dependence between random variables in terms of both their strength and type. Traditionally, the dependence between two random variables is quantified using the linear correlation coefficient. However, this measure can be misleading when the underlying data does not come from an elliptical distribution. In such cases, it is often more appropriate to utilize copula-based measures of dependence, which are scale invariant. A particularly robust copula-based measure is *Kendall's* τ. This measure has gained popularity in the copula literature as an overall measure of dependence and is based on the concept of concordance. Specifically, consider two pairs of observations (x_i, y_i) and (x_j, y_j) drawn from the continuous random variables (X, Y). These pairs are termed *concordant* if $(x_i - x_j)(y_i - y_j) > 0$ and *discordant* if $(x_i - x_j)(y_i - y_j) < 0$. Thus, two random variables are considered concordant when large values of one variable tend to be associated with large values of the other, while similarly small values are associated with each other. Building on the concept of concordance, we can define the dependence measure *Kendall's* τ. The sample version of *Kendall's* τ is given by the difference between the fraction of concordant pairs and the fraction of discordant pairs in the sample. In the population,

Kendall's τ is defined as the difference between the probability of concordance and the probability of discordance:

$$\tau = \tau_{X,Y} = P[(X_1 - X_2)(Y_1 - Y_2) > 0] - P[(X_1 - X_2)(Y_1 - Y_2) < 0]. \quad (1.8)$$

Kendall's τ lies in $[-1, 1]$, where $\tau = -1$ implies perfect negative dependence, $\tau = 1$ perfect positive dependence, and $\tau = 0$ independence. It can be expressed in terms of the expected value of a copula as follows:

$$\tau_C = 4E(C(U, V)) - 1. \quad (1.9)$$

Therefore, we see that Kendall's tau is a copula-based dependence measure, as it only depends on the copula but not on the marginal distributions. Another popular copula-based measure of overall dependence is *Spearman's* ρ, which is defined as

$$\rho_S = Corr(U, V) = \frac{E[UV] - 1/4}{1/12} = 12 \int_0^1 \int_0^1 C(u, v) du dv. \quad (1.10)$$

Due to the fact that it is defined by the linear correlation between U and V, which are proportional to their ranks, it is also known as a *Spearman's rank correlation.* The cases $\rho_S = -1$, $\rho_S = 1$, and $\rho_S = 0$ also imply perfect negative, positive, and independence, respectively.

Perhaps more interesting are dependence measures that allow to measure not only the *strength*, but also the *type* of dependence. *Quantile dependence* for a quantile $q \in (0, 1)$ is defined as

$$\lambda_q = \begin{cases} P[U \leq q | U \leq q] = \frac{C(q,q)}{q}, & \text{for } q \in (0, 0.5] \\ P[U > q | U > q] = \frac{1-2q+C(q,q)}{1-q}, & \text{for } q \in (0.5, 1]. \end{cases} \quad (1.11)$$

Thus, this dependence measure represents the probability that one variable is below (or above) a specific quantile, given that the other variable is also below (or above) the corresponding quantile of its distribution; hence $\lambda_q \in [0, 1]$. To estimate it for a given sample, simply count fraction of pairs of observations for which u and v lie below (above) q and divide this fraction by q $(1 - q)$. A plot of quantile dependence for a grid of values of q ranging from 0 to 1 can give a good impression of whether dependence is symmetric, i.e., similar values for λ_q and λ_{1-q}, or not. Furthermore, for values of q close to 0 or 1 one gets an impression of the strength of dependence in the *tails*, i.e., for *extreme values*. In fact, the limits of λ_q as q goes to 0 and 1, if they exist, are defined as the coefficient of lower and upper *tail dependence*, respectively:

$$\lambda_L = \lim_{q \downarrow 0} \lambda_q$$
$$\lambda_U = \lim_{q \uparrow 1} \lambda_q.$$

The case where the coefficient of tail dependence equals 0 is called tail independence, for positive values we speak of dependent tails. Tail (in)dependence is a property of specific copulas.

1.2.3 Important Bivariate Copulas

Many bivariate copulas exist and are collected in the books Joe (1997) and Nelsen (2006). However, empirical research typically only features a limited amount of models that have proven to provide a good fit to real-world data and these are the models presented here. These models form the basis of most empirical applications and also serve as the building blocks for extending the models to higher dimensions; see, e.g., Patton (2006), Candelon and Manner (2010), Patton (2012), Hakwa et al. (2015), and Jaworski (2017) for papers relying on the below models. Before introducing the most popular copulas, however, we introduce the *independence copula* and the idea of rotating copulas, a very useful approach to transform existing copulas to ones that may be more appropriate for the data one wants to model.

Independence copula

This copula represents the case of independent random variables U and V. It is given by

$$\Pi(u, v) = uv.$$

While this is a very simple model, the case of independence must be allowed for whenever one is interest in modeling and testing for dependence. Note that this model does not have any parameters that need to be estimated.

Survival/rotated copulas

For a specific copula $C(u, v)$, consider the function

$$C^S(u, v) = u + v - 1 + C(1 - u, 1 - v).$$

This is known as the survival or rotated copula, i.e., the copula rotated by 180 degrees. Hence, the roles of the upper and the lower tail are reversed. For example, for a copula with dependence in the upper tail, but independence in the lower tail, the rotated copula has lower tail dependence and upper tail independence. The density is simply given by

$$c^S(u, v) = c(1 - u, 1 - v),$$

where c is the density of the original copula. When $c = c^S$ we call a copula (rotationally) symmetric; otherwise it is asymmetric. This transformation is very popular in applications of copula methods. For example, the Gumbel copula introduced below is, in principle, a very useful model with good properties, but it features stronger dependence for large values of the variables compared to small ones. By rotating the copula by 180° this property is reversed, making it a good model for financial return data that typically feature stronger dependence for losses than for gains.

Gaussian copula

The Gaussian, or normal, copula is derived from the bivariate normal distribution using the relationship in (1.7):

$$C_{Gaussian}(u, v; \rho) = \int_{-\infty}^{\phi^{-1}(u)} \int_{-\infty}^{\phi^{-1}(v)} \frac{1}{2\pi\sqrt{1-\rho^2}} exp\left(- \frac{s^2 - 2\rho st + t^2}{2(1-\rho^2)} \right) dsdt,$$

where ρ denotes the linear correlation coefficient of the associated bivariate normal distribution. A Gaussian copula with Gaussian marginal distributions is simply a bivariate normal distribution, but other margins can be coupled with this copula to obtain "new" bivariate distributions, an approach that has been used a lot to obtain models for pricing credit risk. It is symmetric and can be shown to have independent tails for $\rho < 1$. The model is widely used because of its close relationship with the bivariate normal distribution and is easy to simulate from. It is typically employed when the focus is not on tail dependence but rather on capturing the overall dependence using a simple model.

Note that that there exists a one-to-one relationship between ρ and Kendall's τ, which is given by

$$\tau = \frac{2}{\pi} arcsin(\rho).$$

Student copula

The Student t-copula is also defined using (1.7), but now by starting from a bivariate t-distribution. Its distribution function is given by

$$C_t(u, v; \rho, \nu) = \int_{-\infty}^{t^{-1}(u;\nu)} \int_{-\infty}^{t^{-1}(v;\nu)} \frac{1}{2\pi\sqrt{1-\rho^2}} \left(1 + \frac{s^2 + t^2 - 2\rho st}{\nu(1-\rho^2)} \right)^{-\frac{\nu+2}{2}} dsdt.$$

As before, ρ refers to the linear correlation coefficient of the underlying bivariate t-distribution with ν degrees of freedom. As $\nu \to \infty$, the Student t-copula converges to the Gaussian copula. The relationship between Kendall's τ and ρ remains the same as in the Gaussian copula. The upper and lower tail dependence coefficients, which are identical, are defined as:

$$\lambda = 2\left(1 - t_{\nu+1}\left(\frac{\sqrt{\nu+1}\sqrt{1-\rho}}{\sqrt{1+\rho}} \right) \right),$$

where t_ν denotes the CDF of the standard t-distribution with ν degrees-of-freedom. Consequently, λ is increasing in ρ and decreasing in ν.

The Student copula is an extremely useful copula due to the fact that it does not only capture the overall dependence, but because it allows for tail dependence. In contrast to the Gaussian copula, the tails of the implied distribution are therefore also dependent. In applications, this model typically provides a much better fit due to this property.

The following two copulas are the two most popular models from a class called *Archimedean* copulas. Each member of this class depends on an increasing, convex function φ, called the *generator function*, through $C(u, v) = \varphi^{[-1]}(\varphi(u) + \varphi(v))$, that determines the properties of the copula.

Clayton copula

The Clayton copula is given by

$$C_C(u, v; \theta) = (u^{-\theta} + v^{-\theta} - 1)^{-1/\theta}, \quad \text{with } \theta > 0.$$

This copula is asymmetric with upper tail independence, but positive lower tail dependence equals to $\lambda_L = 2^{-1/\theta}$. Hence, dependence is stronger for small u and v than for large values. This property makes it useful for financial returns where usually losses are more dependent than gains. The relationship between its parameter and Kendall's τ is given by $\tau = \frac{\theta}{\theta+2}$.

Gumbel copula

The Gumbel copula is given by

$$C_G(u, v; \theta) = exp(-[(-ln(u))^\theta + (-ln(v))^\theta]^{1/\theta}), \quad \text{with } \theta \geq 1.$$

It is asymmetric with upper tail dependence given by $\lambda_U = 2 - 2^{1/\theta}$, but the lower tails are independent. For Kendall's τ the relation $\tau = 1 - \frac{1}{\theta}$ holds. Thus, this model can be expected to fit data well where large values are more dependent than small ones.

Mixture copulas

The basic copulas above may not always be flexible enough to be appropriate for modeling that data of interest. However, it is possible to combine them into more general models that allow for a large degree of flexibility generating various types of dependence structures. Let $C_1(u, v; \theta_1)$ and $C_2(u, v; \theta_2)$ be two copulas. Then the mixture copula with mixing weight α is given by

$$C_{mix}(u, v; \theta_1, \theta_2, \alpha) = \alpha C_1(u, v; \theta_1) + (1 - \alpha) C_2(u, v; \theta_2).$$

Hence, an observation comes from C_1 with probability α and from C_2 with probability $1-\alpha$. For example, one could consider the mixture of a Gumbel copula with its rotated version to obtain the following model, called the Gumbel-mixture (GM) model:

$$C_{GM} = \alpha C_G(u, v; \theta_1) + (1-\alpha)(C_G(1-u, 1-v; \theta_2) + u + v - 1).$$

This model is generally asymmetric, but nests symmetry when $\alpha = 0.5$ and $\theta_1 = \theta_2$. The model has both upper tail dependence and lower tail dependence, specifically $\lambda_U = \alpha(2 - 2^{1/\theta_1})$ and $\lambda_L = (1-\alpha)(2 - 2^{1/\theta_2})$. Furthermore, the Gumbel copula tends to fit empirical data reasonably well, but its strict asymmetry property is often too restrictive. The GM model overcomes this restriction and can be considered a highly flexible and recommendable model. One can similarly define a mixture model of the Clayton copula with its rotated version leading to a similar model called the Clayton-mixture (CM) model.

Figure 1.1 presents contour plots of six different distributions based on the copulas introduced above that have standard normal marginal distributions and copulas with an overall dependence of Kendall's $\tau = 0.5$. The Student t-copula has degrees-of-freedom of $\nu = 3$. The Gumbel-mixture model has mixing parameter $\alpha = 0.5$ on the left and $\alpha = 0.7$ on the right, the latter value implying an asymmetric model with tail dependence in both tails. These contour plots nicely show the differences between the copulas, all of which have identical marginal distributions. Both Gaussian and Student copulas show symmetry; however, it can be seen that the Student copula has more probabilities in the tails. The Gumbel copula features stronger upper tail dependence, whereas the Clayton copula has more lower tail dependence. The mixture model can allow for symmetry (lower left plot) or tail asymmetry, in this case in the upper tail.

1.2.4 Estimation, Model Evaluation, and Simulation

Estimation of copulas is typically done using maximum likelihood estimation (MLE). Assuming a sample of independent and identically distributed (iid) observations (u_i, v_i) for $i = 1, \ldots, n$ from a given copula with density $c(u, v; \theta)$, the log-likelihood function is given by

$$\mathcal{L}_C(\theta) = \sum_{i=1}^{n} \ln c(u_i, v_i | \theta). \tag{1.12}$$

Maximizing $\mathcal{L}_C$ with respect to θ gives the maximum likelihood estimator and the asymptotic covariance matrix of $\hat{\theta}$ can be computed using the Fisher information matrix. However, in practice, things are a little more complicated because we do not directly observe the copula data (u_i, v_i), but a sample (x_i, y_i) that is treated as a realization from the joint distribution $H(x, y; \gamma_x, \gamma_y, \theta)$, where γ_x and γ_y are the

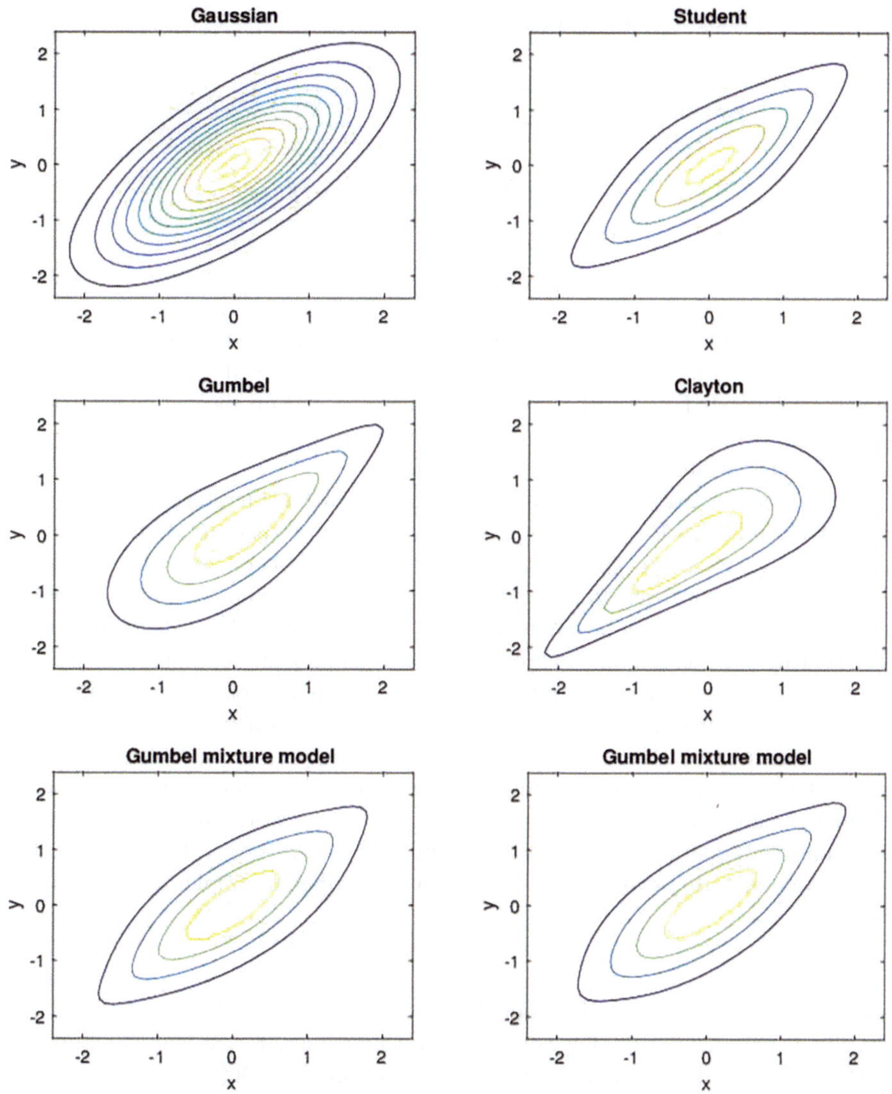

Fig. 1.1 Contour plots of selected copulas with $N(0, 1)$ marginal distributions

parameters of marginal distributions F_X and F_Y, respectively. Using the chain rule, the joint density h based on its copula representation (1.6) is

$$h(x, y; \gamma_x, \gamma_y, \theta) = f_X(x; \gamma_x) \cdot f_Y(y; \gamma_y) \cdot c(F_X(x; \gamma_x), F_Y(y; \gamma_y); \theta).$$

The corresponding log-likelihood function is

$$\begin{aligned}\mathcal{L}_{XY}(\gamma_x, \gamma_y, \theta) &= \sum_{i=1}^{n} \ln f_X(x_i; \gamma_x) + \sum_{i=1}^{n} f_Y(y_i; \gamma_y) \\ &\quad + \sum_{i=1}^{n} \ln c(\underbrace{F_X(x_i; \gamma_x)}_{=u_i}, \underbrace{F_Y(y_i; \gamma_y)}_{=v_i}; \theta) \\ &= \mathcal{L}_X(\gamma_x) + \mathcal{L}_Y(\gamma_y) + \mathcal{L}_C(\theta). \end{aligned} \tag{1.13}$$

Hence, the log-likelihood function consists of three parts, the log-likelihoods of the marginal distributions of X and Y, and the log-likelihood of the copula. The latter, however, depends both on the copula parameter θ and on the parameters of the marginals γ_x and γ_y. If the latter two were known, one could compute u_i and v_i using the PIT and evaluate the copula likelihood alone.

Three popular estimators have been proposed in the literature. The *exact* (or *full information*) MLE maximizes $\mathcal{L}_{XY}(\gamma_x, \gamma_y, \theta)$ with respect to all three (sets of) parameters γ_x, γ_y, and θ. This is the usual maximum likelihood estimator and has all its well-known properties. In particular, the estimator is asymptotically efficient and its covariance matrix can be estimated using the inverse of the Fisher information matrix. However, an alternative approach is computationally and practically easier to implement, namely the estimation in two steps. Two possibilities exist within this framework. The first approach is *fully parametric* and estimates γ_x and γ_y in a first step by maximizing $\mathcal{L}_X(\gamma_x)$ and $\mathcal{L}_Y(\gamma_y)$, respectively. In a second step, the copula parameter θ can be estimated by maximizing the joint likelihood conditional on the estimated marginal parameters, i.e., $\mathcal{L}_{XY}(\hat{\gamma}_x, \hat{\gamma}_y, \theta)$. In practice, this means that first a (conditional) distribution[5] is fit to the samples x_i and y_i for $i = 1, \ldots, n$. The PIT is used to transform the data into the uniformly distributed samples u_i and v_i) via

$$u_i = F_X(x_i; \hat{\gamma}_x)$$

and similarly for v_i. Then, the copula likelihood given in (1.12) is used to estimate the copula parameter. Both the exact and two-step parametric estimators require the marginal distributions of x_i and y_i to be correctly specified. Otherwise, u_i and v_i will not be uniformly distributed and $\hat{\theta}$ will not be a consistent estimator. Therefore, it is recommendable to perform a formal goodness-of-fit test of the null hypothesis; the u_i and v_i are $U(0, 1)$ distributed as mentioned in Section 1.2.1. Alternatively, graphical methods can be used to assess whether u_i and v_i are uniformly distributed. One can plot histograms of u_i and v_i and compare them to a straight line. Alternatively, a

[5] For example, regression models could be considered to model the marginal distributions before fitting a parametric distribution to its residuals. In case of heteroscedastic data, a location-scale model that models both the conditional mean and variance may be appropriate to standardize the data before fitting a parametric distribution.

quantile-quantile (qq) plot plots the empirical quantiles of u_i and v_i against those of the uniform distribution, which is the 45 degree line. Under correct specification, the empirical quantile of the observations should lie close to this line.

The second possibility for two-step estimation is the *semiparametric* estimator that estimates the marginal distributions non-parametrically and the copula parameter by MLE. In particular, the observations are transformed into [0, 1] via the empirical CDF:

$$\hat{u}_i = \hat{F}(x_i) = \frac{1}{n+1}\sum_{j=1}^{n} \mathbf{1}\{x_j \leq x_i\}.$$

In simple terms, it represents the proportion of observations in the sample that are less than or equal to x_i. The copula parameter θ is obtained by maximizing the copula likelihood in (1.12), with u_i and v_i replaced by $\hat{u}_i$ and $\hat{v}_i$. The non-parametric estimation of the marginals avoids any risk of model misspecification, making the procedure more robust and easier to apply. Both two-step approaches are popular in applied research as they are fairly easy to implement. The advantage of the parametric approach over the semiparametric one is that it specifies a fully parametric model that is easy to simulate from and that the properties of the marginal distributions are, in many cases, well understood.

In theory, both approaches are asymptotically less efficient than one-step MLE, but usually this is not a big concern. However, care must be taken when performing inference. Standard error estimates based on the Fisher information matrix corresponding to the copula log-likelihood (1.12) ignore the estimation in the first step and are therefore invalid. In the parametric case, the information matrix must be estimated using the "sandwich" for based on score and hessian of the full likelihood (1.13); see Joe (2005). Things are more difficult for the semiparametric estimator. Details on its theoretical properties can be found in Genest et al. (1995) and a simple step-by-step description on how to compute standard errors in Genest and Favre (2007). A standard iid bootstrap is an alternative approach to estimate the standard errors in both cases.

Two important aspects when modeling dependence with copulas are *model selection* and *goodness-of-fit testing*. Model selection refers to the selection of the best fitting model for a given dataset. A simple yet very reliable way of doing model selection is based on minimizing information criteria, for example the Akaike Information Criterion (AIC). It is calculated as negative two times the maximized log-likelihood statistic, plus two times the number of model parameters:

$$AIC = -2\mathcal{L}_C(\hat{\theta}) + 2\dim(\theta). \tag{1.14}$$

However, often a formal goodness-of-fit test or the graphical evaluation of the fit of a copula is desired. A simple method, analogous to the PIT approach to univariate

GoF testing, is based on the Rosenblatt PIT[6], which is particularly easy to perform for copulas. Consider the conditional copula of v given u:

$$C(v|u) = P[V \leq v|U = u] = \frac{\partial C(u, v)}{\partial u}. \tag{1.15}$$

The Rosenblatt PIT states that the random variable $T = C(V|u)$ follows a $U(0, 1)$ distribution. For a sample (u_i, v_i) for $i = 1 \ldots, n$ this means that $t_i = C(v_i|u_i)$ should be a sample from a $U(0, 1)$ distribution if the copula is correctly specified. Similarly to the univariate case, this can be tested with a formal GoF test or be assessed graphically by, e.g., a qq-plot. A qq-plot graphically displays the theoretical quantiles of a distribution against the empirical (realized) quantiles and a line close to a 45° line indicates an adequate fit.

The Rosenblatt PIT can, again analogously to the univariate case, be used to simulate artificial data from a bivariate copula in the following way:

1. Generate two independent uniform (0,1) variables u and t.
2. Set v = $C^{(-1)}(t|u)$, where $C^{(-1)}$ represents the quasi-inverse of $C(v|u)$.
3. The pair (u, v) then comes from the copula C.

Other approaches exist for simulating data from copulas that are more efficient. Implementations of such functions are readily available for MATLAB and R.

1.3 Advanced Methods

Three advanced copula modeling approaches are presented in this section. Section 1.3.1 shows how copulas can be used for modeling bivariate/multivariate binary choice variables. In Section 1.3.2, copula models for more than two dimensions are discussed and two important model classes are presented.

1.3.1 Copulas for Bivariate Binary Choice Modeling

Copulas are not only useful for modeling continuous random variables, but can also be useful in the discrete case. Here, problem of modeling two binary variables jointly is considered. Consider the binary variable y driven by the latent variable y^* as follows:

$$y = 1 \text{ if } y^* > 0,$$
$$y = 0 \text{ else},$$

---

[6] see Rosenblatt (1952).

where the latent variable is modeled as $y^* = z_1'\beta_1 + \varepsilon_1$. Hence, the latent variable y^* depends linearly on a set of regressors z_1 and on an error term ε_1. Let F_1 be the CDF of ε_1. Then

$$P(y = 1) = P(y* > 0) = P(\varepsilon_1 > -z_1'\beta_1) = 1 - P(\varepsilon < z_1'\beta_1) = 1 - F_1(-z_1'\beta_1)$$

The choice of distribution for F_1 defines the type of binary choice model. For instance, a normal distribution leads to a probit model, whereas a logistic distribution implies a logit model. However, other distributions are possible, in particular asymmetric ones.[7]

Now assume that we observe two binary variables y and x, driven by the latent variables

$$y^* = z_1'\beta_1 + \varepsilon_1 \tag{1.16}$$

$$x^* = z_2'\beta_2 + \varepsilon_2. \tag{1.17}$$

If the error terms ε_1 and ε_2 are independent, then y and x can each be modeled individually, for example, using a logistic regression model. However, if the errors are dependent, a joint model is needed. To specify such a model, we assume that the joint distribution of the error terms is defined using Sklar's theorem:

$$H(\varepsilon_1, \varepsilon_2) = C(F_1(\varepsilon_1), F_2(\varepsilon_2); \theta). \tag{1.18}$$

An appropriate choice of F_1, F_2 (e.g., logistic distribution), and C defines a bivariate binary choice model with a potentially flexible dependence structure for the errors of the underlying latent variables.

There are four possible outcomes that can be observed for the vector (y, x), namely $(1, 1)$, $(0, 0)$, $(1, 0)$, and $(0, 1)$. The four corresponding probabilities can be computed easily for the model. For example,

$$\begin{aligned} P(0,0) = P(y = 0, x = 0) = P(y^* \leq 0, x^* \leq 0) &= P(\varepsilon_1 \leq -z_1'\beta_1, \varepsilon_2 \leq -z_2'\beta_2) \\ &= C(F_1(-z_1'\beta_1), F_2(-z_2'\beta_2); \theta). \end{aligned}$$

The other three probabilities can be shown to be

$$\begin{aligned} P(1,1) &= 1 - F_1(-z_1'\beta_1) - F_2(-z_2'\beta_2) + C(F_1(-z_1'\beta_1), F_2(-z_2'\beta_2); \theta) \\ P(1,0) &= F_2(-z_2'\beta_2) - C(F_1(-z_1'\beta_1), F_2(-z_2'\beta_2); \theta) \\ P(0,1) &= F_1(-z_1'\beta_1) - C(F_1(-z_1'\beta_1), F_2(-z_2'\beta_2); \theta). \end{aligned}$$

---

[7] The usual formula for logit and probit models $P(y = 1) = F(z_1'\beta_1)$ uses the symmetry of F for the logistic and normal distributions.

For a given sample (y_i, x_i) for $i = 1, \dots, n$, $P(y_i, x_i)$ denotes the model implied probability of the corresponding outcome. The model's log-likelihood function is

$$\mathcal{L}(\beta_1, \beta_2, \theta) = \sum_{i=1}^{n} \ln P(y_i, x_i).$$

There are three main advantages of the model compared to a bivariate probit model that assumes a bivariate normal distribution for the joint distribution of the errors. First, different link functions/error distributions F_1 and F_2 (other than the standard normal) can be considered, which may be more appropriate in certain applications. Second, as mentioned above, the dependence implied by the copula may be more suitable than the implicitly assumed Gaussian copula of a bivariate normal distribution. Third, in terms of computational efficiency the advantage of the copula model compared to the classical bivariate probit model is the existence of closed-form expressions for the CDF of the copula. For large datasets, this difference in computation times can be huge.

Applications of this model can be found in the literature, such as in Winkelmann (2012) and Manner et al. (2016). The latter paper presents a generalization to dynamic binary choice models in a time-series context that can be applied in more than two dimensions.

1.3.2 Multivariate Copulas

In Section 1.3.2.1 and Section 1.3.2.2, two important model classes for copulas of more than two variables, applicable for rather large dimensions, are presented: factor copulas and D-vine models. These are important for many empirical applications as some copula models are often quite restrictive when extended to larger dimensions. The simplest approach to model dependence in higher dimension is to rely on the multivariate extensions of Gaussian and Student t-copulas, which can easily be obtained by the inversion method (1.7) applied to the multivariate Normal/ Student distribution. Let X_i for $i = 1, \dots, d$ be random variables with marginal CDF F_i and joint distribution function

$$F(x_1, \dots, x_d) = C(F_1(x_1), \dots, F_d(x_d)) = C(u_1, \dots, u_d).$$

Denote $\boldsymbol{u} = (u_1, \dots, u_d)$, where d denotes the dimension of the copula, and let R be a correlation matrix. The multivariate Gaussian copula is given by

$$C_{Gaussian}(\boldsymbol{u}; R) = \boldsymbol{\Phi}(\Phi^{-1}(u_1), \dots, \Phi^{-1}(u_d); R),$$

where $\boldsymbol{\Phi}$ and Φ denote CDF of the multivariate and univariate standard normal distributions, respectively. The multivariate Student t-copula is defined as

$$C_t(\boldsymbol{u}; R, \nu) = \boldsymbol{t}(t^{-1}(u_1; \nu), \ldots, t^{-1}(u_d; \nu); R, \nu),$$

where $\boldsymbol{t}$ and t denote the multivariate and univariate t-distributions, respectively. These multivariate *elliptical* copulas are parametrized by the $n(n-1)/2$ distinct elements in the correlation matrix R and a single degrees-of-freedom parameter ν in case of the Student copula. These models have three drawbacks. First, the number of parameters increases quadratically with the dimension of the distribution, whereas for the class of factor copula models presented in Section 1.3.2 it only grows linearly. Second, they imply symmetric dependence, a feature often not satisfied by real-world data. Third, although the Student copula allows for tail dependence, the degree-of-freedom parameter is the same for all pairs, which restricts the flexibility of the model. The symmetry property further restricts the tail dependence to be the same for both tails.

1.3.2.1 Factor Copulas

A factor copula is defined as the copula that is implied by a linear factor model and has been introduced and applied by Oh and Patton (2013, 2017).[8] Factor copulas are built upon a set of latent variables that follow a factor structure. Each latent variable is expressed as a linear combination of a smaller set of latent factors and an idiosyncratic term capturing individual characteristics. Consider the following model for the vector of **unobservable variables** Y_i, where $i = 1, \ldots, d$, involving a single common factor:

$$\begin{aligned} Y_j &= \beta_j Z + \epsilon_j, j = 1, 2, \ldots, d, \\ Z &\sim F_Z(\boldsymbol{\theta}_Z), \\ \epsilon_j &\sim \text{iid } F_\epsilon(\boldsymbol{\theta}_\epsilon), \text{ and } \epsilon_j \perp Z, \forall j. \end{aligned} \tag{1.19}$$

The joint distribution of the latent variables Y_j is given by

$$\boldsymbol{Y} = (Y_1, \cdots, Y_d)' \sim F_Y = C_{Factor}(G_1(Y_1; \boldsymbol{\theta}), \ldots, G_d(\boldsymbol{Y_d}; \boldsymbol{\theta}); \boldsymbol{\theta}). \tag{1.20}$$

The crucial part that one is interested in is the underlying copula C_{Factor}. In the model, Z is a common factor, β_j is its factor loading associated with the latent variable $Y_j, j = 1, \ldots, d$, and $\boldsymbol{\theta} = (\boldsymbol{\theta}'_Z, \boldsymbol{\theta}'_\epsilon, \beta_1, \ldots, \beta_d)'$ denotes the vector of the copula parameters. This model defines a multivariate joint distribution F_Y for the latent variables $\boldsymbol{Y}$, which in turn is defined through marginal distributions G_j and a Copula $C_{Factor}(\boldsymbol{u}; \boldsymbol{\theta})$. Hence, the model (1.19) implicitly defines a copula that is used as a model for the dependence of a set of observable variables X_j for $j = 1, \ldots, d$. Interest lies only in this copula C_{Factor} and not in the marginal distributions G_j. The advantage of the model lies in the fact that it is applicable in large dimensions, since

[8] Krupskii and Joe (2013) also introduce a related class of factor copula model that is in a way more general, but less intuitive and transparent than the model based on the linear factor structure.

the number of parameters increases linearly with the dimension d, whereas it grows quadratically in elliptical copulas. The model also offers a large degree of flexibility in terms of the dependence structure it implies due to the modeling choices that can be made regarding the distributions of the common and idiosyncratic components (F_Z and F_ϵ), the number of common factors that are chosen (Z could be a vector of multiple factors), and due to possible restrictions that can be placed on the factor loadings β_i. Choosing different distributions F_Z and F_ϵ, we obtain different copulas with different types of dependence structures. The easiest choice would be standard normal distributions for both. This would imply a Gaussian copula with a specific correlation structure. If t-distributions with common degrees-of-freedom parameter ν are chosen, the resulting copula model is the Student copula. However, if instead each of these t-distributions has a different degrees-of-freedom parameter, the copula is not a standard Student copula any more, but a more flexible model. The most popular choice suggested for the distribution of both the common and idiosyncratic variables is the skewed Student t-distribution defined by Hansen (1994):

$$t(z|\nu,\lambda) = \begin{cases} bc(1+\frac{1}{\nu-2}(\frac{bz+a}{1-\lambda})^2)^{-(\nu+1)/2} & z < -a/b, \\ bc(1+\frac{1}{\nu-2}(\frac{bz+a}{1+\lambda})^2)^{-(\nu+1)/2} & z \geq -a/b, \end{cases}$$

where $2 < \nu < \infty$, $-1 < \lambda < 1$. a, b, and c are constants: $a = 4\lambda c(\frac{\nu+2}{\nu-1})$, $b^2 = 1 + 3\lambda^2 - a^2$ and $c = \frac{\Gamma(\frac{\nu+1}{2})}{\sqrt{\pi(\nu-2)}\Gamma(\nu/2)}$. It is parametrized by two parameters: The parameter ν controls tail thickness, and λ determines skewness. A smaller ν corresponds to fatter tails. When $\lambda = 0$, the distribution reduces to the standard Student t-distribution. A positive λ indicates right skewness, while a negative λ implies left skewness. When this distribution is chosen to construct a factor copula, it implies a model with tail dependence and asymmetric dependence, which are two important features in empirical modeling.

This model has one important disadvantage, which is highly relevant for estimation: Its density is not known in closed form in most cases, and therefore, MLE is not an option. The solution is to rely on a method of moments type estimator. A set of dependence measures $\boldsymbol{m}$ is estimated for a sample and the copula parameters $\boldsymbol{\theta}$ are chosen in such a way that the vector of dependence measures that are implied by the model, denoted by $\boldsymbol{m}(\boldsymbol{\theta})$, is as close as possible to the estimated dependence measures $\hat{\boldsymbol{m}}$. Let $\boldsymbol{g}(\boldsymbol{\theta}) = \hat{\boldsymbol{m}} - \boldsymbol{m}(\boldsymbol{\theta})$ be the difference between the empirical and the model implied dependence measures. Then, the estimator is defined as

$$\hat{\boldsymbol{\theta}} = \arg\min_{\boldsymbol{\theta}} \boldsymbol{g}(\boldsymbol{\theta})'\widehat{\boldsymbol{W}}\boldsymbol{g}(\boldsymbol{\theta}), \tag{1.21}$$

with $\widehat{\boldsymbol{W}}$ an (estimated) weighting matrix. Hence, the estimator resembles a generalized methods of moments (GMM) estimator. Two important issues remain: (i) which dependence measures should be used as moment conditions and (ii) how can you compute $\boldsymbol{m}(\boldsymbol{\theta})$, the dependence measures implied by $C_{Factor}(\boldsymbol{u};\boldsymbol{\theta})$. Concerning (i), the common choice is to use Spearman's ρ defined in (1.10) and quantile dependence defined in (1.11) for a set quantiles spanning both tails of the distribution such as $q = (0.05, 0.1, 0.25, 0.75, 0.9, 0.95)$. Regarding (ii), these dependence measures cannot be computed directly for the factor copula model. The solution is to simulate from the model and compute the empirical dependence measures for the simulated data. This has to be done every time the loss function is evaluated for a different parameter value of $\boldsymbol{\theta}$, making the procedure computationally costly. Due to this simulation step, the estimator is termed simulated method of moments (SMM) estimator. The estimator is consistent and asymptotically normal; details can be found in Oh and Patton (2013).

1.3.2.2 D-vine Copulas

Vine copula models form a large class of copulas that can be constructed in arbitrarily large dimensions. Bivariate copulas are the building blocks in this approach based on sequential conditioning within a multivariate density. There are many different decompositions of a multivariate density and each implies a different kind of vine copula (also called *pair copula construction*). Here the focus lies on a particular class of models called D-vine copulas. In contrast to factor copulas, the joint density of vine copulas is readily available and estimation by maximum likelihood can be applied. In fact, the construction of the copulas through bivariate (conditional) copulas implies that the estimation can also be done in multiple steps in which each of these bivariate copulas can be estimated in a separate step. This makes the estimation and model building particularly easy in practice.

Let $(X_1, \ldots, X_d)$ be a vector of random variable that have joint distribution F and density f, respectively. The density f can be decomposed recursively by

$$f(x_1, \ldots, x_d) = \prod_{k=2}^{d} f(x_k|x_1, \ldots, x_{k-1}) \cdot f(x_1). \tag{1.22}$$

With the help of Sklar's Theorem, the conditional density $f(\cdot|\cdot)$ for dimension $d = 2$ can be rewritten as

$$f(x_1|x_2) = c_{12}(F_1(x_1), F_2(x_2)) \cdot f_1(x_1),$$

with c_{12} being a bivariate copula density. Extending this to dimension $d = 3$, one obtains

$$f(x_1|x_2, x_3) = c_{13|2}(F_{1|2}(x_1|x_2), F_{3|2}(x_3|x_2)|x_2) \cdot c_{12}(F_1(x_1), F_2(x_2)) \cdot f_1(x_1).$$

Here, $c_{13|2}$ denotes the copula of u_1 and u_3, conditional on u_2. The arguments of this conditional copula are the CDFs of X_1 and X_3 conditional on X_2.

Generalizing these decompositions and substituting into (1.22), it is possible to show that the joint density can be decomposed as

$$f(x_1, \dots, x_d) = \left[\prod_{j=1}^{d-1}\prod_{i=1}^{d-j} c_{i,i+j|(i+1):(i+j-1)}\right] \cdot \left[\prod_{k=1}^{d} f_k(x_k)\right], \tag{1.23}$$

where the conditional copulas in this expression are defined as

$$c_{i,j|D} := c_{i,j|D}(F(x_i|x_D), F(x_j|x_D)).$$

$D := \{i_1, \dots, i_k\}$ is a set of indices and $x_D := (x_{i_1}, \dots, x_{i_k})$ the corresponding set of x variables. Hence, the copula can be written as the product of bivariate conditional copulas and the conditioning set are those variables corresponding to the indices between i and $i + j$. Importantly, the arguments of the copulas are the values of the CDFs conditional on this same set of variables.

Vine copulas allow for a nice graphical illustrations in terms of a tree with nodes and edges. The nodes in the D-vine tree correspond to specific pairs of observations or pseudo-observations derived from the preceding trees, while the edges represent the associated pair copula densities with their respective copula parameters. For instance, consider a D-vine copula density with a dimension of $d = 5$:

$$c(u_1, \dots, u_5) = c_{12} \cdot c_{23} \cdot c_{34} \cdot c_{45} \cdot c_{13|2} \cdot c_{24|3} \cdot c_{35|4} \cdot c_{14|23} \cdot c_{25|34} \cdot c_{15|234}.$$

The bivariate unconditional and conditional copula densities in the above factorization can straightforwardly be derived from the corresponding D-vine tree depicted in Figure 1.2.

Taking the natural logarithm its density, it follows that the joint log-likelihood of the D-vine copula is the sum of the log-likelihoods for each pair. Therefore, the estimation is done one tree at a time and each pair copula can be estimated separately. The pseudo-observations in the higher trees are influenced by the pair copulas from the lower trees. Using the pseudo data $u_{j|D_{-i}}$ and $u_{i|D_{-i}}$ obtained from the previous tree (or simply u_i and u_{i+1} in the first tree), the pseudo-observations for the subsequent tree are calculated as:

$$u_{j|D} = h_{ji|D_{-i}}(u_{j|D_{-i}}|u_{i|D_{-i}}; \theta_{ij|D}),$$

where $h_{j|i}(u_j|u_i) := \frac{\partial C_{i,j}(u_i,u_j)}{\partial u_i}$ is the h-function corresponding to the pair copula C_{ij} and $D_{-i} := D \setminus \{i\}$ the appropriate conditioning set. Thus, the pseudo data can be computed recursively, beginning with the copula data from the first tree and subsequently utilizing the bivariate copulas and pseudo data from the lower trees.

More details on vine trees and vine copulas can be found in Aas et al. (2009) or Czado (2019). An application of D-vine copulas is given in Almeida et al. (2016).

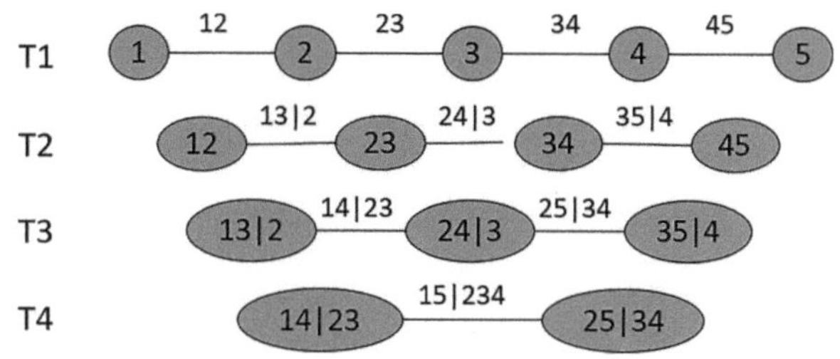

Fig. 1.2 5-dimensional representation of a D-vine tree. Source: Created by the author

1.4 Implementation

This section demonstrates the application of bivariate copula models with a simple copula toolbox for MATLAB going through the steps of the analysis. The provided software can be used for an application of the reader's choice along similar lines as the example below. The toolbox is available for download from the companion webpage of this book.

The toolbox contains the two main scripts `Example_simulated.m` and `Example_empirical.m` that perform the computations presented in Section 1.4.1 and Section 1.4.2, respectively. The reader may run and adapt scripts to get a better feeling for the methods, but also to model her own data of choice or to perform simulation studies. The most important functions are `MLECopula.m`, which allows for estimating copula models and computes the following quantities: $\hat{\theta}$; the dependence measures λ_L, λ_U and Kendall's τ implied by the estimated copula; $\mathcal{L}_C(\hat{\theta})$; AIC; and the p-values of the Kolmogorov-Smirnov test and Neyman's Smooth test based on the Rosenblatt PIT. The function `simCopula.m` can be used to simulate artificial data from a number of bivariate copulas. The folder named `Functions` contains many useful helper function to evaluate copula CDFs, PDFs, conditional CDFs (Rosenblatt PIT), and log-likelihood functions. The following eight copulas are supported: Gaussian, Student, Gumbel, Clayton, rotated Gumbel, rotated Clayton, Gumbel mixture, and Clayton mixture.

Section 1.4.1 presents an example with simulated data, Section 1.4.2 demonstrates the modeling of real-world data, and Section 1.4.3 very briefly discusses other freely available software for copula modeling.

1.4.1 Simulated Data

Consider the script `Example_simulated.m`, which implements the following steps:

1. Simulate artificial data (u_i, v_i) for $i = 1, \ldots, n$ from one of eight possible copulas. This step relies on the function `simCopula.m`, whose input requires a sample size n, a copula parameter (vector) θ, and the name of the copula of choice.

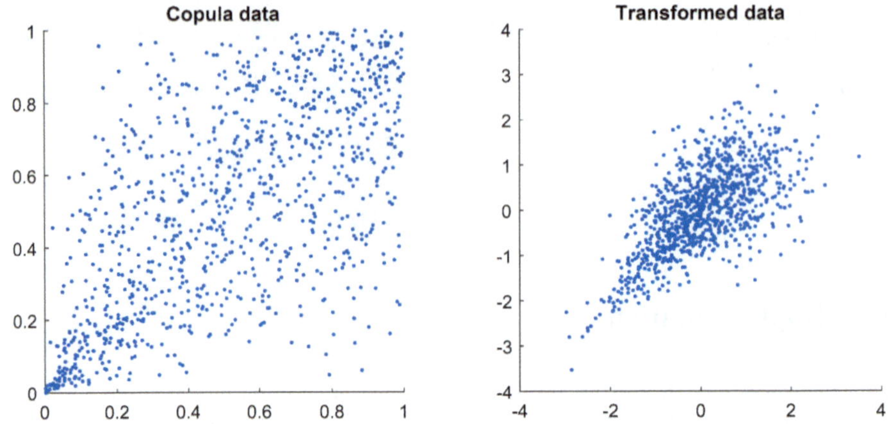

Fig. 1.3 Simulated data: Copula (left) and with standard normal margins (right). Source: Created by the author

2. Transform the data to a marginal distribution of choice via $x_i = F_x^{-1}(u; \gamma_x)$ and $y_i = F_y^{-1}(v; \gamma_y)$. This relies on the in-built functions of MATLAB's statistics toolbox and therefore allows for a large number of distributions. Scatterplots of (u, v) and (x, y) are automatically generated.
3. Estimate the marginal distributions, either parametrically using the function `fitdist.m` from the statistics toolbox or non-parametrically using the function `empiricalCDF`. The PIT is used to compute $(\hat{u}, \hat{v})$.
4. All eight available copula models are estimated using the function `MLECopula.m` and all output of the function is stored.
5. The estimation results are stored in a table that is printed on the screen and converted to a Latex table[9].

Figure 1.3 shows $n = 1000$ simulated data points from a Clayton copula with $\theta = 2$ with uniform margins (left) and standard normal margins (right). Recall that the Clayton copula has dependent lower tails and upper tail independence, which is clearly visible from the figure. Other marginal distributions could be chosen, which may make the visual determination of the type of dependence more difficult.

Table 1.1 presents the estimation results for all eight available copulas based on the two-step parametric estimator assuming knowledge that the marginal distributions are Gaussian. The AIC is lowest for the Clayton copula and it thus selects the true model as the best fitting one. The log-likelihood is slightly larger for the Clayton-mixture model, which nests the Clayton copula when $\alpha = 1$. All models characterized by lower tail dependence such as the rotated Gumbel or the two mixture models appear to have a good fit in terms of the log-likelihood statistics and AIC. This is confirmed by the two goodness-of-fit tests, which reject the models featuring symmetric dependence (Normal and Student) or only upper tail dependence

[9] Using the function `latexTable.m` obtained from the MATLAB file exchange.

Table 1.1 Estimation results for simulated data from a Clayton copula

	$\hat{\theta}_1$	$\hat{\theta}_2$	$\hat{\alpha}$	τ	λ_L	λ_U	LL	AIC	KS test	NS test
Normal	0.668	–	–	0.466	0.000	0.000	272.256	−542.512	0.080	0.000
Student	0.670	6.633	-	0.467	0.256	0.256	280.898	−557.796	0.153	0.002
Gumbel	1.690	–	–	0.408	0.000	0.493	204.300	−406.600	0.054	0.000
Clayton	1.726	–	–	0.463	0.669	0.000	334.270	−666.539	0.726	0.407
rot. Gumbel	1.949	–	–	0.487	0.573	0.000	329.076	−656.153	0.671	0.458
rot. Clayton	0.860	–	–	0.301	0.000	0.447	136.840	−271.681	0.039	0.003
Gumbel mixture	1.906	1.953	0.000	0.488	0.574	0.000	329.074	−652.147	0.702	0.486
Clayton mixture	1.772	1.429	0.961	0.468	0.650	0.026	334.864	−663.728	0.760	0.407

(Gumbel and rotated Clayton). The correctly specified models (Clayton and Clayton mixtures) are not rejected, but neither are the other two models that allow for tail dependence only in the lower tail (rotated Gumbel and Gumbel mixtures), which appears to approximate the true dependence structure reasonably well. Finally, note that the mixture copulas put almost all weight to the component that has lower tail dependence, showing the flexibility of these models.

Simulation exercises:

The following are suggestions for simulation exercises the reader may attempt for improving the understanding of the topic or that can be used as homework/ assignments by a course instructor.

1. Repeat the analysis varying the true copulas, marginal distributions, sample size, and parameters. Interpret your findings.
2. Rewrite the code so that the estimated marginal distribution is misspecified. How does this affect the estimated copula parameter and the selected best fitting model? How does this compare to the case when the margins are estimated non-parametrically?
3. Produce a qq-plot of the Rosenblatt PIT data to evaluate the copula goodness-of-fit. What does it look like for misspecified models?
4. Rewrite the code to perform different kinds of *Monte Carlo simulations*. Some suggestions are:

    - Compare the efficiency of the parametric and semiparametric two-step estimators with the estimator assuming perfect knowledge of the margins (i.e., estimating the copula using the true copula data (u_i, v_i)). Do this by simulating data from a copula of choice for a given sample size n and estimating the model with the three estimators. Repeat this for a large number of Monte Carlo replications (e.g., 1000) and compute the bias, variance, and mean square error of the estimators.

- Study the size and the power of the KS and NS tests for goodness-of-fit. Repeatedly simulating data from a model of choice, estimate the correctly specified and misspecified models. Compute the rejection frequency of the tests under the correct and misspecified models to estimate the size and power of the tests, respectively. Vary the sample size and try this for different models. Which test is more powerful? Do the tests have the correct size? If not, can you find an explanation why this is the case?
- Perform a simulation study to evaluate how reliable minimizing the AIC is for selecting the true model under different scenarios, varying the correct model, the sample size, the strength of dependence, and the number of candidate models.

1.4.2 Illustration Using Real Data

One of the most popular applications of copula methods is for modeling the joint returns of asset prices such as stock (markets), exchange rates (Patton 2006), or CDS spreads (Oh and Patton 2018). The estimated models then serve as a basis for the purpose of asset allocation and portfolio construction (Patton 2004), for studying contagion phenomena in financial markets (Rodriguez 2007; Candelon and Manner 2010), and for the evaluation of the riskiness of portfolio investments (Manner and Reznikova 2012; Hakwa et al. 2015; Almeida et al. 2016; Mensi et al. 2017). The empirical literature has shown that the asymmetric dependence structure captured by the copula models, in particular in the lower tail of the distributions measuring the dependence in unfavorable market conditions, leads to a better performance of these models compared to simpler models such as multivariate Gaussian distributions.

The steps in the empirical literature modeling asset returns typically are along the following lines. Consider the return series y_t for $t = 1, \ldots, T$ and model the conditional mean using an ARMA(p,q) model

$$y_t = \mu + \sum_{j=1}^{p} \phi_j y_{t-j} + \sum_{k=1}^{q} \theta_k \varepsilon_{t-k} + \varepsilon_t. \tag{1.24}$$

The error term ε_t is assumed to have time-varying volatility σ_t featuring volatility clustering that can be modeled using a GARCH(1,1) model (Bollerslev 1986):

$$\begin{aligned} \varepsilon_t &= \sigma_t z_t \\ \sigma_t^2 &= \omega + \alpha \varepsilon_{t-1} + \beta \sigma_{t-1}^2, \end{aligned} \tag{1.25}$$

where z_t is assumed to follow some known distribution F_Z with mean zero and variance one such as a Gaussian or a (standardized) Student t-distribution.[10] The combined ARMA-GARCH model can be estimated using maximum likelihood estimation and easy-to-use routines are available in all econometric software packages that deliver estimates for all model parameters, the time-varying conditional volatility $\hat{\sigma}_t$ and standardized residuals $\hat{z}_t$. Applying the probability integral transformation (1.2), one transforms the data into uniformly distributed copula data.

$$\hat{u}_t = F_Z(\hat{z}_t).$$

Plotting all the values of $\hat{u}_t$ against the fraction of observations being below these respective values (i.e., the empirical distribution function) produces QQ-Plots, where a line close to the 45° indicated a good fit of the model for the marginal distribution.

For an example modeling financial return data, consider the MATLAB script `Example_empirical.m`. A dataset is contained in the file `data.mat`, namely daily returns of the MSCI stock market index and exchange rates against the US dollar for the following 10 countries: Argentina, Brazil, Chile, Mexico, Peru, Australia, Canada, Russia, South Africa, and China. Furthermore, it contains returns of three commodity price indices, namely the CRB total, metal, and energy indices. The data span the period from January 2, 2019, until March 12, 2021. The script performs the following steps of analysis:

1. Load the data and select the variables of interest x and y.
2. Filter the data using a t-GARCH(1,1) model for the conditional heteroscedasticity. Standardize the data and transform to the uniformly distributed variables u and v. Produce qq-plots of u and v against a $U(0, 1)$ distribution. If the fit is unsatisfactory, the function `empiricalCDF.m` may be applied to model the margins non-parametrically.
3. All eight available copula models are estimated using the function `MLECopula.m` and all output of the function is stored.
4. The estimation results are stored in a table that is printed on the screen and converted to a Latex table.

For illustration, the code is applied to the MSCI stock index return of Australia and Canada. Figure 1.4 presents the qq-plots to evaluate the goodness-of-fit of the marginal distributions. We see that the assumed Student t-distribution provides an adequate fit to both standardized return series as the qq-plots are close to the 45 degree line.

Table 1.2 shows the estimation results for the copula models. In terms of the AIC, the Student- and Gumbel-mixture copulas provide the best fit for these data. On the other hand, both goodness-of-fit tests reject the null hypothesis that these two models are correctly specified. The Neyman's smooth test rejects all eight models,

[10] In the latter case, the model is called t-GARCH, which is the model applied in our empirical illustration. This model has degrees-of-freedom parameter $\nu > 2$ measuring the fatness of the tails implying that for smaller values of ν extreme realizations in the tails of z_t are more likely.

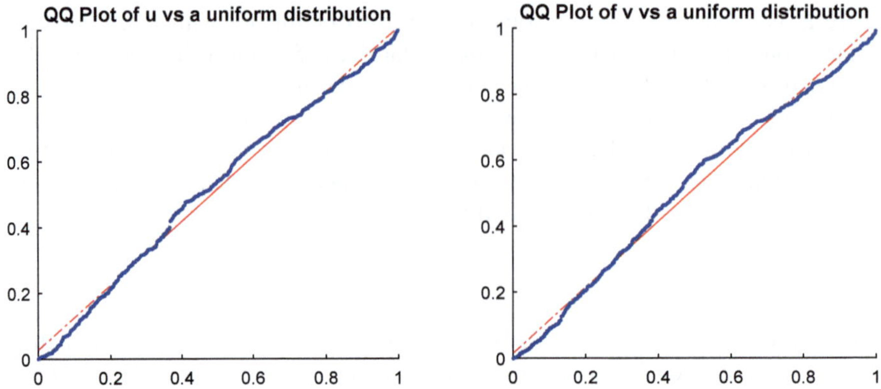

Fig. 1.4 QQ-Plots for margins of MSCI stock market index returns of Australia and Canada. Source: Created by the author

Table 1.2 Copula estimation for MSCI stock market index returns of Australia and Canada

	$\hat{\theta}_1$	$\hat{\theta}_2$	$\hat{\alpha}$	τ	λ_L	λ_U	LL	AIC	KS test	NS test
Normal	0.245	–	–	0.158	0.000	0.000	16.965	−31.929	0.030	0.007
Student	0.234	6.180	–	0.150	0.072	0.072	22.820	−41.640	0.031	0.001
Gumbel	1.210	–	–	0.174	0.000	0.227	19.178	−36.357	0.157	0.030
Clayton	0.240	–	–	0.107	0.056	0.000	16.572	−31.145	0.007	0.000
rot. Gumbel	1.140	–	–	0.123	0.163	0.000	19.029	−36.058	0.015	0.000
rot. Clayton	0.370	–	–	0.156	0.000	0.153	14.587	−27.175	0.120	0.018
Gumbel mixture	1.814	1.064	0.257	0.160	0.061	0.137	22.971	−39.942	0.058	0.004
Clayton mixture	0.422	0.380	0.368	0.165	0.071	0.122	21.789	−37.578	0.054	0.005

whereas the (typically less powerful) Kolmogorov-Smirnov test does not reject the Gumbel and rotated Clayton copula at the 10% level, nor the two mixture models at the 5% level. These results together with the parameter estimates of the two mixture copulas imply that upper tail dependence is stronger than the lower tail dependence. However, it is unclear which model should be selected in this case. Perhaps the Gumbel-mixture model would be recommendable due to its flexibility and overall acceptable fit.

Empirical exercises:

Readers are invited to model some of the other series contained in the dataset or use the code to model the dependence of data of their own choice. The following empirical exercises can serve as a basis for small empirical projects or homework.

1. Select different pairs of variables contained in the file `data.mat` and repeat the analysis. Which copulas seem to be the best fitting ones in general? Are there cases of negative dependence, a feature not supported by most of the copulas available in the software? How could you deal with this issue? Which pair has the strongest overall dependence? How do the results differ if the marginal distribution is modeled using the empirical CDF?
2. Download and import any data of your choice and adapt the code to estimate an appropriate bivariate distribution for that data. Is the data iid or does it require filtering with an appropriate location-scale model? What are appropriate models for the marginal distributions? Use the function `fitdist.m` from the statistics toolbox to estimate the margins. Evaluate their fit using qq-plots or the formal goodness-of-fit tests. Alternatively, use the function `empiricalCDF` to transform the variables to the $U(0, 1)$ distribution. Estimate the different copulas and select the best fitting model. Interpret the implied dependence structure.

1.4.3 *Other Software*

The code in the provided toolbox is not meant to be complete or sufficient for all possible empirical problems. Rather, it is intended for getting familiar with the steps in the analysis of copula-based modeling. A lot of software is available on the internet and a Google search will lead to various useful results. The "Statistics and Machine Learning Toolbox" in MATLAB offers a number of useful functions, which are similar to the ones contained in the toolbox of this chapter; see the documentation of MATLAB. A researcher who has published numerous papers on copula methods in econometrics is Andrew Patton and he provides a lot of useful code on his homepage: http://public.econ.duke.edu/ ap172/.

The toolbox MATVines documented in Coblenz (2021) allows for modeling vine copulas in MATLAB. The R-package CDVine is the most popular toolbox for modeling vine copulas in R, as introduced in Brechmann and Schepsmeier (2013). Much more code is available in R, such as the impressive package "copula" that can easily be found with a Google search.

1.5 Concluding Remarks

Copulas are a useful tool for constructing multivariate distributions and for modeling dependence. Popular dependence measures such as Kendall's τ or the coefficient of tail dependence depend only on the copula and not on the marginal distributions. Copulas are not only useful for modeling continuous variables, but can also be used to construct multivariate binary choice models. In dimensions larger than two vine copulas and factor copulas are recommended approaches. Flexible software for the implementation of copula methods is freely available, and the toolbox that accompanies this chapter may be useful to obtain some experience with this approach.

Applications of copula models are numerous. For example, they are used for financial risk management (Dissmann et al. 2013), pricing options or other complex financial products (van den Goorbergh et al. 2005), portfolio construction (Patton 2004), for constructing quantile regression models (Kraus and Czado 2017), multivariate density forecasts (Dovern and Manner 2020), or to handle endogeneity in regression models without the need for external instruments (Park and Gupta 2012).

Readers interested in learning more about copulas could start by reading the textbooks Joe (1997, 2014) and the research papers in the reference list below. This list is naturally selective and incomplete, because the literature on copula models and their applications is vast and continually growing.

References

Aas K, Czado C, Frigessi A, Bakken H (2009) Pair-copula constructions of multiple dependence. Insur: Math Econ 44(2):182–198
Almeida C, Czado C, Manner H (2016) Modeling high-dimensional time-varying dependence using dynamic d-vine models. Appl Stoch Model Bus Ind 32(5):621–638
Bollerslev T (1986) Generalized autoregressive conditional heteroskedasticity. J Econ 31(3):307–327
Brechmann E, Schepsmeier U (2013) Modeling dependence with c- and d-vine copulas: the r package cdvine. J Stat Softw 52:1–27
Candelon B, Manner H (2010) Testing for asset market linkages: a new approach based on time-varying copulas. Pac Econ Rev 15:364–384
Cherubini G, Luciano E, Vecchiato W (2004) Copula methods in finance. John Wiley and Sons Ltd
Coblenz M (2021) Matvines: a vine copula package for matlab. SoftwareX 14
Czado C (2019) Analyzing dependent data with Vine Copulas. Springer
Dissmann J, Brechmann EC, Czado C, Kurowicka D (2013) Selecting and estimating regular vine copulae and application to financial returns. Comput Stat & Data Anal 59:52–69
Dovern J, Manner H (2020) Order invariant tests for proper calibration of multivariate density forecasts. J Appl Econ 35:440–456
Genest C, Favre AC (2007) Everything you always wanted to know about copula modeling but were afraid to ask. J Hydrol Eng 12:347–368
Genest C, Ghoudi K, Rivest LP (1995) A semiparametric estimation procedure of dependence parameters in multivariate families of distributions. Biometrika 82:543–552
Hakwa B, Jäger-Ambrozewicz M, Rüdiger B (2015) Analysing systemic risk contribution using a closed formula for conditional value at risk through copula. Commun Stoch Anal 9(1):8
Hansen B (1994) Autoregressive conditional density estimation. Int Econ Rev 35(3):705–730
Jaworski P (2017) On conditional value at risk (covar) for tail-dependent copulas. Depend Model 1:1–19
Joe H (1997) Multivariate models and multivariate dependence concepts. CRC Press
Joe H (2005) Asymptotic efficiency of the two-stage estimation method for copula-based models. J Multivar Anal 94:401–419
Joe H (2014) Dependence modeling with copulas. CRC Press
Kraus D, Czado C (2017) D-vine copula based quantile regression. Comput Stat & Data Anal 110:1–18
Krupskii P, Joe H (2013) Factor copula models for multivariate data. J Multivar Anal 120:85–101
Manner H (2007) Estimation and model selection of copulas with an application to exchange rates. Research Memorandum RM/07/056, METEOR

Manner H, Reznikova O (2012) A survey on time-varying copulas: specification, simulations, and application. Econ Rev 31(6):654–687

Manner H, Türk D, Eichler M (2016) Modeling and forecasting multivariate electricity price spikes. Energy Econ 60:255–265

Mensi W, Hammoudeh S, Shahzad SJH, Shahbaz M (2017) Modeling systemic risk and dependence structure between oil and stock markets using a variational mode decomposition-based copula method. J Bank Financ 75:258–279

Nelsen RB (2006) An introduction to Copulas. Springer Science & Business Media, Berlin, Heidelberg

Oh DH, Patton AJ (2013) Simulated method of moments estimation for copula-based multivariate models. J Am Stat Assoc 108(502):689–700

Oh DH, Patton AJ (2017) Modeling dependence in high dimensions with factor copulas. J Bus & Econ Stat 35(1):139–154

Oh DH, Patton AJ (2018) Time-varying systemic risk: evidence from a dynamic copula model of CDS spreads. J Bus & Econ Stat 36(2):181–195

Park S, Gupta S (2012) Handling endogenous regressors by joint estimation using copulas. Mark Sci 31(4):567–586

Patton AJ (2004) On the out-of-sample importance of skewness and asymmetric dependence for asset allocation. J Financ Econ 2:130–168

Patton AJ (2006) Modelling asymmetric exchange rate dependence. Int Econ Rev 47(2):527–556

Patton AJ (2012) A review of copula models for economic time series. J Multivar Anal 110:4–18

Rodriguez JC (2007) Measuring financial contagion: a copula approach. J Empir Financ 14:401–423

Rosenblatt M (1952) Remarks on a multivariate transformation. Ann Math Stat 23(3):470–472

van den Goorbergh RWJ, Genest C, Werker BJM (2005) Bivariate option pricing using dynamic copula models. Insur: Math Econ 37:101–114

Winkelmann R (2012) Copula bivariate probit models: with an application to medical expenditures. Health Econ 21:1444–1455

Manner H, Reznikova O (2012) A survey on time-varying copulas: specification, simulations, and application. Econom Rev 31(6):654–68[illegible]
Manner H, Türk D, Eichler M (2016) Modeling and forecasting multivariate electricity price spikes. Energy Econ 60:[illegible]
[illegible]

Chapter 2
COM-Poisson Regression Model: Application to Bike Sharing in the United States

Suneel Babu Chatla and Galit Shmueli

Abstract Count data prediction is widely used in various domains, including transportation, marketing, and telecommunications. These datasets often exhibit diverse dispersion levels, ranging from overdispersion to equidispersion and underdispersion. The Conway-Maxwell-Poisson (COM-Poisson) distribution provides a flexible framework for modeling such data, encompassing Poisson, Bernoulli, and Geometric distributions within the exponential family. However, its predictive application is limited by computational complexity and the lack of closed-form expressions for the mean and variance. This chapter explores key extensions of the COM-Poisson regression model, including parametric and nonparametric approaches such as additive models, model-based recursive partitioning trees, and gradient boosting. The effectiveness of these methods is demonstrated in R software through a real-world case study on bike-sharing demand in Washington, DC.

2.1 Introduction

With the increasing advances in data collection technologies via sensors and the internet, and the pervasive use of mobile phones and other digital devices, count data are now widely available for many human and social behaviors and events. Examples include counts of the number of accidents, crimes reported, customers visiting a particular store, bicycle rentals, and more. Modeling such data and their relationship

---

Supplementary Information The online version contains supplementary material available at https://doi.org/10.1007/978-981-97-9060-9_2.

S. B. Chatla (✉)
Department of Mathematical Sciences, University of Texas, El Paso, TX, USA
e-mail: sbchatla@utep.edu

G. Shmueli
Institute of Service Science, National Tsing Hua University, Hsinchu, Taiwan
e-mail: galit.shmueli@iss.nthu.edu.tw

F. Hamid and D. Mukherjee (eds.), *Data Science for Modeling Managerial and Socioeconomic Problems*, Contributions to Management Science,
https://doi.org/10.1007/978-981-97-9060-9_2

with other variables is important for advancing knowledge through scientific research and for practical applications.

Poisson regression and negative binomial regression are two common regression methods for count data. The Poisson regression assumes that the data exhibits equidispersion—where the variance of the variable is equal to its mean. Poisson regression is therefore not suitable for capturing either overdispersion (where the variance is larger than mean) or underdispersion (where the variance is smaller than mean). Using the Poisson regression with over- or underdispersed data can therefore lead to erroneous results. The negative binomial regression is more flexible than the Poisson regression, as it can model overdispersion as well as equi-dispersion. However, it too is not suitable for data with underdispersion.

The Conway-Maxwell-Poisson (CMP) distribution (Shmueli et al. 2005) is a two-parameter extension of the Poisson distribution. It is flexible for modeling both overdispersed and underdispersed count data. The Poisson, geometric, and Bernoulli distributions are also included as special cases of the CMP distribution. Furthermore, the CMP belongs to an exponential family and retains the properties and advantages associated with that family. In practice, the type of dispersion is often unknown prior to fitting the model. Also, dispersion level can vary when modeling a count variable across different settings, such as modeling the number of customers at each of several store branches. With such uncertainty and diversity in dispersion levels in practice, the CMP provides an important extension of the count data modeling toolkit as it can handle all the dispersion types. Since the last decade, the CMP distribution has gained an increasing attention given its successes in multiple applications. The recent book by Sellers (2023) provides a comprehensive exposition of the CMP distribution, its historical developments, modeling extensions, and recent advances. In this chapter, we focus on the use of CMP-based regression models for prediction.

In this section, we start by introducing the CMP distribution, its basic characteristics, and alternative formulations. These will be the basis for the next sections, where we describe the parametric CMP regression and then nonparametric extensions that are especially valuable for prediction: additive models, model-based recursive partitioning trees, and gradient boosting.

2.1.1 The CMP Distribution

Suppose a positive and integer valued random variable Y follows CMP(λ, ν), then its probability mass function, for $\nu > 0, \lambda > 0$ and $\nu = 0, \lambda < 1$, is defined as

$$P(Y = y) = \frac{\lambda^y}{(y!)^\nu} \frac{1}{\zeta(\lambda, \nu)}, \qquad y = 0, 1, 2, \ldots, \tag{2.1}$$

where the normalizing constant

$$\zeta(\lambda, \nu) = \sum_{j=0}^{\infty} \frac{\lambda^j}{(j!)^\nu}, \tag{2.2}$$

has the following asymptotic form

$$\zeta(\lambda, \nu) = \frac{\exp(\nu\lambda^{1/\nu})}{\lambda^{(\nu-1)/2\nu}(2\pi)^{(\nu-1)/2}\sqrt{\nu}}(1 + O(\lambda^{-1/\nu})).$$

The asymptotic expression provides a good approximation for large values of $\lambda^{1/\nu}$. In practice, the normalizing constant is approximated by truncating the infinite series in such a way that the truncation error is bounded ($< 10^{-6}$) (Shmueli et al. 2005).

2.1.2 *Moments of the CMP*

The CMP distribution includes three known distributions as special cases: Poisson ($\nu = 1$), geometric ($\nu = 0, \lambda < 1$), and Bernoulli ($\nu \to \infty$ with probability $\lambda/(1+\lambda)$) (Sellers and Shmueli 2010). While each of these three special cases has closed form moments that enable us to easily write the mean and variance of Y, moments for the CMP distribution in general do not have explicit expressions due to the lack of explicit expressions for the normalizing constant $\zeta(\lambda, \nu)$ in (2.1). However, Shmueli et al. (2005) presented the following recursive form of the moments:

$$E(Y) = \begin{cases} \lambda[E(Y+1)]^{1-\nu}, & r = 0, \\ \lambda\frac{\partial E(Y^r)}{\partial\lambda} + E(Y)E(Y^r), & r > 0. \end{cases}$$

The CMP distribution belongs to the exponential family and hence possesses the characteristics and benefits that come with that family. Let $\boldsymbol{\theta} = (\lambda, \nu)$. The likelihood function of CMP distribution takes the following form (Lehmann and Casella 2006):

$$P_Y(y|\boldsymbol{\theta}) = h(y)\exp\left(\sum_{i=1}^{2}\eta_i(\boldsymbol{\theta})T_i(y) - A(\boldsymbol{\theta})\right),$$

where $A(\boldsymbol{\theta}) = \ln\zeta(\lambda, \nu)$, $h(y) = 1$, natural parameters $\eta_1(\boldsymbol{\theta}) = \ln\lambda$ and $\eta_2(\boldsymbol{\theta}) = -\nu$, with corresponding statistics $T_1(y) = y$, and $T_2(y) = \ln(y!)$, respectively (Shmueli et al. 2005; Chatla and Shmueli 2018). As a result, the expected value and variance can be approximated by

$$E(Y) = \frac{\partial\ln\zeta(\lambda, \nu)}{\partial\ln(\lambda)} \approx \lambda^{1/\nu} - \frac{\nu-1}{2\nu}, \text{ and}$$

$$\text{var}(Y) = \frac{\partial E(Y)}{\partial\ln(\lambda)} \approx \frac{1}{\nu}\lambda^{1/\nu}.$$

Several studies have proposed alternative parameterizations for the CMP distribution to provide an explicit expression for the first moment $E(Y)$, which may be modeled using a regression framework. More details may be found in Sellers and Premeaux (2021). We present here two formulations for the link functions which have been successfully considered in multiple studies of count data. The parameterization by Guikema and Goffelt (2008) uses $\mu_* = \lambda^{1/\nu}$ based on the fact that $E(Y^\nu) = \lambda$. This formulation leads to

$$P(Y = y) = \left(\frac{\mu_*^y}{y!}\right)^\nu \frac{1}{\zeta(\mu_*, \nu)}, \qquad y = 0, 1, 2, \ldots,$$

where

$$\zeta(\mu_*, \nu) = \sum_{j=0}^{\infty} \left(\frac{\mu_*^j}{j!}\right)^\nu \qquad \text{for } \nu > 0, \mu_* > 0 \text{ and } \nu = 0, \mu_* < 1. \tag{2.3}$$

This parametrization helps to obtain the following approximations for mean and variance (Sellers and Premeaux 2021)

$$\begin{aligned} E(Y) &= \frac{1}{\nu}\frac{\partial \ln \zeta(\mu_*, \nu)}{\partial \ln(\mu_*)} \approx \mu_* + \frac{1}{2\nu} - \frac{1}{2}, \\ \text{var}(Y) &= \frac{1}{\nu^2}\frac{\partial^2 \ln \zeta(\mu_*, \nu)}{\partial (\ln(\mu_*))^2} = \frac{\mu_*}{\nu}, \end{aligned}$$

where the approximation is accurate for $\mu_* > 10$ (Lord et al. 2008; Barriga and Louzada 2014). Similarly, Huang (2017) reparametrizes (2.1) as a function of the mean, which takes the following form:

$$P(Y = y) = \frac{\lambda(\mu, \nu)^y}{(y!)^\nu}\frac{1}{\zeta(\lambda(\mu, \nu), \nu)}, \qquad y = 0, 1, 2, \ldots,$$

where $\lambda(\mu, \nu)$ satisfies

$$\sum_{y=0}^{\infty} (y - \mu)\frac{\lambda^y}{(y!)^\nu} = 0.$$

This formulation possesses all the key features of the original CMP distribution (2.1). It still has the following three special cases: Poisson (μ) for $\nu = 1$, geometric ($p = 1/(\mu + 1)$) for $\nu = 0$, and Bernoulli ($p = \mu$) for $\nu \to \infty$, provided $\mu \leq 1$. Furthermore, this formulation offers coefficients in the same scale of Poisson and Negative Binomial distributions, allowing for easier comparison of the coefficients.

2.1.3 CMP for Prediction

In this chapter, we will describe a variety of CMP-based models that have been developed with a focus on parameter estimation and statistical inference. For the first time, we will consider these models in terms of their predictive performance—their ability to produce accurate predictions on new data, or *holdout data*, that were not involved in the model fitting step.

After describing each CMP model, we will illustrate its predictive performance (as well as its descriptive insights) on data from a bike-sharing service. Using a real dataset helps elucidate how the CMP can be used and its performance relative to classic models such as Poisson or Negative Binomial regression. We will use the original formulation (2.1) because the methods that will be discussed in this study use the original formulation. However, it is straightforward to implement these methods with other parameterizations.

2.2 Example: Using the CMP for Modeling Data from Bike-Sharing Systems

We consider data from a bike-sharing service to demonstrate the usefulness of CMP distribution for modeling count data in actual applications, similar to Chatla and Shmueli (2020). In bike-sharing systems, bicycles are made available for shared use to individuals on a short-term basis. They are similar to traditional bike rental services with the exception that the entire process including membership, bike rental and return services, is automated. They let individuals rent a bike from one dock and return them to another port that is controlled by a computer. Several studies have been conducted in order to better understand how to operate bike-sharing systems.

In general, bike-sharing systems collect information on each trip (time stamps and locations of rental and return) as well as the user (whether a member or not and information on demographics). We use a dataset from a bike-sharing service in Washington, DC, that includes ride-level data. The data is publicly available from Fanaee-T and Gama (2014) and includes hourly information on the number of rides by casual users and registered users. The data also includes external information such as whether it is a weekday, a working day, or a holiday, as well as weather situation (clear/cloudy/rainy), temperature, and wind speed. A full list of attributes and their descriptions are provided in Table 2.1.

For simplicity of exposition, we consider only the data from the month of November 2012 for estimating the models (*training data*), and we use data from December 2012 as a *holdout*, or *test data*, for evaluating predictive performance. The following `R` code (R Core Team 2021) is used to partition the data and to preprocess the variables.

Table 2.1 Full list of attributes and their description for the bike-sharing data (Chatla and Shmueli 2018)

Name	Description
dteday	date
season	season (1:spring, 2:summer, 3:fall, 4:winter)
yr	year (0: 2011, 1:2012)
mnth	month (1 to 12)
hr	hour (0 to 23)
day	day (1 to 30 or 31)
holiday	whether or not the day is a holiday (extracted from http://dchr.dc.gov/page/holiday-schedule)
weekday	day of the week
workingday	if day is neither weekend nor holiday is 1, otherwise is 0.
weathersit	1= Clear, Few clouds, Partly cloudy
	2= Mist + Cloudy, Mist + Broken clouds, Mist + Few clouds, Mist
	3= Light Snow, Light Rain + Thunderstorm + Scattered clouds
	4= Heavy Rain + Ice Pallets + Thunderstorm + Mist, Snow + Fog
temp	Normalized temperature in Celsius. The values are divided to 41 (max)
atemp	Normalized feeling temperature in Celsius. The values are divided to 50 (max)
hum	Normalized humidity. The values are divided to 100 (max)
windspeed	Normalized wind speed. The values are divided to 67 (max)
casual	count of casual users
registered	count of registered users
cnt	count of total rental bikes including both casual and registered

```
## reading bike-sharing data
hr=read.csv("Bike-Sharing-Dataset/hour.csv",header = T)
head(hr)

########### variable coding
hr$season=with(hr,factor(season,levels = c(1,2,3,4),
                         labels=c("spring", "summer", "fall", "winter")))
hr$yr=with(hr,factor(yr,levels = c(0,1),labels=c( "2011", "2012")))
hr$mnth=with(hr,factor(mnth,levels=c(1:12),
                       labels=c("jan","feb","mar","apr","may","jun",
                                "jul","aug","sep","oct","nov","dec")))
hr$holiday=with(hr,factor(holiday,levels=c(0,1),
                          labels=c("not holiday","holiday")))
hr$weekday=with(hr,factor(weekday,levels = c(0:6),
                          labels = c("sun","mon","tue","wed","thu","fri","sat")))
hr$workingday=with(hr, factor(workingday,levels = c(0,1),
                              labels = c("not working day","working day")))
hr$weathersit=with(hr,factor(weathersit,levels = c(1,2,3,4),
                             labels=c("clear","cloudy","light rain/fog","heavy rain/fog")))
hr$day=as.integer(substr(hr$dteday,9,10))

########### factors to binary coding
weathersit=sapply(unique(hr$weathersit), function(x) as.integer(x == hr$weathersit))
colnames(weathersit)[1:4]=c("clear", "cloudy" ,"lightrain","heavyrain")
holiday=sapply(levels(hr$holiday), function(x) as.integer(x == hr$holiday))
colnames(holiday)=c("notholiday","holidayc")
weekday=sapply(levels(hr$weekday), function(x) as.integer(x == hr$weekday))
## consolidated data
hr=cbind(hr,weathersit,holiday,weekday)
####### Training data: November 2012
nov12=subset(hr,yr=="2012"& mnth=="nov")
####### Validation: For tuning parameters in Boosting, Nov 2011
nov11=subset(hr,yr=="2011"& mnth=="nov")
####### For prediction / holdout: December 2012
dec12=subset(hr,yr=="2012"& mnth=="dec")
```

As an initial data exploration step, we created multiple visualizations to summarize the training data and identify its key patterns. Figure 2.1 shows the average hourly number of rides for both casual and registered users in November 2012. The average number of rides for casual users peaks at 2 p.m. In contrast, registered users' average number of trips has two peaks, at 8 a.m. and 6 p.m., indicating that registered users might be commuting to their offices. The lines placed above the bars represent $\pm$ one standard error. Figure 2.2 depicts the average number of rides per day for both casual and registered users throughout the month of November. The average daily number of rides taken by casual and registered users exhibits contrasting patterns. While for casual users we see peaks on weekends, for registered users we see peaks on weekdays. Figure 2.2 also plots the frequency of rides by weather type, showing that casual users use bikes mostly during clear weather, while registered users use bikes more evenly across all weather types.

Figure 2.3 shows the scatterplots between the number of rides and windspeed, temperature and humidity for weather conditions for both casual and registered users.

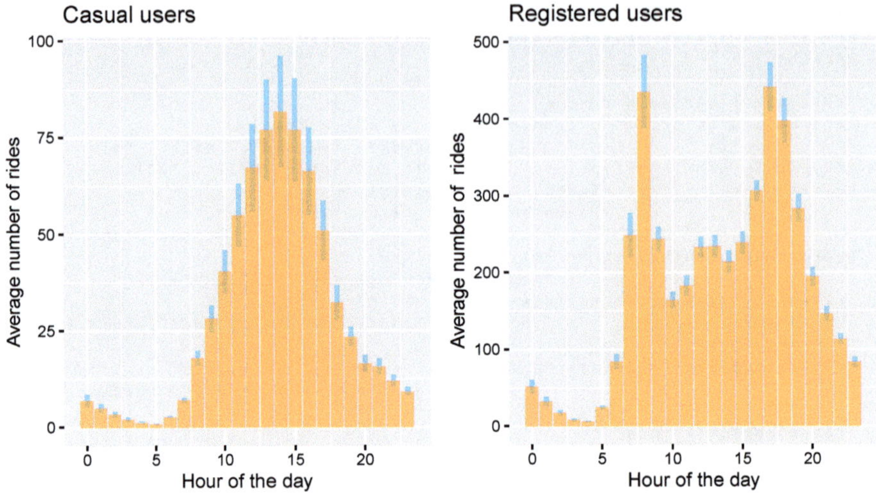

Fig. 2.1 Average number of rides per hour in November 2012. The blue lines indicate ± one standard error

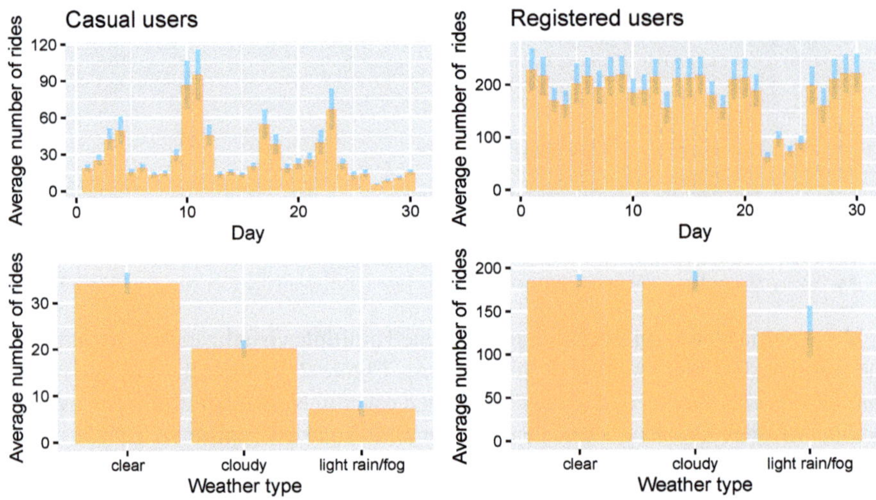

Fig. 2.2 Top: Average number of rides per day in November 2012. Bottom: Average hourly number of rides in November 2012 by weather type. The lines indicate ± one standard error

Temperature seems to have a positive association with the number of casual rentals while humidity displays a negative association. It seems that weather conditions may have little influence on the number of rides by registered users. We also mention that variables *atemp* and *temp* are highly correlated. Therefore, we consider only variable *atemp*. For the sake of simplicity, we mainly focus on casual users in further analysis.

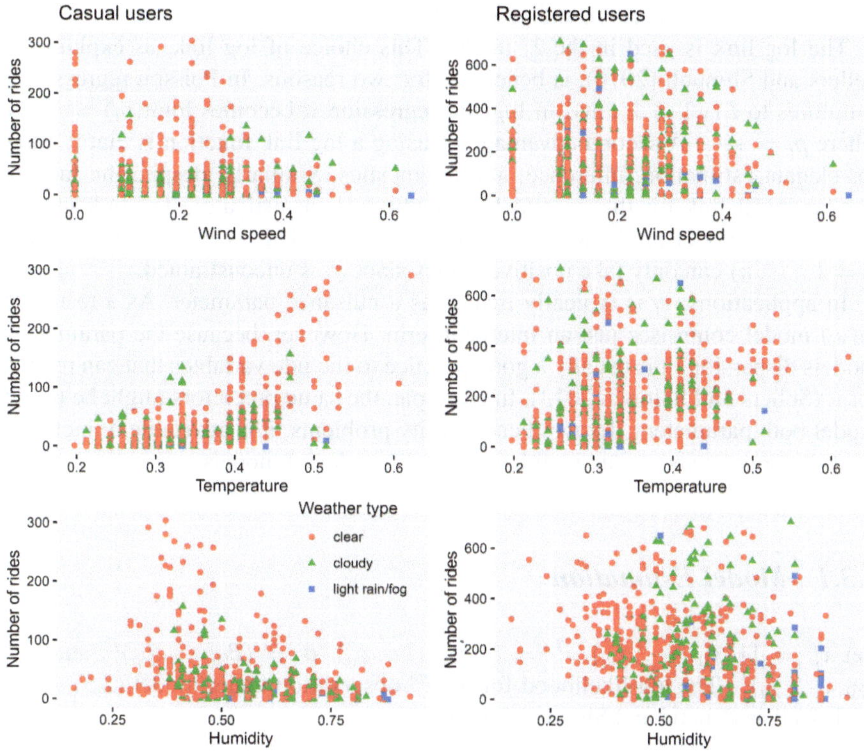

Fig. 2.3 Top: Number of rides vs. wind speed, by weather type. Middle: Number of rides vs. temperature, by weather type. Bottom: Number of rides vs. humidity, by weather types

The dataset and the R code for the data analysis are available at https://github.com/SuneelChatla/CMPTree.

2.3 CMP Regression

Suppose we have a random sample of n observations $\{y_i, x_{i1}, \ldots, x_{ip}, z_{i1}, \ldots, z_{iq}\}_{i=1}^{n}$, where y is the outcome variable and $x_1, \ldots x_p, z_1, \ldots, z_q$ are predictor variables. The CMP regression can be formulated as (Sellers and Shmueli 2010; Chatla and Shmueli 2018)

$$\ln(\lambda_i) = \beta_0 + x_{i1}\beta_1 + \ldots + x_{ip}\beta_p \tag{2.4}$$

$$\ln(\nu_i) = \gamma_0 + z_{i1}\gamma_1 + \ldots + z_{iq}\gamma_q \tag{2.5}$$

where $\beta_j \in \mathbb{R}, j = 0, \ldots, p$, and $\gamma_k \in \mathbb{R}$, $k = 0, \ldots, q$.

The log link is used in the λ_i model. This choice of log link, as explained in Sellers and Shmueli (2010), is beneficial for two reasons. In Poisson regression, it simplifies to $E[y_i] = \lambda_i$, and in logistic regression it becomes $\text{logit}(p_i) = \ln(\lambda_i)$, where $p_i = \frac{\lambda_i}{1+\lambda_i}$. A second advantage of using a log link function is that it allows for elegant estimation, inference, and diagnostics. Similarly, despite the fact that the canonical link for the ν_i model is identity, we investigate a log link in order to constrain model predictions to the range $(0, \infty)$. This is important because ν_i $(i = 1, \ldots, n)$ can only take positive values since γ_i is unconstrained.

In applications, ν_i is typically treated as a nuisance parameter. As a result, the $\ln(\nu_i)$ model comprises just an intercept term. However, because the parameter ν_i models dispersion, it is usually a good practice to include variables that can regulate for it (Sellers and Shmueli 2013). In principle, the same predictors might be used to model both parameters. To avoid collinearity problems, it is advisable to include at least one separate covariate in either the $\ln(\lambda_i)$ or $\ln(\nu_i)$ models.

2.3.1 Model Estimation

Let $\boldsymbol{x}_i^T = [1, x_{i1}, \cdots, x_{ip}]$, $\boldsymbol{z}_i^T = [1, z_{i1}, \cdots, z_{iq}]$, $\boldsymbol{\beta} = (\beta_0, \ldots, \beta_p)^T$, and $\boldsymbol{\gamma} = (\gamma_0, \ldots, \gamma_q)^T$. The log likelihood for the ith observation under model (2.5), can be expressed as (Chatla and Shmueli 2018)

$$\ell_i(y_i, \boldsymbol{\beta}, \boldsymbol{\gamma}) = y_i \boldsymbol{x}_i^T \boldsymbol{\beta} - \ln(y_i!) \exp\{\boldsymbol{z}_i^T \boldsymbol{\gamma}\} - \ln \zeta_i(\exp\{\boldsymbol{x}_i^T \boldsymbol{\beta}\}, \exp\{\boldsymbol{z}_i^T \boldsymbol{\gamma}\}), \tag{2.6}$$

which yields the following score equations:

$$\begin{aligned}
\frac{\partial \ell_i}{\partial \boldsymbol{\beta}^T} &= \boldsymbol{x}_i \left(y_i - \frac{\partial \ln \zeta_i}{\partial \ln \lambda_i} \right) = \boldsymbol{x}_i (y_i - E[y_i]), \\
\frac{\partial \ell_i}{\partial \boldsymbol{\gamma}^T} &= \boldsymbol{z}_i \left[\left(-\ln(y_i!) - \frac{\partial \ln \zeta_i}{\partial \nu_i} \right) \nu_i \right] = \boldsymbol{z}_i \Big[(-\ln(y_i!) + E[\ln(y_i!)]) \Big] \nu_i.
\end{aligned} \tag{2.7}$$

Existing approaches maximize the log-likelihood (2.6) by utilizing nonlinear optimization tools such as *nlminb* or *nlm* in R or using iterative re-weighted least squares (IRLS) algorithm (Chatla and Shmueli 2018; Huang 2017). The IRLS algorithm outperforms gradient-based optimization methods in a number of ways. It is efficient because the Fisher information matrix is somewhat robust to outliers, and it makes modeling extensions like additive models and lasso (Yee 2007) simple to implement. Given its importance, we give further details regarding the IRLS algorithm which is implemented in Chatla and Shmueli (2018).

We use the notation from Chatla and Shmueli (2018). Let $\boldsymbol{u} = \sum_{i=1}^n \frac{\partial \ell_i}{\partial \boldsymbol{\beta}^T}$ and $\boldsymbol{v} = \sum_{i=1}^n \frac{\partial \ell_i}{\partial \boldsymbol{\gamma}^T}$ be the score functions. Define the matrices $\boldsymbol{\Sigma}_Y = \text{diag}(\text{var}[y_1], \ldots, \text{var}[y_n])$, $\boldsymbol{\Sigma}_{Y,\ln(Y!)} = \left[\text{cov}(y_i, \ln(y_j!))\right]_{1 \le i,j \le n}$, and $\boldsymbol{\Sigma}_{\ln(Y!)} =$

diag(var[ln(y_1!)], …, var[ln(y_n!)]). Here, the variance terms are computed using the probability mass function (2.1). We write the full information matrix $\boldsymbol{I}$ using Eq. (2.7) as,

$$\boldsymbol{I} = E\left(\begin{bmatrix}\boldsymbol{u}\\ \boldsymbol{v}\end{bmatrix}[\boldsymbol{u}^T\ \boldsymbol{v}^T]\right) = \begin{pmatrix}E[\boldsymbol{u}\boldsymbol{u}^T] & E[\boldsymbol{u}\boldsymbol{v}^T]\\ E[\boldsymbol{v}\boldsymbol{u}^T] & E[\boldsymbol{v}\boldsymbol{v}^T]\end{pmatrix} = \begin{pmatrix}\boldsymbol{I}_{11} & \boldsymbol{I}_{12}\\ \boldsymbol{I}_{21} & \boldsymbol{I}_{22}\end{pmatrix},$$

where

$$\begin{aligned}
E[\boldsymbol{u}\boldsymbol{u}^T] &= \boldsymbol{I}_{11} = \boldsymbol{X}^T\boldsymbol{\Sigma}_Y\boldsymbol{X},\\
E[\boldsymbol{u}\boldsymbol{v}^T] &= \boldsymbol{I}_{12} = \boldsymbol{I}_{21}^T\ = -\boldsymbol{\nu} * \boldsymbol{X}^T\boldsymbol{\Sigma}_{Y,\ln(Y!)}\boldsymbol{Z},\\
E[\boldsymbol{v}\boldsymbol{v}^T] &= \boldsymbol{I}_{22} = \boldsymbol{\nu}^2 * \boldsymbol{Z}^T\boldsymbol{\Sigma}_{\ln(Y!)}\boldsymbol{Z},
\end{aligned}$$

and where $*$ denotes element-wise multiplication and $\boldsymbol{\nu} = (\nu_1, \ldots, \nu_n)^T$.

The IRLS algorithm has the following form for the mth update:

$$\begin{bmatrix}\boldsymbol{\beta}\\ \boldsymbol{\gamma}\end{bmatrix}^{(m)} = \begin{bmatrix}\boldsymbol{\beta}\\ \boldsymbol{\gamma}\end{bmatrix}^{(m-1)} + \boldsymbol{I}^{-1}\begin{bmatrix}\boldsymbol{u}\\ \boldsymbol{v}\end{bmatrix},$$

which implies the following equations:

$$\begin{aligned}
&\boldsymbol{X}^T\boldsymbol{\Sigma}_Y\boldsymbol{X}\boldsymbol{\beta}^{(m)} - \boldsymbol{X}^T\boldsymbol{\Sigma}_{Y,\ln(Y!)}\boldsymbol{\nu} * \boldsymbol{Z}\boldsymbol{\gamma}^{(m)} =\\
&\qquad \boldsymbol{X}^T\boldsymbol{\Sigma}_Y\boldsymbol{X}\boldsymbol{\beta}^{(m-1)} - \boldsymbol{X}^T\boldsymbol{\Sigma}_{Y,\ln(Y!)}\boldsymbol{\nu} * \boldsymbol{Z}\boldsymbol{\gamma}^{(m-1)} + \boldsymbol{X}^T(\boldsymbol{y} - E[\boldsymbol{y}])
\end{aligned} \tag{2.8}$$

and

$$\begin{aligned}
&-\boldsymbol{\nu} * \boldsymbol{Z}^T\boldsymbol{\Sigma}_{Y,\ln(Y!)}\boldsymbol{X}\boldsymbol{\beta}^{(m)} + \boldsymbol{\nu}^2 * \boldsymbol{Z}^T\boldsymbol{\Sigma}_{\ln(Y!)}\boldsymbol{Z}\boldsymbol{\gamma}^{(m)}\\
&\qquad = -\boldsymbol{\nu}\boldsymbol{Z}^T\Sigma_{Y,\ln(Y!)}\boldsymbol{X}\boldsymbol{\beta}^{(m-1)} + \boldsymbol{\nu}^2 * \boldsymbol{Z}^T\Sigma_{\ln(Y!)}\boldsymbol{Z}\boldsymbol{\gamma}^{(m-1)}\\
&\qquad\qquad + \boldsymbol{\nu} * \boldsymbol{Z}^T(-\ln(\boldsymbol{y}!) + E[\ln(\boldsymbol{y}!)]).
\end{aligned} \tag{2.9}$$

Here, $\boldsymbol{y} = (y_1, \ldots, y_n)^T$ and $\ln(\boldsymbol{y}!) = (\ln(y_1!), \ldots, \ln(y_n!))^T$. Since the two equations (2.8) and (2.9) may not be solved simultaneously we fix one parameter in each equation (Chatla and Shmueli 2018). Fixing $\boldsymbol{\gamma}$ (i.e., $\boldsymbol{\gamma}^{(m-1)} = \boldsymbol{\gamma}^{(m)}$) in Eq. (2.8) yields

$$\boldsymbol{X}^T\boldsymbol{\Sigma}_Y\boldsymbol{X}\boldsymbol{\beta}^{(m)} = \boldsymbol{X}^T\boldsymbol{\Sigma}_Y\boldsymbol{X}\boldsymbol{\beta}^{(m-1)} + \boldsymbol{X}^T(\boldsymbol{y} - E[\boldsymbol{y}]), \tag{2.10}$$

which, for a given $\boldsymbol{\beta}^{(m-1)}$, is a normal equation for the regression model of $\boldsymbol{X}$ on $\boldsymbol{y}$ with weights $\boldsymbol{\Sigma}_Y$, and can be solved directly. Similarly, if we fix $\boldsymbol{\beta}$, (i.e., $\boldsymbol{\beta}^{(m-1)} = \boldsymbol{\beta}^{(m)}$) Eq. (2.9) yields

$$\boldsymbol{\nu}^2 * \boldsymbol{Z}^T \boldsymbol{\Sigma}_{\ln(Y!)} \boldsymbol{Z}\boldsymbol{\gamma}^{(m)} = \boldsymbol{\nu}^2 * \boldsymbol{Z}^T \boldsymbol{\Sigma}_{\ln(Y!)} \boldsymbol{Z}\boldsymbol{\gamma}^{(m-1)} + \boldsymbol{\nu} * \boldsymbol{Z}^T(-\ln(\boldsymbol{y}!) + E[\ln(\boldsymbol{y}!)]). \tag{2.11}$$

For a given $\boldsymbol{\gamma}^{(m-1)}$, this is a normal equation for the regression model of $\boldsymbol{\nu} * \boldsymbol{Z}$ on $\ln(\boldsymbol{y}!)$ with weights $\Sigma_{\ln(\boldsymbol{y}!)}$.

The two updates (2.10) and (2.11) can be performed using two weighted least square (WLS) regression models. The pseudo algorithm and the convergence properties of the above two-step IRLS algorithm are provided in Chatla and Shmueli (2018).

2.3.2 *Model Estimation Using R*

The discussed CMP regression method is implemented in R programming language (R Core Team 2021). The computation routines are available in the github package *cmp* (https://github.com/SuneelChatla/cmp) which can be installed as follows:

```
devtools::install_github("SuneelChatla/cmp")
```

This package can be used to fit CMP regression and the CMP generalized additive model (GAM) which will be discussed in Sect. 2.4. It mainly includes the following functions:

1. `gam.cmp(formula.list, data, family=cmp)`: The formula.list object can be a list of two formulas: one for the $\ln(\lambda)$ model and another for the $\ln(\nu)$ model. It can also be a single formula for the $\ln(\lambda)$ model and in this case the $\ln(\nu)$ model is fitted with only an intercept.
2. `summary.gam.cmp(object)`: Provides the summary of fit results. Object is the output from the function in step 1. It provides the coefficient values, standard errors, and p-values.
3. `predict.gam.cmp(object, newdata, type="response")`: Provides the out-of-sample predictions for the fitted object in step 1. The argument type takes two values. If type="link" the predictions are for the linear predictors; and if type="response" the predictions are obtained for the response variable.

In the following, we analyze the bike-sharing data using the aforementioned functions.

2.3.3 *Predicting Bike-Sharing Rides with CMP Regression*

We use the following model for predicting the number of rides of casual/registered users, similar to Chatla and Shmueli (2018):

$$\ln(\lambda) = \beta_0 + \beta_1\text{day} + \beta_2\text{hour} + \beta_3\text{holiday} + \beta_4\text{weekday} + \beta_5\text{weathersit} + \beta_6\text{atemp} + \beta_7\text{hum} + \beta_8\text{windspeed},$$
$$\ln(\nu) = \gamma_0.$$

2.3.3.1 Model Estimation (Training)

We estimate the model using the original formulation (package: *cmp*, Chatla and Shmueli [2018]) and the mean formulation (package: *mpcmp*, Huang [2017] and Fung et al. [2020]). For comparison, Poisson regression and negative binomial regressions are also fitted. The following code is used to fit the above models:

```
library(cmp)

# model formula
regform.cas <- as.formula(casual~day+hr+holiday+weekday+weathersit+
                 atemp+hum+windspeed)

# original formulation: cmp
cmp.reg <- gam.cmp(regform.cas,family = cmp, data = nov12)

# Poisson
pos.reg <- glm(regform.cas,family = poisson(),data = nov12)

# Negative Binomial
nb.reg <- glm.nb(regform.cas, data = nov12)

# mean formulation : mpcmp
mpcmp.reg <-mpcmp::glm.cmp(regform.cas, data = nov12)
```

The results for the training data and for the casual users are presented in Table 2.2. The results indicate that both CMP and negative binomial regression models are able to capture the overdispersion in the training data. Poisson regression, on the other hand, failed to model the overdispersion. Consequently, it is inferior for purposes of statistical inference and estimation, as it provides misleading coefficient significance results. However, our focus is on out-of-sample prediction. We therefore look at the training AIC values and later at performance on the holdout data. In terms of AIC, the negative binomial was best (5564.4) followed by cmp (5665.56) and mpcmp (6103.96).

From the models' fit to the training data, we can observe a few insights about relationships with predictors: We see that *atemp*, *hum*, *hr*, and *weekday* are statistically significant for modeling the number of rentals from casual users. Rentals are positively correlated with temperature, weekend, and holidays.[1]

While we do not share the findings from the model with registered users, similar results are seen, and the AIC value from the mpcmp formulation is slightly smaller

[1] If we're interested in comparing the magnitudes of the coefficients across models, as in descriptive or explanatory modeling, then note that cmp model coefficients are different from the magnitudes of coefficients from other models. We can divide the coefficients by $\hat{\nu}$ to obtain the coefficients with magnitudes that are comparable with other models (Sellers and Shmueli 2010).

than the AIC value from the cmp formulation. The discrepancies in results across different parameterizations may be attributed to numerical errors. In practice, we recommend fitting both cmp and mpcmp models and deciding which is better for the data.

2.3.3.2 Predictive Performance (Holdout)

To assess the predictive performance of the CMP regression model, we use December 2012 data (*dec12*) as a test or holdout data. For evaluation, we compute the root mean square error (RMSE) which is defined as

$$\text{RMSE} = \sqrt{\frac{1}{n}\sum_{i=1}^{n}(y_i - \widehat{y}_i)^2},$$

where y_i is the ith observation of the response in *dec12* data and $\widehat{y}_i$ is its predicted value from the regression model. The following code is used to compute the predictions from the above four models and to compute RMSE values for the casual users.

```
# rmse
rmse <- function(y,yhat)
{
  if(length(y) != length(yhat)) stop("both vectors are not of same size")
  sqrt(mean((y-yhat)^2))
}

# original formulation: cmp
cmp.pred <- predict.gam.cmp(cmp.reg,dec12,type="response")$fit
rmse(dec12$casual,cmp.pred)

# Poisson
pos.pred <- predict(pos.reg,dec12, type="response")
rmse(dec12$casual, pos.pred)

# Negative Binomial
nb.pred <- predict(nb.reg, dec12, type="response")
rmse(dec12$casual, nb.pred)

# mean formulation : mpcmp
mp.pred <- predict(mpcmp.reg, dec12, type="response")
rmse(dec12$casual,mp.pred)
```

The predictions and RMSE values for the registered users can be computed similarly. The RMSE values are reported in Table 2.3. Except for the negative binomial regression, the RMSE values for the regression models are very similar. While CMP regression yields the smallest RMSE values for model with casual users, Poisson regression has the smallest RMSE for the model with registered users. Between the two CMP regression models, the original formulation yielded a smaller RMSE than the mpcmp model, for both models with casual and registered users. Boxplots for the prediction errors are presented in Fig. 2.4. The distribution of the prediction errors across the four regression models is very similar.

Table 2.2 Coefficients table, fit statistics, and AIC for the four models: cmp (*gam.cmp()* in R [Chatla and Shmueli 2018]), Poisson (*glm()* in R), negative binomial (*glm.nb()* in R), and mpcmp (*glm.cmp()* function of *mpcmp* package in R)

	Dependent variable: casual			
	cmp	*Poisson*	*negative binomial*	*mpcmp*
	(1)	(2)	(3)	(4)
day	−0.0004	−0.004***	−0.006	−0.004
	(0.0003)	(0.001)	(0.004)	(0.002)
hr	0.0015***	0.009***	0.038***	0.009**
	(0.0004)	(0.001)	(0.005)	(0.002)
holiday: yes	0.053***	0.512***	0.369**	0.512***
	(0.01)	(0.034)	(0.159)	(0.077)
weekday: mon	−0.076***	−1.083***	−1.028***	−1.082***
	(0.010)	(0.031)	(0.134)	(0.07)
weekday: tue	−0.032**	−0.502***	−0.525***	−0.50***
	(0.01)	(0.032)	(0.133)	(0.07)
weekday: wed	−0.052***	−0.701***	−0.577***	−0.70***
	(0.01)	(0.032)	(0.130)	(0.07)
weekday: thu	−0.078***	−1.027***	−0.835***	−1.026***
	(0.01)	(0.031)	(0.130)	(0.069)
weekday: fri	−0.029***	−0.516***	−0.547***	−0.515***
	(0.006)	(0.023)	(0.120)	(0.052)
weekday: sat	0.015*	0.160***	0.049	0.159**
	(0.006)	(0.021)	(0.126)	(0.05)
weathersit: cloudy	0.001	−0.009	−0.088	−0.008
	(0.006)	(0.019)	(0.078)	(0.042)
weathersit: light rain/fog	−0.075*	−0.525***	−0.534***	−0.525***
	(0.033)	(0.075)	(0.204)	(0.150)
atemp	0.552***	8.734***	9.777***	8.733***
	(0.033)	(0.106)	(0.569)	(0.243)
hum	−0.178***	−2.206***	−1.778***	−2.20***
	(0.023)	(0.069)	(0.297)	(0.156)
windspeed	0.036	0.325·	1.220***	0.324*
	(0.02)	(0.065)	(0.334)	(0.151)
Constant	0.035	1.655***	0.528	1.65***
	(0.027)	(0.082)	(0.359)	(0.186)
Observations	718	718	718	718
Log Likelihood	−2816.8	−6277.8	−2767.2	−3035.95

(continued)

Table 2.2 (continued)

	Dependent variable: casual			
	cmp	*Poisson*	*negative binomial*	*mpcmp*
	(1)	(2)	(3)	(4)
θ or $\ln \nu$	−2.99*** (0.010)		1.456*** (0.089)	−1.783
Akaike Inf. Crit.	5665.6	12585.6	5564.4	6103.96

Note *p < 0.1; **p < 0.05; ***p < 0.01

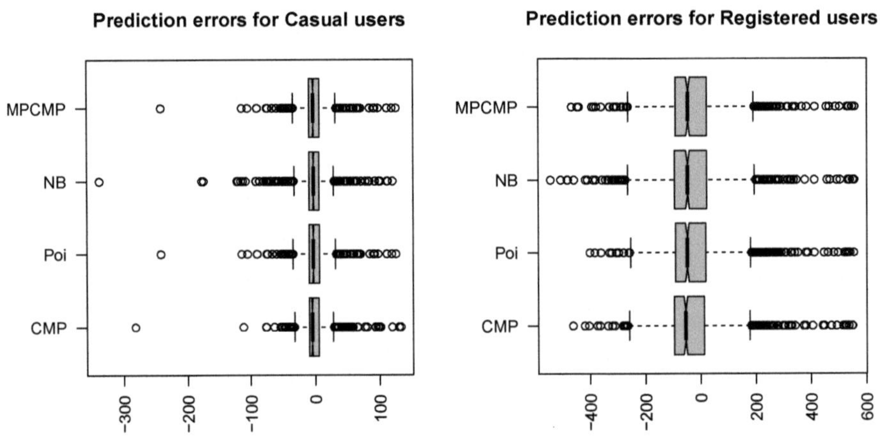

Fig. 2.4 Prediction errors for GLMs on the holdout data from December 2012

Table 2.3 Computed RMSE values for models with casual and registered users based on their predictions on the holdout data from December 2012

Model	RMSE for casual	RMSE for registered
CMP	23.7366	136.1823
Poisson	23.8755	133.577
Negative Binomial	29.2997	144.6394
MPCMP	23.8756	138.0379

2.4 CMP Generalized Additive Model

Additive models are well known for their flexibility and ability to circumvent the curse of dimensionality. Generalized additive models are generalized linear models with a linear predictor including smooth functions of covariates. Similar to CMP regression (2.5), Chatla and Shmueli (2018) formulate the CMP generalized additive model (CMPGAM) as

$$\ln(\lambda_i) = \boldsymbol{x}_i^{*^T} \boldsymbol{\theta}^* + \sum_{j=1}^{p} f_j(x_{ij}), \tag{2.12}$$

$$\ln(\nu_i) = \boldsymbol{z}_i^{*^T} \boldsymbol{\delta}^* + \sum_{l=1}^{k} m_l(z_{il}), \tag{2.13}$$

for $i = 1, \ldots, n$, where $\boldsymbol{\theta}^*$ and $\boldsymbol{\delta}^*$ are the parameter vectors for the parametric parts $\boldsymbol{x}_i^*$ and $\boldsymbol{z}_i^*$, respectively. The functions f_j, $1 \le j \le p$, and m_l, $1 \le l \le k$, are smooth and identifiable subject to the constraints $\sum_{i=1}^{n} f_j(x_{ij}) = 0$ and $\sum_{i=1}^{n} m_l(z_{il}) = 0$, respectively.

Chatla and Shmueli (2018) consider the penalized splines approach which is one of the most popular approaches for estimation of additive models (Wood 2006a). This approach uses intermediate rank spline-type basis functions to approximate each f_j and m_l. If $b_r(x)$ and $c_s(z)$ are rth, $1 \le r \le n_j$, and sth, $1 \le s \le n_l$, basis functions of f_j and m_l, respectively, then

$$f_j(x) = \sum_{r=1}^{n_j} b_r(x)\beta_r, \qquad m_l(z) = \sum_{s=1}^{n_l} c_s(z)\gamma_s,$$

where n_j and n_l are the number of basis functions required for the approximation of f_j and m_l. After some rearrangement of terms, we write in vector form

$$\ln(\boldsymbol{\lambda}) = \boldsymbol{X}\boldsymbol{\beta},$$
$$\ln(\boldsymbol{\nu}) = \boldsymbol{Z}\boldsymbol{\gamma}.$$

While this makes the model a generalized linear model (GLM), penalized likelihood is considered to avoid overfitting. In practice, this is achieved by the penalized re-weighted least squares (P-IRLS). Wood (2006a, 2008) proposed two approaches to estimate to maximize the penalized likelihood.

- *Performance iteration:* A working penalized linear model is estimated at each step of the P-IRLS and the smoothing parameter is computed.
- *Outer iteration:* The P-IRLS algorithm is iterated until convergence and a set of smoothing parameters are estimated at convergence using GCV or AIC scores.

The performance iteration method is considered in Chatla and Shmueli (2018) as it is faster and more stable than the outer iteration method, which needs some numerical derivatives that may be error-prone with CMP distribution. Each P-IRLS iteration includes minimizing

$$\|\sqrt{\boldsymbol{W}_1^{(k)}}(T_1^{(k)} - \boldsymbol{X\beta})\|^2 + \boldsymbol{\beta}^T\boldsymbol{H}_1\boldsymbol{\beta} + \sum_j \omega_{1j}\boldsymbol{\beta}^T\boldsymbol{S}_{1j}\boldsymbol{\beta} \text{ w.r.t } \boldsymbol{\beta} \text{ and}$$

$$\|\sqrt{\boldsymbol{W}_2^{(k)}}(T_2^{(k)} - \boldsymbol{X\gamma})\|^2 + \boldsymbol{\gamma}^T\boldsymbol{H}_2\boldsymbol{\gamma} + \sum_j \omega_{2j}\boldsymbol{\gamma}^T\boldsymbol{S}_{2j}\boldsymbol{\gamma} \text{ w.r.t } \boldsymbol{\gamma}$$

where $\boldsymbol{W}_1^{(k)}$, $\boldsymbol{W}_2^{(k)}$, $T_1^{(k)}$, and $T_2^{(k)}$ are defined as in Algorithm 1, and ω_{1j} and ω_{2j} are smoothing parameters, $\boldsymbol{S}_{1j}$ and $\boldsymbol{S}_{2j}$ are roughness penalty matrices, and $\boldsymbol{H}_1$ and $\boldsymbol{H}_2$ are fixed positive semidefinite penalty matrices that allow for multiple extensions of GAMs (Wood 2004). These objective functions are solved using penalized least squares estimation methodology for fixed smoothing parameter values ω_1 and ω_2. On the other hand, smoothing parameters are chosen by minimizing the following GCV criterion

$$V_{1g} = \frac{n\|T_1 - \boldsymbol{A}_1 T_1\|^2}{[\mathrm{tr}(\boldsymbol{I} - \varrho_1 \boldsymbol{A}_1)]^2}, \qquad V_{2g} = \frac{n\|T_2 - \boldsymbol{A}_2 T_2\|^2}{[\mathrm{tr}(\boldsymbol{I} - \varrho_2 \boldsymbol{A}_2)]^2},$$

where $\boldsymbol{A}_1$ and $\boldsymbol{A}_2$ are influence or hat matrices for both objective functions and ϱ_1 and ϱ_2 are used to inflate the GCV function to ensure that the models are smoother Wood (2004) and Chatla and Shmueli (2018). Model parameters and smoothing parameters are estimated using the *magic* function in *mgcv* package in R (Wood 2017).

As in Wood (2006a), a Bayesian smoothing framework is considered for the inference of the spline regression coefficients. In this framework, the smoothing parameters are induced by improper Gaussian priors on $\boldsymbol{\beta}, \boldsymbol{\gamma}$. The estimates $\hat{\boldsymbol{\beta}}$ and $\hat{\boldsymbol{\gamma}}$ are also the modes of the posterior densities of $\boldsymbol{\beta}$ and $\boldsymbol{\gamma}$, respectively (Wahba 1983; Silverman 1985; Nychka 1988). More details are found in Wood (2006b, 2012) and Chatla and Shmueli (2018). The large sample posterior distribution for the regression spline coefficients in CMP GAM is (Chatla and Shmueli 2018),

$$\boldsymbol{\beta}|\eta_1, \boldsymbol{y} \sim \mathcal{N}(\hat{\boldsymbol{\beta}}, \boldsymbol{\Sigma}_\beta), \quad \boldsymbol{\gamma}|\eta_2, \boldsymbol{y} \sim \mathcal{N}(\hat{\boldsymbol{\gamma}}, \boldsymbol{\Sigma}_\gamma),$$

where $\boldsymbol{\Sigma}_\beta = (\boldsymbol{X}^T\boldsymbol{W}_1\boldsymbol{X} + \sum_{j=1}^{p}\eta_{1j}\boldsymbol{S}_{1j})^{-1}\phi_1$, $\boldsymbol{\Sigma}_\gamma = (\boldsymbol{Z}^T\boldsymbol{W}_2\boldsymbol{Z} + \sum_{j=1}^{q}\eta_{2j}\boldsymbol{S}_{2j})^{-1}\phi_2$ with $\boldsymbol{W}_1$ and $\boldsymbol{W}_2$ as weight matrices and ϕ_1 and ϕ_2 as the estimated scale parameters at the convergence of the P-IRLS algorithm.

2.4.1 Predicting Bike-Sharing Rides with CMP Generalized Additive Model

We use the following model for predicting the number of rides of casual and registered users:

$$\begin{aligned}\ln(\lambda) =&\beta_0 + \beta_1\text{day} + \beta_2\text{holiday} + \beta_3\text{weekday} + \beta_4\text{weathersit}+\\ &\beta_5\text{hum} + \beta_6\text{windspeed} + s(\text{hr}) + s(\text{atemp})\\ \ln(\nu) =&\gamma_0,\end{aligned}$$

where $s(\cdot)$ indicates the variable that is modeled nonparametrically. The variables *atemp* and *hr* are chosen to be modeled nonparametrically while the variables *holiday*, *weekday*, *day*, hum, windspeed, and *weathersit* are chosen to be parametric effects. We assume constant dispersion. Therefore, the $\ln(\nu)$ model contains just an intercept.

2.4.1.1 Model Estimation (Training)

The additive model implementation for the CMP distribution uses the original formulation (Chatla and Shmueli 2018). For comparison, we also provide the results from the Poisson and negative binomial additive models which are estimated using *mgcv* package (Wood 2017) in R. The following code is used to estimate these models:

```
library(cmp)
library(mgcv)

# model formula
gamform.cas <-as.formula(casual~day+holiday+weekday+weathersit
                        +hum+windspeed +s(hr)+ s(atemp))

# original formulation: cmp
cmp.gam <-gam.cmp(gamform.cas,family = cmp,data = nov12)

# Poisson
pos.gam <- gam(gamform.cas,family = poisson(),data = nov12)

# Negative Binomial
nb.gam <- gam(gamform.cas,data = nov12,family = nb())
```

The results for the additive functions are shown in Table 2.4. Because our focus is on prediction rather than inference, the coefficient significance results for the parametric components are not included. We do note, for the purpose of insights about predictor importance, that all of the additive components are significant for casual users and for registered users. To explore these effects further, the partial plots for casual users and registered users are presented in Fig. 2.5. According to Fig. 2.5, hour appears to have a positive effect during day time and a negative effect during night time on the number of rentals from casual users. At the same time, Fig. 2.5 also shows two hourly peaks in the number of rentals from registered users. It might be due to users commuting to their offices and homes. Furthermore, temperature appears to have a favorable influence on the number of rentals made by registered users.

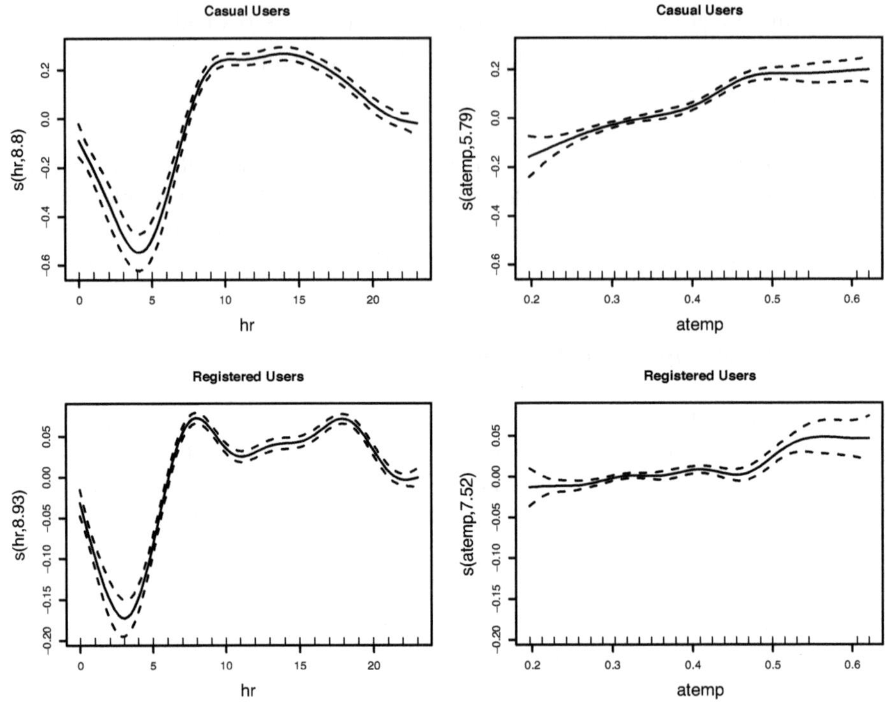

Fig. 2.5 Partial plots for the nonparametric functions estimated from the model with # rentals

Table 2.4 Comparison of CMP, Poisson, and negative binomial GAMs in terms of coefficient function significance and fit, for # rentals for casual and registered users in November 2012

	Casual Users		
	cmp − gam	*poisson − gam*	*nb − gam*
	(edf)	(edf)	(edf)
s(hr)	8.80**	8.95***	8.73***
s(atemp)	5.79***	7.55***	5.35***
γ_0 or θ	−1.85		5.49
AIC	4863.8	6596.76	4820.49
$^{***}p < 0.001$, $^{**}p < 0.01$, $^{*}p < 0.05$			
	Registered Users		
	cmp − gam	*poisson − gam*	*nb − gam*
	(edf)	(edf)	(edf)
s(hr)	8.92***	8.99***	8.90***
s(atemp)	7.52***	8.95***	1.00***
γ_0 or θ	−2.99		3.41
AIC	7977.2	27647.77	7905.70
$^{***}p < 0.001$, $^{**}p < 0.01$, $^{*}p < 0.05$			

2.4.1.2 Predictive Performance (Holdout)

We computed the RMSE values for the three additive models based on their predictions on the December 2012 holdout data. The following code is used to compute these values for casual users. Similar code can be used to compute them for the registered users.

```
# original formulation: cmp
cmpgam.pred <- predict.gam.cmp(cmp.gam,dec12,type="response")$fit
rmse(dec12$casual,cmpgam.pred)

# Poisson
posgam.pred <- predict(pos.gam,dec12,type="response")
rmse(dec12$casual,posgam.pred)

# Negative Binomial
nbgam.pred <- predict(nb.gam,dec12,type="response")
rmse(dec12$casual,nbgam.pred)
```

The RMSE values for both casual and registered users are presented in Table 2.5. Similar to GLMs in Sect. 2.3.3.2, CMPGAM yielded the lowest RMSE for casual users while Poisson GAM yielded the lowest for registered users. As expected, the RMSE values for the additive models are smaller than the RMSE values for GLMs. Further, boxplots for the prediction errors are presented in Fig. 2.6. According to results, the distribution of the prediction errors across the three models is very similar.

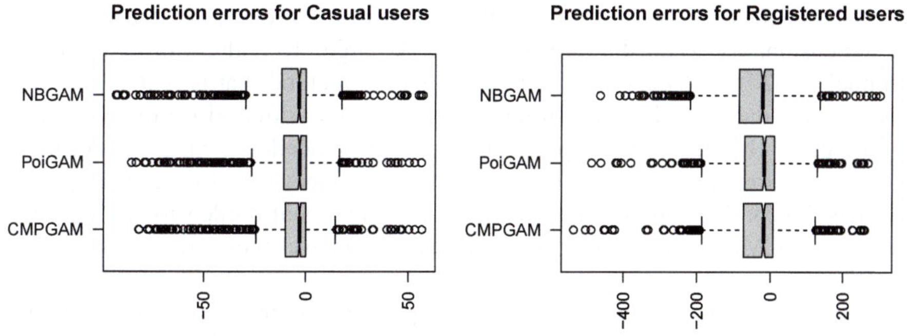

Fig. 2.6 Prediction errors for additive models on the data from December 2012

Table 2.5 Computed RMSE values for additive models with casual and registered users based on their predictions on the data from December 2012

Model	RMSE for casual	RMSE for registered
CMPGAM	18.41	98.86
PoiGAM	18.76	96.38
NBGAM	20.73	104.23

2.5 A Tree-Based CMP Varying Coefficient Model

The *varying coefficient model* is a flexible nonparametric model that allows for nonlinear interactions between the variables. Moreover, it includes the additive model as a special case. The varying coefficient model is similar to an ordinary regression model, except that the regression coefficients are free to vary based on the values of other covariates (Hastie and Tibshirani 1993; Park et al. 2015) known as *moderator variables*.

For a random sample of n iid observations $\{y_i, \boldsymbol{x}_i^T, \boldsymbol{z}_i^T, \boldsymbol{w}_i^T\}_{i=1}^n$, where $\boldsymbol{x}_i^T = [x_{i1}, \cdots, x_{iP}]$, $\boldsymbol{z}_i^T = [z_{i1}, \cdots, z_{iL}]$ and $\boldsymbol{w}_i^T = [1, w_{i1}, \ldots, w_{iQ}]$, Chatla and Shmueli (2020) considers the following CMP varying coefficient model:

$$\ln(\lambda_i) = \eta_{1i} = \boldsymbol{x}_{1i}^T \boldsymbol{\beta}(z_i) + \boldsymbol{x}_{2i}^T \boldsymbol{\phi}_1, \tag{2.14}$$

$$\ln(\nu_i) = \eta_{2i} = \boldsymbol{w}_{1i}^T \boldsymbol{\gamma}(z_i) + \boldsymbol{w}_{2i}^T \boldsymbol{\phi}_2, \tag{2.15}$$

where the covariate vectors $\boldsymbol{x}_i^T = (\boldsymbol{x}_{1i}^T, \boldsymbol{x}_{2i}^T)$, $\boldsymbol{w}_i^T = (\boldsymbol{w}_{1i}^T, \boldsymbol{w}_{2i}^T)$ and the parameter vectors $\boldsymbol{\beta}(\cdot)$, $\boldsymbol{\gamma}(\cdot)$ are varying coefficients and $\boldsymbol{\phi}_1$, $\boldsymbol{\phi}_2$ are global parameter vectors which do not vary. The invariant part in both models may be parametric or nonparametric or both. As discussed in Sect. 2.4, we can always represent the smooth functions as $\boldsymbol{x}_{2i}^T \boldsymbol{\phi}_1$ or $\boldsymbol{w}_{2i}^T \boldsymbol{\phi}_2$ using finite basis expansion.

The methods described in Sect. 2.4 may be used to estimate models (2.14) and (2.15) when the dimension of z_i is small. They, however, break down due to the curse of dimensionality when the dimension of z_i is large. To address this, Chatla and Shmueli (2020) considers a tree-based estimation method that is scalable to high-dimensional data and does not suffer from the curse of dimensionality. They use a model-based (MOB) recursive partitioning algorithm (Zeileis et al. 2008) because it is simple to implement and employs coefficient constancy tests for variable splitting (Zeileis 2005). This approach reduces the computational burden for fitting a tree because the CMP distribution is computationally intensive owing to the normalization constant.

The MOB framework employs a vectorial piecewise constant function to approximate the varying coefficients $\boldsymbol{\beta}(\cdot)$, $\boldsymbol{\gamma}(\cdot)$ in models (2.14) and (2.15). Consider a partition $\mathcal{Z}_1 \times \cdots \times \mathcal{Z}_L$ of L splitting variables $Z_1, \ldots, Z_L$ into M (terminal) nodes $\mathcal{B}_1, \ldots, \mathcal{B}_M$. Chatla and Shmueli (2020) express that at node m the linear predictors are

$$\eta_{1i} = 1(z_i \in \mathcal{B}_m)\left\{\boldsymbol{x}_{1i}^T \boldsymbol{\beta}(z_i) + \boldsymbol{x}_{2i}^T \boldsymbol{\phi}_1\right\}, \tag{2.16}$$

$$\eta_{2i} = 1(z_i \in \mathcal{B}_m)\left\{\boldsymbol{w}_{1i}^T \boldsymbol{\gamma}(z_i) + \boldsymbol{w}_{2i}^T \boldsymbol{\phi}_2\right\}, \tag{2.17}$$

where $1(\cdot)$ is an indicator function that takes 1 if the argument is true and 0 otherwise. The MOB algorithm uses generalized M-fluctuation tests (Zeileis 2005) known as coefficient constancy tests for the selection of a splitting variable at each node of the CMP tree using the following score functions:

$$\begin{aligned}\hat{s}_{1i} &= 1(z_i \in \mathcal{B}_m)\boldsymbol{x}_{1i}(y_i - \widehat{E[y_i]}),\\ \hat{s}_{2i} &= 1(z_i \in \mathcal{B}_m)\boldsymbol{w}_{1i}(-\ln(y_i!) + \widehat{E[\ln(y_i!)]})\nu_i.\end{aligned} \tag{2.18}$$

It determines the best splitting variable at each node based on the p-values obtained from the test, after applying the Bonferroni or a similar correction for multiple testing.

The MOB algorithm employs an exhaustive search for split point selection, fitting two models for each potential split point candidate. While this is computationally intensive, the existing implementations such as *lmtree*, *glmtree* functions in R package *partykit* (Hothorn and Zeileis 2013) are generally fast.

2.5.1 *Split Point Estimation*

Because regression with the CMP distribution is computationally intensive, the exhaustive search for split point estimation can be extremely intensive. We therefore describe a solution using tools borrowed from change-point estimation, introduced by Chatla and Shmueli (2020).

The optimal split point for the exhaustive search method is estimated using the following objective function (Chatla and Shmueli 2020):

$$\underset{1\le k\le n_z}{\operatorname{argmin}}\Big(-2\ell(\widehat{\boldsymbol{\beta}}, \widehat{\boldsymbol{\gamma}}; \boldsymbol{y}, \boldsymbol{x}, z \le z_{(k)}) - 2\ell(\widehat{\boldsymbol{\beta}}, \widehat{\boldsymbol{\gamma}}; \boldsymbol{y}, \boldsymbol{x}, z > z_{(k)})\Big), \tag{2.19}$$

where n_z is the number of unique values in z, $z_{(k)}$ is the kth ordered statistic, and $\ell(\cdot)$ is the log-likelihood function. To assess the two likelihood functions, we may need to fit two models for every k: $\ell(\widehat{\boldsymbol{\beta}}, \widehat{\boldsymbol{\gamma}}; \boldsymbol{y}, \boldsymbol{x}, z \le z_{(k)})$ and $\ell(\widehat{\boldsymbol{\beta}}, \widehat{\boldsymbol{\gamma}}; \boldsymbol{y}, \boldsymbol{x}, z > z_{(k)})$, where $z > z_{(k)} := (z_1 > z_{(k)}, \ldots, z_n > z_{(k)})^T$. As previously stated, this procedure is computationally heavy, particularly for the CMPMOB tree. Because the purpose is to compare the total of -2ℓ values for different values of k, calculating the exact value may not be essential; an approximate value will suffice. For each model, the algorithm may run for a few fixed number of iterations instead of running until convergence to approximate the -2ℓ values.

Alternatively, Chatla and Shmueli (2020) consider a change-point estimation approach for split point estimation. For illustration, we consider a simplified model of (2.14). Suppose z is a potential splitting variable, and $z_{(k)}$ is the kth order statistic of z. Assume that the *true model* takes the following form:

$$\eta_{1i} = 1(z \le z_{(k)})\boldsymbol{x}_{1i}^T\boldsymbol{\beta}_L + 1(z > z_{(k)})\boldsymbol{x}_{1i}^T\boldsymbol{\beta}_R + \boldsymbol{x}_{2i}^T\boldsymbol{\phi}_1, \tag{2.20}$$

where $\boldsymbol{\beta}_L, \boldsymbol{\beta}_R$ are the corresponding parameter vectors when $z \leq z_{(k)}$ and $z > z_{(k)}$. The *working model*, on the other hand, may take the following form

$$\widetilde{\eta}_{1i} = \boldsymbol{x}_{1i}^T \boldsymbol{\beta} + \boldsymbol{x}_{2i}^T \boldsymbol{\phi}_1. \tag{2.21}$$

Combination of (2.20) and (2.21) yields that

$$\begin{aligned} \eta_{1i} =& \boldsymbol{x}_{1i}^T \boldsymbol{\beta} + \boldsymbol{x}_{2i}^T \boldsymbol{\phi}_1 + 1(z \leq z_{(k)}) \boldsymbol{x}_{1i}^T \boldsymbol{\beta}_L + 1(z > z_{(k)}) \boldsymbol{x}_{1i}^T \boldsymbol{\beta}_R - \boldsymbol{x}_{1i}^T \boldsymbol{\beta} \\ =& \widetilde{\eta}_{1i} + \epsilon_i, \end{aligned}$$

where

$$\epsilon_i = 1(z \leq z_{(k)}) \boldsymbol{x}_{1i}^T \boldsymbol{\beta}_L + 1(z > z_{(k)}) \boldsymbol{x}_{1i}^T \boldsymbol{\beta}_R - \boldsymbol{x}_{1i}^T \boldsymbol{\beta},$$

is the omitted part from the *working model*. In other words,

$$\epsilon_i = \begin{cases} \boldsymbol{x}_{1i}^T \boldsymbol{\beta}_L - \boldsymbol{x}_{1i}^T \boldsymbol{\beta}, \text{ if } z \leq z_{(k)} \\ \boldsymbol{x}_{1i}^T \boldsymbol{\beta}_R - \boldsymbol{x}_{1i}^T \boldsymbol{\beta}, \text{ if } z > z_{(k)} \end{cases}, \quad i = 1, \ldots, n. \tag{2.22}$$

Chatla and Shmueli (2020) discuss that the omitted parts ϵ_i's may be captured in the score functions of *working model* (2.21). The formulation (2.22) resembles the change-point problem (e.g. Hinkley 1970; Boukai 1993; Antoch et al. 1995; Hawkins and Zamba 2005; Brodsky and Darkhovsky 2013) as the significance of the omitted parts may be determined by identifying whether ϵ_i shows deviation in either mean or variance or both for values $z > z_{(k)}$.

Although they share some similarities, change point and split point are two distinct concepts Banerjee et al. (2007). For change-point analysis, the underlying true model must have a discontinuity, but for split-point analysis, the underlying true function might be a smooth curve. As a result, change-point estimation approaches may not be applicable for split-point analysis. However, Chatla and Shmueli (2020) proposes to use tools from change-point analysis to at least identify some potential split points, thereby allowing us to reduce the search space. This is very useful in practice.

Assume a change point happens at some location k. Change-point estimation involves splitting the residuals, $\widehat{\epsilon}_i$, for $i = 1, \ldots, n$, into two samples $\{\widehat{\epsilon}_{z(1)}, \ldots, \widehat{\epsilon}_{z(k)}\}$ and $\{\widehat{\epsilon}_{z(k+1)}, \ldots, \widehat{\epsilon}_{z(n)}\}$, and testing whether the both sets are identically distributed or not (Hinkley 1970). Here $z_{(k)}$ is the kth order statistic of z. Formally,

$$H_0 : \widehat{\epsilon}_{z(i)} \sim F_0, \quad i = 1, \ldots, n \tag{2.23}$$

$$H_1 : \widehat{\epsilon}_{z(i)} \sim \begin{cases} F_0 & \text{if } \; i \leq k, \\ F_1 & \text{if } \; i > k. \end{cases} \tag{2.24}$$

In general, a two-sample hypothesis test may be constructed to test for a change point at $z_{(k)}$. Let $D_{k,n}$ be an appropriately chosen test statistic. Then, Chatla and Shmueli

(2020) use the test statistic $D_n = \max_{1 \le k \le n} D_{k,n}$ to test the above hypothesis problem (Ross et al. 2013). If both F_0 and F_1 are Gaussian, then $D_{k,n}$ is similar to generalized likelihood ratio (GLR) test. This testing approach is intended mostly for continuous and ordinal variables. Chatla and Shmueli (2020) recommends transforming categorical variables into ordinal variables using the CRIMCOORD procedures defined in Loh and Shih (1997). The aforementioned approach is then used to determine the split point. The above test is being implemented using the `R` package *cpm* (Ross et al. 2013). This testing procedure has the benefit of reducing the number of potential candidates for the default exhaustive search by choosing the locations of top 5% or 10% of $D_{k,n}$ values instead of all the possible splitting locations. We refer to Chatla and Shmueli (2020) for more details about the estimation.

2.5.2 CMPMOB Model Estimation Using R

The computation routines for the proposed tree-based method are available at https://github.com/SuneelChatla/CMPTree. These routines can be used to fit the CMPMOB tree. The following are the main functions:

1. `cmpmob(formula, fformula, vcformula, data, sflag=TRUE)`: The `formula` object is for a model estimated in terminal nodes. The `fformula` is a formula object for fixed/global terms. The `vcformula` object contains the model formula for partitioning variables. All of them can be a list of two formulas: one for the $\ln(\lambda)$ model and another for the $\ln(\nu)$ model. They can also contain a single formula. Moderator variables for both $\ln(\lambda)$ and $\ln(\nu)$ models are considered to be the same. The keyword sflag=TRUE uses an exhaustive search for split selection; or otherwise the change point estimation approach is used. Categorical variables need to be converted into dummies for `fformula`.
2. `Partial.mob(object, name.vars)`: Here `object` is the fitted CMPMOB tree and `name.vars` is the names of variables for which to estimate partial effects.
3. `predict.mob.cmp(object, newdata, type="response")`: Provides the out-of-sample predictions for the fitted object in step 1. The argument type takes two values. If type="link" the predictions are for the linear predictors, and if type="response" the predictions are obtained for the response variable.

2.5.3 Predicting Bike-Sharing Rides with CMPMOB Tree

For modeling the number of rentals from casual users, we consider the following semi-varying coefficient model:

$$
\begin{aligned}
\ln \lambda = {} & \beta_0(\text{atemp, hum, windspeed, weathersit, holiday}) \\
& + \text{hum} \times \beta_1(\text{atemp, hum, windspeed, weathersit, holiday}) \\
& + \text{weekday} \times \beta_2(\text{atemp, hum, windspeed, weathersit, holiday}) \\
& + f_1(\text{hr}) \\
\ln \nu = {} & \gamma_0(\text{atemp, hum, windspeed, weathersit, holiday}) \\
& + \text{hum} \times \gamma_1(\text{atemp, hum, windspeed, weathersit, holiday}) + \gamma_2\text{atemp} + \gamma_3\text{day},
\end{aligned} \tag{2.25}
$$

where γ_2, γ_3 and function $f_1(\cdot)$ are global or fixed in the sense that their estimation is outside the tree. The functions $\beta_0(\cdot)$, $\beta_1(\cdot)$, and $\beta_2(\cdot)$ are the varying coefficient functions for $\ln \lambda$ model, and $\gamma_0(\cdot)$ and $\gamma_1(\cdot)$ are the varying coefficients functions for the $\ln \nu$ model.

2.5.3.1 Model Estimation (Training)

The following code is used to fit the tree-based model in (2.25) and to plot the estimated tree and global additive function.

```
### input model formulation
formula <- list(as.formula(casual~ hum+weekday),as.formula(casual~ hum))
fformula <- list(as.formula(~s(hr)),as.formula(~atemp+day))
vcformula <- list(as.formula(~   atemp+ hum+ windspeed+ weathersit  +
    holiday), as.formula(~ atemp+ hum+ windspeed+ weathersit  + holiday))

### CMP MOB tree
cmp.mob <- cmpmob(formula,fformula,vcformula,data = nov12,family = cmp,
    sflag = TRUE)
### plot for the tree
plot(cmp.mob$tree)
### partial residual plot for the global effect
plot.gam(cmp.mob$fixed,rug=TRUE)
```

The estimated tree is shown in the left plot of Fig. 2.7. It seems that weather type plays a significant role in the rentals from casual users. Interestingly, the humidity seems to have a positive effect on rentals on a clear day, but a negative effect when the weather is not clear. Irrespective of weather, casual users have preference for renting bikes on Saturday.

From the partial residual plots for the global effect, shown in the right plot of Fig. 2.7, we see an apparent increase in rentals at noon, which then drops in the evening. This model is very flexible since it allows global effects and varying dispersion levels.

2.5.3.2 Predictive Performance (Holdout)

The following code is used to compute the predictions on the holdout data from December 2012 and to compute the RMSE value.

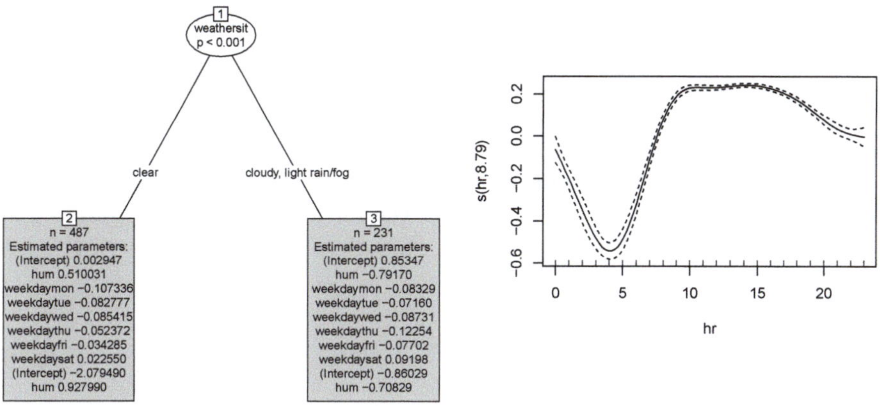

Fig. 2.7 Left: CMPMOB tree with global effects for variables *day*, *hr*, and *hum*, for modeling # rentals from casual users. The local likelihood value (−2432.04) is higher than the global likelihood value without varying coefficients (−2465.94). Right: Partial plot for the global effect $s(\text{hr})$ in Model (2.25)

```
### Predictions on holdout
cmpmob.pred <- predict.mob.cmp(cmp.mob, dec12, type="response")
rmse(dec12$casual,cmpmob.pred)
```

The RMSE values for the casual users and registered users are 20.76 and 100.14, respectively. They are slightly larger than the RMSE values of the CMPGAM model. Nonetheless, with fixed or global effects, the CMPMOB tree is very competitive while providing more insights into the data. It is worth mentioning that the best model during our simulations did not have any splits in the tree. However, for illustration, we chose the single-split model.

2.6 CMPBoost Semi-Varying Coefficient Model

2.6.1 Background

Because a single tree may not be able to recover the underlying smooth function in the data, tree-based models produce discontinuous estimates of the coefficient functions $\boldsymbol{\beta}(z_i)$ or $\boldsymbol{\gamma}(z_i)$. Smooth estimates are favored in practice because they give considerably greater insights into the data and models. In this case, ensemble methods are advantageous because they integrate predictions from multiple models to obtain smoother estimates of the coefficient functions. Boosting, also known as gradient boosting (Freund et al. 1996; Freund and Schapire 1997; Friedman 2001), is a popular ensemble method that produces smoother estimates than a single tree. The basic concept of boosting is to combine a large number of base learners (such as simple trees) to improve prediction accuracy. For a thorough review, we refer to Friedman (2001) and Buehlmann (2006).

Wang and Hastie (2014) propose a gradient boosting approach for the tree-based varying coefficient model. Chatla and Shmueli (2020) expands this concept to CMP distribution. Because of the complicated nature of the CMP distribution, the implementation is not straightforward. For illustration, consider the varying coefficient model: $y_i = f(\boldsymbol{x}_i, z_i) + \epsilon_i = \boldsymbol{x}_i^T \boldsymbol{\beta}(z_i) + \epsilon_i$. The function $f(\cdot)$ may be estimated by specifying an empirical loss function $\phi(y, f)$, such as least square error or absolute error loss, as follows:

$$\widehat{f} = \arg\min_{f \in \mathcal{F}} \frac{1}{n} \sum_{i=1}^{n} \phi(y_i, f(\boldsymbol{x}_i, z_i)), \tag{2.26}$$

where the function space $\mathcal{F} = \{f(\boldsymbol{x}, z) \mid f(\boldsymbol{x}, z) = \boldsymbol{x}^T \boldsymbol{\beta}(z)\}$. At each iteration, boosting finds an incremental model $T(\boldsymbol{x}_i, z_i)$ that minimizes empirical risk,

$$\widehat{T} = \arg\min_{f \in \mathcal{F}} \frac{1}{n} \sum_{i=1}^{n} \phi\left(y_i, \widehat{f}^{[b-1]}(\boldsymbol{x}_i, z_i) + T(\boldsymbol{x}_i, z_i)\right),$$

where $\widehat{f}^{[b-1]}$ is the model fit from the $[b-1]$th iteration. The boosting procedure begins with a basic fit $\widehat{f}^{(0)}$ and iteratively updates it by adding the incremental model fitted on "residuals" from the current iteration, $u_i = -\frac{\partial}{\partial f}\phi(y_i, f)|_{f=\widehat{f}}$. At the Bth iteration, boosting model takes the following form:

$$\widehat{f}^{(B)} = \widehat{f}^{(0)} + \xi \sum_{b=1}^{B} \sum_{m=1}^{M} \boldsymbol{x}_i^T \widehat{\boldsymbol{\beta}}_m^{(b)} \mathbf{1}_{(z_i \in \widehat{C}_m^{(b)})}, \tag{2.27}$$

where $\widehat{C}_m^{(b)}$ is the terminal node for the tree fitted in the bth boosting iteration and B is the total number of boosting iterations. The four tuning parameters of the boosting algorithm, according to Wang and Hastie (2014) and Chatla and Shmueli (2020), are the number of boosting iterations B, the learning rate ξ, the minimal node size, and the size of base learner M. To achieve better predictive accuracy, it is advisable to use small values for $0 < \xi < 1$, for example $\xi = 0.005$ (Friedman 2001).

The gradient boosting approach may be extended to nonlinear and generalized linear models by employing various loss functions such as absolute error loss, Huber-loss (Bühlmann and Hothorn 2007), or likelihood. In general, when there is a nuisance parameter that is orthogonal to the main parameter, gradient boosting methodology for the main parameter and the profile likelihood method for the nuisance parameter (Yang et al. 2017) may be utilized. Because the parameters in the CMP distribution are not orthogonal, Chatla and Shmueli (2020) considers updating both parameters in the gradient boosting algorithm.

2.6.2 CMPBoost Estimation

Chatla and Shmueli (2020) present an approach for boosting that is similar to the IRLS Algorithm 1 for estimating CMP regression (see Sect. 2.3) and requires updating both parameters λ and ν as follows:

$$\ln(\lambda_i) = \eta_{1i} = \boldsymbol{x}_{1i}^T \boldsymbol{\beta}(z_i) + \boldsymbol{x}_{2i}^T \boldsymbol{\phi}_1, \tag{2.28}$$

$$\ln(\nu_i) = \eta_{2i} = \boldsymbol{w}_{1i}^T \boldsymbol{\gamma}(\boldsymbol{u}_i) + \boldsymbol{w}_{2i}^T \boldsymbol{\phi}_2, \tag{2.29}$$

where $\boldsymbol{u}_i = (u_{i1}, \ldots, u_{iM})^T$ is an another set of moderator variables. In practice, the same set or different sets of variables may be used for $\boldsymbol{z}_i$ and $\boldsymbol{u}_i$.

The adjusted dependent variable for the $\ln(\lambda_i)$ model in (2.28) for the bth iteration is expressed as follows, using notation from Sect. 2.3:

$$\hat{t}_{i1}^{(b)} = \hat{\eta}_{i1}^{(b-1)} + \frac{y_i - \widehat{E[y_i]}^{(b-1)}}{\widehat{V[y_i]}^{(b-1)}},$$

where $\widehat{E[y_i]}^{(b-1)}$ and $\widehat{V[y_i]}^{(b-1)}$ are the estimated mean and variance in the bth iteration. The linear predictor $\hat{\eta}_{i1}^{(b-1)} = \hat{f}_{i1}^{(b-1)} = \boldsymbol{x}_{1i}^T \hat{\boldsymbol{\beta}}^{(b-1)} + \boldsymbol{x}_{2i}^T \hat{\boldsymbol{\phi}}_1$ is updated using the baseline learners fitted to the estimated residual $\frac{y_i - \widehat{E[y_i]}^{(b-1)}}{\widehat{V[y_i]}^{(b-1)}}$. The adjusted dependent variable for the $\ln(\nu_i)$ model in (2.29) for the bth iteration is

$$\hat{t}_{i2}^{(b)} = \hat{\nu}_i^{(b-1)} \hat{\eta}_{i2}^{(b-1)} + \frac{-\ln(y_i!) + \widehat{E[\ln(y_i!)]}^{(b-1)}}{\widehat{V[\ln(y_i!)]}^{(b-1)}},$$

where $\widehat{E[\ln(y_i!)]}^{(b-1)}$ and $\widehat{V[\ln(y_i!)]}^{(b-1)}$ are the estimated mean and variance for $\ln(y_i!)$ from the bth iteration. Similarly, we can fit the base learners to the estimated residual $\frac{-\ln(y_i!) + \widehat{E[\ln(y_i!)]}^{(b-1)}}{\widehat{V[\ln(y_i!)]}^{(b-1)}}$ for updating the linear predictor similar to Eq. (2.27).

Because the estimated residual is a continuous variable, any regression tree technique may be used to fit the base learner, as mentioned in Chatla and Shmueli (2020). Still the "PartReg" algorithm (Wang and Hastie 2014) is considered because it is easier to evaluate model predictions and easier to regulate its tuning parameters. The PartReg algorithm employs a breadth-first search cycle, as opposed to regular trees, such as MOB tree, which grows a larger-than-necessary tree and then trims it back to the desired size. As a result, the PartReg makes it easy to limit the number of terminal nodes. On the flip side, it has a selection bias for variables having a large number of split points (Strobl et al. 2008) and involves some computational challenges for estimating split points in ordinal and nominal variables (see Wang and Hastie [2014] for a detailed discussion). Nonetheless, Chatla and Shmueli (2020) still considers the PartReg algorithm as it is convenient.

The boosting algorithm, in general, includes four tuning parameters: ξ, M, B, and the minimal node size. Chatla and Shmueli (2020) uses small values for ξ such as 0.01, 0.05, 0.1, or 0.05, and a minimal node size of 20 in their simulations and real-world applications. Both ξ and B are determined using data-driven methods such as cross-validation. Care should be given here because a smaller value of ξ leads to a large B and vice versa. At the same time, a large value for ξ may cause the algorithm to fail to converge. As a result, they advocated starting with a few values and observing the convergence path (sequence of $-2l$ values). They consider the -2*log likelihood* as an evaluation criterion and stop the algorithm when there is not much improvement. Finally, the parameter M, the number of terminal nodes, is chosen based on the performance of the test data.

One of the key constraints of the CMPMOB model, as described in Chatla and Shmueli (2020), is computational complexity. Nonetheless, the CMPBoost model is computationally simpler than the CMPMOB since the base learners are regular regression trees, which are faster to compute. The findings of the boosting model may be interpreted using a variable importance plot and a partial dependence plot (Wang and Hastie 2014). The importance of variable z_j in a single tree (T) is defined as

$$\mathcal{I}_j^2(T) = \sum_{l=1}^{M-1} \Delta SSE_l \mathbf{1}_{(v(l)=z_j)},$$

where $v(l)$ denotes the chosen splitting variable in the lth step, M is the number of terminal nodes, and ΔSSE_l is the improvement in the lth iteration of the model fit (Friedman 2001; Hastie et al. 2015). Denote T_b as the bth tree then the relative importance of z_j can be computed as the average of $\mathcal{I}_j^2(T_b)$ among all iterations, namely,

$$\mathcal{I}_j^2 = \frac{1}{B}\sum_{b=1}^{B} \mathcal{I}_j^2(T_b).$$

As discussed in Chatla and Shmueli (2020) coefficient interpretation has always been a difficult task with CMP models since λ is not the mean parameter. Furthermore, because of the nonorthogonality of the two models, $\ln(\lambda_i)$ and $\ln(\nu_i)$, it is not easy to construct partial dependence plots for the CMPBoost.

2.6.3 *CMPBoost Model Estimation Using R*

The computation routines for the CMPBoost method are available at https://github.com/SuneelChatla/CMPTree. The following are the main functions:

1. `Boost.Tree(formula, fformula, vcformula, data, M, nBoost, eta)`: The three formula arguments are defined similar to `cmpmob()` function. The formula object consists of the model fitted at terminal nodes. The `fformula` is a formula object for fixed/global terms. the `vcformula` object contains the model formula for partitioning variables. Each one of them can be a list of two formulas: one for the $\ln(\lambda)$ model and another for the $\ln(\nu)$ model. They can also contain a single formula. Unlike the `cmpmob()` function, moderator variables for both $\ln(\lambda)$ and $\ln(\nu)$ models need not be the same. Currently, `vcformula` does not support categorical variables, so categorical variables must first be converted into dummies (1/0). Here, M is the number of terminal nodes in the base learner (tree), `nBoost` is the number of boosting iterations, and `eta` is the learning rate which is usually set to 0.01.
2. `Partial.Boost(object, name.vars)`: Here `object` is the fitted CMPBoost model and `name.vars` is the names of variables for which to estimate partial effects.
3. `predict.boost.cmp(object, newdata, type="response")`: Provides the out-of-sample predictions for the fitted object in step 1. The argument type takes two values. If type="link" the predictions are for the linear predictors, and if type="response" the predictions are obtained for the response variable.

2.6.4 *Predicting Bike-Sharing Rides with CMPBoost Model*

We consider the following CMPBoost model for modeling the number of rentals from casual users:

$$\begin{aligned}
\ln\lambda &= \beta_0(\text{windspeed, weathersit, holiday}) \\
&\quad + \text{hum} \times \beta_1(\text{windspeed, weathersit, holiday}) \\
&\quad + f_1(\text{hr}) + f_2(\text{atemp}) + \beta_2\text{weekday} \\
\ln\nu &= \gamma_0(\text{weathersit, holiday, weekday}) \\
&\quad + \text{windspeed} \times \gamma_1(\text{weathersit, holiday, weekday}) + \gamma_2\text{atemp} + \gamma_3\text{day},
\end{aligned} \tag{2.30}$$

where $f_1(\cdot)$ and $f_2(\cdot)$ are fixed effects or global functions, $\beta_0(\cdot)$, $\beta_1(\cdot)$, and $\gamma_0(\cdot)$, $\gamma_1(\cdot)$ are the additive functions for $\ln\lambda$ and $\ln\nu$ models, respectively.

2.6.4.1 Model Estimation (Training)

We employed binary coding for the categorical variables (*weathersit*, *weekday*, and *holiday*) and used them instead of categorical variables in the model. The learning rate ξ has been set to 0.01. We use data from November 2011 to fine-tune the number of terminal nodes M and number of boosting iterations (B). The following code is used to find the optimal M and B for the model with casual users. We can find the optimal values for registered users similarly.

```
### model formulae
formula <- list(as.formula(casual~ hum),as.formula(casual~ windspeed))
fformula <- list(as.formula(~s(hr) + s(atemp)+weekday),as.formula(~day))
vcformula <- list(as.formula(~    windspeed+ clear + cloudy + lightrain +
    heavyrain + notholiday + holidayc),as.formula(~ clear + cloudy +
    lightrain + heavyrain + notholiday + holidayc + sun + mon + tue + thu
    + fri + sat ))

# grid for the terminal nodes M
grid.M <- c(15,20,25)
# grid for the number of boosting iterations
nboost <- seq(1,40,3)
# grid search
prmse <- matrix(0,length(nboost),length(grid.M))
for(i in 1:length(nboost))
{
  for(j in 1:length(grid.M))
  {
  # Fitting CMPBoost model for November 2012 data
  tres <-   Boost.Tree(formula=formula,fformula=fformula,vcformula=
    vcformula, data= nov12,eta=0.01,mini.size=20,M=grid.M[j],nBoost =
    nboost[i])
  # Predictions for the validation set: November 2011
  tpred <-predict.boost.cmp(tres,test,type = "response")
  # RMSE
  prmse[i,j] <- rmse(test$registered,tpred)
  }
 }
```

The RMSE values (`prmse`) from the validation data for different values of terminal nodes and different number of boosting iterations from models with casual and registered users are provided in Fig. 2.8. The minimum values are achieved for ($M = 25, B = 19$) and ($M = 20, B = 22$) for casual and registered users, respectively. With this set of tuning parameters, we fit the CMPBoost model. The variable importance plot for the CMPBoost model is shown in Fig. 2.9. The findings suggest that the most important variables are *lightrain* and *thu* for the $\ln \lambda$ and $\ln \nu$ models, respectively. Although not presented here, we find that *notholiday* is the most important variable for registered users. This validates our findings from Sect. 2.2 that most of the registered users might be renting bicycles to commute to work.

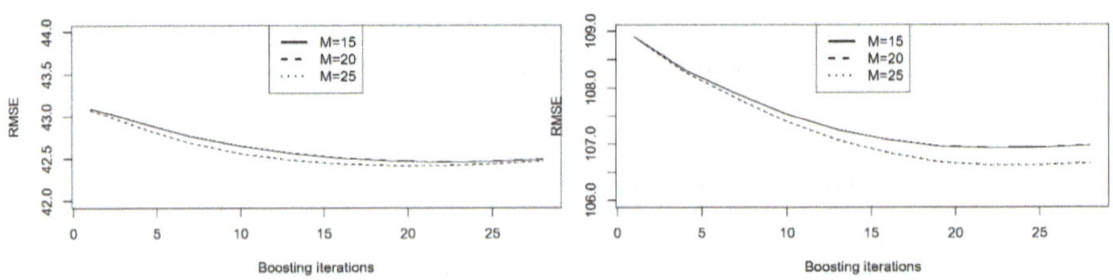

Fig. 2.8 RMSE values for casual (left) and registered (right) users models on the validation data (November 2011) for different values of M and different number of boosting iterations (B)

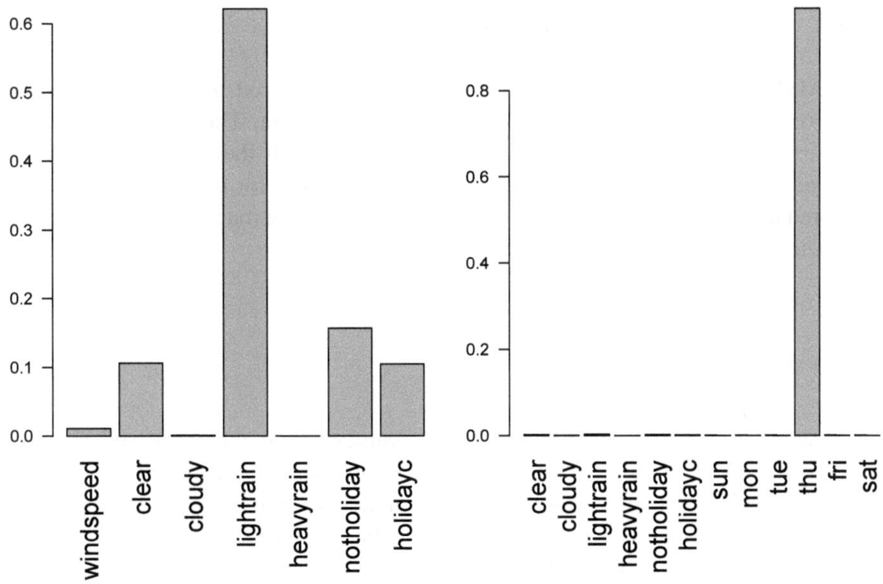

Fig. 2.9 Variable importance plots for $\ln \lambda$ and $\ln \nu$ models in (2.30) for casual users

To provide additional insights, we provide the partial dependence plots for variable *lightrain* in Fig. 2.10. Similar plots may be provided for other significant moderator variables such as *thu* and *notholiday*. Figure 2.10 illustrates that when it rains a little, humidity negatively affects the rentals from casual users.

To summarize, the findings demonstrate that CMPBoost is quite flexible for modeling the bike-sharing data and provides some helpful insights.

Fig. 2.10 Partial plot for the variable *lightrain*

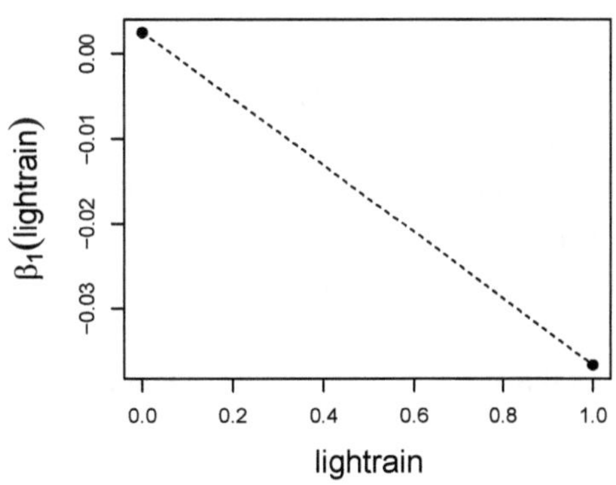

2.6.4.2 Predictive Performance (Holdout)

The RMSE values for both models of casual and registered users are 18.518 and 93.07, respectively, on the holdout (December 2012) data. The CMPBoost model is so far the best for registered users and is on par with the additive models for casual users. Here, for simplicity, we use the same model for both casual users and registered users. We believe that the predictive performance may be improved further if we select the best model individually.

2.6.5 Summary

For easy reference, we provide the RMSE values of all the discussed methods in Table 2.6. We see that the ordinary regression models performed the poorest, whereas all the more flexible models (additive model, MOB tree, and boosting model) performed better, and very similarly to each other. In terms of the magnitude, the CMP additive model (CMPGAM) achieved the lowest RMSE for the casual users, with the boosting model (CMPBoost) a very close second best. For registered users, CMPBoost achieved the lowest RMSE. In order to validate this finding statistically, we would need to obtain standard errors for the RMSEs, for example, by using a bootstrap approach.

Table 2.6 Computed RMSE values for all the discussed models with casual and registered users based on their predictions on the holdout data from December 2012.

		Holdout RMSE	
Model	Model Type	Casual	Registered
CMPGLM	Regression	23.7366	136.1823
Poisson GLM		23.8755	133.577
Negative Binomial GLM		29.2997	144.6394
MPCMP GLM		23.8756	138.0379
CMPGAM	Additive model	**18.41**	98.86
Poisson GAM		18.76	96.38
Negative Binomial GAM		20.73	104.23
CMPMOB	Tree	20.76	100.14
CMPBoost	Boosting model	18.52	**93.07**

2.7 Concluding Remarks

This chapter described several regression models based on the Conway-Maxwell-Poisson distribution that provide further flexibility for modeling count data. The discussed models include generalized linear models, additive models, tree-based models, and boosting models. We introduced these models and showed how to apply them to data by using a dataset from a bike-sharing service in Washington, DC. We further showed how to evaluate the predictive performances of the different regression models using holdout data. The overall predictive performances of all the models were compared, showing the advantage of the more flexible CMP models over models such as Poisson, Negative Binomial, and even the ordinary CMP generalized linear model. We note that while the predicted values across the CMP, Poisson, and Negative Binomial models are similar, the prediction intervals that these different regression models generate for each record may differ. Prediction intervals can be obtained empirically by using bootstrapping (closed-form formulas are not easily available due to the lack of explicit expressions for CMP moments and percentiles). For further details on prediction intervals, see Sellers and Shmueli (2010).

In conclusion, and as we have illustrated in this chapter, we believe that the CMP regression with its additional family of flexible models provides a valuable addition to the modeling toolkit for count data. The CMP distribution has been successfully used in multiple applications and there has been a growing interest in proposing further modeling extensions. Sellers (2023) provides a comprehensive introduction to the CMP, its various extensions, and recent advances. The great majority of these models are aimed at descriptive modeling, where estimation and statistical inference are focal. In this chapter, we investigated CMP regression from the angle of prediction. We hope this chapter motivates further research and applications on using CMP for prediction.

References

Antoch J, Hušková M, Veraverbeke N (1995) Change-point problem and bootstrap. J Nonparametric Stat 5(2):123–144

Banerjee M, McKeague IW et al (2007) Confidence sets for split points in decision trees. Ann Stat 35(2):543–574

Barriga GD, Louzada F (2014) The zero-inflated conway-maxwell-poisson distribution: Bayesian inference, regression modeling and influence diagnostic. Stat Method 21:23–34

Boukai B (1993) A nonparametric bootstrapped estimate of the change-point. J Nonparametric Stat 3(2):123–134

Brodsky E, Darkhovsky BS (2013) Nonparametric methods in change point problems, vol 243. Springer Science & Business Media

Buehlmann P (2006) Boosting for high-dimensional linear models. Ann Stat: 559–583

Bühlmann P, Hothorn T (2007) Boosting algorithms: regularization, prediction and model fitting. Stat Sci: 477–505

Chatla SB, Shmueli G (2018) Efficient estimation of com-poisson regression and additive model. Comput Stat Data Anal 121:71–88

Chatla SB, Shmueli G (2020) A tree-based semi-varying coefficient model for the com-poisson distribution. J Comput Graphical Stat 29(4):827–846
Fanaee-T H, Gama J (2014) Event labeling combining ensemble detectors and background knowledge. Prog Artif Intell 2:113–127
Freund Y, Schapire RE (1997) A decision-theoretic generalization of on-line learning and an application to boosting. J Comput Syst Sci 55(1):119–139
Freund Y, Schapire RE et al (1996) Experiments with a new boosting algorithm. Icml, Bari, Italy 96:148–156
Friedman JH (2001) Greedy function approximation: a gradient boosting machine. Ann Stat: 1189–1232
Fung T, Alwan A, Wishart J, Huang A, Fung MT (2020) Package 'mpcmp'
Guikema SD, Goffelt JP (2008) A flexible count data regression model for risk analysis. Risk Anal Int J 28(1):213–223
Hastie T, Tibshirani R (1993) Varying-coefficient models. J Roy Stat Soc Series B (Methodological): 757–796
Hastie T, Tibshirani R, Wainwright M (2015) Statistical learning with sparsity: the lasso and generalizations. CRC Press
Hawkins DM, Zamba K (2005) A change-point model for a shift in variance. J Qual Technol 37(1):21
Hinkley DV (1970) Inference about the change-point in a sequence of random variables. Biometrika 57(1):1–17
Hothorn T, Zeileis A (2013) partykit: a toolkit for recursive partytioning, 2013. R package version 01-6 66
Huang A (2017) Mean-parametrized conway-maxwell-poisson regression models for dispersed counts. Stat Model 17(6):359–380
Lehmann EL, Casella G (2006) Theory of point estimation. Springer Science & Business Media
Loh WY, Shih YS (1997) Split selection methods for classification trees. Statistica Sinica 7:815–840
Lord D, Guikema SD, Geedipally SR (2008) Application of the Conway-Maxwell-Poisson generalized linear model for analyzing motor vehicle crashes. Accid Anal Prev 40(3):1123–1134
Nychka D (1988) Bayesian confidence intervals for smoothing splines. J Am Stat Assoc 83(404):1134–1143
Park BU, Mammen E, Lee YK, Lee ER (2015) Varying coefficient regression models: a review and new developments. Int Stat Rev 83(1):36–64
R Core Team (2021) R: a language and environment for statistical computing. R Foundation for Statistical Computing, Vienna, Austria. https://www.R-project.org/
Ross GJ, et al (2013) Parametric and nonparametric sequential change detection in R: The cpm package. J Stat Softw 78
Sellers KF (2023) The Conway-Maxwell-Poisson distribution, vol 8. Cambridge University Press
Sellers KF, Premeaux B (2021) Conway-maxwell-poisson regression models for dispersed count data. Wiley Interdisc Rev Comput Stat 13(6):e1533
Sellers KF, Shmueli G (2010) A flexible regression model for count data. Ann Appl Stat 4(2):943–961
Sellers KF, Shmueli G (2013) Data dispersion: now you see it- now you don't. Commun Stat-Theor Methods 42(17):3134–3147
Shmueli G, Minka TP, Kadane JB, Borle S, Boatwright P (2005) A useful distribution for fitting discrete data: revival of the Conway-Maxwell-Poisson distribution. J Roy Stat Soc: Series C (Appl Stat) 54(1):127–142
Silverman BW (1985) Some aspects of the spline smoothing approach to non-parametric regression curve fitting. J Roy Stat Soc Series B (Methodological): 1–52
Strobl C, Boulesteix AL, Kneib T, Augustin T, Zeileis A (2008) Conditional variable importance for random forests. BMC Bioinf 9(1):1–11
Wahba G (1983) Bayesian "confidence intervals" for the cross-validated smoothing spline. J Roy Stat Soc Series B (Methodological): 133–150

Wang JC, Hastie T (2014) Boosted varying-coefficient regression models for product demand prediction. J Comput Graphical Stat 23(2):361–382
Wood SN (2004) Stable and efficient multiple smoothing parameter estimation for generalized additive models. J Am Stat Assoc 99(467):673–686
Wood SN (2006a) Generalized additive models: an introduction with R. CRC Press
Wood SN (2006b) On confidence intervals for generalized additive models based on penalized regression splines. Aust NZ J Stat 48(4):445–464
Wood SN (2008) Fast stable direct fitting and smoothness selection for generalized additive models. J Roy Stat Soc: Series B (Stat Method) 70(3):495–518
Wood SN (2012) On p-values for smooth components of an extended generalized additive model. Biometrika 100(1):221–228
Wood SN (2017) mgcv. R package version 18-18. https://CRAN.R-project.org/package=mgcv
Yang Y, Qian W, Zou H (2017) Insurance premium prediction via gradient tree-boosted tweedie compound poisson models. J Bus Econ Stat: 1–15
Yee TW (2007) Vector generalized linear and additive models. Springer
Zeileis A (2005) A unified approach to structural change tests based on ml scores, f statistics, and ols residuals. Econometric Rev 24(4):445–466
Zeileis A, Hothorn T, Hornik K (2008) Model-based recursive partitioning. J Comput Graphical Stat 17(2):492–514

Chapter 3
Generalized Ordered Logit Model: Application to Airline Demand Modeling in the United States

Sudipta Dey Tirtha and Naveen Eluru

Abstract Traditional ordered outcome models, such as ordered logit and probit, are widely employed to analyze the effects of independent variables on ordered response data. These models assume fixed thresholds across observations, limiting their flexibility. The generalized ordered response (GOR) model overcomes this constraint by allowing thresholds to vary as a function of independent variables. Additionally, it can be extended to account for unobserved heterogeneity, similar to mixed ordered logit/probit models. This chapter presents a detailed formulation of the mixed generalized ordered response (MGOR) model and its application to US airport-level airline demand. Analyzing quarterly passenger demand data from 300 airports (2010–2018) in GAUSS, we demonstrate how the mixed generalized ordered logit (MGOL) model captures variations in air travel demand.

3.1 Introduction

Discrete outcome variables can be broadly classified as ordered or unordered variables. Examples of ordered discrete variables include household vehicle ownership, crash injury severity and three- or five-point Likert scale ratings. Example of unordered discrete variables includes travel mode choice, activity type selection and crash type outcomes. Unordered model frameworks including multinomial logit model (MNL) and other variants can be employed to analyze all discrete variables.

Supplementary Information The online version contains supplementary material available at https://doi.org/10.1007/978-981-97-9060-9_3.

S. D. Tirtha
Whitman, Requardt and Associates, LLP, Tampa, FL, USA
e-mail: stirtha@wrallp.com

N. Eluru (✉)
Department of Civil, Environmental & Construction Engineering, University of Central Florida, Orlando, FL, USA
e-mail: naveen.eluru@ucf.edu

F. Hamid and D. Mukherjee (eds.), *Data Science for Modeling Managerial and Socioeconomic Problems*, Contributions to Management Science,
https://doi.org/10.1007/978-981-97-9060-9_3

However, for modeling ordered discrete variables a specific class of discrete outcome models referred to as ordered response models is widely employed. In this chapter, we provide a discussion on ordered response frameworks including mathematical details of ordered response model, generalized ordered response model and mixed generalized ordered response model with an example application for airline demand analysis (for discussion on unordered models please see Bhat et al. 2008).

The ordered response approaches accommodate the inherent ordering of the chosen/dependent variable while partitioning a uni-dimensional latent propensity equation using a set of fixed thresholds to compute the probability of the ordered alternatives. Further, the uni-dimensional propensity approach allows for identification of factors affecting the dependent variable in a parsimonious manner relative to the unordered model systems. In traditional ordered response systems, the number of parameters/coefficients estimated is tied to the number of independent variables (and is not impacted by the number of alternatives). While this ability to incorporate the impact of independent variables on the dependent variable parsimoniously through a single equation is useful, in some cases this can also result in restrictive specifications. The restrictive structure only allows for a single parameter to represent the impact of each independent variable. The structure does not allow for a variable to impact partial group of alternatives. For example, consider an analysis of household vehicle ownership variable defined as zero, one, two and three or more vehicles. Household income available in the data is defined as < 30,000$, between 30,000$ and 60,000$ and greater than 60,000$. Let us consider the possibility that households with income greater than 60,000$ are not likely to own zero cars while the exact number of vehicles owned is not affected by income. Incorporating this behavior in the traditional ordered model explicitly is not possible as reducing the probability of zero vehicles is also associated with a corresponding increase in the probability of three or more vehicles (see Eluru et al. 2008 for discussion in another empirical context). The restrictive assumption in traditional ordered models can be addressed by allowing for the impact of independent variables to vary across the alternative space in two forms. First, the change is accommodated by allowing the thresholds to vary across observations as a function of observed attributes. Second, the traditional approach is modified to allow for the impact of exogenous variables to vary across alternatives. The first approach gives rise to the model referred to as the generalized ordered response model (Terza 1985; Eluru et al. 2008; Srinivasan 2002; Yasmin and Eluru 2013) and the second approach results in a model referred to as the partial proportional odds model (Peterson and Harrell 1990).

Eluru and Yasmin (2015) established that these two variants are mathematically identical. Further, they highlight the specific advantages of the generalized ordered response approach for accommodating for the presence of unobserved heterogeneity. Henceforth, in this chapter we refer to these models as generalized ordered response (GOR) models. In literature, GOR models have been widely employed for modeling ordered response variables in different domains such as crash severity analysis, transportation demand modeling, social science, tourism, medical science, agriculture and education (see Sect. 3.3 for more details).

The objective of the current chapter is to offer the reader a detailed description of the generalized ordered response model from first principles. An overview of the chapter organization follows: An overview of the different variations of the ordered response model including traditional ordered response model, GOR model and mixed GOR model are presented in Sect. 3.2. Section 3.3 provides a summary of ordered response framework applications from various research domains. Section 3.4 provides a real-world application of the generalized ordered model system. Finally, Sect. 3.5 concludes the chapter.

3.2 Mathematical Formulation of Ordered Response Models

In this section, we present detailed mathematical formulations for traditional OR model and its advanced variants including GOR model and mixed GOR model. In OR model, probabilities of ordered categories are estimated by partitioning a continuous latent propensity using a set of fixed thresholds. In GOR model, thresholds are parametrized to allow them to vary across observations. Finally, mixed GOR model is formulated to accommodate for the potential influence of unobserved factors on the dependent variable.

3.2.1 Traditional Ordered Response Model

Let $i\ (i = 1, 2, \ldots\ldots\ldots, N)$ and $j\ (j = 1, 2, \ldots\ldots\ldots, J)$ be indices to denote the observation and the different levels of an ordinal dependent variable, respectively. The observation-specific dependent variable y_i is postulated to be linked with an underlying continuous latent propensity variable y_i^*. This variable is parameterized as a linear function as follows:

$$y_i^* = X_i\beta + \varepsilon_i, \text{ for } i = 1, 2, \ldots\ldots\ldots, N \tag{3.1}$$

where X_{i} is a vector of variables (excluding a constant), β is a vector of unknown parameters/coefficients to be estimated and ε_i is an idiosyncratic error term. The exact formulation of the model is determined based on the distribution of the error term. If ε_i is assumed to be identically and independently standard logistic (normal) distributed across observation i, then the model will result in ordered logit (probit) model. The continuous latent variable y_i^* is mapped to actual ordered categories y_i using a set of fixed thresholds, $\tau\ (-\infty < \tau_1 < \tau_2 < \ldots < \tau_{J-1} < +\infty)$. The variable y_i^* is linked to the observed variable y_i in the following manner:

$$y_i = j, \textit{ if } \tau_{j-1} < y_i^* < \tau_j, \text{ for } j = 1, 2, \ldots, J \tag{3.2}$$

Now, the probability expression for alternative j is computed as follows:

$$\Pr(y_i = j) = \Lambda(\tau_j - X_i\beta) - \Lambda(\tau_{j-1} - X_i\beta) \tag{3.3}$$

where $\Lambda(\cdot)$ represents the standard logistic (normal) cumulative distribution function. The standard logistic (normal) assumption results in an ordered logit (probit) model. The unknown parameters in Eq. (3.3) are τ and β which can be estimated by maximizing the log-likelihood function presented below:

$$LL = \sum_{i=1}^{N} \sum_{j=1}^{J} \ln\left[\Pr(y_i = j|X_i)^{m_{ij}}\right] \tag{3.4}$$

In Eq. (3.4), m_{ij} is a binary variable where $m_{ij} = 1$ if j is the observed level and 0, otherwise. The reader would note that the ordering of thresholds must be maintained to ensure non-negativity of probability values.

3.2.2 *Generalized Ordered Response Model*

The generalized ordered response framework allows the thresholds to vary across the observations. This flexibility is achieved by the expression of thresholds as a linear function of independent variables (Terza 1985; Srinivasan 2002; Eluru and Yasmin 2015) using the following parametric form:

$$\tau_{i,j} = \alpha_j + \delta_j Z_i \tag{3.5}$$

where α_j is a constant specific to category j, Z_i is a set of exogenous variables excluding a constant and δ_j is a vector of parameters to be estimated. Therefore, the probability expression in Eq. (3.3) can be modified as follows:

$$\Pr(y_i = j) = \Lambda(\tau_{i,j} - X_i\beta) - \Lambda(\tau_{i,j-1} - X_i\beta) \tag{3.6}$$

The unknown parameters, α, δ and β, are estimated maximizing the same log-likelihood function presented in Eq. (3.4) with probability values generated employing Eq. (3.6).

3.2.3 *Mixed Generalized Ordered Response Framework*

The GOR framework described in Sect. 3.2.2 can be employed to examine the influence of observed variables on the decision process. However, in many contexts, it is also important for an analyst to recognize that variables that are missing or unobserved factors for a specific observation (or a set of observations) can impact the decision process. For example, individual characteristics and vehicle safety equipment specifications are often missing in police/hospital reported crash dataset. Therefore, in crash severity analysis using traditional OR and GOR model, it is possible that the presence of unobserved factors can modify the effect of independent variables on outcome variables (Eluru and Yasmin 2015). In another context, consider that an analyst has access to a multi-year air passenger demand data at the airport level and is interested in identifying the key factors affecting the demand. In this context, airport-specific factors such as number of terminals, service quality of terminals and intermodal transportation services affecting the demand at an airport are unobserved or missing in the dataset. The unobserved factors can affect airline passenger demand at several resolutions such as airport, year and airport-year (Tirtha et al. 2023). The mixed GOR (MGOR) framework includes additional parameters to account for the presence of such unobserved effects. Several research efforts considered the impact of unobserved factors in GOR modeling approach (please see Eluru et al. 2008; Marcoux et al. 2018; Yasmin et al. 2015; Balusu et al. 2018; Balan and Paleti 2018; Rahman et al. 2019; Tirtha et al. 2023). In this section, we will present MGOR model structure that allows us to incorporate the potential effect of unobserved factors/variables on the dependent variable.

Let μ_i and γ_{ij} be two column vectors representing the vectors of unobserved factors modifying the effect of independent variables for observation i in Eq. (3.1) and (3.5), respectively. η_k represents the vector of parameters/coefficients representing the impact of these unobserved factors influencing outcome variable across repetition level k. Therefore, we can rewrite the propensity and threshold functions as follows:

$$y_i^* = (X_i + \mu_i)\beta + (\eta_k)X_i + \varepsilon_i, \text{ for } i = 1, 2, \ldots\ldots\ldots, N \tag{3.7}$$

$$\tau_{i,j} = \alpha_j + (\delta_j + \gamma_{ij})Z_i \tag{3.8}$$

However, in ordered model framework, it is important to maintain non-negativity of the probabilities of ordered outcomes. To ensure non-negativity condition, it is necessary that $(\alpha_j + (\delta_j + \gamma_{ij})Z_i) > (\alpha_{j-1} + (\delta_{j-1} + \gamma_{i,j-1})Z_i)$. . In GOR model (without any unobserved heterogeneity consideration), it is very unlikely that this condition will be violated. However, in MGOR model estimation ensuring that the thresholds ordering remains as required is not straightforward (see detailed discussion in Srinivasan 2002; Eluru et al. 2008). To avoid such potential negative probability scenarios, Eluru et al. (2008) proposed the following alternative formulation for threshold parameters:

$$\tau_{i,j} = \tau_{i,j-1} + \exp[\alpha_j + (\delta_j + \gamma_{ij})Z_i] \tag{3.9}$$

The reader would note that all thresholds are not identifiable. Hence, we need to set either the constant in the propensity or one of the thresholds to 0. In our model structure, we include a constant in latent propensity, and we set $\tau_{i,1} = 0$. For probability computation, t is necessary to outline the structure for the different unobserved vectors μ_i, γ_{ij} and η_k. We consider that these three vectors are drawn independently from normal distributions as follows: $N\left(0, \sigma^2\right), N\left(0, \vartheta^2\right)$ and $N\left(0, \rho^2\right)$. Thus, conditional on σ, ϑ and ρ, the probability formula for observation i and alternative j in MGOL model is computed as shown below:

$$\begin{aligned} \Pr\left(y_i = j|\mu_i, \gamma_{ij}, \eta_k\right) = & \Lambda\left(\tau_{i,j} - ((X_i + \mu_i)\beta + (\eta_k)X_i)\right) \\ & - \Lambda\left(\tau_{i,j-1} - ((X_i + \mu_i)\beta + (\eta_k)X_i)\right) \end{aligned} \tag{3.10}$$

The unconditional probability can be expressed as below (Eluru and Yasmin 2015):

$$\Pr(y_i = j) = \int\limits_{\mu_i, \gamma_{ij}, \eta_k} \Pr\left(y_i = j|\mu_i, \gamma_{ij}, \eta_k\right) * dF\left(\mu_i, \gamma_{ij}, \eta_k\right) d\left(\mu_i, \gamma_{ij}, \eta_k\right) \tag{3.11}$$

To estimate the above MGOR model, we apply Quasi-Monte Carlo simulation techniques based on scrambled Halton sequences for the maximum simulated likelihood approach (see Bhat 2001; Eluru et al. 2008 for more details). The approach followed in this study employs Halton draws to calculate the multidimensional integrals presented in Eq. (3.11). The model systems are programmed and estimated in GAUSS. As GAUSS is not readily accessible to most readers, we provide our model estimation procedures for OR and GOR models using statistical software SAS in the supplementary materials. Advanced readers with GAUSS access can contact the authors for GAUSS codes of advanced GOR variants.

3.3 Ordered Response Framework Applications

Several research studies have employed ordered response frameworks. In this section, we provide a review of ordered response framework applications including traditional ordered models, generalized ordered models and mixed generalized ordered models. Table 3.1 provides a summary of a sample of earlier studies with information on study, study objective, data source, dependent variable (number of alternatives), alternative definitions and the exact model structure considered. Given the dominance of ordered response model applications in the transportation field, studies in the table are organized by transportation and other fields. Table 3.1 provides a summary of 21 studies from the transportation domain and 11 studies from other domains. The reader would note that literature review was not intended to serve as an exhaustive review—it is developed to provide an overview of different fields of application as well as a representation of different ordered response variants employed across wide range of research domains.

Table 3.1 Summary of literature review

Study	Study Objective	Data Source (Region; Year)	Dependent variable (No of alternatives)	Alternatives	Model (logit/ probit)
Transportation Studies					
Eluru et al. 2008	To study injury severity of non-motorists	General Estimates System (GES) database (US; 2004)	Injury severity (4)	No injury, non-incapacitating, incapacitating and fatal injury	MGOR (logit)
Yasmin et al. 2014	To estimate an econometric model of driver injury severity	Victorian Crash Database (Australia; 2006–2010)	Injury severity (4)	No injury, minor, serious and fatal injury	GOR (logit)
Kaplan and Prato 2012	To analyze the injury severity in bus accidents	GES database (US; 2005–2009)	Injury severity (5)	Property damage, light, severe non-incapacitating, incapacitating injury and fatality	PPO (logit)
Wang et al. 2018	To study e-bicyclists' injury severity	Police Department Database (China; 2008–2014)	Injury severity (4)	No injury, no treatment injury, treatment injury and fatality	PPO (logit)
Abegaz et al. 2014	To characterize the impact of speeding and falling asleep on crash injury severity	Crash detector data (Ethiopia; 2012–2013)	Injury severity (4)	Property damage, minor injury, serious injury and fatal	PPO (logit)
Marcoux et al. 2018	To quantify the temporal variation in exogenous factor/variable impact on crash severity	GES database (US; 1989–2014)	Injury severity (5)	No injury, possible injury, non-incapacitating, incapacitating and fatal injury	MGOR (logit)
AlKheder 2023	To identify how built environment affects traffic safety	Police reported crash data (UAE; 2008–2013)	Injury severity (4)	Minor, moderate, severe, and fatal injury	MGOR (logit)

(continued)

Table 3.1 (continued)

Study	Study Objective	Data Source (Region; Year)	Dependent variable (No of alternatives)	Alternatives	Model (logit/ probit)
Aidoo and Ackaah 2021	To study factors/variables influencing driver injury severity	National Road Traffic Accident Database (Ghana; 2014–2016)	Injury severity (3)	No injury, minor and fatal or seriously injured	PPO (logit)
Yasmin et al. 2012	To study the factors the severity of a crash involving emergency vehicles	Alberta Transportation and Infrastructure crash database (Canada; 1999–2008)	Injury severity (3)	Property damage, minor injury, and serious injury	PPO (logit)
Yasmin et al. 2015	To identify the various risk factors for driver fatalities	Fatality Analysis Reporting System dataset (US; 2010)	Fatality timeline (7)	6–30 days, 2–5 days and so on	MGOR (logit)
Balusu et al. 2018	To discuss potential limitations of the MGOR model	Simulated data	Injury severity (4)	No injury, non-incapacitating, incapacitating and fatal injury	MGOR (logit)
Chang et al. 2022	To analyze e-bike safety data	Incident Case Data Report (China; 2014–2016)	Injury severity (3)	Minor, major, and fatal injury	MGOR (probit)
Chen and Shen 2016	To determine built environment effects on bicyclist injury severity	Seattle Department of Transportation (US; 2004–2013)	Injury severity (4)	Property damage, possible, evident, severe injury or fatality	PPO (logit)
Balan and Paleti 2018	To develop a modified MGOR model of crash severity	GES database (US; 2014)	Injury severity (4)	No injury, possible injury, non-incapacitating and incapacitating or fatal injury	MGOR (probit)

(continued)

Table 3.1 (continued)

Study	Study Objective	Data Source (Region; Year)	Dependent variable (No of alternatives)	Alternatives	Model (logit/ probit)
Castro et al. 2013	To identify factors/variables affecting crash severity while accommodating spatial dependency	TxDOT Crash Records Information System (Texas, US; 2009)	Injury severity (4)	No injury, possible injury, non-incapacitating, incapacitating or fatal injury	MGOR (logit)
Fountas et al. 2018	To accommodate for the influence of unobserved heterogeneity on crash injury severity	Single-vehicle crash data (Washington, US; 2011–2013)	Injury severity (4)	No evident injury, non-incapacitating, incapacitating and fatality	OR (probit)
Michalaki et al. 2015	To identify the factors/variables affecting crash severity	Police reported crash data (Great Britain; 2005–2011)	Injury severity (3)	Slight injury, serious injury and fatality	PPO (logit)
Rahman et al. 2019	To study how new public transportation infrastructure affects transit infrastructure usage	Lynx bus ridership (Orlando, US; 2013–2015)	Bus ridership: average weekday boarding and alighting (13)	Discretized log-transformed ridership (0–5, 5–10, so on)	MGOR (logit)
Tirtha et al. 2023	To identify and quantify the impact of independent variables on airline demand	Bureau of Transportation Statistics (US; 2010–2018)	Air passenger departures and arrivals (14)	Discretized log-transformed demand (1–2, 2–4 and so on)	MGOR (probit)
Bellizzi et al. 2018	To identify the factors influencing air passenger satisfaction	Survey data (Italy; 2012–2014)	Passenger satisfaction (5)	Very poor to Excellent	OR (logit)
Alemi et al. 2019	To identify the factors/variables impacting the frequency of ride-hailing service usage	Survey data (California, US)	Ride-hailing service usage (3)	Non-user, Infrequent user, and frequent user	OR (probit)

(continued)

Table 3.1 (continued)

Study	Study Objective	Data Source (Region; Year)	Dependent variable (No of alternatives)	Alternatives	Model (logit/ probit)
Studies in other disciplines					
Guagnano et al. 2016	To identify the association between social capital and subjective poverty	EU-SILC survey and Eurostat statistics database (Europe; 2009)	Subjective poverty (6)	Very easily, easily, fairly easily, with some difficulty, with difficulty and with great difficulty	PPO (logit)
Millán et al. 2013	To determine the factors influencing job satisfaction	European Community Household Panel (Europe; 1994–2001)	Job satisfaction (3)	Dissatisfied, neither dissatisfied nor satisfied and satisfied	PPO (logit)
Abrudan et al. 2020	To study how hotel facilities impact customer rating	Online data (Romania; 2019)	Customer rating (4)	Acceptable, medium, good, and excellent	PPO (logit)
Brida et al. 2012	To study resident's preferences in cruise tourism investment	Survey data (Italy; 2011)	Level of investment (5)	Very low, low, medium, high and very high	OR (logit)
Khedhiri et al. 2010	To examine the risk to develop bladder cancer due to smoking and metabolizing enzyme genotypes	Patient data (Tunisia)	Stage of bladder tumors (3)	3 stages	PPO (logit)
McHugh and Lake 2010	To study the factors affecting clinical nursing expertise	Statewide survey of registered nurses (Pennsylvania, US; 1999)	Level of Expertise (4)	Advanced beginner, competent, proficient and expert	PPO (logit)
Cyrenne and Chan 2022	To determine the factors determining student success in university	Administrative data (Canada; 1997)	Ordered Outcomes: Success (3)	Withdrawn, continuing and graduated	OR and PPO (logit)

(continued)

Table 3.1 (continued)

Study	Study Objective	Data Source (Region; Year)	Dependent variable (No of alternatives)	Alternatives	Model (logit/ probit)
Lin and Zhang 2020	To represent how COVID-19 affected agricultural exports	Survey data (China; 2020)	Change of exports (5)	Increased or no change, decreased by less than 10%, decreased between 10 and 30% and so on	OR (logit)
Islam et al. 2017	To identify the determinants of crop production loss severity	Survey data (Bangladesh)	Loss severity (3)	Low, medium, and high	OR (probit)
Zaurin et al. 2020a	To investigate how incorporating an Adaptive Learning Module (ALM) in Statics course affected student learning	Class performance data (US; 2019)	Binary outcome (2)	Withdraw/fail and pass	OR (logit)
Zaurin et al. 2020b	To compare between different instructional delivery methods	Class performance data (US; 2019)	Students' grade (6)	A through F	OR (logit)

Several observations can be made from our review. First, ordered response frameworks have been employed in transportation, social sciences, tourism, medical science, agriculture and education. Second, among transportation studies, large number of studies focus on transportation safety examining injury severity. In fact, the reader would note that majority of seminal work on mixed generalized ordered response frameworks arose from injury severity modeling. Other fields of transportation where ordered response models are employed include public transport usage, ride-hailing demand and airline demand. Third, among the different fields employing ordered response models the variables of interest include poverty levels, nurse expertise levels, bladder tumor stages and student grades. Fourth, several variants of ordered response frameworks are implemented in literature including ordered logit and probit, generalized ordered logit (including PPO) and probit, and mixed generalized ordered logit and probit model. Among alternative model structures considered, OR, GOR, PPO and MGOR formulations have been considered in 9, 1, 13 and 10 studies, respectively. While GOR formulation is commonly employed in transportation literature, studies in other domains mostly considered PPO variant. Fifth, the number of alternatives of ordered variables considered in the studies summarized in Table 3.1 ranges between 2 and 14. Finally, we observe that error term is assumed to follow logistic and normal distributions across these studies. Among the studies presented in Table 3.1, we observe that logistic (normal) distribution was employed in 26 (6) studies out of 32 indicating a preference for logit variants.

3.4 Model Application

In this section, we present an application of MGOR model system using domestic airline demand data for US airports. Domestic aviation sector contributes to more than 5% of US Gross Domestic Product (FAA 2023). Analyzing the factors affecting airport-level passenger demand is useful for informing facility-level (e.g., runways, terminals and intermodal transportation system) improvement and operational decisions (e.g., crew management).

Previous studies analyzed air passenger demand/flows at regional/country level (Chang 2014; Chi 2014; Kalić et al. 2014; Tsui et al. 2014; Valdes 2015; Mostafaeipour et al. 2018; Albayrak et al. 2020) and airport level (Suryani et al. 2010; Zhou et al. 2018; Li and Wan 2019; Tirtha et al. 2022). Among different mathematical approaches, linear regression model and its variants (Chang 2014; Chi 2014; Valdes 2015; Albayrak et al. 2020; Tirtha et al. 2022), autoregressive integrated moving average (Tsui et al. 2014), artificial neural networks (Mostafaeipour et al. 2018), fuzzy models (Kalić et al. 2014), gravity models (Zhou et al. 2018) and optimization techniques (Li and Wan 2019) are commonly employed to model air passenger demand. Tirtha et al. (2023) developed a joint panel generalized ordered probit model of domestic airline demand at US airports and offered a formulation to predict continuous demand using the developed model. Independent variables considered in airline demand analysis include socio-demographic factors (population, education, age distribution, median income, employment and GDP), built environment factors (number of business centers and tourism), airport classification, level of service factors (average airfare and distance), temporal lag variables and spillover effects (see Tirtha et al. 2023 for more details). In this chapter, we apply MGOR framework to model domestic airline demand to illustrate how the framework can be applied in practice.

The airline demand data is sourced from origin and destination survey dataset for 5 years (2010, 2012, 2014, 2016 and 2018) hosted by Bureau of Transportation Statistics (BTS) for 300 airports. The dataset provides detailed information on 10% tickets of domestic flights operated in the United States. Airport-level quarterly passenger departures are processed from the dataset by aggregating the departing passengers for year, quarter and airport ID.[1] The resulting demand dataset consists of 20 demand measures for each airport (5 years $\times$ 4 quarters). Therefore, the estimation sample

[1] As the dataset represents 10% tickets of domestic flights operated in the US, we appropriately scale the aggregated numbers.

consists of 6,000 demand records. To generate the dependent variable, we categorize airline demand into 6 equal bins based on their quarterly ranking as follows: Bin 1: rank 251–300, Bin 2: 201–250, Bin 3: 151–200, Bin 4: 101–150, Bin 5: 51–100 and Bin 6: 1–50. Therefore, grouping of the airports has an inherent ordering: 1—the lowest demand group—and 6—the highest demand group. This ordered dependent variable is supplemented with a number of independent variables including metropolitan statistical area (MSA)-specific socio-demographic characteristics, built environment factors, and spatial and temporal factors. Socio-demographic characteristics include population, household median income, employment and educational attainment status in the corresponding MSA. Built environment factors include the number of airports in close proximity and tourism ranking of the corresponding state. Spatial factors include airport location by region (including south, north-east, west, mid-west and pacific). Temporal factors include the quarter of the passenger demand data considered for the analysis. A summary of the variables considered in the specification of models is provided in Table 3.2. Using the processed dataset, we develop airline demand models using logistic error distribution resulting in traditional ordered logit (OL), generalized ordered logit (GOL) and mixed GOL formulations. In OL and GOL models, we do not consider the effect of unobserved factors on the outcome variable. In the MGOL model, we consider the potential effect of unobserved factors at different repetition levels and also consider the factors modifying the impact of independent variables (random effect).

3.4.1 Model Development

The parameter estimates and t-statistics (in parenthesis) from OL, GOL and MGOL models are presented in Tables 3.3, 3.4 and 3.5, respectively. To be consistent across the alternative models, we employ the formulation presented in Sect. 3.2.3. We consider a constant in the demand propensity and set the first threshold $\tau_{i,1} = 0$. The model specification begins with the estimation of an OL model. Subsequently, a GOL model is estimated allowing for additional parameters in the thresholds. Finally, a MGOL model is estimated where the parameters are more flexible as they are allowed to vary across observations through random parameters or unobserved factors impact accommodated at the airport level (random parameter for the constant). In our estimation process, the OL model with 11 parameters offered the best fit with a LL value of -9347. The GOL model with 13 parameters offered the best fit with a LL value of -8405. The improvement in GOL model can be attributed to the two additional threshold variables for the population variable. The additional degrees of freedom allowed within the GOL framework allow for improving the model flexibility. Finally, the MGOL model with 13 parameters offered the best with a LL value of -7408. The improvement in the MGOL model can be attributed to the random effect estimated for the population variable. For the sake of completeness, we also computed the Bayesian Information Criterion (BIC) values. The BIC measures for the three models are: OL 18789, GOL 16924 and MGOL 14929. Clearly, the BIC

Table 3.2 Descriptive Statistics of Independent Variables (Number of airports = 300)

Variables	Description	Mean	Standard Deviation
Continuous variables			
Demographic characteristics			
Population	MSA population in million	1.430	3.241
Median Income	MSA level median income in 100K	0.544	0.112
Employment	Ln (MSA employment in thousand + 1)	5.223	1.503
Built environment factors			
No. of airports	Ln (No. of airports in 50-mile buffer area + 1)	1.803	0.737
Categorical variables			
Variables	**Description**	**Freq**	**Percentage**
Demographic characteristics			
Educational Status			
Low	% of adults without a high school degree > 12%	2748	45.80
High	% of adults without a high school degree ≤ 12%	3252	54.20
Built environment factors			
Tourism ranking			
Top 10	Airport is located in the top 10 tourist attraction states	1900	31.67
Other	Airport is located in other states	4100	68.33
Spatial factors			
Region			
South	Airport lies in the south region in the US	2160	36.00
North-East	Airport lies in the north-east region in the US	660	11.00
West	Airport lies in the west region in the US	1500	25.00
Mid-West	Airport lies in the mid-west region in the US	1340	22.33
Pacific	Airport lies in the pacific region in the US	340	5.67
Temporal factors			
Quarter 1	From January to March	1500	25.00
Quarter 2	From April to June	1500	25.00
Quarter 3	From July to September	1500	25.00
Quarter 4	From October to December	1500	25.00

values confirm that the MGOL model offers a superior fit for our data. We will only discuss the results for MGOL model in this chapter.

Table 3.3 Traditional ordered logit model

Variables	Latent propensity	Threshold 2–3	Threshold 3–4	Threshold 4–5	Threshold 5–6
Constant	−1.483 (−8.329)	0.084 (2.847)	−0.120 (−4.114)	−0.039 (−1.278)	0.345 (10.081)
Demographic characteristics					
Population	0.417 (6.524)	–	–	–	–
Median Income	2.977 (9.105)	–	–	–	–
Education status (Base: High)					
Low	−0.547 (−10.627)	–	–	–	–
Built environment factors					
No. of airports	0.773 (15.410)	–	–	–	–
Tourism ranking (Base: Other)					
Top 10	0.220 (3.838)	–	–	–	–
Spatial factors					
Region (Base: Other regions)					
South	0.910 (15.179)	–	–	–	–

3.4.2 MGOL Model Results

The results from the MGOL model are shown in Table 3.5. Positive (negative) value of a propensity parameter (β) indicates that an increase of the variable increases (decreases) the propensity of higher outcome levels. The threshold parameters (δ) indicate shifting of thresholds depending on sign of the parameters. Positive (negative) value of a threshold parameter indicates that an increase of the variable shifts the threshold to the right (left). For threshold formulation presented in Eq. (3.9), an increase (decrease) of a threshold increases (decreases) the probability of the alternative immediately to the left of the threshold and decreases (increases) the probability of alternatives to the right of the threshold. In the following sections, we discuss the results from MGOL model by variable group.

3.4.2.1 Socio-Demographic Characteristics

Among the socio-demographic characteristics considered, MSA-specific population, median income and education status are found to have significant impact on the

Table 3.4 Generalized ordered logit model

Variables	Latent propensity	Threshold 2–3	Threshold 3–4	Threshold 4–5	Threshold 5–6
Constant	−1.143 (−4.974)	1.066 (27.335)	0.342 (2.333)	0.135 (4.418)	0.516 (14.783)
Demographic characteristics					
Population	0.119 (5.658)	−2.762 (−25.335)	−0.614 (−2.580)	–	–
Median Income	3.581 (7.751)	–	–	–	–
Education status (Base: High)					
Low	−0.682 (−12.121)	–	–	–	–
Built environment factors					
No. of airports	0.653 (13.626)	–	–	–	–
Tourism ranking (Base: Other)					
Top 10	0.274 (4.087)	–	–	–	–
Spatial factors					
Region (Base: Other regions)					
South	0.705 (12.224)	–	–	–	–

airport-level demand variable. The results suggest that the population in the corresponding MSA significantly influences air passenger demand at US airports. As expected, it is found that an increase in population is associated with an increase in the propensity of airline demand (see Zhou et al. 2018; Tirtha et al. 2022, 2023 for similar findings). However, the effect of population in MGOL framework is modified by unobserved effects as illustrated by the statistically significant standard deviation parameter. The estimated parameter distribution indicates that more than 99% of the distribution is higher than zero. We also found that MSA population influences threshold parameters. Positive coefficient of the parameter demarcating Bin 3 and Bin 4 indicates that an increase in MSA population increases the likelihood of demand category 3 and decreases the likelihood of higher-order demand groups (Bin 4 and higher). Further, the results show that the air demand has significant income association. An increase in MSA median income contributes to higher demand. This result may suggest that higher affordability of the residents associated with increased median income may result in higher demand. From the results, we also observe that airline demand is significantly low if the overall education status of a MSA is lower.

Table 3.5 Mixed generalized ordered logit model

Variables	Latent propensity	Threshold 2–3	Threshold 3–4	Threshold 4–5	Threshold 5–6
Constant	−2.106 (−9.125)	0.396 (12.667)	0.351 (9.796)	0.865 (20.509)	1.996 (33.209)
Demographic characteristics					
Population	8.425 (24.679)		0.236 (128.211)	–	–
SD population	*3.637 (14.048)*	–	–	–	–
Median Income	3.002 (7.103)	–	–	–	–
Education status (Base: High)					
Low	−1.057 (−14.036)	–	–	–	–
Built environment factors					
No. of airports	0.366 (7.022)	–	–	–	–
Tourism ranking (Base: Other)					
Top 10	0.743 (9.139)	–	–	–	–
Spatial factors					
Region (Base: Other regions)					
South	0.183 (2.262)	–	–	–	–

3.4.2.2 Built Environment Factors

From the table, we also observe that built environment factors also determine air passenger demand at US airports. If the number of airports near an airport of interest increases, the probabilities of higher demand categories increase. The result might suggest that the higher number of airports in an area indicates overall higher airline demand in that area. We also consider the impact of tourism on demand. From the result, we found that if an airport is located in the top 10 tourist attraction states, the propensity of airline demand increases.

3.4.2.3 Spatial Factors

We consider spatial variation in airline demand using US region indicators and found that airports in the south experience higher airline demand compared to other airports.

3.4.3 Policy Exercise

As is evident from the results section, the effect of independent variables in MGOL model is complex and can only be determined by examining the variable impacts in propensity and thresholds collectively. Hence, to allow for an easier understanding of variable impacts across the various alternatives, we conducted a policy exercise. In this policy exercise, we examine the changes in alternative-level aggregate probability due to changes in independent variables. For our exercise, we quantify the impact of a subset of independent variables including continuous and categorical variables. Among continuous variables, we consider a 25% increase in population, median income and number of airports. Among categorical variables, we select education status—low indicator variable—and modify 25% of high-education status (indicator value = 0) records and set them to 1. The changes to alternative probabilities are computed in two steps. In step one, for each record, we compute the predicted probability for each alternative for the original independent variables and aggregate the probabilities across all records. In step two, we repeat the same procedure for modified variables (as described above) and compute the new aggregate values. A percentage change measure is computed using the alternative-specific aggregate values from the two computations. The result of the analysis is presented in Fig. 3.1. From the figure, we can see that 25% increase of population, median income and number of airports increases the probabilities of higher-order categories. Therefore, as expected, the overall impact of population, median income and number of airports on airline demand are positive. Conversely, the impact of low education status is negative as aggregate probabilities of higher-order alternatives decrease due to increase in proportion of education status—low. The analysis illustrates how the results from the MGOL model can be employed for policy exercises.[2]

3.5 Concluding Remarks

Ordered response models, a specific class of discrete outcome models, are widely employed for modeling ordered discrete variables. In ordered response frameworks, a uni-dimensional latent propensity equation is partitioned using a set of fixed thresholds to compute the probability of the ordered alternatives. The restrictive assumption

[2] The dependent variable was characterized as a ranking among airports—the 300 airports were ranked into 6 equal bins of 50 airports each. The policy exercise is conducted examining how each airport probability of belonging to a particular group changes with changes to different variables. In this process, we recognize that we are examining changes assuming a static cut-off demand based on initial ranking for the original set of independent variables. In our policy exercise, we are not ranking them anymore. We are examining if the airport demand can cross the corresponding ranking threshold employed during dependent variable creation. So, after employing the model with new policy variables it is possible that all 6 categories of dependent variables will not have equal numbers (as was the case in the model estimation).

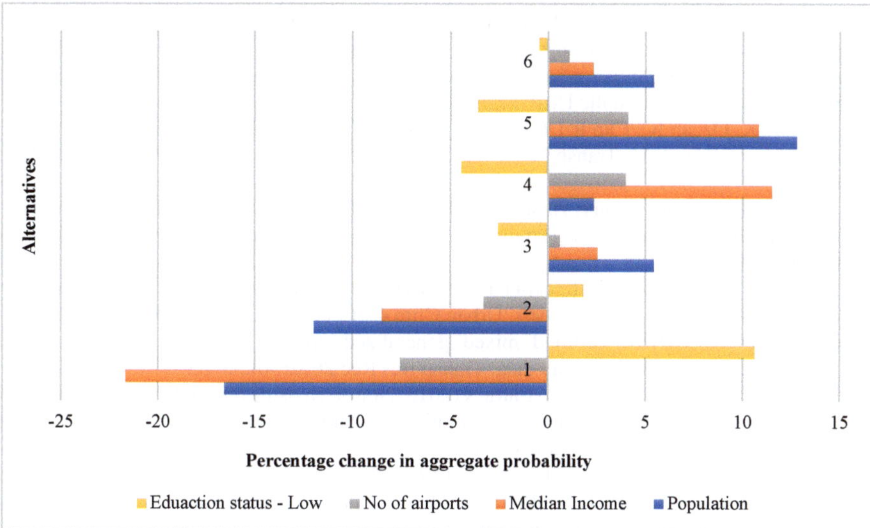

Fig. 3.1 Results of policy exercise

in traditional ordered models can be relaxed by allowing for the impact of exogenous/independent variables vary across the alternative space using two formulations: generalized ordered response model and partial proportional odds model. In this chapter, we provide a discussion on ordered response frameworks including mathematical details of traditional ordered response model, generalized ordered response model and mixed generalized ordered response model. We also provide an application of MGOL model using air passenger demand data at US airports for multiple years. In model development, we consider different variants of order response model including traditional ordered logit model, generalized ordered logit model and mixed generalized ordered logit model. Among the alternative models, MGOL offers the best data fit in terms of LL and BIC measures. The results from MGOL model highlight that MSA-specific socio-demographic characteristics, built environment factors and spatial factors are strongly associated with passenger demand at US airports. Moreover, we found population standard deviation parameter significant in the MGOL model highlighting how unobserved factors influence the effect of population on airline demand. The book chapter provides an illustration of how various ordered response models can be applied in practice.

References

Abegaz T, Berhane Y, Worku A, Assrat A, Assefa A (2014) Effects of excessive speeding and falling asleep while driving on crash injury severity in Ethiopia: A generalized ordered logit model analysis. Accid Anal Prev 71:15–21

Abrudan IN, Pop CM, Lazăr PS (2020) Using a general ordered logit model to explain the influence of hotel facilities, general and sustainability-related, on customer ratings. Sustain 12(21):1–21
Aidoo EN, Ackaah W (2021) A generalized ordered logit analysis of risk factors associated with driver injury severity. J Public Heal 29(2):471–477
Albayrak MBK, Özcan İÇ, Can R Dobruszkes F (2020) The determinants of air passenger traffic at Turkish airports. J Air Transp Manag 86:101818. https://doi.org/10.1016/j.jairtraman.2020.101818
Alemi F, Circella G, Mokhtarian P, Handy S (2019) What drives the use of ridehailing in California? Ordered probit models of the usage frequency of Uber and Lyft. Transp Res Part C Emerg Technol 102(May 2018):233–248.
AlKheder S (2023) Studying the effect of built environment on traffic safety with random parameter and generalized ordered logit models. Urban Clim 47(December 2022):101388.
Balan L, Paleti R (2018) Modified mixed generalized ordered response model to handle misclassification in injury severity data. Transp Res Rec 2672(30):53–63
Balusu SK, Pinjari AR, Mannering FL, Eluru N (2018) Non-decreasing threshold variances in mixed generalized ordered response models: a negative correlations approach to variance reduction. Anal Methods Accid Res 20(October):46–67
Bellizzi MG, Eboli L, Forciniti C, Mazzulla G (2018) Air Transport Passengers' Satisfaction: An Ordered Logit Model. Transp Res Procedia 33:147–154
Bhat CR (2001) Quasi-random maximum simulated likelihood estimation of the mixed multinomial logit model. Transp Res Part B Methodol 35(7):677–693
Bhat CR, Eluru N, Copperman RB (2008) Flexible model structures for discrete choice analysis. In: D.A. Henser, Button KJ (eds) Handbook of transport modelling, 2nd edn. Elsevier Science, pp 75–104
Brida JG, Del Chiappa G, Meleddu M, Pulina M (2012) Cruise tourism externalities and residents' support: A mixed approach. Economics 6. https://doi.org/10.5018/economics-ejournal.ja.2012-40
Castro M, Paleti R, Bhat CR (2013) A spatial generalized ordered response model to examine highway crash injury severity. Accid Anal Prev 52:188–203
Chang F, Haque MM, Yasmin S, Huang H (2022) Crash injury severity analysis of E-Bike Riders: A random parameters generalized ordered probit model with heterogeneity in means. Saf Sci 146(April 2021):105545.
Chang LY (2014) Analysis of bilateral air passenger flows: a non-parametric multivariate adaptive regression spline approach. J Air Transp Manag 34:123–130
Chen P, Shen Q (2016) Built environment effects on cyclist injury severity in automobile-involved bicycle crashes. Accid Anal Prev 86:239–246
Chi J (2014) A cointegration analysis of bilateral air travel flows: the case of international travel to and from the United States. J Air Transp Manag 39:41–47
Cyrenne P, Chan A (2022) The determinants of student success in university: a generalized ordered logit approach. Can J High Educ 52(1):84–107
Eluru N, Bhat CR, Hensher DA (2008) A mixed generalized ordered response model for examining pedestrian and bicyclist injury severity level in traffic crashes. Accid Anal Prev 40(3):1033–1054
Eluru N, Yasmin S (2015) A note on generalized ordered outcome models. Anal Methods Accid Res 8:1–6
FAA (2023) Air traffic by the numbers. https://www.faa.gov/air_traffic/by_the_numbers/. Accessed 02 Feb 2023
Fountas G, Anastasopoulos PC, Mannering FL (2018) Analysis of vehicle accident-injury severities: A comparison of segment- versus accident-based latent class ordered probit models with class-probability functions. Anal Methods Accid Res 18:15–32
Guagnano G, Santarelli E, Santini I (2016) Can social capital affect subjective poverty in Europe? An empirical analysis based on a generalized ordered logit model. Soc Indic Res 128(2):881–907
Islam Z, Alauddin M, Sarker MAR (2017) Determinants and implications of crop production loss: An empirical exploration using ordered probit analysis. Land Use Policy 67(May):527–536

Kalić M, Kuljanin J, Dožić S (2014) Air travel demand fuzzy modelling: trip generation and trip distribution. Adv Intell Syst Comput 223:279–289

Kaplan S, Prato CG (2012) Risk factors associated with bus accident severity in the United States: a generalized ordered logit model. J Safety Res 43(3):171–180

Khedhiri S, Stambouli N, Ouerhani S, Rouissi K, Marrakchi R, Gaaied AB, Slama MB (2010) The impact of smoking and polymorphic enzymes of xenobiotic metabolism on the stage of bladder tumors: A generalized ordered logistic regression analysis. J Cancer Res Clin Oncol 136(7):1111–1116

Li T, Wan Y (2019) Estimating the geographic distribution of originating air travel demand using a bi-level optimization model. Transp Res Part E Logist Transp Rev 131(October):267–291

Lin B-xi, Zhang YY (2020) Impact of the COVID-19 pandemic on agricultural exports. J Integr Agric 19(12):2937–2945

Marcoux R, Yasmin S, Eluru N, Rahman M (2018) Evaluating temporal variability of exogenous variable impacts over 25 years: An application of scaled generalized ordered logit model for driver injury severity. Anal Methods Accid Res 20(September):15–29

McHugh MD, Lake ET (2010) Understanding clinical expertise: nurse education, experience, and the hospital context. Res Nurs Health 33(4):276–287

Michalaki P, Quddus MA, Pitfield D, Huetson A (2015) Exploring the factors affecting motorway accident severity in England using the generalised ordered logistic regression model. J Safety Res 55:89–97

Millán JM, Hessels J, Thurik R, Aguado R (2013) Determinants of job satisfaction: a European comparison of self-employed and paid employees. Small Bus Econ 40(3):651–670

Mostafaeipour A, Goli A, Qolipour M (2018) Prediction of air travel demand using a hybrid artificial neural network (ANN) with Bat and Firefly algorithms: a case study. J Supercomput 74(10):5461–5484

Peterson B, Harrell FEJ (1990) Partial proportional odds models for ordinal response variables author(s): Bercedis Peterson, Frank E . Harrell and Jr Published by: Wiley for the Royal Statistical Society Stable URL: https://www.jstor.org/stable/2347760 Partial Proportional Odds M. Wiley R Stat Soc 39(2):205–217

Rahman M, Yasmin S, Eluru N (2019) Evaluating the impact of a newly added commuter rail system on bus ridership: a grouped ordered logit model approach. Transp A Transp Sci 15(2):1081–1101

Srinivasan KK (2002) Injury severity analysis with variable and correlated thresholds: Ordered mixed logit formulation. Transp Res Rec 1784:132–142

Suryani E, Chou SY, Chen CH (2010) Air passenger demand forecasting and passenger terminal capacity expansion: A system dynamics framework. Expert Syst Appl 37(3):2324–2339

Terza JV (1985) Ordinal probit: a generalization. Communications in Statistics-Theory and Methods 14(1):1–11

Tirtha SD, Bhowmik T, Eluru N (2022) An airport level framework for examining the impact of COVID-19 on airline demand. Transp Res Part A Policy Pract 159(March):169–181

Tirtha SD, Bhowmik T, Eluru N (2023) Understanding the factors affecting airport level demand (arrivals and departures) using a novel modeling approach. J Air Transp Manag 106(September 2022):102320.

Tsui WHK, Ozer Balli H, Gilbey A, Gow H (2014) Forecasting of Hong Kong airport's passenger throughput. Tour Manag 42(2014):62–76

Valdes V (2015) Determinants of air travel demand in Middle Income Countries. J Air Transp Manag 42:75–84

Wang T, Chen J, Wang C, Ye X (2018) Understand e-bicyclist safety in China: crash severity modeling using a generalized ordered logit model. Adv Mech Eng 10(6):1–11

Yasmin S, Anowar S, Tay R (2012) Effects of drivers' actions on severity of emergency vehicle collisions. Transp Res Rec 1(2318):90–97

Yasmin S, Eluru N (2013) Evaluating alternate discrete outcome frameworks for modeling crash injury severity. Accid Anal Prev 59:506–521

Yasmin S, Eluru N, Bhat CR, Tay R (2014) A latent segmentation based generalized ordered logit model to examine factors influencing driver injury severity. Anal Methods Accid Res 1:23–38

Yasmin S, Eluru N, Pinjari AR (2015) Analyzing the continuum of fatal crashes: A generalized ordered approach. Anal Methods Accid Res 7:1–15

Zaurin R, Tirtha SD, Eluru N (2020a) A comparative analysis of the students' performance in two statics courses due to the inclusion of an Adaptive Learning Module (ALM) to review the mathematics pre-requisite knowledge. ASEE Annu Conf Expo Conf Proc 2020-June. https://doi.org/10.18260/1-2--33982

Zaurin R, Tirtha SD, Eluru N (2020b) A comparison between mixed-mode and face-to-face instructional delivery approaches for engineering analysis: Statics. ASEE Annu Conf Expo Conf Proc 2020-June. https://doi.org/10.18260/1-2--33985

Zhou H, Xia J (Cecilia), Luo Q, Nikolova G, Sun J, Hughes B, Kelobonye K, Wang H, Falkmer T (2018) Investigating the impact of catchment areas of airports on estimating air travel demand: A case study of regional Western Australia. J Air Transp Manag 70(May):91–103

Chapter 4
Conjoint Analysis: Application to Consumer Preference for Organic Rice in the Philippines

Shihomi Ara Aksoy

Abstract This chapter explores Conjoint Analysis as a tool for valuing non-market goods within environmental evaluation and providing critical inputs for cost-benefit analyses. It begins with a brief literature review and theoretical foundations, followed by practical guidance on survey design, profile construction, model estimation, and result interpretation. Using R programming, the chapter demonstrates the estimation of random parameter logit models. A case study on consumer preferences for organic rice in the Philippines illustrates the application of Conjoint Analysis, assessing willingness to pay for various attributes, including price, health benefits, environmental impact, eating quality, organic certification, and fair-trade practices. By offering a structured approach, this chapter serves as a practical guide for researchers and practitioners seeking to apply Conjoint Analysis.

4.1 Introduction

Imagine conducting a cost–benefit analysis (CBA) for a project aimed at reducing air pollution. The expenditures required to improve environmental quality represent accounting costs, which are relatively straightforward to obtain and calculate. These costs might include the price of additional filters or monitoring devices that must be installed. But what about the benefits of such projects? You might list several items, such as improved health, longer lifespans, reduced mortality rates, and clearer scenic views. However, for a cost–benefit analysis, every benefit must be monetized and expressed in monetary terms. This raises questions: What are the monetary values of reduced mortality, fewer cases of respiratory diseases, fewer days with severe headaches, or improved clarity of scenic views? If these benefits are not

Supplementary Information The online version contains supplementary material available at https://doi.org/10.1007/978-981-97-9060-9_4.

S. Ara Aksoy (✉)
Department of Economics, Hacettpe University, Ankara, Turkey
e-mail: sara@hacettepe.edu.tr

F. Hamid and D. Mukherjee (eds.), *Data Science for Modeling Managerial and Socioeconomic Problems*, Contributions to Management Science,
https://doi.org/10.1007/978-981-97-9060-9_4

appropriately valued and end up underestimated, costs may exceed benefits, leading to the rejection of the project and, consequently, no improvement in environmental quality. It is essential to assign accurate monetary values to goods that lack price tags because they are not traded in markets. Only by doing so can we ensure that environmental projects are evaluated fairly and their true benefits recognized.

Environmental evaluation methods are crucial in assessing the monetary worth of environmental goods such as clean air, water, and soil, a calm environment without undesired noise, beautiful scenic views without smog from air pollution, renewable energy, and improved human health status (reduced mortality and morbidity). It also reveals people's preferences regarding energy efficiency, water conservation, adaptation of new green technologies, green-certified buildings, and more. Successful implementation of the evaluation makes it possible for us to protect the right of the public to enjoy a clean environment without health hazards.

As shown in Fig. 4.1, environmental evaluation methods are categorized into two sub-branches: stated and revealed preference methods. While stated preference methods include conjoint analysis and contingent valuation method (CVM), travel cost methods and hedonic price methods are categorized as revealed preference methods. Stated preference methods are survey-based methods that attempt to reveal the value of non-market goods by asking respondents directly, while revealed preference methods rely on the actual behavior and expenditures of individuals. The significant advantage of the stated preference method is that it allows us to evaluate the values of use and non-use values. It also allows us to evaluate the hypothetical goods/services/policies that do not exist at the point of the survey. Since researchers often wish to evaluate how the public values the potential policy changes in the future before actually implementing them, the stated preference method provides decision-makers with the prediction of potential welfare changes and entries to be listed in environmental cost–benefit analysis. The general procedure of stated preference methods is shown in Fig. 4.2.

Conjoint analysis originated in the 1960s within the field of Psychometrics and has since been developed further in marketing research and transportation engineering. The theory of household production by Lancaster (1966) laid the foundation for

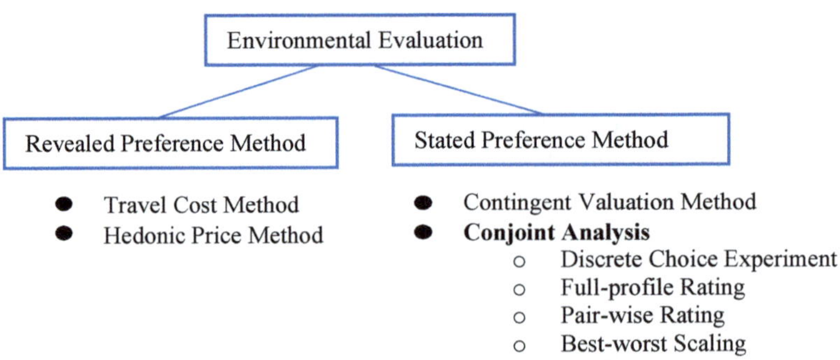

Fig. 4.1 Environmental evaluation methods. Source: Generated by Author

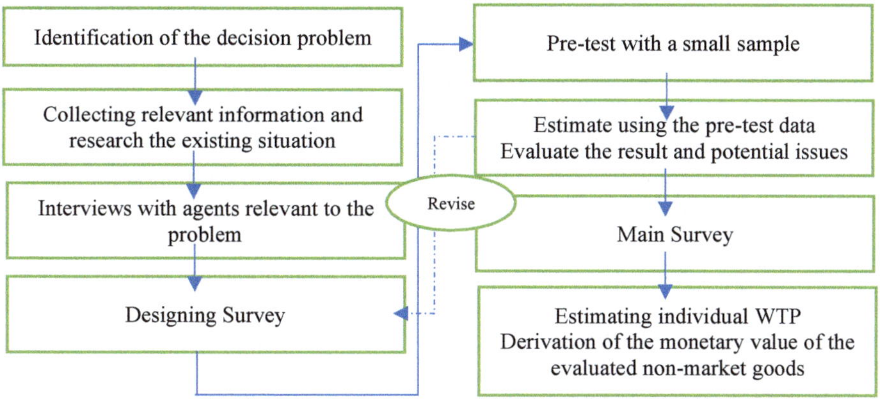

Fig. 4.2 Stated preference method procedure. Source: Inspired by Washida (1999), modified and recreated by author

the multi-attribute utility function. The basic idea behind the multi-attribute utility function is that consumers derive utility from each attribute of a good. Key references for conjoint analysis include Louviere et al. (2000), Train (2003), and Holmes and Adamowicz (2003).

While there are different kinds of conjoint analysis (compositional model vs. decompositional model, self-explicated vs. profile data (Green and Krieger 1991), we focus mainly on the discrete choice experiment (DCE) in this chapter. DCE and pair-wise evaluation methods discussed in Sect. 4.2 are the conjoint analysis methods based on random utility theory. Conjoint analysis used in environmental evaluation is mainly utilizing DCE. Since (McFadden 1974) integrated the random utility model (Thurstone 1927) with a discrete choice model, various discrete choice models have been developed, such as conditional logit model, nested logit model (McFadden 1978), random parameter logit (Revelt and Train 1998; Mcfadden and Train 2000) and latent class model (McFadden 1986). A well-described history of conjoint analysis can be found in Holmes and Adamowicz (2003).

Suppose the environmental good, program, project, or policy can be expressed with multiple attributes, including one expressed in monetary terms. In that case, it is possible to apply DCE to evaluate the monetary values of each attribute as well as the good itself. DCE applications in environmental evaluation are extensively used in the fields of environmental and resource economics (Scarpa et al. 2009; Cameron et al. 2023; Kruse and Atkinson 2022; Uehara et al. 2023; Khan et al. 2020; Meyer 2013; Tang and Zhang 2016; De Marchi et al. 2020; Ndunda and Mungatana 2013; Gundlach et al. 2018), energy economics (Longo et al. 2008; Kuhn et al. 2022; Roe et al. 2001; Paudel et al. 2023; Plum et al. 2019), transportation economics (Will et al. 2022; Plananska and Gamma 2022; Huber and Wicki 2021; Yap et al. 2016), agricultural and food economics (Sayadi et al. 2009; Grunert et al. 2014; Lusk et al. 2003; Darby et al. 2008; Onozaka and McFadden 2011; Uchida et al. 2014; Latacz-Lohmann and Schreiner 2019; Onozaka et al. 2023; Aoki and Akai 2023) and health economics (Tsuge and Takeuchi 2005; Alberini and Ščasný 2011; Clark et al. 2014; Pereira et al. 2011; Bonner et al. 2023; Kreps et al. 2020). The topics include water, air, soil, and noise pollution management, various climate change policies, national parks and wildlife management, consumer preference for recreational activities, tourism, electric vehicles, renewable energy, sustainable housing, alternative fuels, and many more. There are also many applications in health economics, estimating the value of statistical life and morbidities and uncovering individuals' preferences for various types of vaccines and treatment plans.

Conjoint analysis assumes that consumers value the attributes of goods or services when making decisions. Imagine you are considering buying an electric vehicle (EV). What characteristics of EVs might be essential to you when making a purchasing decision? These characteristics, referred to as attributes in conjoint analysis, may include the manufacturer, range, charging speed, battery life, resale value, performance, safety features, maintenance cost, design, size, color, and environmental impact. Each attribute contributes to the overall value of the good, and the sum of these individual contributions, known as part-worth, determines the total utility of the good. The set of essential attributes and their importance can vary among consumers. However, what is generally assumed in decision-making is that a consumer makes a purchasing decision based on the features and levels of a set of attributes that matter to them. Conjoint analysis enables the estimation of the value of each attribute.

In other words, the marginal utility of each attribute can be estimated. If one of the attributes is expressed in monetary terms, such as price, fee, or tax, the marginal willingness to pay (MWTP) for each attribute can also be calculated. Estimating MWTP is particularly useful for evaluating future products, programs, and policies, as it allows us to predict the added value of a product given a marginal change in one or more attributes. For example, if an EV offers a 100 km longer range, how much additional value would a customer attribute to it? How much extra are people willing to pay if the charging time is reduced by 10 min? If the lifetime greenhouse gas (GHG) emissions of an EV are included as an attribute, we can assess whether consumers care about the GHG emission level of an EV and how much they are willing to pay for a one-ton reduction in GHG emissions. This information is invaluable for EV producers and policymakers in designing strategies and policies to promote EV adoption and reduce GHG emissions.

There are three representative styles of question types: full-profile rating, pair-wise rating, and choice experiment. Full-profile rating directly asks for the purchase probability for the presented goods, one by one (Fig. 4.3). If a policy is evaluated, the probability of agreeing with the policy is asked. Since consumers do not usually have experience rating their purchase probability specifically, answering this type of question could be demanding for respondents. Full-profile rating does not rely on a random utility model. It can be estimated using linear regression analysis, where the given probability is treated as the dependent variable, and the attributes and their levels are used as independent variables.

Pair-wise rating asks a respondent to compare two products and rate the relative attractiveness of each pair. As shown in Fig. 4.4, each respondent is asked to mark

	Price: 1 million TL Range: 300 km Warranty: 3 years VAT exemption: 25% Lifetime GHG Emission: 10 ton	Purchase Probability (Answered by each respondent) 45
	Price: 2 million TL Range: 500 km Warranty: 10 years VAT exemption: 50% Lifetime GHG Emission: 20 ton	62

Fig. 4.3 Full-profile rating example. Source: Generated by Author

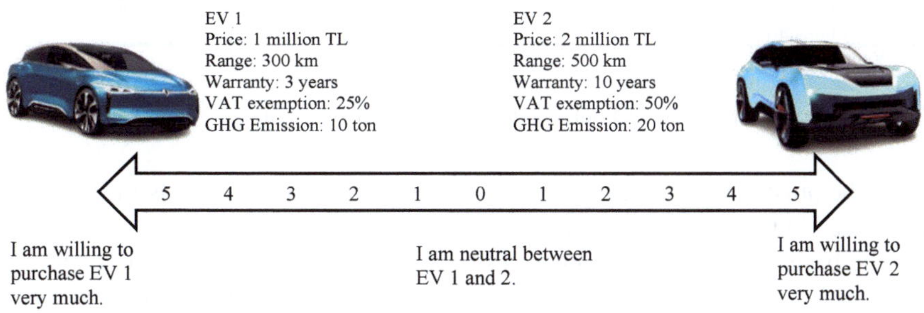

Fig. 4.4 Pair-wise rating example. Source: Generated by Author

the relative rating of each combination of two products or policies. This type is more manageable for respondents to answer compared to a full-profile rating.

Figure 4.5 illustrates the basic structure of discrete choice experiment (DCE). The process of asking respondents to rank all the alternatives from most to least preferred is known as the full-ranking method, while if the best two or three alternatives to be chosen, it is called the partial-ranking method. The choice experiment is specifically asked to choose the best alternative (Holmes and Adamowicz 2003). The estimation procedure of DCE using R-program is discussed in Sect. 4.4.

4.2 Theory

4.2.1 Random Utility Theory

Random utility theory (RUT) is the economic theoretical background of choice experiment, ranking, and pair-wise rating methods. In this section and onward, we focus on the discrete choice experiment. Application of choice experiment depends on Lancaster's multi-attribute utility function (Lancaster 1966) and random utility theory (McFadden 1974). Lancaster introduced the concept that utility is derived

	Alternative 1 (EV 1)	Alternative 2 (EV 2)	Alternative 3 (non- EV)

Attribute 1 (Price)	1 million TL	2 million TL	0.8 million TL
Attribute 2 (Range)	300 km	500 km	N.A.
Attribute 3 (Warranty)	3 years	10 years	N.A.
Attribute 4 (VATExemption)	25%	50%	0%
Attribute 5 (GHG Emission)	10 ton / lifetime	20 ton / lifetime	40 ton/ lifetime
Your Choice			

Fig. 4.5 Choice experiment example. Source: Generated by Author

from the attributes (characteristics) of goods rather than the goods themselves, as illustrated in the EV example above.

The basis of RUT is to express utility function with observable (V_{ij}) and unobservable (ε_{ij}) parts for researchers, where i indicates i th response[1] and j represents profile out of the choice set $C = \{1,2, ..J\}$ where J is the number of profiles presented in a choice set. Now utility is derived from multiple attributes of a good, $x_1, x_2 \ldots x_A$, where A is the number of attributes. For EV case, A is five and $x_1 = price, x_2 = range, x_3 = warranty, x_4 = vat, and, x_5 = emission$, where C = {EV1, EV2, non-EV}.

$$U_{ij} = V_i(x_{1j}, x_{2j} \ldots x_{nj}) + \varepsilon_{ij} \tag{4.1}$$

Figure 4.5, since there are three alternatives presented at a time, $J = 3$. If the first respondent selects the second alternative out of three alternatives, it can be expressed as $U_{1,2} = V_{1,2} + \varepsilon_{1,2}$. Now we consider the probability of a respondent selecting jth profile against $k(j \neq k)$ and express it as:

$$P_{ij} = \Pr[U_{ij} > U_{ik}, \forall j \in C, j \neq k] \tag{4.2}$$

When we combine (4.1) and (4.2), the choice probability can be expressed as (4.3) below:

$$P_{ij} = \Pr[V_{ij} + \varepsilon_{ij} > V_{ik} + \varepsilon_{ik}, \forall j \in C, j \neq k] \tag{4.3}$$

By collecting the observed part of the utility to the left and the difference of the unobserved part of the utility to the left, we obtain (4.4).

[1] Since each respondent is asked to decide on a multiple choice set, i goes from 1 through N, where N = # respondents * # choice set answered per respondent.

$$P_{ij} = \Pr[V_{ij} - V_{ik} > \varepsilon_{ik} - \varepsilon_{ij}, \forall j \in C, j \neq k] \tag{4.4}$$

4.2.2 Conditional Logit Model

Assuming each unobserved error term follows type I extreme value distribution (a.k.a. Gumbel distribution), the difference between the two random variables with type I extreme value distribution follows a logistic distribution. Given this assumption, Eq. (4.4) can be rewritten as follows:

$$P_{ij} = \frac{e^{\mu v_{ij}}}{e^{\mu v_{i1}} + e^{\mu v_{i2}} + \cdots + e^{\mu v_{iJ}}} \tag{4.5}$$

where the denominator contains the number of terms equal to the number of profiles presented each time. P_{ij} is the probability of the respondent i (more specifically, i th response) of selecting j th profile. For EV example, it simply becomes $e^{\mu v_{i1}} + e^{\mu v_{i2}} + e^{\mu v_{i3}}$, and $V_1(price_1, range_1, warranty_1, vat_1, emission_1)$ for the first profile presented. μ is a scale parameter of the distribution and is generally set to 1. By assuming the linearity, the linear utility function is expressed as:

$$V_j\left(x_{1j}, x_{2j} \ldots x_{Aj}\right) = \beta_1 x_{1j} + \beta_2 x_{2j} + \cdots + \beta_A x_{Aj} + \gamma_j{}' z_i \tag{4.6}$$

where this is the observed part of the utility function defined with A alternatives ($a = 1 \ldots A$, index for attributes) for j th profile and z_i is the choice invariant characteristics of the individual i, such as income, education level, or age. Note that if the coefficients for the choice invariant variables $\gamma_j{}'$ are zero, it simply becomes a multinomial logit model. With EV example, it can be written as $V_j\left(price_j, range_j, warranty_j, vat_j, emission_j\right) = \beta_0 + \beta_1 price_j + \beta_2 range_j + \beta_3 warranty_j + \beta_4 vat_j + \beta_5 emission_j$. We estimate β_j by using a maximum log-likelihood function:

$$lnL = \sum_{i=1}^{N} \sum_{j=1}^{J} d_i^j lnP_{ij} \tag{4.7}$$

where d_i^j is a dummy variable set to 1 if the individual i selects the profile j and 0 if the profile is not selected. That is, if the second alternative is selected out of three profiles, the function simply becomes $lnP_{i2}(= 0lnP_{i1} + 1lnP_{i2} + 0lnP_{i3})$. Note that N is the total number of choice experiment answered, not a total number of respondents. If there are 100 respondents asked five choice sets to select from three alternatives each, N becomes $100 \times 5 = 500$, and J is 3. With EV example, the log-likelihood function can be expressed as:

$$lnL = \sum_{i=1}^{N}\sum_{j=1}^{3} d_i^j lnP_{ij} = \sum_{i=1}^{N}\left[d_i^1 lnP_{i1} + d_i^2 lnP_{i2} + d_i^3 lnP_{i3}\right] \quad (4.8)$$

For the first respondent, who repeated five choice experiments (hence $i = 1 \sim 5$), and selected 1st, 3rd, 2nd, 3rd, and 1st profiles for each experiment, this part becomes $lnP_{11} + lnP_{23} + lnP_{32} + lnP_{43} + lnP_{51}$ and for each probability function can be stated as follows. Note that P_{11} is the probability for the respondent 1 (answering the first choice experiment) to select the first profile out of three alternatives.

$$P_{11} = \frac{e^{v_{11}}}{e^{v_{11}} + e^{v_{12}} + e^{v_{13}}}, P_{23} = \frac{e^{v_{23}}}{e^{v_{21}} + e^{v_{22}} + e^{v_{23}}}, P_{32} = \frac{e^{v_{32}}}{e^{v_{31}} + e^{v_{32}} + e^{v_{33}}}$$

$$P_{43} = \frac{e^{v_{43}}}{e^{v_{41}} + e^{v_{42}} + e^{v_{43}}}, P_{51} = \frac{e^{v_{51}}}{e^{v_{51}} + e^{v_{52}} + e^{v_{53}}}$$

Since McFadden (1974) integrated the discrete choice model with the random utility model, the conditional logit model has become widely used. However, its limitations have led to the development of more flexible alternatives, including the nested logit model, random parameter logit model, and latent class model. They are currently used widely in environmental evaluation. The following sections will discuss conditional logit and random parameter logit models in detail.

Although the conditional logit model is computationally less demanding, there are two substantial restrictions: the assumption of IIA (indifference from irrelevant alternative) and the homogeneity of preference. Indifference from irrelevant alternative (IIA) indicates that the likelihood of selecting one option over another (provided both have a non-zero chance of being chosen) remains unchanged regardless of whether additional alternatives are included or excluded from the choice set (Louviere et al. 2000). Note that the probabilities will change once new alternatives are introduced. However, the ratio of the probabilities stays the same under IIA. Suppose two alternatives are initially available and the probability of selecting either can be expressed as $P_1 = P_2 = \frac{1}{2}$ if $U_1 = U_2$. Here, the ratio of probabilities is $\frac{P_1}{P_2} = \frac{1/2}{1/2} = 1$. Now consider adding the third option. Under IIA, $P_1 = P_2 = P_3 = \frac{1}{3}$ if $U_1 = U_2 = U_3$, the ratio of the probabilities ($\frac{P_1}{P_2} = \frac{1/3}{1/3} = 1$) does not change even after the introduction of a new alternative. This is a convenient assumption for estimation since it is unnecessary to re-estimate the model every time a new alternative is added or an existing alternative is removed.

However, IIA could cause the so-called blue bus, red bus problem. Consider a transportation mode choice between blue bus and train. If $U_{train} = U_{bluebus}$, then $P_{train} = P_{bluebus} = \frac{1}{2}$. Now, introduce the perfect substitute for the blue bus, that is a red bus. If IIA holds under the current situation, then $P_{train} = P_{bluebus} = P_{redbus} = \frac{1}{3}$. Now, the probability of selecting a bus becomes 2/3 ($P_{bus} = P_{bluebus} + P_{redbus} = \frac{2}{3}$) from ½ without the red bus option. As a result, adding the red bus resulted in the overvalued probability of buses since blue and red buses are perfect substitutes. This example indicates that if any set of alternatives are related and can be considered as partial substitutes, the probabilities of profiles with similar profiles are overvalued and the rest is undervalued. Since the conditional logit model assumes IIA, a researcher has to make sure that IIA is not violated if the conditional logit model is used. (Hausman and McFadden 1984) suggested such test of IIA.

The second constraint of the conditional logit model is the homogeneity of parameters across individuals. When we estimate β s in Eq. (4.6), we assume these β s are the same for all respondents. Since it is pretty challenging to consider the situation where all the individuals share the same evaluation of the attributes, especially for environmental goods, this constraint is quite restrictive. With the EV example, while for Range, Warranty, and VAT reduction, consumers' preferences could be similar, for GHG emission, the preferences could vary across individuals. For the organic rice case that we will discuss as a case study, environmental quality (EVN), organic certification (CER), and fair trade (FAIR) are likely to display heterogeneous preferences while we may assume homogeneous parameters for Eating Quality (EAT). The estimated result could be biased and may provide inaccurate information if a conditional logit model is used regardless of IIA or heterogeneous preference.

4.2.3 *Random Parameter Logit Model (Mixed Logit Model)*

The random parameter logit model, also referred to as the mixed logit model, is a more flexible approach that relaxes the IIA assumption and accounts for heterogeneous preferences. In order to incorporate heterogeneous parameters, Eq. (4.6) needs to be modified by adding an individual index i to β s and the utility function is now expressed as follows:

$$\begin{aligned} U_{ij}\left(x_{1j}, x_{2j}\ldots x_{Aj}\right) = V_{ij}\left(x_{1j}, x_{2j}\ldots x_{Aj}\right) + \varepsilon_{ij} = \beta_{i1}x_{1j} + \beta_{i2}x_{2j} \\ + \ldots + \beta_{iA}x_{Aj} + \gamma_j{'}z_i + \varepsilon_{ij} \end{aligned} \tag{4.9}$$

where i represents an individual, j is the index for the profile (if three profiles are presented at a time, $j = 1, 2, 3$ (=J)) and a is for attributes (if there are five attributes, a=1,2,3,4,5 (=A)).

By assuming type I extreme value distribution for ε_{ij}, the logit probability for i evaluated at β to select profile k out of $j = 1, \ldots, J$ becomes

$$L_{ij} = \frac{\exp(\beta_{i1}x_{1k} + \beta_{i2}x_{2k} + \cdots + \beta_{iA}x_{Ak})}{\sum_{j=1}^{J}\exp(\beta_{i1}x_{1j} + \beta_{i2}x_{2j} + \cdots + \beta_{iA}x_{Aj})} \tag{4.10}$$

and its choice probability can be expressed as follows:

$$P_{ij} = \int L_{ij}(\beta)f(\beta)d\beta \tag{4.11}$$

Since we cannot observe β, we have to derive the integral of the density function of $\beta, f(\beta)$. Note that the individual-specific choice invariant terms drop in the choice probability function since these do not change between profiles. The density function is usually set at normal distribution with two parameters, mean (b) and the covariance (W), $f(\beta) \sim \Phi(\beta|b, W)$. The parameters b and W need to be estimated. For the details on the random parameter logit model, refer to Train (2003).

4.2.4 Alternative Specific Constants

There are three methods of handling the status quo or opt-out option. 1. Treat the status quo profile as a status quo dummy variable, 2. treat as one of the profiles, 3. assume a nested structure of deciding first either status quo or new options, then among new options. If we adopt the first treatment, the utility functions look like the below, where β_0 is considered as a status quo constant.

$$\begin{aligned} U_{i1} &= \beta_1 x_{i1} + \beta_2 x_{i2} + \ldots + \beta_J x_{iJ} + \gamma_j{}' z_i + \varepsilon_1 \\ U_{i2} &= \beta_1 x_{i1} + \beta_2 x_{i2} + \ldots + \beta_J x_{iJ} + \gamma_j{}' z_i + \varepsilon_2 \\ U_{i3} &= \beta_0 + \gamma_j{}' z_i + \varepsilon_3 \end{aligned} \tag{4.12}$$

If the first method is used, we include (J−1) alternative specific constants (ASCs) in our model. Under this treatment, we can rewrite the utility functions for each profile when there are three alternatives, and the last one is the status quo as below:

$$\begin{aligned} U_{i1} &= ASC_{i1} + \beta_{i1} x_{i11} + \beta_{i2} x_{i21} + \ldots + \beta_{iJ} x_{iA1} + \gamma_j{}' z_i + \varepsilon_1 \\ U_{i2} &= ASC_{i2} + \beta_{i1} x_{i12} + \beta_{i2} x_{i22} + \ldots + \beta_{iJ} x_{iA2} + \gamma_j{}' z_i + \varepsilon_2 \\ U_{i3} &= \beta_{i1} x_{i13} + \beta_{i2} x_{i23} + \ldots + \beta_{iJ} x_{iA3} + \gamma_j{}' z_i + \varepsilon_3 \end{aligned} \tag{4.13}$$

where x_{iaj} represents the attribute a in j th profile presented to the individual i.

If treatment 3 is appropriate, then a nested logit model would be used under the specification of ASCs with Eq. (4.13).

4.2.5 Calculation of Willingness to Pay for Each Attribute

Given the estimated coefficients, marginal willingness to pay (MWTP) for attribute k, the willingness to pay (WTP) for a one-unit change in attribute k can be determined using the following formula:

$$MWTP_k = -\frac{dv}{dx_a} \Big/ \frac{dv}{dp} = -\frac{\beta_a}{\beta_p} \tag{4.14}$$

where x_a is one of the attributes except for price, and p is the price attribute. β_a is the estimated parameters for the attribute "a" while β_p is the estimated coefficient for PRICE. The actual calculation with the estimated coefficients is shown in Sect. 4.5.

4.2.6 Sample Size Calculation

Multiple studies and formulas for determining minimum sample size for discrete choice experiments exist. The most cited among all would be the one based on (Louviere et al. 2000) to measure a choice probability, and the formula is as follows:

$$n \geq \frac{q}{rpc^2}\left[\Phi^{-1}\left(1-\frac{\alpha}{2}\right)\right]^2 \tag{4.15}$$

where n is the minimum required sample size, p is the estimated proportion of the population choosing the preferred option, $q = 1 - p$, r is the number of repeated choice sets asked each respondent, c is the relative accuracy (10%, 20%), α is the significance level (1%, 5%, 10%), and Φ^{-1} is the inverse cumulative normal distribution function of standard normal distribution. c represents the degree of accuracy we aim for when estimating p from a sample size of n, centered around the actual mean. If we set c to 0.1, it indicates our intention for the average estimated p derived from the sample to deviate by no more than ± 10% from the true mean. For example, if α is 0.05, $\left(1-\frac{\alpha}{2}\right)$ = (1–0.05/2) = 0.975. Φ^{-1} finds the z value which corresponds to F(z) = 0.975. It can be calculated as qnorm(0.975) = 1.96 in R or simply find from the standard normal table. If we set the choice probability p as 0.6, c as 0.1, α as 0.05, then the minimum sample size is calculated as follows:

$$n \geq \frac{0.4}{6\times 0.6\times 0.1^2}\left[\Phi^{-1}\left(1-\frac{0.05}{2}\right)\right]^2 = 42.68 \tag{4.16}$$

#R code for sample size calculation using Eq. (4.15).

```
p = 0.4;q = 1-p;r = 6;a = 0.1;k = 0.05;
g = q/(r*p*(a**2))
n<-g*(qnorm(1-(k/2)))**2
n
```

The rule of thumb suggested by Orme (1998) is that if only the main effect is estimated, it is

$$n \geq \frac{500\times L}{J\times r}$$

where L is the maximum number of levels per attribute, J is the number of alternatives in a choice set, and r is the number of choice sets answered by each respondent. In our case, $L = 7$ for PRICE attribute (others are 4 or 2), $J = 3$ and $r = 6$. Hence the minimum sample size is suggested as 195. Other rules of thumb and derivation methods depend on the survey design and objectives. Refer to (de Bekker-Grob et al. 2015) and (Rose and Bliemer 2013) for the further discussion.

4.3 Survey Design

4.3.1 Description of the Case Study

In this section, steps into survey design and implementation are discussed. Since discussion with a specific example is expected to help understand the concepts clearly, a case study on consumer preference for organic rice, conducted in Naga City in the Philippines, is used to illustrate the details henceforward. The estimation of the models will also use the dataset collected by the study. The study's objective was to help newly emerging organic rice farmers back in 2001 to understand consumer preference for organic rice feature by feature. The study details are available in (Ara 2002), and all the monetary terms are expressed in the 2001 Philippine peso (\$1 = 51.8 peso at the time of survey). This study might be considered a marketing study since the price premium for organic rice is estimated at the end of the study. However, the scope of the study is much more comprehensive, from personal tastes of rice to the perceptions of health risks and environmental pollution, as well as preferred administrative structure regarding organic certification and social factors revealing people's support toward fair payment mechanisms for organic farmers.

4.3.2 Attributes and Levels

The procedure of survey design for conjoint analysis is illustrated in Fig. 4.6. Once an environmental or health good/service/policy is identified for evaluation, attributes must be selected carefully. Usually, there are more than six potential attributes in constructing a good or service. However, including more than six attributes could significantly increase the respondents' burden, so we have to carefully select the most important/relevant attributes for the research purpose. Typically, literature reviews of existing relevant articles, interviews with experts and people from the potential target population, and evaluation of related reports could help identify the candidate attributes. A critical criterion for selecting an attribute is whether it is seriously considered by respondents when evaluating the good or service. Focus-group discussion and pre-test results could help determine the importance of each attribute.

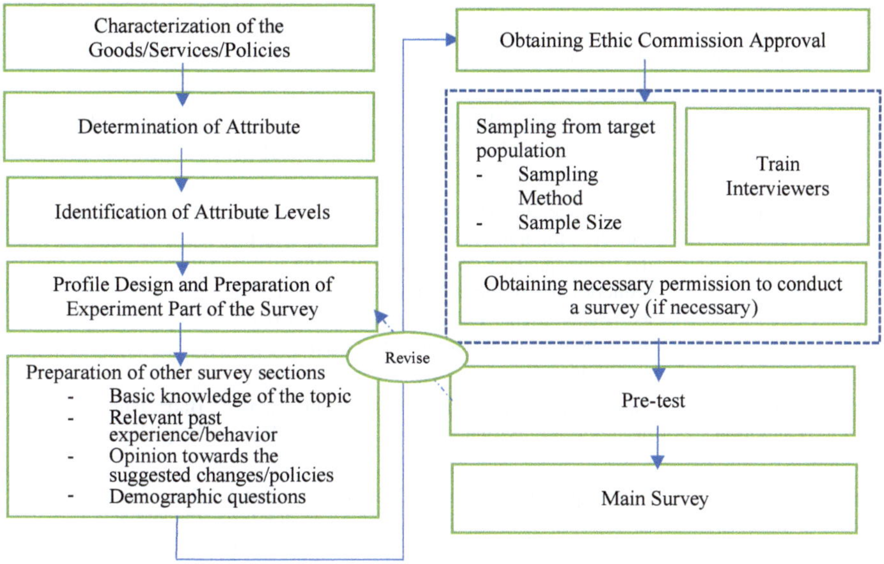

Fig. 4.6 Conjoint analysis survey design procedure. Source: Generated by Author

Although attributes have to be realistic, it is possible to include an attribute that does not exist at the point of the survey if the purpose of the research is to reveal the value of new features of the good or service. As discussed earlier, conjoint analysis allows us to evaluate a hypothetical good with hypothetical characteristics. However, if respondents are not familiar with any of the attributes, a clear explanation of such attributes is necessary. All the respondents must answer questions by understanding exactly the same thing. Attributes and levels used for organic rice study are summarized in Table 4.1.

4.3.3 Profile Design

The total number of profiles that can be generated based on the specified attributes and their levels is calculated as $\prod_{j=1}^{K} level_j$ where $j \in K$ is an index for an attribute with a total of K attributes included. For instance, if there are five attributes and each has four levels, the total number of profiles is $4 \times 4 \times 4 \times 4 \times 4 = 4^5 = 1024$. If the attributes have varying numbers of levels, the calculation adjusts accordingly. For example, with three attributes having four levels, one attribute having six levels, and two attributes having two levels, the total number of profiles is calculated as $4^3 \times 6 \times 2^2 = 1536$.

Since it is nearly impossible to present all profiles to the respondents and ask them to select the best option or rank them from the most favorable to the least, we usually select a subset of these profiles. The most commonly used method of determining the

Table 4.1 Attributes and levels for organic rice study

Attributes (Variable Names)	Description	Levels
Price (PRICE)	Price differences between the rice the respondent regularly buys and the other types of rice	0[a],2,4,6,10,15,20
Reduced Health Risk (RISK)	% of the health risk which can be reduced by choosing the non-status quo rice compared to the status quo rice. We assume 50, 80% reduction as "reduced pesticide rice" and 99% as "organic rice." The different levels of risks are explained with risk ladder to each respondent	0[a],50,80,99%
Environmental Quality (ENV_(Fair,Good, VGood))	Lake water pollution level on farm site due to pesticide use. Explained with a ladder index	Bad[a], Fair, Good, Very Good
Eating Quality (EAT_(Fair, Good, Exc))	Eating quality of rice is evaluated in terms of 1. softness, 2. white grain, 3. good smell and 4. high purity. Fair with 1 & 2 features, Good with 1, 2, & 3, Excellent with 1, 2, 3 & 4 features	Bad, Fair, Good[a], Excellent
Certification (CER_(COOP, NGO, DA))	Whether the rice is certified as reduced pesticide or organic by one of the potential certification bodies (Farmers' Cooperative, NGO, Department of Agriculture)	None[a], Farmers' Cooperative, NGO, Dept. of Agriculture
Fair Trade (FAIR)	If a trader who buys rice from an organic/ reduced pesticide farmer pay the appropriate price of the product, 1, 0 otherwise	No Fair Tarde[a], With Fair Trade

[a]Levels indicating status quo option
Source: Ara (2002)

profiles to be included in the survey is orthogonal main-effect design using orthogonal array. The orthogonal array guarantees that the correlation between attributes in all combinations becomes zero by setting all attributes as orthogonal. Suppose the levels of some profiles are moving in the same direction in many profiles. In that case, we cannot separate the preferences for two different attributes, and a problem similar to the multicollinearity problem in multiple regression analysis occurs.

4.3.3.1 Profile Design with Orthogonal Array Tables

Orthogonal main-effect design can be conducted by using orthogonal array tables. Orthogonal array is constructed to have Balance (all columns have the same number of levels appearing in each column) and Orthogonality (zero correlation coefficient for any two columns). Therefore, instead of experimenting with all possible level combinations, conjoint analysis can be performed using a significantly smaller set of profiles. Depending on the number of attributes and levels, the necessary table

Table 4.2 L16 orthogonal array and label assignment

Profile No.	Att1	Att 2	Att3	Att 4	Att 5
1	1	1	1	1	1
2	1	2	2	2	2
3	1	3	3	3	3
4	1	4	4	4	4
5	2	1	2	3	4
6	2	2	1	4	3
7	2	3	4	1	2
8	2	4	3	2	1
9	3	1	3	4	2
10	3	2	4	3	1
11	3	3	1	2	4
12	3	4	2	1	3
13	4	1	4	2	3
14	4	2	3	1	4
15	4	3	2	4	1
16	4	4	1	3	2

Profile No.	PRICE	RISK	ENV	EAT	CER
1	5	0	Bad	Bad	None
2	5	50	Fair	Fair	Coop
3	5	80	Good	Good	NGO
4	5	99	Very Good	Excellent	DA
5	10	0	Fair	Good	DA
6	10	50	Bad	Excellent	NGO
7	10	80	Very Good	Bad	Coop
8	10	99	Good	Fair	None
9	15	0	Good	Excellent	Coop
10	15	50	Very Good	Good	None
11	15	80	Bad	Fair	DA
12	15	99	Fair	Bad	NGO
13	20	0	Very Good	Fair	NGO
14	20	50	Good	Bad	DA
15	20	80	Fair	Excellent	None
16	20	99	Bad	Good	Coop

Labels	PRICE	RISK	ENV	EAT	CER
1	5	0	Bad	Bad	None
2	10	50	Fair	Fair	Coop
3	15	80	Good	Good	NGO
4	20	99	Very Good	Excellent	DA

Source: Generated by Author

changes. For example, the following table is used for the case with five attributes with four levels each. Instead of 4^5=1024 profiles, 16 profiles are generated for this case. Table 4.2 illustrates how the orthogonal array generates 16 profiles with each attribute level using partial attributes and levels of organic rice example.

4.3.3.2 Profile Design Using SPSS

Using SPSS Conjoint Package to generate profiles using an orthogonal array is also common. Especially if the attributes contain different numbers of levels, it could be convenient to use this program. By selecting Data => Orthogonal Design => nerate tabs, the input window appears. Here, a researcher enters the Factor Name (the name of one of the attributes) and Factor Label (optional), then hit Add. Click on the added attribute in the list, and click Define Values. In the Define Values window, Values and Labels of attributes are entered. Once all the attributes and levels are entered, choose "Create a new dataset" or "Create new data file" depending on your purpose. Once OK is clicked, the generated profiles are displayed in the Data window. SPSS generates the minimum number of cases necessary for the orthogonal design. Figure 4.7 illustrates the entries of attributes and levels, and Fig. 4.8 displays the output of the generated 32 profiles, out of $7 \times 4^4 \times 2 = 3{,}584$ possible profiles by using organic rice case.

Each row in Fig. 4.8 represents one profile. Since the attributes and levels are entered as shown in Table 4.3, for example, the first profile means PRICE = 3 (4 peso price premium), RISK = 2 (50% health risk reduction), ENV = 1 (bad environmental quality), EAT = 1 (bad eating quality), CER = 4 (certified by Department of Agriculture) and FAIR = 2 (with fair trade).

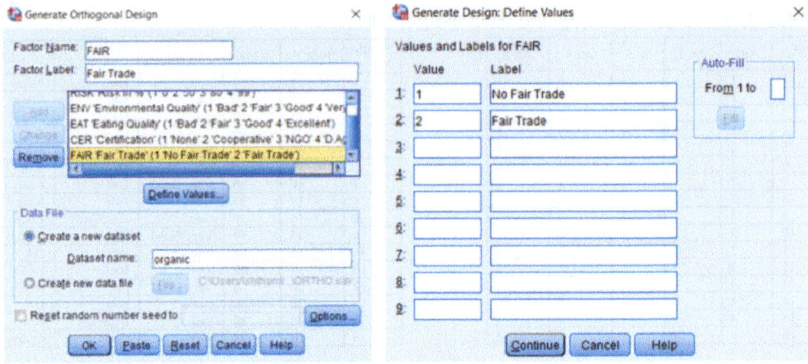

Fig. 4.7 Orthogonal design in SPSS. Source: Generated by Author

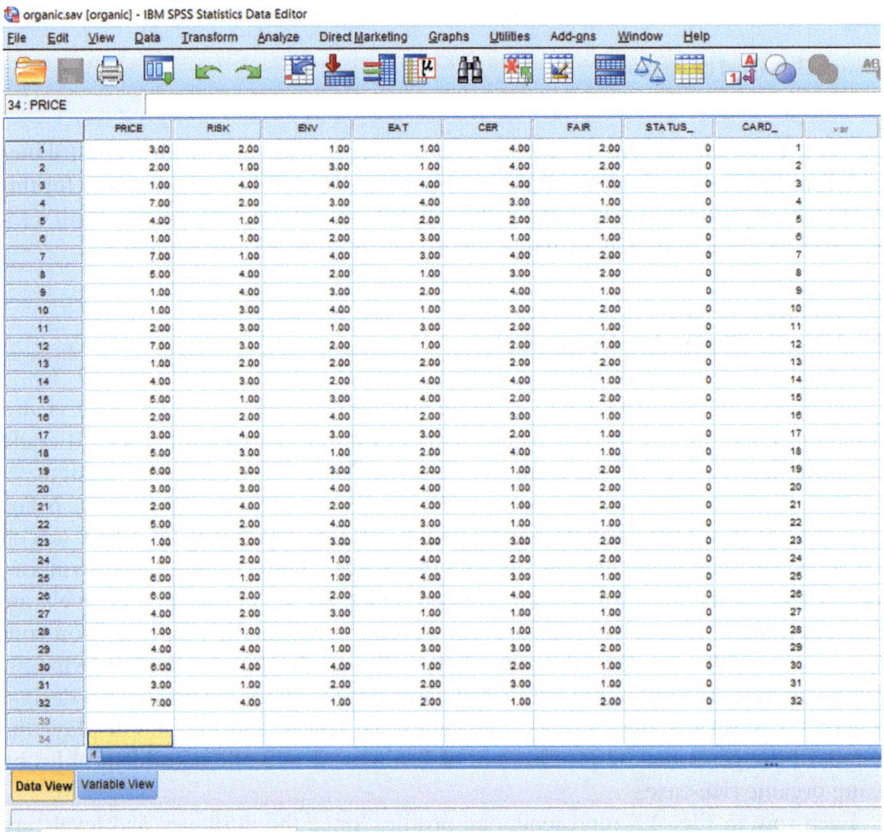

	PRICE	RISK	ENV	EAT	CER	FAIR	STATUS_	CARD_	var
1	3.00	2.00	1.00	1.00	4.00	2.00	0	1	
2	2.00	1.00	3.00	1.00	4.00	2.00	0	2	
3	1.00	4.00	4.00	4.00	4.00	1.00	0	3	
4	7.00	2.00	3.00	4.00	3.00	1.00	0	4	
5	4.00	1.00	4.00	2.00	2.00	2.00	0	5	
6	1.00	1.00	2.00	3.00	1.00	1.00	0	6	
7	7.00	1.00	4.00	3.00	4.00	2.00	0	7	
8	5.00	4.00	2.00	1.00	3.00	2.00	0	8	
9	1.00	4.00	3.00	2.00	4.00	1.00	0	9	
10	1.00	3.00	4.00	1.00	3.00	2.00	0	10	
11	2.00	3.00	1.00	3.00	2.00	1.00	0	11	
12	7.00	3.00	2.00	1.00	2.00	1.00	0	12	
13	1.00	2.00	2.00	2.00	2.00	2.00	0	13	
14	4.00	3.00	2.00	4.00	4.00	1.00	0	14	
15	5.00	1.00	3.00	4.00	2.00	2.00	0	15	
16	2.00	2.00	4.00	2.00	3.00	1.00	0	16	
17	3.00	4.00	3.00	3.00	2.00	1.00	0	17	
18	5.00	3.00	1.00	2.00	4.00	1.00	0	18	
19	6.00	3.00	3.00	2.00	1.00	2.00	0	19	
20	3.00	3.00	4.00	4.00	1.00	2.00	0	20	
21	2.00	4.00	2.00	4.00	1.00	2.00	0	21	
22	5.00	2.00	4.00	3.00	1.00	1.00	0	22	
23	1.00	3.00	3.00	3.00	3.00	2.00	0	23	
24	1.00	2.00	1.00	4.00	2.00	2.00	0	24	
25	6.00	1.00	1.00	4.00	3.00	1.00	0	25	
26	6.00	2.00	2.00	3.00	4.00	2.00	0	26	
27	4.00	2.00	3.00	1.00	1.00	1.00	0	27	
28	1.00	1.00	1.00	1.00	1.00	1.00	0	28	
29	4.00	4.00	1.00	3.00	3.00	2.00	0	29	
30	6.00	4.00	4.00	1.00	2.00	1.00	0	30	
31	3.00	1.00	2.00	2.00	3.00	1.00	0	31	
32	7.00	4.00	1.00	2.00	1.00	2.00	0	32	
33									
34									

Fig. 4.8 Orthogonal design output in SPSS. Source: Generated by Author

Table 4.3 Attributes and levels entries in SPSS

Factor Name	Factor Label	Values and Labels
PRICE	'Price in peso'	(1 '0' 2 '2' 3 '4' 4 '6' 5 '10' 6 '15' 7 '20')
RISK	'Risk in %'	(1 '0' 2 '50' 3 '80' 4 '99')
ENV	'Environmental Quality'	(1 'Bad' 2 'Fair' 3 'Good' 4 'Very Good')
EAT	'Eating Quality'	(1 'Bad' 2 'Fair' 3 'Good' 4 'Excellent')
CER	'Certification'	(1 'None' 2 'Cooperative' 3 'NGO' 4 'D.A')
FAIR	'Fair Trade'	(1 'No Fair Trade' 2 'Fair Trade')

Source: Ara (2002)

This profile reveals one potential problem of orthogonal design. Some generated profiles could contain very unrealistic or conflicting contexts. For example, if we select organic rice over conventional rice, we expect the environmental quality to improve. However, the first profile states that a consumer will pay 4 pesos extra to obtain bad environmental and quality rice, although the health risk is reduced by 50%, certified by the government, and fair-traded. While it is possible to remove unrealistic profiles to avoid confusion among respondents, the orthogonality of the profiles decreases by doing so. Hence, unless it is a hundred percent impossible profile such as a 20 peso price premium with 0% risk reduction, bad environmental and eating qualities, without certification or fair trade, slightly unrealistic ones are kept not to jeopardize the orthogonality.

4.3.3.3 Profile Design Using R-program

If SPSS is not readily available, a conjoint package of the R-program can also be used to generate profiles. The below code is an example of five attributes with four levels each, using orthogonal design. Other available types include "full," "fractional," "ca," and "aca."

#R code for profile design

```
library(conjoint)
experiment<-expand.grid(
  price=c("5","10","15","20"),
  risk=c("0","50","80","99"),
  env=c("bad","fair","good","very good"),
```

```
  eat=c("bad","fair","good","excellent..."),
  cer=c("no","coop","ngo","da"))
design=caFactorialDesign(data=experiment,type="orthogonal")
print(design)
print(cor(caEncodedDesign(design)))
```

4.3.3.4 Creating Choice Sets

Once profiles are generated, we ask respondents to rank the choice probability for each profile if the full-profile rating is used. For pair-wise rating, 16 choice sets with two profiles each are generated. Since it is demanding for respondents to evaluate all 16 pairs meaningfully, we usually generate several versions of the same survey, containing less than or equal to ten pairs each. In our case, two versions containing eight pairs each are reasonable.

Once 32 profiles are generated, choice sets have to be determined. The number of profiles included in a choice set varies. However, if many profiles are to be evaluated and compared in one choice set, the respondent's burden could grow, resulting in fatigue, and the credibility of the responses could drop. Usually, 3 or 4 profiles are included in each choice set. Our case study included three profiles per choice set, one being the status quo. Let us proceed with the case of 3 profiles per choice set. (Note that inclusion of status quo/prefer not to select (purchase) to each choice set decreases the efficiency of the profiles since the status quo profiles appear in all the choice sets.)

Louviere et al. (2000) introduce four different methods to generate choice sets (pages 114–115). Here, the simplest two methods are introduced. Let L represent the total number of generated profiles and M the number of profiles per choice. While the fourth method is the most efficient, it may not be practical when the generated profiles L increases. We apply two profiles ($M = 2$) + 1 status quo case with $L = 32$ case to the described methods.

Method 1: Given the total profiles generated ($L = 32$), copy-paste the same profiles $M-1$ times. Since $M = 2$ in our case, we copy-paste one more set of profiles, and in the end, we have two identical sets of 32 profiles. Let's call them Right and Left profiles. We randomly pick one profile from the Right and do the same from the Left. This operation generates our first pair of choice sets. We repeat this procedure without replacement of the already picked profiles for $L = 32$ times. Make sure that all of the pair is different. Copy-paste the original profiles twice if three profiles per choice set are included. This time, randomly pick one profile from Right, Middle, and Left to generate a choice set with three different profiles. The procedure is illustrated in Fig. 4.9. A status quo option can be added to all of the generated choice sets. This is the most straightforward, but least statistically efficient procedure.

Method 2: Create M distinct profiles of size L as described in the previous section. If SPSS is used for orthogonal design, repeat the process twice using the same set of

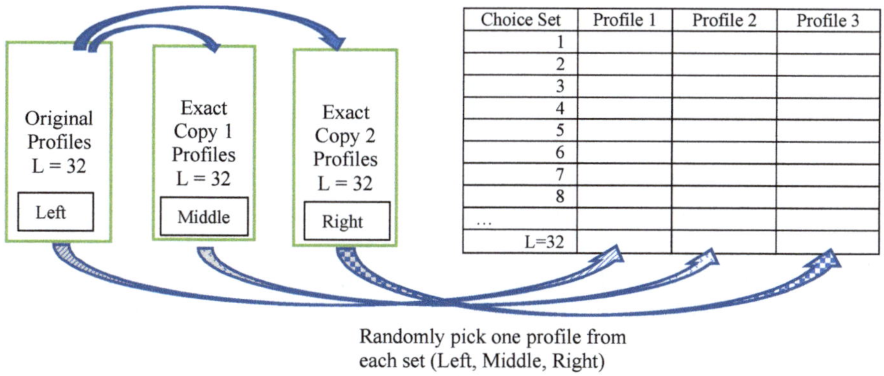

Fig. 4.9 Illustration of binding choice sets, Method 1. Source: Generated by Author

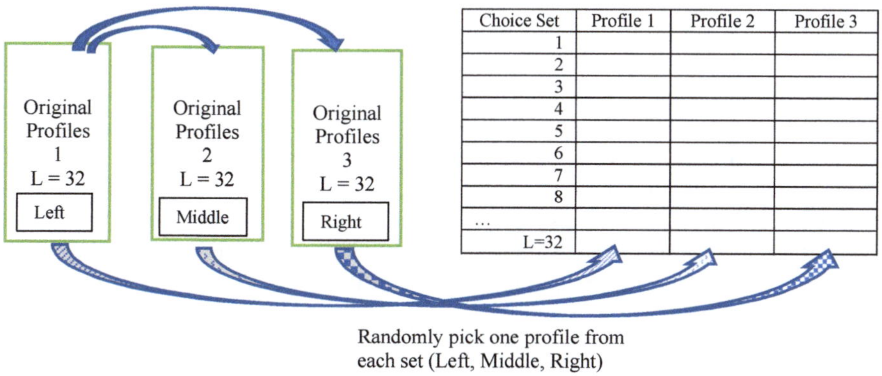

Fig. 4.10 Illustration of binding choice sets, Method 2. Source: Generated by Author

attributes and levels. The resulting profiles differ if the seed values are different (or not set) each time. Once M profiles are generated by repeating the exact procedure M times, randomly pick one from each L. When M = 2 and L = 32, we generate 32 profiles twice, again allocating them to the Right and the Left. Then, randomly pick from each side one at a time to generate 32 pairs (Fig. 4.10). For the remaining two methods, refer to (Louviere et al. 2000).

4.3.4 Explaining Choice Experiment to Respondents

Once the choice sets are determined, a researcher should control for the dominant profile; one of the profiles dominates other profiles in all attributes (Terawaki et al. 2003). For example, if the organic rice's price premium is 0 while the health risk reduction is 99%, environmental and eating quality are very good/excellent, certified by one of the alternative agencies, and is fair-traded such a profile will likely be preferred regardless of the other profiles in the choice set. As the result, no information is collected from the choice set if every respondent selects the same profile.

The number of choice experiments asked per respondent generally varies between 5 and 10. There is a trade-off between the amount of information collected from a respondent and the burden imposed by asking too many choice experiments for a single respondent. Several factors influence the burden on respondents, including the number of profiles per choice set, the number of attributes per profile, the number of levels per attribute, respondents' familiarity with the topics, and the overall length of the survey. To finalize the number of choice sets in a survey, it is essential to conduct focus group discussions and pre-tests.

Using the organic rice example, suppose that the 8th and 14th profiles are matched to be presented in one of the choice sets. As shown in Table 4.4, a table needs to be prepared with a clearly and easily understood format.

Before choice experiment questions are asked, it is critical to describe each attribute clearly so that all the respondents have the same understanding of the meanings, impacts, and scales. The organic rice study was conducted in face-to-face format. The interviewers used verbal descriptions of each attribute and panel displays. When we state levels as subjective expressions "bad" or "good," we have to define each level as clearly as possible so that when we say "good" eating quality, all the respondents imagine the rice with the same or very similar eating quality. Note that we have set EAT_GOOD as the level of inorganic rice. During the survey design phase, these levels have been discussed with the rice consumers in Naga City in the Philippines at the point of research, and they evaluated EAT_GOOD as the eating level of the inorganic rice.

Especially when a researcher conducts conjoint analysis in developing countries, the expressions with percentages (such as an 80% reduction in risk) may not be understood clearly. If such a situation is observed in pre-tests, the attribute description can be supplemented with a risk ladder compared with the familiar risks to the respondents. Examples from (Ara 2002) are shown in Figs. 4.11 and 4.12.

Table 4.4 Choice set display example

Factors	Alternative 1	Alternative 2	Alternative 3
Price	P 10 Higher	P 6 Higher	Current Price
Health Risk	99% Reduced (0.1/100,000)	80% Reduced (2/100,000)	Current Risk (10/100,000)
Impact on the Environment	Fair	Fair	Bad
Eating Quality	Bad	Excellent	Good
Certification	NGO	DA	None
Fair Trade	Fair Trade	No	No
Your Choice	❑	❑	❑

Source: Ara (2002)

1. **Farm Water Quality**: Imagine rice fields next to a lake. If rice farmers use pesticides and chemical fertilizers on their farms, those chemicals can run into the lake and pollute the water. The pollution level depends on the amount of chemical inputs the farmers use. Now take a look at Panel 2.

<Point Panel 2>

Four levels of water quality will be shown in the rice package you are going to assess. These levels are:

A: Very good, B: Good, C: Fair, D: Bad

Assume the water quality level depends on the amount of chemical fertilizer and pesticide the farmers use. If farmers use fewer chemical inputs and manage their farms in an environmentally friendly way, water quality can be improved. However, if they use pesticides, fish living in the water can be polluted, which may not be good for your health if you eat such fish.

---

2. **Eating Quality**: Now assume there are four levels of eating quality: Excellent, Good, Fair, and Bad.

<Point Panel 3>

- Excellent: Soft texture, white grains, good smell, high purity.
- Good: Soft texture, white grains, good smell, but no high purity.
- Fair: Soft texture, white grains, but lacks other good features.
- Bad: None of these good features.

Assume that the eating quality of the rice you eat every day is "Good."

---

3. **Certification**: If the rice you are going to buy is organic or a combination, you may wonder if it is truly grown organically or with reduced pesticide and chemical fertilizer use. There are three possible institutions that can test and certify whether it is organic or a combination:

1. Farmer's Cooperative
2. Non-Governmental Organization (NGO)
3. Department of Agriculture

Including the option of no certification, there are four different options for certification.

---

4. **Fair Trade**:

<Point Panel 3>

- Fair Trade: Producers receive a proper selling price when they sell their rice.
- No Fair Trade: Producers receive a regular or prevailing purchasing price. With Fair Trade, producers can obtain 10% to 15% higher purchasing prices.

Fig. 4.11 Example of attribute descriptions. Source: Ara (2002)

4.4 Estimation Procedures

In this section, the estimation of random parameter logit model using R-program is demonstrated. Although the NLOGIT program is also commonly used to estimate choice experiments and derive willingness to pay, R-program has the advantage of being readily available online; anyone can download the program and conduct the conjoint analysis right away for free. If one does not have an R-program, download the latest version at https://www.r-project.org/ or https://cran.r-project.org/, and also download RStudio, which is user-friendly R interface, for convenience at https://posit.co/.

In our example, two packages are used to estimate the random parameter logit model, namely mlogit and gmnl packages. The structure of these packages is quite similar to each other. Here, the estimation is demonstrated with the mlogit package.

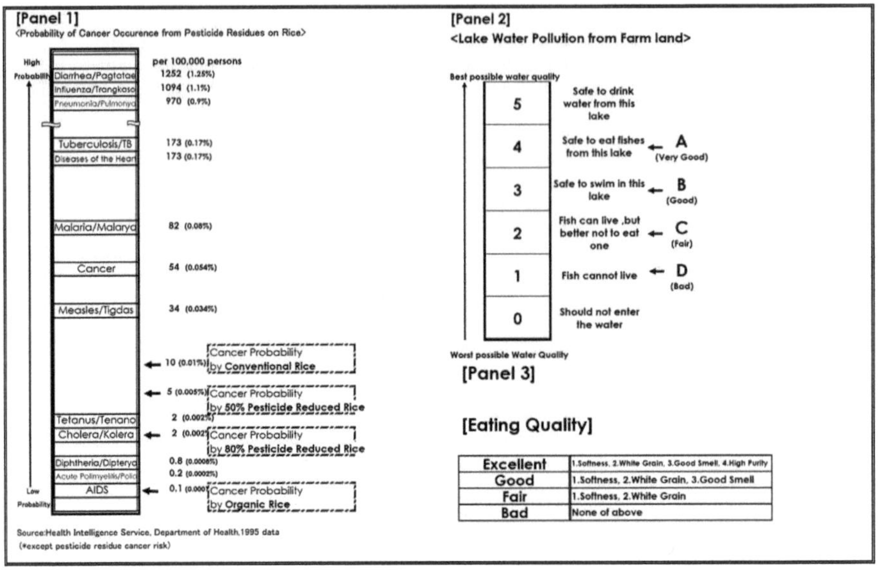

Fig. 4.12 Example of attribute illustrations. Source: Ara (2002)

Importing the mlogit package can be done from RStudio -> Tools -> Install Packages -> {type mlogit in Packages}.

```
library(mlogit)
library(gmnl)
```

The structure of the data is shown in Fig. 4.13. The data is organized in a "long" format.[2] The "long" format is used in the NLOGIT program, which is a paid program also used to conduct DCE), as well. In mlogit.data, both long and wide formats can be used. In "wide" format, each choice set is coded in one row, while in "long" format, each profile from a choice set is coded in one row, and a choice set is coded in three (in our case) or the number of alternatives included in one choice set. Note that the data has been slightly modified from the original data from Ara (2002) for illustration purposes. The descriptions of the variables in Fig. 4.13 are as follows:

idnaga: individual-specific id. Since each respondent answered six choice experiments with three alternatives each, we have 6x3=18 rows belonging to the same respondent. Hence, this column contains 18 cells with number 1, followed by 18 cells with number 2, … till the end with the total number of observations.

[2] The discussion of "wide" versus "long" data format can be found at https://cran.r-project.org/web/packages/mlogit/vignettes/c2.formula.data.html.

	idnaga	case	alt	cset	optout	choice	price	risk	envfair	envgood	envvgood	eatfair	eatgood	eatexc	coop	ngo	da	fair	inc
2	1	1	1	1	0	0	20	50	0	1	0	0	1	0	1	0	0	1	5000
3	1	1	2	1	0	1	6	50	0	1	0	0	0	1	0	0	0	0	5000
4	1	1	3	1	1	0	0	0	0	0	0	0	1	0	0	0	0	0	5000
5	1	2	1	2	0	1	4	50	1	0	0	1	0	0	1	0	0	0	5000
6	1	2	2	2	0	0	10	80	0	0	1	1	0	0	0	0	0	0	5000
7	1	2	3	2	1	0	0	0	0	0	0	0	1	0	0	0	0	0	5000
8	1	3	1	3	0	0	20	80	0	0	0	0	0	1	1	0	0	0	5000
9	1	3	2	3	0	0	2	0	0	0	0	1	0	0	1	0	0	1	5000
10	1	3	3	3	1	1	0	0	0	0	0	0	1	0	0	0	0	0	5000
11	1	4	1	4	0	0	6	50	0	0	1	1	0	0	0	0	1	0	5000
12	1	4	2	4	0	1	2	99	0	1	0	0	0	0	1	0	0	0	5000
13	1	4	3	4	1	0	0	0	0	0	0	0	1	0	0	0	0	0	5000
14	1	5	1	5	0	0	10	0	1	0	0	0	0	1	0	1	0	1	5000
15	1	5	2	5	0	0	6	99	0	0	0	1	0	0	0	1	0	1	5000
16	1	5	3	5	1	1	0	0	0	0	0	0	1	0	0	0	0	0	5000
17	1	6	1	6	0	0	4	99	0	1	0	1	0	0	0	0	0	1	5000
18	1	6	2	6	0	1	2	80	0	1	0	0	0	1	0	0	1	0	5000
19	1	6	3	6	1	0	0	0	0	0	0	0	1	0	0	0	0	0	5000
20	2	7	1	1	0	0	20	50	0	1	0	0	1	0	1	0	0	1	7500
21	2	7	2	1	0	1	6	50	0	1	0	0	0	1	0	0	0	0	7500
22	2	7	3	1	1	0	0	0	0	0	0	0	1	0	0	0	0	0	7500
23	2	8	1	2	0	1	4	50	1	0	0	1	0	0	1	0	0	0	7500
24	2	8	2	2	0	0	10	80	0	0	1	1	0	0	0	0	0	0	7500
25	2	8	3	2	1	0	0	0	0	0	0	0	1	0	0	0	0	0	7500
26	2	9	1	3	0	0	20	80	0	0	0	0	0	1	1	0	0	0	7500
27	2	9	2	3	0	1	2	0	0	0	0	1	0	0	1	0	0	1	7500
28	2	9	3	3	1	0	0	0	0	0	0	0	1	0	0	0	0	0	7500

Fig. 4.13 Example data structure. Source: This study

case: id for each choice set asked. The first respondent answered six choice set (hence case goes from 1 to 6), and then the responses of the second respondent are coded from 7 to 12 and continue till the end ($6 \times 272 = 1632$).

alt: id for each alternative, repeated as 1, 2 and 3 for 272 times. In our study, 1 and 2 are alternative 1 and 2 while 3 is status quo (non-organic) profile.

cset: id for individual choice set, repeated 1 ~ 6 for each choice set for each individual.

optout: dummy variable set to 1 for status quo or opt-out option. Since the third profile is the status quo option in our case, the sequence goes 0, 0, 1 for each choice set.

choice: set as 1 if the profile is selected, 0 otherwise. For example, if the profile 2 is selected, code the choice as 0, 1, 0 for the choice set. "choice" is the left-hand side variable in the formula in the mlogit function.

From **price** to **fair** are the attributes and their levels. The discrete variables (ENV, EAT, CER, and FAIR) are converted to dummy variables. One level of each attribute is set at a baseline, hence dropped from the formula setting.

inc: household income, an example of a socio-demographic variable. A researcher can list the collected socio-demographic information here. Since this data is organized in a "long" format, each piece of information has to be repeated for (# alternatives per choice set × # choice set per respondent) times. Here, for each respondent, socio-demographic variables are repeated $3\times6=18$ times. More individual-specific variables could be included.

Once the collected survey data is organized in the above format, import the data to RStudio, and define the dataset using mlogit.data. The italic terms need to be changed according to the file's names and variables of each researcher.

org<-mlogit.data(*organic1*,choice="*choice*", shape="long", alt.var="*alt*",alt.levels=c("*1*","2","*3*"),id.var="*idnaga*", chid.var="*case*")

After deciding which type of constant structure is appropriate (refer to Sect. 4.2.4. Alternative Specific Constants), the code can be organized one of three ways: (i) a model without constant, (ii) a model with a status quo constant, and (iii) a model with alternative specific constants (if you have status quo option, the constants are added to the non-status quo option).

As shown in the R code below, in "rpar" setting of mlogit function, the distributions for random parameters are set where "n" is normal, "l" is log-normal, "t" is truncated normal, and "u" is a uniform distribution. Here, we have selected "n" for all parameters.[3] Here, we have selected Haltons draws, "bhhh" or Berndt-Hall-Hall-Hausman method for the maximization method ("bfgs" (Broyden-Fletcher-Goldfarb-Shanno) and "nr" (Newton–Raphson) are also available). R is the number of draws of pseudo-random numbers in the simulation. The discussion of Halton draws can be found in Train (2003), Bhat (2001), and Ellis et al. (2019). According to Ellis et al. (2019), the stable random parameter logit model estimation requires more than 1000 draws. The use of more draws is recommended as insufficient draws may cause biases. The larger R gets, the more time-consuming the estimation will be. Hence, initially, using a smaller R like 40 as the default of the mlogit function is recommended and set higher when one finalizes the result. Note that the coefficient monetary attribute (=PRICE) is set as a non-random parameter, not appearing in the ramp definition here. It is possible to include PRICE as a random parameter. However, the calculation of WTP becomes slightly complex. Since each respondent repeated to answer six choice sets, we set panel = TRUE to incorporate the fact that the random parameters of an individual are the same for all six choice settings.

#Model 1: Random Parameter Logit Model without constant.

```
rplnc<- mlogit(choice ~ price + risk + envfair + envgood+envvgood+
     eatfair+eatgood+eatexc+coop+ngo+da+fair|-1, org,
                              rpar = c(risk = "n", envfair="n",
envgood="n",envvgood="n",eatfair="n",eatgood="n",
          eatexc="n",coop="n",ngo="n", da="n",fair="n"),
          R=1000, halton = NA, method="bhhh", panel = TRUE)
summary(rplnc)
```

In order to include the status quo constant (Model 1 of Sect. 4.2.4), the only modification needed is to add the "opt-out" term in the function.

#Model 2: Random Parameter Logit Model with status quo constant.

[3] See other specifications at https://cran.r-project.org/web/packages/gmnl/index.html.

```
rplsqc <- mlogit(choice ~ optout + price + risk + envfair +
envgood+envvgood+
    eatfair+eatgood+eatexc+coop+ngo+da+fair|-1, org,
                          rpar = c(risk = "n", envfair="n",
envgood="n",envvgood="n",eatfair="n",eatgood="n",
eatexc="n",coop="n",ngo="n", da="n",fair="n"),
    R=1000, halton = NA, method="bhhh", panel = TRUE)
summary(rplsqc)
```

and as for the alternative specific constants added to non-status quo options (alternative 1 and 2) by setting the reference level as alternative 3 or status quo option, the function has to be changed slightly by adding the bold part. The default setting is to drop alternative 1 and include ASCs for alternatives 2 and 3.

#Model 3: Random Parameter Logit Model with ASCs.

```
rplasc <- mlogit(choice ~ 1+price + risk + envfair + envgood+envvgood+
    eatfair+eatgood+eatexc+coop+ngo+da+fair, reflevel="3", org,
                          rpar = c(risk = "n", envfair="n",
envgood="n",envvgood="n",eatfair="n",eatgood="n",
      eatexc="n",coop="n",ngo="n", da="n",fair="n"),
      R=1000, halton = NA, method="bhhh",panel=TRUE)
summary(rplasc)
```

4.5 Estimated Results and Interpretations

In this section, the estimated results of random parameter logit models with different constant terms are presented and interpreted.

4.5.1 Estimated Results of Random Parameter Logit Model

The estimated outcomes are listed in Table 4.5. While Model 1 does not contain any constant term, Model 2 and Model 3 are corresponding to Eq. (4.12) and (4.13), respectively.

If ASCs are positive and significant in Model 3, alternative profiles, instead of the status quo option, are selected with a statistically significantly higher probability. This suggests that respondents are inclined to choose a non-status quo option, regardless of the attribute levels. If the estimates are negative and significant, it indicates that

Table 4.5 Estimated result of random parameter logit model for organic rice case

Mean Parameters	Model 1		Model 2		Model 3	
	Mean Parameters	St. Dev. Parameters	Mean Parameters	St. Dev. Parameters	Mean Parameters	St. Dev. Parameters
ASC1					-0.090	
					(0.301)	
ASC2					-0.018	
					(0.260)	
SQC			-0.010			
			(0.258)			
RISK	0.014 ***	0.019 ***	0.014 ***	0.019 ***	0.014 ***	0.019 ***
	(0.002)	(0.002)	(0.002)	(0.002)	(0.002)	(0.002)
ENV_FAIR	0.925 ***	-0.241	0.922 ***	-0.242	0.949 ***	-0.244
	(0.163)	(0.767)	(0.178)	(0.774)	(0.185)	(0.769)
ENV_GOOD	0.894 ***	-0.938 *	0.891 ***	-0.938 *	0.917 ***	-0.962 *
	(0.217)	(0.507)	(0.234)	(0.507)	(0.237)	(0.504)
ENV_VGOOD	1.368 ***	-1.487 ***	1.363 ***	-1.486 ***	1.356 ***	-1.455 ***
	(0.228)	(0.429)	(0.257)	(0.432)	(0.258)	(0.440)
EAT_FAIR	0.825 ***	0.037	0.825 ***	0.036	0.863 ***	0.038
	(0.176)	(1.406)	(0.176)	(1.409)	(0.188)	(1.409)
EAT_GOOD	0.899 ***	-1.187 ***	0.901 ***	-1.187 ***	0.917 ***	-1.196 ***
	(0.173)	(0.231)	(0.177)	(0.232)	(0.181)	(0.233)
EAT_EXC	1.584 ***	0.918 ***	1.582 ***	0.918 ***	1.598 ***	0.923 ***
	(0.190)	(0.300)	(0.198)	(0.301)	(0.200)	(0.301)
CER_COOP	0.735 ***	0.393	0.733 ***	0.394	0.726 ***	0.412
	(0.137)	(0.568)	(0.147)	(0.567)	(0.148)	(0.548)
CER_NGO	0.678 ***	0.695 *	0.676 ***	0.694 *	0.681 ***	0.678
	(0.197)	(0.415)	(0.204)	(0.415)	(0.204)	(0.425)
CER_DA	0.723 ***	1.211 ***	0.720 ***	1.211 ***	0.721 ***	1.203 ***
	(0.152)	(0.275)	(0.161)	(0.275)	(0.161)	(0.275)
FAIRTRADE	0.650 ***	0.902 ***	0.646 ***	0.901 ***	0.653 ***	0.902 ***
	(0.138)	(0.286)	(0.167)	(0.290)	(0.167)	(0.289)
Fixed Parameters						
PRICE	-0.0645 ***		-0.0646 ***		-0.0597 ***	
	(0.010)		(0.010)		(0.013)	
Log likelihood	-1426.5		-1426.5		-1426.3	
N. Observations	1632		1632		1632	

Source: This study

respondents prefer to remain with the status quo, irrespective of the profile characteristics. Similarly, a positive and significant status quo constant in Model 2 implies that respondents tend to favor the status quo option. In contrast, a negative and significant constant indicates that respondents prefer acting by selecting the non-status quo. Although our SQC, ASC1, and ASC2 estimates are not statistically significant, the signs indicate that our respondents are willing to act by selecting organic rice instead of conventional rice.

The likelihood ratio test among the three models indicates that all three models are indifferent regarding the model fit (Table 4.6). Since neither SQC nor ASCs are not statistically significant, we adopt Model 1 without any constant term and interpret the results in terms of Model 1 henceforward. The second column of each model estimate is the standard deviation of the distribution of the random parameters.

lrtest(rplnc,rplsqc,rplasc)

All the attributes are statistically significant, at least at a 5% level with expected signs, except for the CER_NGO variable. Since the baseline for the CER variable is no certification, this result indicates that organic certification by NGOs is different from no organic certification. If the estimated coefficients in Table 4.6 are positive and statistically significant, the probability of the profile selection increases, while the probability decreases if the coefficient is negative. In our case, all the attributes except for PRICE carry positive features which increase utility. As we set ENVBAD

Table 4.6 Likelihood ratio test result

	Df	LogL	Df	Chisq P	r(>Chisq)
1	23	– 1426.5			
2	24	– 1426.5	1	0.0017	0.9672
3	25	– 1426.3	1	0.3964	0.5289

Source: This study

and EATBAD as the baseline variables, other levels for ENV and EAT variables are considered better than the BAD level. Hence, we observe all the positive estimated coefficients. Since an increase in PRICE decreases the utility, respondents tend to select lower-priced profiles. Hence, the sign of the PRICE coefficient is negative, as expected.

As stated earlier, to derive WTP estimates, we have to have price or other attributes expressed in monetary terms. Depending on the context, this attribute can be any of the "Payment Vehicles," such as tax payments, user fees, entrance fees, license fees, donations, or expenditures. Note that the choice of Payment Vehicle could influence respondents' WTP. Refer to the discussion in Sect. 4.6.2. RISK reduction is evaluated positively, indicating that respondents prefer selecting the profile with the higher risk reduction. Similarly, organic certification, especially by cooperatives and the Department of Agriculture, is evaluated positively. At the same time, consumers do care about fair trade mechanisms, supporting the fair payment to the organic rice producers when they make their purchasing decisions.

Although we assumed all attributes except PRICE are random parameters, we can re-estimate the model by setting the variables whose standard deviation parameters are not statistically significant. Using the estimated coefficients, we can calculate the choice probability for each set presented given the assumed choice. Given the profile in Table 4.4, estimated coefficients, and assuming Alternative 2 is selected, the choice probability is calculated with the following steps.

First, we can express the observed part of the utility functions can be expressed as:

$$
\begin{aligned}
V_1 &= \beta_{PRICE}PRICE_1 + \beta_{RISK}RISK_1 + \ldots + \beta_{FAIR}FAIR_1 \\
V_2 &= \beta_{PRICE}PRICE_2 + \beta_{RISK}RISK_2 + \ldots + \beta_{FAIR}FAIR_2 \\
V_3 &= \beta_{PRICE}PRICE_3 + \beta_{RISK}RISK_3 + \ldots \beta_{EATGOOD}EATGOOD_3 + \beta_{FAIR}FAIR_3
\end{aligned}
\tag{4.17}
$$

After plugging in both profiles (Table 4.4) and estimated coefficients (Table 4.5) values, each utility is calculated as follows:

$$
\begin{aligned}
V_{i1} &= -0.0645\times 10 + 0.014\times 99 + \ldots + 0.650\times 1 = 3.005 \\
V_{i2} &= -0.0645\times 6 + 0.014\times 80 + \ldots + 0.650\times 0 = 3.974 \\
V_{i3} &= -0.0645\times 0 + 0.014\times 0 + \ldots 0.899\times 1 + 0.650\times 0 = 0.899
\end{aligned}
\tag{4.18}
$$

The choice probability of selecting Alternative 2 out of these three alternatives can be expressed as:

$$P[2|1,2,3] = \frac{\exp(V_2)}{\exp(V_1) + \exp(V_2) + \exp(V_3)}$$
$$= \frac{1}{1 + \exp(-(V_2 - V_1)) + \exp(-(V_3 - V_1))} = 0.7014$$

meaning that the probability of selecting Alternative 2 out of these three alternatives is calculated as 70.14%. Calculating the similar probability for any selected profiles out of a given choice set is possible.

4.5.2 MWTP Calculation

Marginal willingness to pay (MWTP) is calculated as shown in Table 4.7. For example, the estimated coefficient for RISK variable from Model 1 is 0.014 and it is –0.0645 for PRICE variable. Hence, MWTP is calculated as $-\frac{\beta_{RISK}}{\beta_{PRICE}} - \frac{0.014}{-0.0645} = 0.217$. The same calculation can be done by using the following commands in R. The following command calculates the MWTP. The denominator is set as β_p the estimated price coefficient.

```
-coef(rplnc.mxlu)[- 1] / coef(rplnc.mxlu)[1]
```

Table 4.7 MWTP calculation

	MWTP	Std. Err	
RISK	0.22	0.03	***
ENV_FAIR	14.34	2.65	***
ENV_GOOD	13.86	3.27	***
ENV_VGOOD	21.20	3.49	***
EAT_FAIR	12.79	3.00	***
EAT_GOOD	13.94	3.05	***
EAT_EXC	24.55	3.93	***
CER_COOP	11.39	2.56	***
CER_NGO	10.51	3.37	***
CER_DA	11.20	2.57	***
FAIR	10.07	2.35	***

Source: This study

It is also possible to use wtp.gmnl function in gmnl package below for the same purpose. This function uses the Delta Method to derive standard errors for marginal effects, not the Krinsky and Robb Method. Note that MWTP is derived without multiplying by (−1). Hence, interpret the derived MWTP after multiplying by −1.

```
wtp.gmnl(rplnc.mxlu, wrt=“price”)
```

Although we investigate the signs and statistical significance of the estimated coefficients in Table 4.5, the interpretation is based on the calculated MWTP as shown in Table 4.7. For RISK, a one percent decrease in health risk corresponds to 0.22 pesos. Since we assumed organic rice could result in a 99% health risk reduction, respondents valued the health risk reduction aspect of organic rice as 0.22 peso/% × 99% = 21.78 peso. Note that the average price of inorganic rice paid at the point of survey among the respondents was 18.22 pesos. Hence, this result indicates that health risk reduction is given significant importance. ENV variables need to be interpreted by taking the ENV_BAD as the baseline. Hence, 14.34 pesos for ENV_FAIR means that people are willing to pay 14.34 pesos more to upgrade from ENV_BAD to ENV_FAIR situation, that is, to improve lake water pollution on farm site due to pesticide use from “fish cannot live” in the lake level (BAD) to “fish can live, but better not to eat” (FAIR) situation.

Although we expect MWTP(ENVFAIR)<MWTP(ENVGOOD)<MWTP(ENVVery Good), ENV_GOOD lower than MWTP(ENV_FAIR) with our data. This is difficult to interpret. An improvement from BAD to VERY GOOD (safe to eat fish from the lake), MWTP jumps up to 21.20 pesos. This result may indicate that people did not perceive the difference between ENVFAIR and ENVGOOD levels of lake water quality correctly or evaluate them as very similar. In contrast, improving to the “Very Good” level is highly appreciated.

As for EAT variables, we have set EAT_BAD as the baseline level. Hence, the other levels can be interpreted as the change from BAD to {Fair, Good, Excellent} quality. As illustrated in Fig. 4.12, the eating quality is defined as Excellent level and has all the desired features of rice (softness, white grain, good smell, and high purity), and each level below has lesser preferred features. EAT_BAD does not have any of these features. MWTP for FAIR is 12.79 pesos, meaning people attach this amount to rice's “softness” and “white grain” features. MWTP for GOOD is 13.94; the extra feature added from FAIR to GOOD is good smell. Hence, the value for good smell corresponds to 13.94–12.79 = 1.15 peso per kilo. MWTP attached for Excellent eating quality of rice is estimated as 24.55 pesos, meaning that having all four good features of rice is worth 24.55 pesos. This result indicates that consumers place a very high value on the eating quality of rice.

Interpretation of other attributes is straightforward. The baseline level of organic certification is “no certification.” Hence, we can observe that potential consumers attach the highest value (11.39 pesos) to the certification by the Farmers' cooperative, followed by the Department of Agriculture (11.20 pesos). Certification by NGO was valued the least, and it is 10.51 pesos. Lastly, 10.07 pesos was attached to fair trade. Although consumers do not obtain any tangible benefits by consuming fair-trade rice, they place significant value on organic rice producers getting proper selling prices.

4.6 Conducting Conjoint Analysis in Developing Countries

If one compares the number of studies conducted in developed countries against developing countries, the latter is much smaller, although the latter tends to have more severe environmental issues. When there is no primary study in a country, the benefit transfer method is typically used to derive the value of environmental goods. Although benefit transfer controls the income difference between study and policy sites, the transfer error could be significant due to the differences between two sites beyond the national income level (if a study in a developed country is transferred to a site in a developing country). Because we are in a Global Boiling era, and the developing world is more vulnerable to the damages caused by Climate Change, more studies are needed in developing countries. However, researchers encounter specific difficulties when conducting survey-based research in developing countries. In this section, some examples and potential solutions are discussed.

4.6.1 Understanding the Complex Survey Contents

Compared to developed countries, the share of university graduates is usually lower in the developing world. In some countries, the majority could be elementary school graduates or lower. If the environmental goods a researcher attempts to evaluate influence everyone, including those with lower education levels, it is essential to include their perspectives. Hence, it is critical to design the survey in the format people are familiar with. The problem we encountered in the past is the expression of probabilities and percentages. Probabilities and percentages are often used to describe risks, such as mortality and morbidity, or to express the changes in the current situation of any environmental goods. Both concepts could be challenging to capture the exact situation clearly and correctly. Hence, it is critical to pre-test the use of probabilities and percentages beforehand. Instead of an expression with percentage, an illustration with figures to show how much, for example, is 8 in 10,000 risk could be used as a substitute as illustrated in Fig. 4.14 (Tekeşin and Ara 2014).

As in the organic rice study, the risk description and already familiar risks could also help. It is also recommended to include one or a few straightforward "probability/percentage" quiz questions, such as "Which of the following two conditions has a higher risk of contracting a disease? A. The risk of getting the disease is in 5 out of 10,000 people. B. What is the risk of getting the disease in 10 out of 10,000 people? Which would you rather be in? A or B?" If this kind of question is not answered correctly, it indicates that the respondents do not understand the

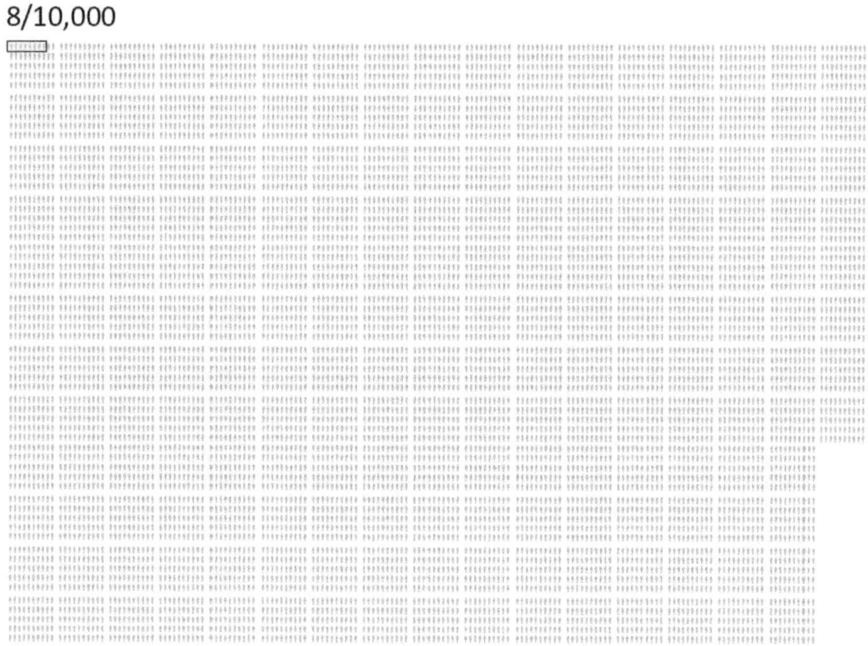

Fig. 4.14 Mortality risk illustration example. Source: Tekeşin and Ara (2014)

risk correctly. A researcher may consider dropping these observations since their evaluation of conjoint analysis may not reveal the true preference.

4.6.2 Selection of Payment Vehicle

As discussed in the earlier section, the choice of payment vehicle is critical. While it has to be realistic, the vehicle should not cause any biased responses outside the context of the choice experiment. For example, the use of tax as a payment vehicle could result in protest votes, especially in developing countries since tax is associated with the government, and if the government is responsible for the changes suggested, respondents refuse to reveal their actual preference for the attributes or profiles because they consider their government should be the one paying. At the same time, if people perceive that the existing (already high) tax collected is not used properly (e.g., environmental tax collected is not used to reduce pollution or environmental hazards), then this tendency could be strengthened. Hence, carefully selecting the attributes and wordings of the attributes' descriptions is required and must be tested before the primary survey.

4.6.3 Conducting Survey in Developing Countries

It is recommended to conduct conjoint analysis face to face in developing countries. As discussed earlier, it is somewhat complex to understand the structure of conjoint analysis clearly. Given the population with shorter average years of education, it is likely not to obtain reliable data through online surveys in developing countries. It is critical to train interviewers for possible questions from the respondents and provide them with standardized clarifying answers. Adapt necessary visual aids such as Que Cards to help respondents capture each concept as researchers expect by incorporating the familiar decision mechanisms for respondents. Although it may be costly, hiring multiple interviewers with desirable education levels for face-to-face interviews within the project's budget in developing countries is still feasible. At the same time, conducting face-to-face interviews in the developed world is usually extremely expensive. Given the determined random sample, it is necessary to establish a control mechanism for interviewers to visit their assigned places to collect responses. Otherwise, researchers could have difficulty conducting data analysis based on the data randomly filled by interviewers.

4.7 Further Exploration of Conjoint Analysis

For those who wish to explore the world of conjoint analysis and discrete choice experiments, here are some suggestions regarding the directions one might take.

The first set of research could be the models that attempt to account for the heterogeneity in preference across individuals by uncovering underlying classes and/or hierarchical data structure. The latent class model (LCM) is used to identify latent (unobserved) classes or segments in a population based on observed choice behavior. The model assumes that the stated preference within a class is similar while between classes is heterogeneous. LCM makes it possible to estimate different parameters for different classes separately. The hierarchical Bayesian model (HBM) is used to incorporate hierarchical data structures and estimate parameters at different levels of the hierarchy. Parameters are estimated for both the individual and the group levels. The choice between these methods depends on the nature of the data and research objectives. Empirical studies using LCM include, but are not limited to, Boxall and Adamowicz (2002), Birol et al. (2006), Campbell et al. (2011), Hensher and Greene (2010), Ortega et al. (2011), Hidrue et al. (2011), Axsen et al. (2015), Choi et al. (2023), and Kyoi et al. (2023). Literature on DCE with HBM contains Huber and Train (2001), Layton and Levine (2003), Regier et al. (2009), Byun and Lee (2017), Antonio et al. (2018), and Hein et al. (2022).

The second set of research could be combining the stated preference method with the revealed preference method. Since there are advantages and disadvantages to each method, the integration and its methodology have been discussed and developed since the late 1980s (Morikawa 1989; Ben-Akiva et al. 1994; Morikawa et al. 2002;

Whitehead et al. 2008; Hensher et al. 2008). The hierarchical Bayesian approach was initially applied in studies such as Adamowicz et al. (1994, 1997), Brownstone et al. (2000), Earnhart (2001), and Boxall and Englin (2008), and has been employed in more recent research by Bottero et al. (2023) and Li et al. (2023).

Another exciting integration of methods is combining DCE with agent-based modeling. Agent-based modeling is often used to study the diffusion of new technologies and innovations. Zhang et al. (2011) developed a multi-agent model by feeding the parameters from the results of a discrete choice experiment to model the diffusion process of alternative fuel vehicles. Since then, studies such as Klein et al. (2020), Chappin et al. (2022), Huber et al. (2023), and Trinh and Munro (2023) have applied similar concepts in different technologies and contexts.

4.8 Concluding Remarks

The intention of this chapter is a detailed introduction to conjoint analysis, focusing on the discrete choice experiment (DCE) so that anyone can conduct a basic DCE using the free software R-program. Although the estimation procedure is vital to learn, survey design has even greater importance since if it is done wrong, there is no meaning in analyzing the collected data. Since, in any country, conducting a survey costs a lot of money and time, it is essential to give extra attention to adequately designing a survey before conducting one. Hopefully, the examples shown by using the organic rice case study would help design your future survey and estimate your DCE models to reach policy implications and recommendations. We also hope that more conjoint analyses and more DCE will be conducted in developing countries to help design future environmental and health policies for the countries under the fast-changing climate.

References

Adamowicz W, Louviere J, Williams M (1994) Combining revealed and stated preference methods for valuing environmental amenities. J Environ Econ Manag 26(3):271–292

Adamowicz W, Swait J, Boxall P, Louviere J, Williams M (1997) Perceptions versus objective measures of environmental quality in combined revealed and stated preference models of environmental valuation. J Environ Econ Manage 32:65–84

Alberini A, Ščasný M (2011) Context and the VSL: Evidence from a stated preference study in Italy and the Czech Republic. Environ Resource Econ 49(4):511–538. https://doi.org/10.1007/s10640-010-9444-8

Antonio ALM, Weiss RE, Saigal CS, Dahan E, Crespi CM (2018) A Bayesian hierarchical model for discrete choice data in health care. Stat Methods Med Res 27(12):3544–3559. https://doi.org/10.1177/0962280217704226

Aoki K, Akai K (2023) A comparison between Spain and Japan with respect to the color, expected taste scale, and sustainability of strawberries: a choice experiment. Food Qual Prefer 103(104671):1–13. https://doi.org/10.1016/j.foodqual.2022.104671

Ara S (2002) Environmental evaluation of organic rice: a case study in the Philippines. Kobe University

Axsen J, Bailey J, Castro MA (2015) Preference and lifestyle heterogeneity among potential plug-in electric vehicle buyers. Ener Econ 50:190–201. https://doi.org/10.1016/j.eneco.2015.05.003

Ben-Akiva M, Bradley M, Morikawa T, Benjamin J, Novak T, Oppewal H, Rao V (1994) Combining revealed and stated preferences data. *Marketing Letters* 5(4):335

Bhat CR (2001). Quasi-random maximum simulated likelihood estimation of the mixed multinomial logit model. *Transportation Research* 35:677–693. www.elsevier.com/locate/trb

Birol E, Karousakis K, Koundouri P (2006) Using a choice experiment to account for preference heterogeneity in wetland attributes: The case of Cheimaditida wetland in Greece. Ecol Econ 60(1):145–156. https://doi.org/10.1016/j.ecolecon.2006.06.002

Bonner KE, Chyderiotis S, Sicsic J, Mueller JE, Ulrich AK, Toomey T, Horvath KJ, Neaton JD, Basta NE (2023) What motivates adults to accept influenza vaccine? An assessment of incentives, ease of access, messaging, and sources of information using a discrete choice experiment. SSM—Population Health 22(101384):1–9. https://doi.org/10.1016/j.ssmph.2023.101384

Bottero M, Bravi M, Caprioli C, Dell'Anna F (2023) Combining Revealed and Stated Preferences to design a new urban park in a metropolitan area of North-Western Italy. Ecol Model 483:110436. https://doi.org/10.1016/j.ecolmodel.2023.110436

Boxall PC, Adamowicz WL (2002) Understanding heterogeneous preferences in random utility models: a latent class approach. Environ Resource Econ 23:421–446

Boxall PC, Englin JE (2008) Fire and recreation values in fire-prone forests: exploring an intertemporal amenity function using pooled RP-SP data. J Agric Resour Econ 33(1):19–33

Brownstone D, Bunch DS, Train K (2000) Joint mixed logit models of stated and revealed preferences for alternative-fuel vehicles. Trans Res Part B 34:315–338. www.elsevier.com/locate/trb

Byun H, Lee CY (2017) Analyzing Korean consumers' latent preferences for electricity generation sources with a hierarchical Bayesian logit model in a discrete choice experiment. Ener Poli 105:294–302. https://doi.org/10.1016/j.enpol.2017.02.055

Cameron DB, Ray I, Parida M, Dow WH (2023) Product preferences and willingness to pay for potable water delivery: experimental evidence from rural Bihar, India. *PLoS One* 18(4):e0283892. https://doi.org/10.1371/journal.pone.0283892

Campbell D, Hensher DA, Scarpa R (2011) Non-attendance to attributes in environmental choice analysis: A latent class specification. J Environ Planning Manage 54(8):1061–1076. https://doi.org/10.1080/09640568.2010.549367

Chappin EJL, Schleich J, Guetlein MC, Faure C, Bouwmans I (2022) Linking of a multi-country discrete choice experiment and an agent-based model to simulate the diffusion of smart thermostats. Technol Forecast Soc Chang 180(121682):1–26. https://doi.org/10.1016/j.techfore.2022.121682

Choi J, Kee D, Lee J, Kim JJ (2023) Understanding heterogeneous consumer preferences for residential zero-energy buildings (ZEBs) in South Korea: a latent class approach. Sustain Cities Soc 97:104747. https://doi.org/10.1016/j.scs.2023.104747

Clark MD, Determann D, Petrou S, Moro D, de Bekker-Grob EW (2014) Discrete choice experiments in health economics: a review of the literature. PharmacoEconomics 32(9):883–902. Springer International Publishing. https://doi.org/10.1007/s40273-014-0170-x

Darby K, Batte MT, Ernst S, Roe B (2008) Decomposing local: a conjoint analysis of locally produced foods. Am J Agr Econ 90(2):476–486. https://doi.org/10.1111/j.1467-8276.2007.01111.x

de Bekker-Grob EW, Donkers B, Jonker MF, Stolk EA (2015) Sample size requirements for discrete-choice experiments in healthcare: a practical guide. Patient 8(5):373–384. https://doi.org/10.1007/s40271-015-0118-z

De Marchi E, Pigliafreddo S, Banterle A, Parolini M, Cavaliere A (2020) Plastic packaging goes sustainable: an analysis of consumer preferences for plastic water bottles. Environ Sci Policy 114:305–311. https://doi.org/10.1016/j.envsci.2020.08.014

Earnhart D (2001) Combining revealed and stated preference methods to value environmental amenities at residential locations. Land Econ 77(1):12–29

Ellis A, de Bekker-Grob E, Howard K, Thomas K, Lancsar E (2019) Number of Halton Draws required for valid random parameter estimation with discrete choice data. Patient 14(4):432–432

Green PE, Krieger AM (1991) Segmenting markets with conjoint analysis. Source J Mark 55(4)

Grunert KG, Hieke S, Wills J (2014) Sustainability labels on food products: consumer motivation, understanding and use. Food Poli 44:177–189. https://doi.org/10.1016/j.foodpol.2013.12.001

Gundlach A, Ehrlinspiel M, Kirsch S, Koschker A, Sagebiel J (2018) Investigating people's preferences for car-free city centers: a discrete choice experiment. Transp Res Part D Transp Environ 63:677–688. https://doi.org/10.1016/j.trd.2018.07.004

Hausman J, Mcfadden D (1984) Specification tests for the multinomial logit model. Econometrica 52(5):1219–1240

Hein M, Goeken N, Kurz P, Steiner WJ (2022) Using Hierarchical Bayes draws for improving shares of choice predictions in conjoint simulations: A study based on conjoint choice data. Eur J Oper Res 297(2):630–651. https://doi.org/10.1016/j.ejor.2021.05.056

Hensher DA, Greene WH (2010) Non-attendance and dual processing of common-metric attributes in choice analysis: a latent class specification. Emp Econ 39(2):413–426. https://doi.org/10.1007/s00181-009-0310-x

Hensher DA, Rose JM, Greene WH (2008) Combining RP and SP data: biases in using the nested logit "trick"—Contrasts with flexible mixed logit incorporating panel and scale effects. J Transp Geogr 16(2):126–133. https://doi.org/10.1016/j.jtrangeo.2007.07.001

Hidrue MK, Parsons GR, Kempton W, Gardner MP (2011) Willingness to pay for electric vehicles and their attributes. Res Ener Econ 33(3):686–705. https://doi.org/10.1016/j.reseneeco.2011.02.002

Holmes TP, Adamowicz WL (2003) Attribute-based methods. In A primer on nonmarket valuation. Kluwer Academic Publishers, pp 171–219

Huber J, Train K (2001) On the similarity of classical and Bayesian estimates of individual mean partworths. Mark Lett 12(3)

Huber RA, Wicki M (2021) What explains citizen support for transport policy? The roles of policy design, trust in government and proximity among Swiss citizens. Energy Res Soc Sci 75(101973):1–10. https://doi.org/10.1016/j.erss.2021.101973

Huber R, Späti K, Finger R (2023) A behavioural agent-based modelling approach for the ex-ante assessment of policies supporting precision agriculture. Ecol Econ 212. https://doi.org/10.1016/j.ecolecon.2023.107936

Khan RAJ, Thaheem MJ, Ali TH (2020) Are Pakistani homebuyers ready to adopt sustainable housing? An insight into their willingness to pay. Ener Poli 143(111598):1–10. https://doi.org/10.1016/j.enpol.2020.111598

Klein M, Lüpke L, Günther M (2020) Home charging and electric vehicle diffusion: agent-based simulation using choice-based conjoint data. Transp Res Part D Transp Environ 88(102475):1–20. https://doi.org/10.1016/j.trd.2020.102475

Kreps S, Prasad S, Brownstein JS, Hswen Y, Garibaldi BT, Zhang B, Kriner DL (2020) Factors associated With US adults' likelihood of accepting COVID-19 vaccination. JAMA Netw Open 3(10):e2025594. https://doi.org/10.1001/jamanetworkopen.2020.25594

Kruse T, Atkinson G (2022) Understanding public support for international climate adaptation payments: evidence from a choice experiment. Ecol Econ 194(107321):1–19. https://doi.org/10.1016/j.ecolecon.2021.107321

Kuhn S, Kutzner F, Thøgersen J (2022) How to make energy efficiency labels more effective: insights from discrete choice experiments in Ghana and the Philippines. Energy Res Soc Sci 84. https://doi.org/10.1016/j.erss.2021.102320

Kyoi S, Kuriyama K, Hashimoto S (2023) People's heterogeneous preferences for future development scenarios: a case study of Ishikawa Prefecture, Japan. Sustain Sci 18(4):1907–1924. https://doi.org/10.1007/s11625-023-01334-z

Lancaster KJ (1966) A new approach to consumer theory. J Polit Econ 74:132–157

Latacz-Lohmann U, Schreiner JA (2019) Assessing consumer and producer preferences for animal welfare using a common elicitation format. J Agric Econ 70(2):293–315. https://doi.org/10.1111/1477-9552.12297

Layton DF, Levine RA (2003) How much does the far future matter? A hierarchical bayesian analysis of the public's willingness to mitigate ecological impacts of climate change. J Am Stat Assoc 98(463):533–544. https://doi.org/10.1198/016214503000000341

Li Z, Washington SP, Zheng Z, Prato CG (2023) A Bayesian hierarchical approach to the joint modelling of revealed and stated choices. J Choice Model 47(100419):1–16. https://doi.org/10.1016/j.jocm.2023.100419

Longo A, Markandya A, Petrucci M (2008) The internalization of externalities in the production of electricity: willingness to pay for the attributes of a policy for renewable energy. Ecol Econ 67(1):140–152. https://doi.org/10.1016/j.ecolecon.2007.12.006

Louviere JJ, Hensher DA, Swait JD (2000) Stated choice methods: analysis and applications. Cambridge University Press

Lusk JL, Roosen J, Fox JA (2003) Demand for beef from cattle administered growth hormones or fed genetically modified corn: a comparison of consumers in France, Germany, the United Kingdom and the United States. Am J Agr Econ 85(1):16–29

McFadden D (1974) Conditional logit analysis of qualitative choice behavior. In: Frontiers in Econometrics. Academic Press, pp 105–142

McFadden D (1978) Modeling the choice of residential location. In: Zarembka P (ed) Spatial interaction theory and planning models. North Holland, pp 75–96

McFadden D (1986) The choice theory approach to market research. Mark Sci 5(4):275–297

Mcfadden D, Train K (2000) Mixed MNL models for discrete response. Source J Appl Econo 15(5)

Meyer A (2013) Intertemporal valuation of river restoration. Environ Resource Econ 54(1):41–61. https://doi.org/10.1007/s10640-012-9580-4

Morikawa T (1989) Incorporating stated preference data in travel demand analysis [Dissertation]. MIT

Morikawa T, Ben-Akiva M, McFadden D (2002) Discrete choice models incorporating revealed preferences and psychometric data. In: Franses PH, Montgomery AL (eds) Econometric models in marketing: advances in econometrics, vol 16, pp 29–56.

Ndunda EN, Mungatana ED (2013) Evaluating the welfare effects of improved wastewater treatment using a discrete choice experiment. J Environ Manage 123:49–57. https://doi.org/10.1016/j.jenvman.2013.02.053

Onozaka Y, Honkanen P, Altintzoglou T (2023) Sustainability, perceived quality and country of origin of farmed salmon: Impact on consumer choices in the USA, France and Japan. Food Pol 117(102452):1–12. https://doi.org/10.1016/j.foodpol.2023.102452

Onozaka Y, McFadden DT (2011) Does local labelling complement or compete with other sustainable labels? A conjoint analysis of direct and joint values for fresh produce claim. Am J Agr Econ 93(3):689–702. https://doi.org/10.1093/ajae/aar005

Orme B (1998) Sample size issues for conjoint analysis studies. Sequim: Sawtooth software technical paper

Ortega DL, Wang HH, Wu L, Olynk NJ (2011) Modeling heterogeneity in consumer preferences for select food safety attributes in China. Food Pol 36(2):318–324. https://doi.org/10.1016/j.foodpol.2010.11.030

Paudel J, Sharifi A, Khan GD (2023) What are the drivers of sustainable energy transition? Insights from an empirical analysis of household preferences for electric induction cooking in Nepal. J Clean Prod 417(138021):1–10. https://doi.org/10.1016/j.jclepro.2023.138021

Pereira CCA, Mulligan M, Bridges JFP, Bishai D (2011) Determinants of influenza vaccine purchasing decision in the US: a conjoint analysis. Vaccine 29(7):1443–1447. https://doi.org/10.1016/j.vaccine.2010.12.027

Plananska J, Gamma K (2022) Product bundling for accelerating electric vehicle adoption: A mixed-method empirical analysis of Swiss customers. Renew Sustain Energy Rev 154(111760):1–14. https://doi.org/10.1016/j.rser.2021.111760

Plum C, Olschewski R, Jobin M, van Vliet O (2019) Public preferences for the Swiss electricity system after the nuclear phase-out: A choice experiment. Ener Pol 130:181–196. https://doi.org/10.1016/j.enpol.2019.03.054

Regier DA, Ryan M, Phimister E, Marra CA (2009) Bayesian and classical estimation of mixed logit: an application to genetic testing. J Health Econ 28(3):598–610. https://doi.org/10.1016/j.jhealeco.2008.11.003

Revelt D, Train K (1998) Mixed logit with repeated choices: households' choices of appliance efficiency level. Rev Econ Stat 80:647–657

Roe B, Teisl MF, Levy A, Russell M (2001) US consumers' willingness to pay for green electricity. Ener Pol 29:917–925

Rose JM, Bliemer MCJ (2013) Sample size requirements for stated choice experiments. Transportation 40(5):1021–1041. https://doi.org/10.1007/s11116-013-9451-z

Sayadi S, González-Roa MC, Calatrava-Requena J (2009) Public preferences for landscape features: the case of agricultural landscape in mountainous Mediterranean areas. Land Use Pol 26(2):334–344. https://doi.org/10.1016/j.landusepol.2008.04.003

Scarpa R, Gilbride TJ, Campbell D, Hensher DA (2009) Modelling attribute non-attendance in choice experiments for rural landscape valuation. Eur Rev Agric Econ 36(2):151–174. https://doi.org/10.1093/erae/jbp012

Tang C, Zhang Y (2016) Using discrete choice experiments to value preferences for air quality improvement: the case of curbing haze in urban China. J Environ Planning Manage 59(8):1473–1494. https://doi.org/10.1080/09640568.2015.1079518

Tekeşin C, Ara S (2014) Measuring the value of mortality risk reductions in Turkey. Int J Environ Res Public Health 11(7):6890–6922. https://doi.org/10.3390/ijerph110706890

Terawaki T, Kuriyama K, Yoshida K (2003) The importance of excluding unrealistic alternatives in choice experiment designs. http://www.taku-t.com/

Thurstone LL (1927) A law of comparative judgment. Psychol Rev 34(4):273–286. https://doi.org/10.1037/h0070288

Train KE (2003) Discrete choice methods with simulation. Cambridge University Press, Cambridge

Trinh TT, Munro A (2023) Integrating a choice experiment into an agent-based model to simulate climate-change induced migration: The case of the Mekong River Delta, Vietnam. J Choi Model 48(100428):1–15. https://doi.org/10.1016/j.jocm.2023.100428

Tsuge T, Takeuchi K (2005) A choice experiment approach to the valuation of mortality. J Risk Uncert 31(1):73–95

Uchida H, Onozaka Y, Morita T, Managi S (2014) Demand for ecolabeled seafood in the Japanese market: A conjoint analysis of the impact of information and interaction with other labels. Food Pol 44:68–76. https://doi.org/10.1016/j.foodpol.2013.10.002

Uehara T, Nakatani J, Tsuge T, Asari M (2023) Consumer preferences and understanding of bio-based and biodegradable plastics. J Clean Prod 417(137979):1–15. https://doi.org/10.1016/j.jclepro.2023.137979

Washida T (1999) Introduction to environmental evaluation (in Japanese). Keisou-Publishing

Whitehead JC, Pattanayak SK, Van Houtven GL, Gelso BR (2008) Combining revealed and stated preference data to estimate the nonmarket value of ecological services: an assessment of the state of the science. J Econ Surv 22(5):872–908. https://doi.org/10.1111/j.1467-6419.2008.00552.x

Will C, Lehmann N, Baumgartner N, Feurer S, Jochem P, Fichtner W (2022) Consumer understanding and evaluation of carbon-neutral electric vehicle charging services. Appl Energy 313(118799):1–20. https://doi.org/10.1016/j.apenergy.2022.118799

Yap MD, Correia G, van Arem B (2016) Preferences of travellers for using automated vehicles as last mile public transport of multimodal train trips. Trans Res Part A Pol Prac 94:1–16. https://doi.org/10.1016/j.tra.2016.09.003

Zhang T, Gensler S, Garcia R (2011) A study of the diffusion of alternative fuel vehicles: an agent-based modeling approach. J Prod Innov Manag 28(2):152–168. https://doi.org/10.1111/j.1540-5885.2011.00789.x

Chapter 5
Spatial Data Analysis: Application to Child Malnutrition in India

Zakir Husain and Mousumi Dutta

Abstract Integrating spatial information with conventional datasets enhances data analysis by uncovering geographic patterns, identifying regional disparities, and detecting spatial clusters. These insights help policymakers in designing targeted interventions through visually intuitive representations. This chapter introduces key steps in spatial data analysis, starting with exploratory techniques such as geo-visualization, choropleth mapping, and statistical distributions. It then examines spatial dependence using global (Moran's I) and local (Gettis-Ord) correlation measures to identify clusters. The analysis is extended to confirmatory modeling by incorporating spatial dependencies into regression models, improving the explanatory power of traditional linear regression. Using GeoDa software, the chapter illustrates these concepts through a case study on the relationship between child marriage and child malnutrition in India.

5.1 Introduction

Data refers to information on a group of variables that may be continuous, discrete and categorical in nature. Implicitly, the data matrix contains information, and the purpose of statistical analysis is to bring out this information to identify patterns and confirm hypotheses. In recent times, the incorporation of locational information has opened up new vistas in statistical analysis. Conventional statistical techniques have been revised to incorporate the locational dimension, leading to the creation of new statistical tools and methods which enrichen the process of information extraction from the data set.

A simple example is the histogram. While histograms reveal the distribution of a variable, incorporating locational information means that the histogram can be

Z. Husain (✉) · M. Dutta
Department of Economics, Presidency University, Kolkata, India

M. Dutta
e-mail: dmousumi1970@gmail.com

F. Hamid and D. Mukherjee (eds.), *Data Science for Modeling Managerial and Socioeconomic Problems*, Contributions to Management Science,
https://doi.org/10.1007/978-981-97-9060-9_5

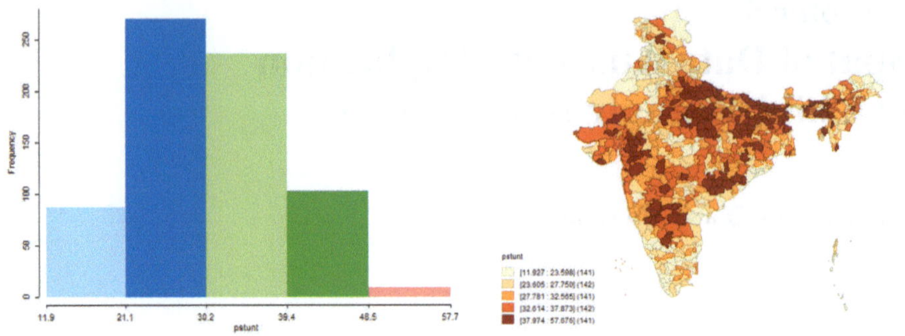

(a) Histogram (b) Choropleth map

Fig. 5.1 Analysis of stunting levels in Indian districts—2019–2021. Source: Estimated by authors using NFHS data (2019–2021)

redrawn as a choropleth map, which brings to the forefront perspectives that cannot be obtained using conventional statistics. An example is Fig. 5.1 which analyzes the percentage of children below five years who are stunted in Indian districts. On the left hand side, we have used a histogram to analyze the variable. The highest proportion of districts have stunting levels between 21.1% and 30.2%, followed by the second bin (30.2% to 39.4%). Simultaneously, we observe that very few districts have stunting levels above 48.5%. The right hand side panel (Fig. 5.1b) draws a map using the same data, adding the locational dimension. The map reveals new information. Stunting is high in the districts of Uttar Pradesh, Bihar, Jharkhand and northern West Bengal and relatively low in Southern India and in the Punjab-Haryana belt.

This example shows us how geovisualizing the data can yield interesting insights. While the mapping of social phenomenon has a long lineage,[1] its theoretical origin lies in the recognition of social interaction and its influence on individual behavior. It was articulated in the large-scale empirical studies on crime and communities undertaken by the sociologists of the Chicago school. The basic premise of the Chicago school was that:

> one cannot understand social life without understanding the arrangements of particular social times and places. Another way of stating this is to say that Chicago felt that no social fact makes any sense abstracted from its context in social (and often geographic) space and social time. Social facts are *located*. This means a focus on social relations and spatial ecology …. Every social fact is situated, surrounded by other contextual facts and brought into being by a process relating to its past contexts. (Abbott 1997: 1152)

The thrust on social place and interaction between actors was echoed in other domains of social science. In economics, there was a recognition of the inadequacies of the neo-classical model of a rational optimizing individual and a recognition of

[1] An early use of maps to study the spatial distribution of disease was by Dr. John Snow during the cholera epidemic of London in 1854 (Tulchinsky 2018).

various forms of external influences on the behavior of the consumer. For instance, Leibenstein pointed out that an increase in demand for a product can have a bandwagon or snob effect, increasing or reducing demand of an individual, respectively (Leibenstein 1950). The explicit acknowledgment of the need to take into interaction between individuals led to a study of diffusion, peer effects and spatial externality. The development of social network theory (Moreno 1932; White 1963) was another influence.

On the empirical side, the advances in technologies such as geographic information systems (GIS), the Global Positioning System (GPS), satellite remote sensing and computer cartography were the other factors encouraging the development of spatial statistical analysis (Anselin and Getis 1992). Such analysis enabled researchers to examine regional disparities, detect spatial patterns, identify chronic hotspots accurately and predict areas where threats were emerging. Spatial analysis also enabled the assessment of the regional applicability and relevance of intervention strategies, thereby preventing maladaptation. It was also realized that spatial analysis provided a visually attractive and easy-to-understand means to disseminate information to policymakers, activists and the general public and make a strong case for policy intervention and community action (Douglas et al. 2009; Kamel Boulos and Wilson 2023; Crespi et al. 2023). Further, the development of low-cost technologies to capture real-time facilitated participatory GIS, having a social transformative potential.

5.2 Exploratory Analysis

Tukey argues that Exploratory Data Analysis (EDA) comprises an important step in data analysis (Tukey 1963). By summarizing the main statistical properties of the data set being analyzed, it helps the researcher to get a feel of the data set and identify general pattern. This helps to understand the nature of relationship between variables (viz., no relationship, linear relationship or non-linear relationship) and facilitates the design of statistical analysis yielding meaningful results. EDA, therefore, comprises a crucial preliminary step for causal analysis. Typically, histograms, box plots and graphs are used in EDA. In spatial data science, they are replaced by choropleth maps and statistical maps.

Maps have been defined as a collection of spatially defined objects (Monmonier 2018). They are used to present facts or as a device of analysis. Like any other statistical tool, maps are used to generate information about the regional distribution of a phenomenon. For instance, the map in Fig. 5.1(b) reveals the spatial distribution of stunting in India. Such geovisualization, viz. the creation and use of visual representation to facilitate thinking, understanding and creating knowledge, helps the researcher to detect expected patterns and uncover unexpected variations (Thomas and Cook 2006). In particular, Exploratory Spatial Data Analysis (ESDA) describes and visualizes spatial distribution, identifies spatial clusters and outliers, and reveals spatial regimes, forming the basis for analysis based on spatial regression (Becker and Cleveland 1987; Anselin 1996; Wall and Devine 2000; Bivand 2010).

5.2.1 Choropleth Maps

Choropleth maps plot the values of a variable at different locations to visualizes the spatial distribution based on a color mapping symbology technique. It displays geographical regions that are colored based on the value of the variable being analyzed. There are three types of choropleth maps:

a. *Quantile maps*: Categories are formed so that the frequency in each category or bin) is equal.[2] The lower and upper limits, and hence intervals, will vary across categories.
b. *Equal interval maps*: As the name suggests, the categories are formed so that intervals are equal. A feature of such maps is that the frequency of each bin will vary, and some can even be very small or very large. They are the closest to the histogram.
c. *Natural break maps*: In natural break maps, an algorithm is used to form the categories (Jenks 1977). The algorithm maximizes the dissimilarity between categories, while minimizing the dissimilarity with each group. Thus, both intervals and frequencies will vary across categories. Equal interval maps are best for data displaying wide variation.

In GeoDa click on Maps, choose Quantile or Equal Intervals or Natural Break maps and select the number of categories. Maps can be saved by right clicking and choosing SAVE IMAGE AS option.

Examples of these three types of maps are given in Fig. 5.2. It should be seen how the regional pattern of stunting varies across each map. Stunting appears a serious concern in the quantile map, but less in the equal interval and natural break maps. This raises the question: Which map to use? If the distribution of the variable being studied varies widely, we should sue a natural breaks map; otherwise we should use the quantile map (even distribution with few ties) or equal interval maps (the distribution is clustered around several values).

Another important issue in drawing the maps is the number of categories to be used. While there is no hard and fast rule, a rule of thumb suggests that 3–5 categories should be used, with the final choice being determined by the number of observation.

[2] Ideally if there are n observations and m categories, each bin should have n/m observations. In reality, there may be a large number of ties. In such cases, there will be differences in frequencies of each category.

Fig. 5.2 Choropleth maps showing prevelence of stunting among 0–5 year children in India in 2019–21. Source: Estimated by authors using NFHS data (2019–2021)

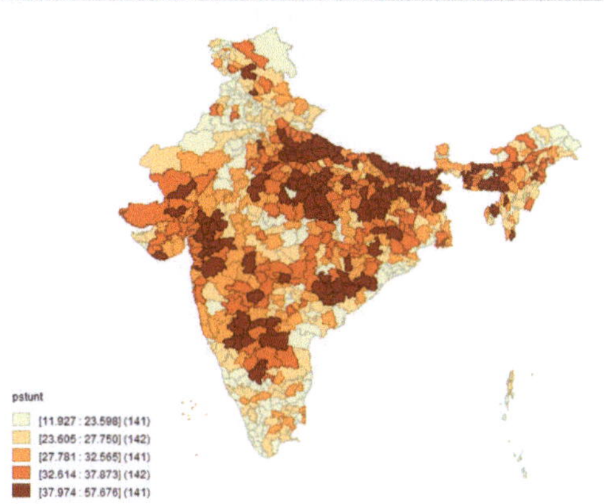

(a) Quantile maps

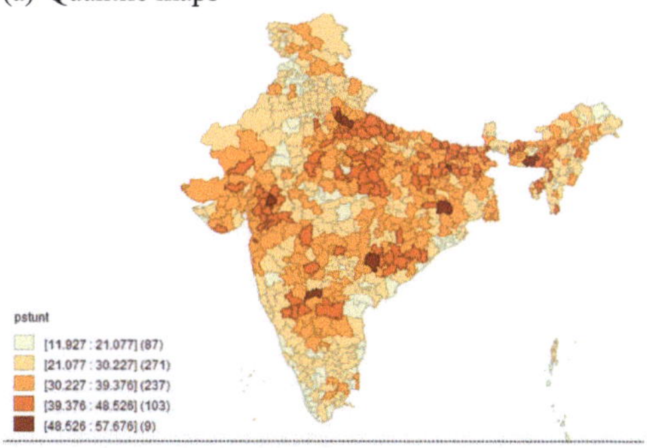

(b) Equal interval maps

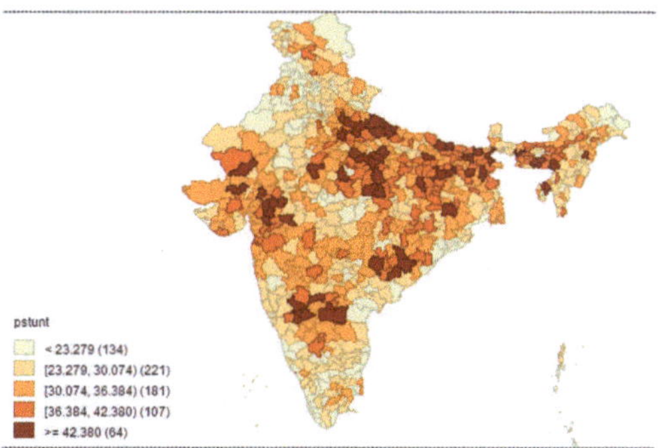

(c) Natural break intervals

5.2.2 *Statistical Maps*

Statistical maps explore the nature of data. The focus is on the distribution of data, and the central question is: How is the variable distributed around the measure of central tendency (generally median)?

In conventional statistics, box plots are used to examine the distribution of values of the variable being studied. Three important values, forming the box, are Q_1, Q_2 (or median) and Q_3. The whiskers and fences are formed as follows: $Q_1 - 1.5 \times IQR$ and $Q_3 - 1.5 \times IQR$, when IQR is the inter-quartile range ($= Q_3 - Q_1$). Any observation below the lower fence ($Q_1 - 1.5 \times IQR$) is called a lower outlier, while upper outliers are values lying above the upper fence ($Q_3 - 1.5 \times IQR$). Some statisticians suggest that the hinges should be 3 (instead of 1.5).

In spatial analysis, box maps are the counterparts of the box plots. They are a type of choropleth maps. The categories are formed as follows:

Below $Q_1 - 1.5 \times IQR$
Between $Q_1 - 1.5 \times IQR$ and Q_1
Between Q_1 and Q_2
Between Q_2 and Q_3
Between Q_3 and $Q_3 - 1.5 \times IQR$
Above $Q_3 - 1.5 \times IQR$

> Click on Map, select Box Maps 1.5 Hinges, or Box Maps 3 Hinges. In case of percentile maps, choose Percentile maps instead. Since categories are predetermined, there is no option to select number of bins.

A box plot and box map are drawn in Fig. 5.3. While the box plot reveals whether the distribution is symmetric and whether there are any outliers, the box map additionally identifies the location of districts falling in each category. They also reveal whether very high values of stunting are located close by in a particular region.

A less used map is the percentile map. They are used when the data size is large and has very few ties. They are similar to the quantile map. Only, instead of 3–5 categories, 100 groups are formed. Values are ranked sequentially in ascending order, and bins are formed containing the first 1% observations, the second 1% observations ..... The large number of bins are reclassified to form six categories: <1%, 1–10%, 10–50%, 50–90%, 90–99% and >99%. The emphasis is to identify extreme values and find out whether they are in the same place, or whether they are scattered.

The discussion so far has focused on continuous variables. But a variable can be categorical in nature. In such cases, particularly if it does not have any natural ordering, the color coding does not have any intuitive meaning like in the choropleth maps.[3] Unique value maps may be drawn in such cases. The fundamental difference

[3] In choropleth, maps higher valued bin are generally coded with darker variants of the base color.

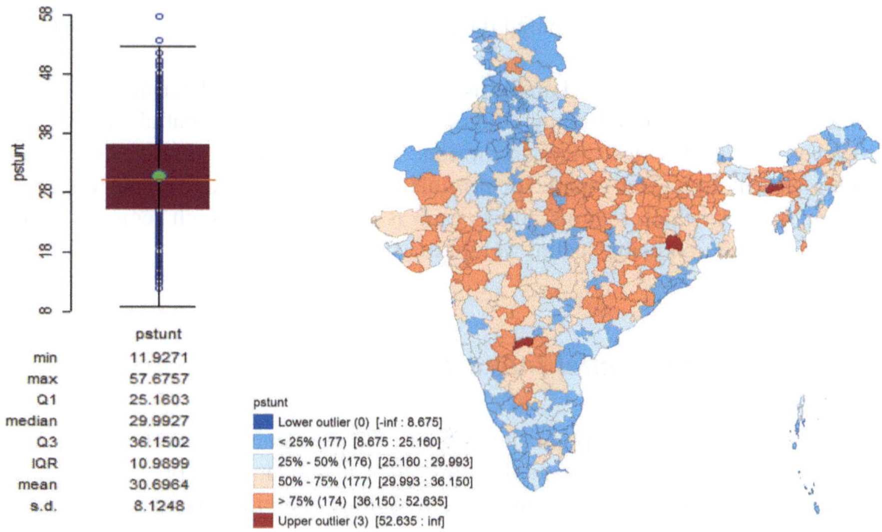

(a) Box plot **(b) Box map**

Fig. 5.3 Box plots and box maps. Source: Estimated by authors using NFHS data (2019–2021)

between categorical maps and other choropleth map types is that the numerical legend values do not imply any ordering. So whether we label the same group as 5 or 4 is irrelevant—only the grouping matters. Furthermore, the values in the legend are represented as integers for convenience, but they could just as well have been A, B, C, …., or other descriptors.

5.3 Spatial Weights

When we undertake spatial analysis we contrast two alternative perspectives of the locational data, spatial random pattern and spatial deterministic pattern. Spatial randomness assumes that the location of values can be reshuffled without changing its information content. It implicitly assumes that the observed spatial pattern is equally likely as any other spatial pattern. In contrast, if the data displays a spatially deterministic distribution, then it has two properties. Firstly, the reshuffling of locational information modifies the information content of the data. In Fig. 5.1, for instance, if we had distributed the locational content of the stunting data, we would have got the same histogram but a different choropleth map. Secondly, value of the attribute at one location depends upon its values at other (neighboring) locations. The latter implies the presence of interactions or spatial feedbacks between regions. While similar patterns are observed for time series data, in case of spatial data we have a two-way flow of effects and not one way. Thus, in time series data x_t affects x_{t+1} but not vice versa. In the case of spatial data, however, x_i affects x_j and is also affected by x_j.

So spatial statistics is based on the assumption of a spatially deterministic pattern in the variable or attribute being studied and the negation of spatial randomness. However, we need to test this assumption. To testing the hypothesis of spatially random distribution versus spatial deterministic pattern, we need to create a structure that can capture which regions affects (and is affected by) which regions. Such a structure is captured using spatial weight matrices.

5.3.1 Capturing Spatial Interaction

Spatial weight matrices capture the degree of dependence between spatial units formally. If two locations are neighboring, then it indicates the presence of interaction between regions. Now, if there are n regions, then the spatial weight matrix should capture $n \times (n - 1)/2$ interaction between the regions.[4] Given n observations, however, the incidental parameter problem will occur, making it impossible to capture all possible interactions even by increasing the number of regions.

The solution is to impose a structure and limit number of interactions on which we focus on. The structure is based on Tobler's law (Anselin and Getis 1992). In his study of population growth in Detroit, Tobler (1970) found that the population in a cell depends upon population in the neighboring cell. Further, the influence of a cell decreased as distance between them increased. Referred to as distance decay, this argues that the strength of interaction will decrease with distance and, beyond a distance, the feedback is eliminated. This law, summed up neatly as 'Everything depends upon everything else, but closer things more so' (Tobler 1970), allows us to solve the incidental parameter problem by focusing on a sub-set of possible interactions.

How is this sub-set defined? There are two broad methods, based on geographical contiguity and distance, respectively. Distance-based weights are formed by focusing on interactions between regions whose centroids fall within the perimeter around centroid of a particular region. In this study, however, we will discuss the more common geographical contiguity or neighborhood-based weights.

The spatial weight matrix (W) is defined as a $n \times n$ matrix, where an element w_{ij} indicates whether region j is the neighbor of region i. The element w_{ij} is binary. If the two regions are neighbors, then $w_{ij} = 1$; if they are not neighbors, $w_{ij} = 0$. In case of the diagonal elements (w_{ii}), we impose the condition $w_{ii} = 0$. While it may seem counter-intuitive, there is a statistical reason underlying this structure. Many statistical tests are based on the trace of the weight matrix, and if the diagonal elements are 1, the trace will become very large when the data set has a large number of regions. So the sample size will determine the statistic value, which is undesirable.

[4] Assuming a symmetric relationship between two regions i and j. If the relations are asymmetric, the number of interactions to be estimated will increase, compounding the problem.

Fig. 5.4 Defining neighbors

	B	R	B	
	R	A	R	
	B	R	B	

The next question is then how do we define neighbors? Chess fans have an advantage here, as the concept of geographical contiguity or common boundary is based on the moves of chess pieces. There are three possible ways of defining neighbors:

a. *Bishop criterion*: Two regions are neighbors if they touch each other at a single point. In Fig. 5.4, the four diagonal cells around cell A (labeled as B) are neighbors.
b. *Rook criterion*: Two regions are neighbors if they share a common set of points. In Fig. 5.4, the two vertical cells and two horizontal cells labeled as R are A's neighbors.
c. *Queen criterion*: Two regions are neighbors if they share a common boundary, irrespective of whether it is a single point or set of points. Thus, all eight cells around A are neighbors using the Queen criterion.

The weight matrix for a map with four regions will look like:

$$W = \begin{bmatrix} 0\,1\,0\,0 \\ 1\,0\,1\,0 \\ 0\,1\,0\,1 \\ 0\,0\,1\,0 \end{bmatrix}$$

when A and B, B and C, and B and D are neighbors.

When constructing spatial weights, we use either the Queen criterion or the Rook criterion. In case of a chess board like demarcation of regions, there are differences between the spatial weight matrix created using the Queen criterion and the Rook criterion. Such situations, however, are rare. In general, both yield the same neighbors. In India, for instance, the weight matrix will be identical for both the criterion. The Bishop criterion, in contrast, is rarely used as few regions have a single boundary point.

Apart from contiguity-based weights, we can also construct distance-based binary weight matrix. It is used for point data, or after creating centroids of polygons. A critical distance (d*) is defined, and if the centroid of any region j is within a distance of less than d* from region i (i.e., $d_{ij} < d^*$), it is considered a neighbor of region i and assigned a value of $w_{ij} = 1$.[5] Distance-based weights have three problems. The first relates to the defining of the optimal distance, or bandwidth. If the bandwidth is too small, we may end up having too many isolates. On the other hand, a large bandwidth will result in regions having too many neighbors. A convenient solution is to take the maximum of the minimum distance between regions. The second problem relates to the strength of the spatial dependence between regions. We are assigning the same weight to all regions whose centroid is within the bandwidth. However, some regions may be further off and others close to the region being considered. This problem is tackled by weighting w_{ij} by the inverse of distance. So the weights will be $(\frac{w_{ij}}{d_{ij}})$ or $(\frac{w_{ij}}{d_{ij}})^k$. The third problem occurs f we have large regions. In that case, even when a region is contiguous to the region being considered, its centroid may lie outside the bandwidth.

It is also possible to have non-geography-based weights capturing the idea that remote regions may have strong ties (Case et al. 1993; Conley and Copa 1999; Conley 1999) or are similar in some respects (Doreian 1980; F. Dormann et al. 2007; Duncan et al. 2017). Examples are matrices created based on migration patterns, capital inflows, etc. Such matrices are exogenous to the model and discouraged by statisticians. However, it is possible to combine economic or other forms of linkages with distance-based links. In that case $wij = \frac{w_{ij}}{d_{ij}\delta_{ij}}$ when δ_{ij} is the non-geography-based distance (Earnest et al. 2007).

The discussion on spatial weights indicates that there are several ways to construct spatial weights. Which method is chosen will depend upon the research objective, and will vary across studies.

5.3.2 *Some Technical Issues*

After creating the contiguity weights, we should always check its properties. The quality of analysis depends upon the accuracy of the shp file, and mistakes in the locational information show up in the weight matrix. Possible errors are detected using a histogram. The histogram depicts the frequency of regions with a certain number of regions. If a region does not have any neighbor, then it is called an isolate. The histogram in Fig. 5.5 reveals that there are three isolates in India. Normally, the presence of such isolates indicates possible errors in the creation of the shp file. But, in this case, should we treat the presence of isolates as errors? The answer is no, and we leave it to the reader to work out the reason.

After ruling out possible errors in the shp file, we often normalize the weight matrix. Row normalization is undertaken by dividing each row by the row sum, to ensure that the row sum ($\Sigma_j w_{ij} = 1$) equals unity (Moscone et al. 2007). Thus, the revised components are: $w^*_{ij} = w_{ij} / \Sigma_j w_i$. The row normalized weight matrix will

[5] We estimate d_{ij} using either the Euclidean distance, $= \sqrt{(x_i - x_j)^2 + (y_i - y)^2}$, or the Minkowski distance, $= p\sqrt{(x_i - x_j)^p + (y_i - y)^p}$.

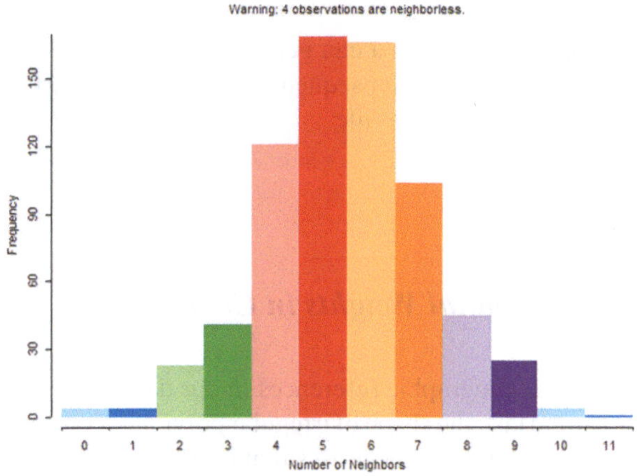

Fig. 5.5 Connectivity histogram. Source: Estimated by authors using NFHS data (2019–2021)

look like:

$$W = \begin{bmatrix} 0 & 1 & 0 & 0 \\ 0.5 & 0 & 0.5 & 0 \\ 0 & 0.5 & 0 & 0.5 \\ 0 & 0 & 1 & 0 \end{bmatrix}$$

as the row elements are divided by 1, 2, 2 and 1, respectively. Row normalization ensures that spatial weight matrices (and spatial autocorrelations) are comparable. However, they also destroy the symmetric nature of the weight matrix.

Some statisticians suggest that we divide by the total sum of w_{ij}. In this case, the division will be by six. It converts the elements into stochastic components with the sum of weights ($\Sigma_{ij}\ w_{ij} = 1$) equaling to unity. When there are few regions, such normalization does not create much problems. But if n is very large, the number of contiguous regions will also be very large. Division of unity by a large number will result in very small w_{ij} values, which may be close to zero, moderating the strength of feedbacks between spatial units. The underlying premise of the spatial framework is weakened if we assume that spatial interactions tend to zero.

Spatial weights are created using the menu option Tools > Weights. A window opens up, asking you to create or construct the matrix. Click on CONSTRUCT. In the next window, we should check that the Contiguity Weight tab is selected. Click on the pop down options in SELECT ID VARIABLE and select the variable identifying regions. In our example, it is REGCODE. Ensure that the order of contiguity is 1. Click on OK. A window opens up asking you to name the weight matrix (*.gal) in the folder where the shp files are located. After clicking on save, the weight matrix window is populated with the details of the weight matrix. You can get the connectivity histogram by clicking the button HISTOGRAM. Subsequently, you only have to click Tools > Weight > Load and choose the gal file.

5.3.3 *Spatial Weights in GeoDa*

Since the geographic references in our dataset are defined by administrative boundaries and take the shape of irregular polygons, contiguity-based weights following the queen criterion are the most suitable choice (O'Sullivan and Unwin 2010). Following standard practice, we also undertake row standardization (Moscone et al. 2007).

5.4 Spatial Autocorrelation

A statistic, calculated from the data (i.e., the sample), is any value that summarizes the characteristics of a distribution. In order to assess whether the statistic estimated from the sample holds for the population or not, we use statistical tests. We estimate a test statistic, calculated from the data and compare it to a reference distribution. In spatial analysis, we examine the likelihood of the statistic taking the estimated value for a spatially random distribution. Thus, the null hypothesis is that the data is spatially random. When the probability of the null hypothesis being accepted is very low, we reject the null hypothesis, opening the way to the search for spatial regimes (Anselin and Amaral 2024). So we need to find out statistic which will not only capture the distribution, but also the spatial component.

In conventional statistics, the Pearson correlation coefficient indicates the similarity of two variables. In time series, autocorrelation estimates the similarity between x_t and its preceding value (x_{t-1}). In the case of spatial statistics, autocorrelation has two important features:

i. *Attribute similarity*: The autocorrelation for a variable x at two points i and j is depicted by $f(x_i, x_j)$. It looks at the similarity of x_i with other values of the attribute.[6] Alternately autocorrelation can estimate the dissimilarity.[7] The expression $f(x_i, x_j)$ is generally measured using either $(x_i - x_j)^2$ or $|x_i - x_j|$.

[6] Under the assumption of spatially randomness, the value of $f(y_i, y_j)$ will not be systematically large or small. When large values are systematically together, the value of $f(y_i, y_j)$ will be large, and vice versa.

[7] Under the assumption of spatially randomness, the value of $f(y_i, y_j)$ will be systematically large or small. When large values are systematically together, the value of $f(y_i, y_j)$ will be small, and vice versa.

ii. *Locational similarity*: In addition to attribute similarity, spatial autocorrelation also looks at the locational similarity. In other words, we should look at the similarity of x_i with its neighbors. Obviously, we have to take the weighted mean of the attribute for the neighbors; the weight is the spatial weight matrix. The spatial weight matrix, therefore, is extremely important in helping to estimate spatial autocorrelation.

Spatial autocorrelation is the sum of the product of an attribute similarity measure with a neighbor indicator (spatial weight) over all pairs of observations. Given that $f(x_i\ x_j)$ is attribute similarity between i and j for x, while w_{ij} is a spatial weight between i and j, the spatial autocorrelation statistic is $\Sigma_{ij}\ f(x_i\ x_j)\ w_{ij}$.

There are two types of autocorrelation measures: global and spatial. Global autocorrelation is a single indicator capturing the general similarity of regions. They indicate the presence of clusters, but not the location of such clusters, nor the nature of such clusters. Local autocorrelation measures are estimated for each region and indicate the nature of association between the value of the attribute in region i and the values of the attribute in the surrounding regions.

5.4.1 Global Moran's I

Global Moran's I is the most widely used autocorrelation statistic. Originally proposed by Moran (1948), it gained prominence through the seminal work on spatial autocorrelation by Cliff and Ord (Getis 1995). The statistic is estimated to test whether the spatial distribution has a deterministic pattern, or it is random. It is the cross-product statistic between a variable and its spatial lag, with the variable expressed in deviations from its mean. Moran's I is given by:

$$I = \frac{\sum_i^n \sum_j W_{ij}(x_i - \overline{x})(x_j - \overline{x}) \big/ S_0}{\sum_i (x_i - \overline{x})^2 \big/ n} \tag{5.1}$$

where n is the number of spatial units indexed by i and j; x is the variable of interest; w_{ij} is a matrix of row standardized spatial weights with zeros along the diagonal; and S_0 is the sum of all w_{ij}. The range of Global Moran is $[-1, 1]$.

In our example, the value of Global Moran's I is found to be 0.49, which is quite high. There is thus reason for doubting the assumption of spatial randomness. However, we need to assess whether the computed value of Moran's I is significantly different from its value for a spatially random distribution. If we assume a normal distribution (independent normal random variates) or so-called randomization (i.e., each value is equally likely to occur at any location), it is easy to derive expressions for the mean and the variance of the statistic under the null hypothesis. We can conduct an inference test based on the mean and variance, and employing an approximation to a standard normal distribution.[8] A z-test is conducted comparing the estimated Moran's I with the mean of this reference distribution. In our example, the mean under the assumption of spatial randomness is −0.0009, with a standard deviation of 0.0228.[9] The z-value is 21.4764. As the p-value is 0.001, we cannot accept the null hypothesis of spatial randomness and conclude that there are clusters in the spatial distribution of stunting.

5.4.2 Local Spatial Autocorrelation

Although Moran's I indicates whether there is clustering, it does not identify the location of statistically significant spatial clusters; nor does it indicate the type of autocorrelation (Anselin 1995, 2010). Therefore, we use local spatial autocorrelation statistics. While there are several possible local statistics, we discuss two of the most commonly used statistics—the Local Indicators of Spatial Autocorrelation (LISA) and the Getis-Ord G_i statistic.

The LISA statistics was proposed by Anselin (Anselin 1995) and is given by:

$$I_i = \frac{(x_i - x)\sum_{i=1}^{n} w_{ij}\left(x_j - x\right)}{\sum_{i=1}^{n}(x_i - x)^2/n}. \tag{5.2}$$

This statistic examines whether a region is surrounded by neighboring regions with relatively similar values of the variable of interest, or whether the variable values are different. If the region I and its neighbors have high values of the variable of interest, it the cluster is referred to as a hotspot. Such points are located in the north-west quadrant of Fig. 5.6. On the other hand, if the values are low, the cluster is a cold spot and lies on the south-eastern quadrant. Spatial outliers are cases where a region has a high/low value, but its neighbors have dissimilar (low/high) values. They lie on the south-western and north-eastern quadrants, respectively.

Statistical testing of the hypotheses that $I_i = 0$, against the alternative hypotheses of $I_i \neq 0$, is possible. It is undertaken through conditional randomization or permutation. Bao states that a low p-value indicates that the region I is surrounded by relatively high values, while a high p-value indicates that the region I is surrounded by relatively low values (Bao 2000). The values of the local autocorrelation statistics and their statistical significance are captured through the significance and cluster maps (Fig. 5.7). Hotspots, indicated by the color red, are located in the states of Uttar Pradesh, Bihar and West Bengal, on the eastern side, and in the Gujarat and northern Maharashtra side, on the western side of India. Cold spots, indicated by blue, are mainly located in Southern India and in the Rajasthan, Punjab and Haryana belt.

---

[8] In operational terms, we randomly shuffle the observations repeatedly based on a Monte Carlo simulation and recomputing Moran's I each time to get the distribution of Moran's I for a spatially random attribute.

[9] The distribution was generated after 999 permutations.

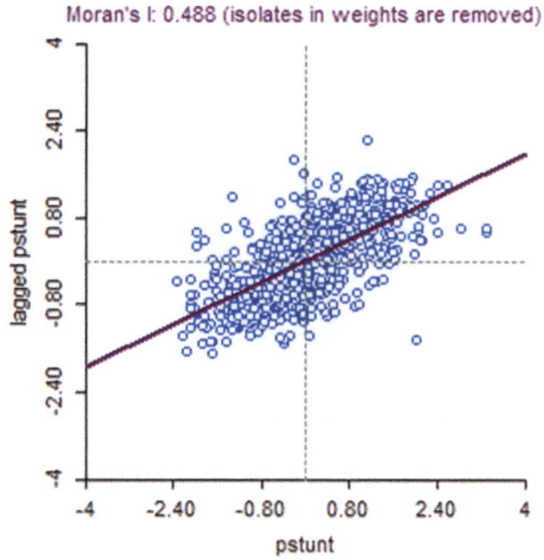

Fig. 5.6 Global Moran's I for stunting. Source: Estimated by authors using NFHS data (2019–2021)

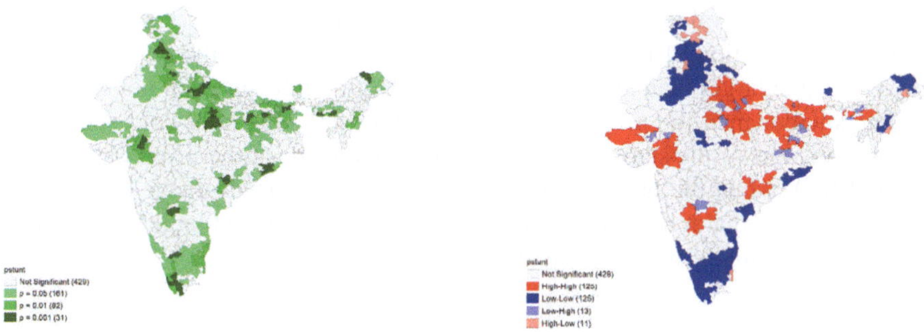

(a) Significance map

(b) Cluster map

Fig. 5.7 LISA maps for stunting. Source: Estimated by authors using NFHS data (2019–2021)

The Getis-Ord $G_i^*(d)$ statistic is used to identify hotspots or clusters with high intensity of transmission (Getis and Ord 1992; Ord and Getis 1995). The statistic is given by:

$$G_i^*(d) = \frac{\sum_j^N W_{ij} x_j - \overline{X} \sum_j^N W_{ij}}{S \times \sqrt{\frac{N \sum_j^N W_{ij}^2 - \left(\sum_j^N W_{ij}\right)^2}{N-1}}} \quad (5.3)$$

when $\overline{X}$ and S are the mean and standard deviation of x, respectively; the other notations have been defined before. This statistic is derived from a point pattern analysis logic and initially was stated as the ratio of the number of observations

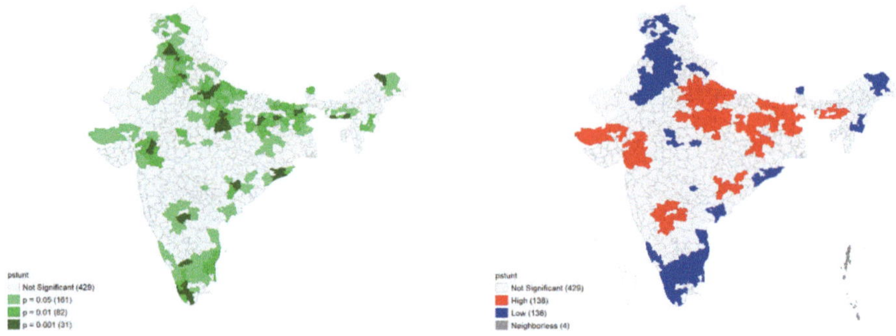

(a) Significance map (b) Cluster map

Fig. 5.8 Getis-Ord statistic maps for stunting. Source: Estimated by authors using NFHS data (2019–2021)

within a given bandwidth from a point to the total count of points. Subsequently, it was applied to the values of the variable of interest in the neighboring regions.

The interpretation of the Getis-Ord statistics is easier than the interpretation of LISA. A value larger than the mean indicates a High-High cluster or hot spot, while a value smaller than the mean demarcates a Low-Low cluster or cold spot. There are two versions of the statistic, with one taking the value at the given location into account (G_i), while the other version does not (G_i*). A disadvantage of the Getis-Ord approach is that it does not indicate spatial outliers. By and large both the local statistics give similar pictures (Fig. 5.8).

5.4.3 *Multivariate Spatial Correlation*

The Multivariate Local Geary statistic is used to examine whether the spatial distribution of the different categories of IPV is overlapping (Anselin 2019). This statistic is given by:

$$C_i^M = \sum_h^k \sum_j^N w_{ij} \left(x_{hi} - x_{hj} \right)^2, \tag{5.4}$$

when k indicates the number of variables (types of IPV) in the analysis indexed by h, and the other symbols have the same meanings as before. Figure 5.9 reveals substantive overlap between child marriage and stunting, with a significant positive association in 62% in districts.

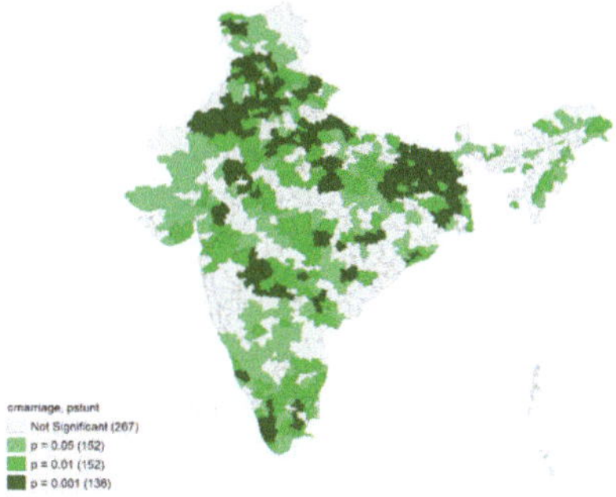

(a) Significance map

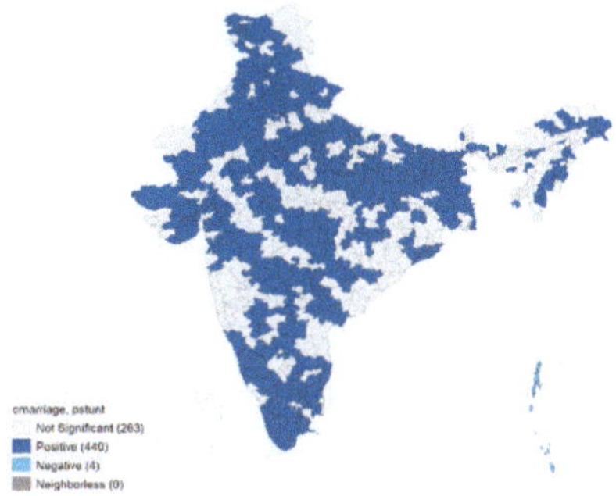

(b) Cluster map

Fig. 5.9 Multivariate Geary statistics for stunting and child marriage. Source: Estimated by authors using NFHS data (2019–2021)

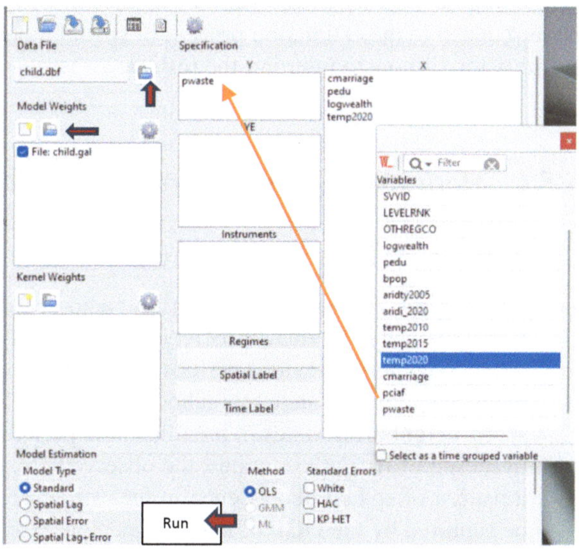

Fig. 5.10 Execution of an OLS regression model using GeoDaSpace

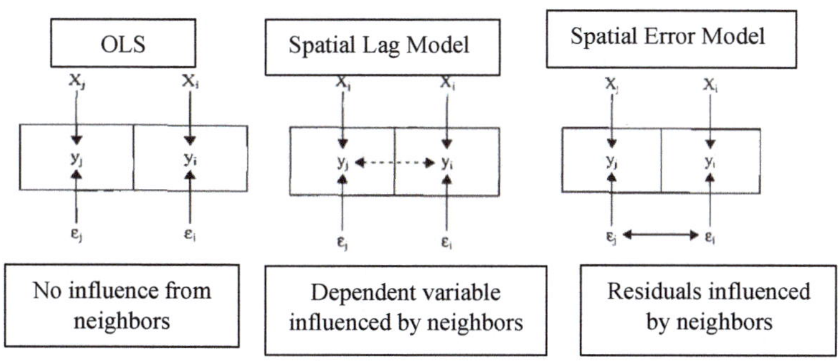

Fig. 5.11 Non-spatial and spatial functional forms. Source: Kopczewska (2021: 214)

5.5 Spatial Regression Motivation and Interpretation

In this section, our objective is to present the paradigm of perception and modeling of spatial interactions introducing the preliminaries of spatial regression analysis. Section 5.1 illustrates how different motivations may generate a host of alternative spatial regression structures. Section 5.2 briefly introduces a family of conventional spatial regression models, namely spatial autoregressive (SAR), spatial error model (SEM) and the spatial autoregressive moving average (SARMA) models, popularized by Anselin in his influential text on spatial econometrics (Anselin 1988). This section also provides the *data generating process* (DGP), the method of estimation and the diagnostics for these spatial regression models. Spatial lags are a hallmark of spatial regression models and can be used to provide extended versions of SAR model. One such extension is the spatial Durbin model (SDM). This model includes spatial lags of explanatory variables as well as the dependent variable. SDM along with its *data generating process* has been described in Sect. 5.3. In Sects. 5.2 and 5.3, readers will be introduced to the estimation process of spatial models using GeoDa, followed by a discussion of how to interpret the results.

5.5.1 Motivating Spatial Regression Models

Spatial models are motivated by different concerns:

(a) *A time-dependence motivation:* In any socio-economic context, often the present behavior of an individual or society or a region may be influenced by the behavior of other agents in previous periods. For example, suppose the education department in a particular state sets school fees after taking into consideration the fees set by neighboring states in previous time periods. Although there is a time lag in setting of fees across states, the observed cross-sectional fees would exhibit a pattern of spatial dependence in fees structure. This spatial dependence can be captured by the DGP of a SAR model as expressed in Eqs. (5.7) and (5.7a).

(b) *An omitted variable motivation*: Omitted variables are likely problems in applied work with regional data. For example, let us consider selling prices of houses in region i (y_i) as a function of past period selling prices of neighboring houses (y_j) and several characteristics of those houses (X_i). The characteristics of houses may consist of the number of bedrooms, bathrooms, etc., which are observable, along with several unobserved latent factors like local amenities, highway accessibility, or neighborhood prestige and so on. The impact of y_j on y_i may be captured introducing a spatial lag of the dependent variable as stated in (5.7) and (5.7a). The latent variables, however, are likely to be correlated with the past period selling prices of both region i and j and also likely to exhibit spatial dependence. These unobserved counterfactuals will remain present in the residuals of house price equations of both the regions i and j. A SEM and its DGP as stated in equations (5.8) and (5.8a) may be introduced to capture the spatial lag of the error terms in this context. However, Lee (2004), Davidson and MacKinnon (2004) and LeSage et al. (2009) using Hausman tests (Hausman 1978) show that SEM captures the effect of omitted variables, but these are not correlated with the included explanatory variables. In this context, we introduce SDM which matches the implied DGP that arises from the presence of omitted variables and spatially dependent explanatory variables.

(c) *A spatial heterogeneity motivation*: In a spatial context where each region has a single observation, the intercept α in the regression,

$$\mathbf{y} = \alpha + \beta\mathbf{X}$$

, can be treated as a spatially structured random effect vector. A fundamental assumption in spatial analysis is that neighboring observational units are likely to exhibit similar effect levels, which serve as a basis for modeling spatial heterogeneity. One way to incorporate this spatial dependence is by introducing a spatial autoregressive process, represented as:

$$\alpha = \lambda\mathbf{W}\alpha + \varepsilon$$

$$\text{or } \alpha = (\mathrm{I} - \lambda\mathbf{W})^{-1}\varepsilon,$$

where $\mathbf{W}$ is the spatial weight matrix. Plugging this in the original regression framework we get:

$$\mathbf{y} = \beta\mathbf{X} + (\mathrm{I} - \lambda\mathbf{W})^{-1}\varepsilon.$$

The above equation yields the DGP of the SEM provided later in equations (5.8) and (5.8a). Thus, spatial heterogeneity provides a way of motivating spatial dependence. In this discussion, we have assumed that α is not affected by the $\mathbf{X}$ vector.

If α is a function of $\mathbf{X}$, i.e., say $\alpha = \lambda\mathbf{W}\alpha + \mathbf{X}\gamma + \boldsymbol{\varepsilon}$, in that case:

$$\alpha = (\mathrm{I} - \lambda\mathbf{W})^{-1}\mathbf{X}\gamma + (\mathrm{I} - \lambda\mathbf{W})^{-1} + \text{ resulting in}$$

$$\mathbf{y} = \beta\mathbf{X} + (\mathrm{I} - \lambda\mathbf{W})^{-1}(\mathbf{X}\gamma + +).$$

$$\text{or, } \mathbf{y} = \gamma\mathbf{W}\,\mathbf{y} + \mathbf{X}(\beta + \gamma) + \mathbf{W}\mathbf{X}(-\lambda\beta) + +.$$

this takes form of the SDM where we have spatial lag for both the dependent and the independent variables.

(d) *An externalities-based motivation*: In a spatial context, externalities (both positive and negative) arising from neighborhood characteristics often have substantial impacts. For instance, filthy surroundings may attract rats and snakes in contiguous yards and reduce the property values. Conversely, a home surrounded by beautifully landscaped yards with flower garden would have positive effects on house values. In terms of modeling, the spatial lags of neighboring home characteristics (WX) will be able to capture these externalities.

(e) *A model uncertainty motivation*: There is always an inherent element of uncertainty regarding the type of model selection, conventional parameter uncertainty and uncertainty regarding specification of the appropriate explanatory variables. This can be incorporated through the Bayesian model. The Bayesian approach focusses on both the parameters and the distribution of the data. A likelihood function, reflecting the distribution of the data, is combined with a prior distribution assigned by the researcher to derive the posterior distribution. The posterior distribution contains all relevant information (information about the sample data and the prior distribution) required for estimating the model. The complex estimation process is decomposed into simpler problems that are based on the conditional distributions of the model parameters (known as the Markov Chain Monte Carlo) (Gelfand and Smith 1990). The latter, however, is beyond the scope of this chapter.

5.5.2 *SAR, SEM and SARMA*

In this section, we will introduce the readers to the host of spatial models where the spatial components have been combined with the conventional regression model. As the model building exercise can be seen as a deviation from or an extension of a classical linear regression model, we start with a brief overview of the classical model under finite sample and large sample properties. We consider both the least square and the maximum likelihood estimation techniques to present the estimated values of the parameters of the model. We briefly touch the inference and the regression diagnostics as these will help us to understand different test statistics and diagnostics related to spatial models.

5.5.2.1 A Review of the Classical Linear Regression Model

Let's start with a regression equation without any aspect of spatial association among the explained (y) and explanatory variables (X), β's are regression coefficients and u_i is the random error component of the regression model:

$$\mathbf{y}_{nx1} = \mathbf{X}_{nxk}\beta_{kx1} + \mathbf{u}_{nx1}. \tag{5.5}$$

(i) Regularity conditions

The standard set of regularity conditions for equation (5.5) includes

(a) The correct functional specification of the regression (e.g., functional relation is not non-linear)

(b) $E[u_i] = 0$, i.e., no systematic bias,
(c) $E[u_i^2] = \sigma^2$, constant error variance (no heteroscedasticity),
(d) $E[u_i, u_j] = 0$, no error correlation (including no spatial autocorrelation),
(e) $E[X_i, u_i] = 0$, no endogeneity (every explanatory variable is uncorrelated with the error terms and
(f) $\mathbf{u} \sim N(\mathbf{0}, \sigma^2 \mathbf{I}_{nxn})$, i.e., the error terms follow a multivariate (homoscedastic and uncorrelated) normal distribution. Here I is an identity matrix of order n.

All these assumptions are conditioned on the **X** matrix.

(ii) Estimation and inference

Under set of regularity conditions, both ordinary least square (OLS) and maximum likelihood (ML) estimation techniques yield the best (minimum variance) unbiased estimator (Gauss-Markov theorem):

$$\hat{\beta} = [\mathbf{X}'\mathbf{X}]^{-1}\mathbf{X}'\mathbf{y} \tag{5.6}$$

with $E(\hat{\beta}) = \beta$ and $\mathrm{Var}(\hat{\beta}) = \sigma^2 [\mathbf{X}'\mathbf{X}]^{-1}$

Estimators for error variance: $s^2_{OLS} = e'e/n - k$ [unbiased] and

$$s^2_{ML} = e'e/n \text{ [consistent, but not unbiased]}$$

Here s^2 indicates the sample variance, e is the residual[10] and $e'e$ is the residual sum of squares (RSS). Inference on the coefficients consists of null hypothesis, H_0: $\beta_i = 0$. To test this null hypothesis, we consider 't' test (with assumption of normality), where $t = \hat{\beta}_i/SE(\hat{\beta}_i)$.

For joint significance, H_0: $\beta_2 = \beta_3 = \ldots\ldots\ldots = \beta_k = 0$ we use F test with $F = \frac{[RRSS - URSS]/k-1}{URSS/n-k}$, where RRSS is the RSS of the restricted equation (the equation consisting of only the intercept term under H_0) and URSS is the RSS of the original unrestricted regression equation.

The tests based on ML are Likelihood Ratio (LR) test, Wald test and Lagrange Multiplier test. The relevant test statistics are as follows:

(a) Likelihood Ratio (LR) test is based on Likelihood Ratio $(\lambda) = \frac{L(\tilde{\beta},\tilde{\sigma}^2)}{L(\hat{\beta},\hat{\sigma}^2)}$ where

$L(\hat{\beta}, \hat{\sigma}^2)$: unrestricted ML function with the residual, e
$L(\tilde{\beta}, \tilde{\sigma}^2)$: restricted ML function with the residual, e_*.
Intuitively, we expect to reject H_0 if λ is small and this can be captured by:

$$\mathrm{LR} = -2\ \ln\lambda = \frac{n(e_*'e_* - e'e)}{e_*'e_*} + \frac{n}{2}\left(\frac{e_*'e_* - e'e}{e_*'e_*}\right)^2 \sim \chi^2(q)$$

[10] A residual 'e' is the difference between observed and predicted values of y_i and used as some sort of estimator of u_i which is the difference between true and observed values of y_i, thus unobservable.

(b) Wald test is done using the statistic, W $= \frac{n(e_*' e_* - e'e)}{e'e}$ ~ χ^2 (q)

(c) Lagrange Multiplier (LM) test is based on the test statistic, LM $= \frac{n(e_*' e_* - e'e)}{e_*' e_*}$ ~ χ^2 (q)

For all three test statistics, 'e' refers to the residual of the unrestricted equation, while e_* refers to the residual of the restricted equation. The null hypothesis for all three tests is that the restricted model is the 'true' model.

(iii) Measures of fit

The goodness of fit of OLS estimators is given by:
$(Multiple)\, R^2 = 1 - \frac{RSS}{TSS}$; Adjusted$(Multiple)\, R^2 = 1 - \frac{RSS/n-k}{TSS/n-1}$.
Measures based on ML

$$\text{Akaike Information Criterion (AIC)}: \ln\frac{e'e}{n} + \ln n,\ \text{where } RSS = e'e.$$

$$\text{Schwartz Criterion (SC)}: \ln\frac{e'e}{n} + \frac{2k}{n}.$$

(iv) Regression diagnostics and some associated detection mechanisms

Multicollinearity

Multicollinearity occurs when two or more independent variables have a high correlation with one another in a regression model. There are several techniques by which we may detect multicollinearity:

i. *Condition number*: This is not a test, but an indication of potential problem. The rule of thumb is to consider it to be excessive when its value exceeds 50.
ii. *Variance inflation factor (VIF)*: It is defined as $VIF_h = 1/(1 - R_h^2)$, where R_h^2 is the multiple correlation of x_k with other explanatory variables. The rule of thumb is that a VIF less than 10 is desirable.

Checking Normality Assumption

The *Jarque–Bera test* statistic is estimated considering the distribution of regression residuals.

Based on the skewness (b_1) and kurtosis (b_2) matching a normal distribution:

$$JB = (n/6)\, b_1^2 + (n/24)\, b_2^2 \sim \chi^2(2)$$

$$\text{where } b_1 = (m_3/\sigma^3) \text{and } b_2 = (m_4 - 3)/\sigma^4.$$

The null hypothesis of normality is tested against non-normality. In general, normality test has become less relevant with larger data.

Heteroscedasticity

Initial regularity conditions related to the distribution of u_i conditional on X_i state that it has a mean of zero. If, furthermore, the variance of conditional distribution does not depend on X_i, then the errors are said to be homoscedastic. However, if $var(u_i | X_i)$ depends on X_i that is known as heteroscedasticity.

To detect heteroscedasticity, we consider the null hypothesis of no heteroscedasticity. Several possible tests to detect heteroscedasticity are:

i. *Breusch-Pagan Lagrange Multiplier test*, 1979 (based on normality assumption): Here we regress the squared residual of the original regression equation on the explanatory variables. Multiple R^2 generated out of this equation is used to form the test statistic $nR^2 \sim \chi^2$ (with k degrees of freedom). If the test statistic has a p-value below an appropriate threshold (e.g., $p < 0.05$), then the null hypothesis of homoscedasticity is rejected.
ii. *Koenker and Bassett test*, 1982: A new class of tests for heteroscedasticity in linear models based on the regression quantile statistics of Koenker and Bassett. A large numerical difference between BP and KB statistics suggests that normality may be an issue.
iii. *White test*, 1983: Residual square of the original regression equation is regressed on explanatory variables, their squares and cross products. Multiple R^2 generated out of this equation is used to form the test statistic $nR^2 \sim \chi^2$.

We estimate a *Robust Coefficient Variance* of the least square estimates in presence of heteroscedasticity. The General Covariance Structure of the error in a regression in the presence of *heteroscedasticity* and *autocorrelation* is given by:

$$\mathrm{E}\left[\mathrm{UU}'\right] = \Omega, \quad \text{where } \Omega \neq \sigma^2\mathbf{I}.$$

The variance–covariance matrix of the regression coefficients then becomes:

$$\mathrm{Var}\ (\hat{\beta}) = \left[\mathbf{X}'\mathbf{X}\right]^{-1}[\mathbf{X}'\Omega\mathbf{X}]\,[\mathbf{X}'\mathbf{X}]^{-1}.$$

While the individual elements in the matrix Ω cannot be estimated separately, White (1983) suggests a consistent estimator for $V = [\mathbf{X}'\ \Omega\ \mathbf{X}]$. Specifically for heteroscedasticity, he proposes:

$$\hat{V} = (1/\mathrm{n})\mathbf{X}'\mathrm{S}\mathbf{X},$$

where S is a diagonal matrix with elements $e_i{}^2$, the squared residuals.

The coefficient variance matrix then becomes:

$$\widehat{Var}(\hat{\beta}) = \left[\mathbf{X}'\mathbf{X}\right]^{-1}\left[\mathbf{X}'\mathrm{S}\mathbf{X}\right][\mathbf{X}'\mathbf{X}]^{-1}.$$

The square root of the above estimator is known as *robust standard error*.

Kelejian and Prucha (2007) provide an estimator covering heteroscedasticity and spatial autocorrelation of unknown form using kernel estimator for covariance terms (HAC standard error): $\hat{V} = (1/n) \sum_i \sum_j x_i x_j' e_i e_j k\left(\frac{d_{ij}}{d}\right)$, where K() is the kernel function with bandwidth d. The coefficient variance matrix then estimated as:

$$\widehat{Var}(\hat{\beta}) = n\left[\mathbf{X}'\mathbf{X}\right]^{-1}\left[\mathbf{X}'\mathbf{X}\right]^{-1}.$$

We now illustrate this using our data set. We will first estimate an OLS model. The weight matrix is constructed using the queen's contiguity method.[11] We construct a regression model based on district-level information on the percentage of wasted children. We hypothesize that this percentage depends upon contextual factors like district-level educational status of women (pedu), cultural factor like prevalence of child marriage (cmarriage) and environmental factor like temperature (temp2020) in India. The equation has been estimated using GeoDaSpace.

The regression output in Table 5.1 provides information on measures of goodness of fit based on least square (R^2 and Adjusted R^2) and based on ML (AIC and SC). Overall significance of the model has been captured by 'F' statistic along with p-value. Student's 't' statistics and their p-values along with estimated coefficient values provide information on whether null hypothesis of insignificant coefficients have been rejected in favor of alternative specifications or not. In this example, all three regression coefficients are significant at 1% level of significance. The rise in proportion of child marriage and temperature increases the proportion of children wasted. However, increase in percentage of women with higher level of education reduces child malnutrition.

In the regression diagnostics part, information on condition number suggests that there is no evidence of multicollinearity. JB-statistic shows that null hypothesis of normality of error distribution has been violated. Outcomes of Breusch-Pagan Lagrange Multiplier test and Koenker and Bassett test suggest that homoscedasticity prevails in this model. Diagnostics for spatial dependence will be discussed in the following subsection.

5.5.2.2 Spatial Dependence

Two primary types of spatial dependence:

(a) Spatial lag model or spatial autoregressive (SAR) model: The dependent variable y in place i is affected by the independent variables in both place i and j.
(b) Spatial error model (SEM): The error terms across different spatial units are correlated.

[11] Although we do not require the spatial weight matrix to estimate the OLS, it is needed to obtain the spatial diagnostics as part of the output.

Table 5.1 Outcome of an OLS regression model using GeoDaSpace

```
REGRESSION
----------
SUMMARY OF OUTPUT: ORDINARY LEAST SQUARES
-----------------------------------------
Data set            :   child.dbf
Weights matrix      :File: child.gal
Dependent Variable  :      pwaste                Number of Observations:         707
Mean dependent var  :     16.5360                Number of Variables   :           4
S.D. dependent var  :      5.7397                Degrees of Freedom    :         703
R-squared           :      0.1154
Adjusted R-squared  :      0.1116
Sum squared residual:   20573.682                F-statistic           :     30.5767
Sigma-square        :      29.266                Prob(F-statistic)     :   1.378e-18
S.E. of regression  :       5.410                Log likelihood        :   -2194.745
Sigma-square ML     :      29.100                Akaike info criterion :    4397.490
S.E of regression ML:      5.3944                Schwarz criterion     :    4415.734

------------------------------------------------------------------------------------
            Variable     Coefficient       Std.Error     t-Statistic     Probability
------------------------------------------------------------------------------------
            CONSTANT      13.4292401       1.9426917       6.9126974       0.0000000
           cmarriage       0.0780223       0.0178092       4.3810150       0.0000136
                pedu      -0.0465501       0.0174174      -2.6726176       0.0077007
            temp2020       0.1831428       0.0517660       3.5378979       0.0004298
------------------------------------------------------------------------------------
REGRESSION DIAGNOSTICS
MULTICOLLINEARITY CONDITION NUMBER            22.455

TEST ON NORMALITY OF ERRORS
TEST                             DF        VALUE           PROB
Jarque-Bera                       2       49.591          0.0000

DIAGNOSTICS FOR HETEROSKEDASTICITY
RANDOM COEFFICIENTS
TEST                             DF        VALUE           PROB
Breusch-Pagan test                3        0.962          0.8104
Koenker-Bassett test              3        0.811          0.8469

DIAGNOSTICS FOR SPATIAL DEPENDENCE
TEST                           MI/DF       VALUE           PROB
Lagrange Multiplier (lag)         1      249.042          0.0000
Robust LM (lag)                   1        0.277          0.5984
Lagrange Multiplier (error)       1      277.263          0.0000
Robust LM (error)                 1       28.499          0.0000
Lagrange Multiplier (SARMA)       2      277.540          0.0000

================================ END OF REPORT =====================================
```

Spatial Autoregressive (SAR) Model

In our example, it is possible that child malnutrition in neighboring districts affects malnutrition level in the concerned district. In this case, we consider a spatial lag model presented as:

$$\mathbf{y} = \rho\mathbf{W}\mathbf{y} + \mathbf{X}\beta + \mathbf{u}, \tag{5.7}$$

where Wy is the spatially lagged dependent variable with associate autoregressive coefficient ρ. This is also known as spatial autoregressive (SAR) model. Now the question is why two y's are related? Contextual factors like education level of women, cultural factor like prevalence of child marriage and environmental factor like temperature may be similar in neighboring regions. The DGP of the SAR model is given by:

$$\mathbf{y} = (\mathbf{I} - \rho\mathbf{W})^{-1}\mathbf{X}\beta + (\mathbf{I} - \rho\mathbf{W})^{-1}\mathbf{u}. \tag{5.7a}$$

An ignored spatial lag term corresponds to an omitted variable and will make the estimates of other regression coefficients biased. To understand spatial dependence, we consider a test on setting the null hypothesis, H_0: $\rho = 0$ (spatial independence) against H_1: $\rho \neq 0$ (spatial dependence). We use the Lagrange Multiplier or Rao Score test statistic:

$$\mathrm{LM}_\rho = \frac{d_\rho^2}{D} \sim \chi^2(1),\ \text{where } \mathrm{d}\rho = \frac{e'\mathrm{Wy}}{s_{ML}^2}.$$

e: vector of OLS residuals,
Wy is the spatial lag and $s_{ML}^2 = \frac{e'\mathrm{e}}{\mathrm{n}}$,
D consists of two parts:

$$\mathrm{D} = (\mathbf{WX}\hat{\beta})/\left[\mathrm{I}{-}\mathbf{X}(\mathbf{X'X})^{-1}\mathrm{X'}\right](\mathbf{WX}\hat{\beta})s_{ML}^2 + \mathrm{T}.$$

i. The first part is the sum of squared residuals in a regression of the spatially lagged predicted values ($\mathbf{WX}\ \hat{\beta}$) on the **X,** and
ii. The second part is the trace expression, T = [tr (WW + **W**′W)].

Spatial Error Model (SEM)

It is also possible that omitted variables in a region and their values in the surrounding areas are correlated. For example, awareness and dietary practices, empowerment/agency of women may spill over and diffuse across regions. The spatial error model (SEM) will be an appropriate regression in this setting, and it is presented as:

$$\mathbf{y} = \mathbf{X}\beta + \mathbf{u} \text{ and}$$

$$\mathbf{u} = \lambda\mathbf{Wu} + +.$$

Combining the above two equations we get:

$$\mathbf{y} = \mathbf{X}\beta + \lambda\mathbf{Wu} + +, \tag{5.8}$$

where $\boldsymbol{\varepsilon}$ can be heteroscedastic or homoscedastic; however, the errors are non-autocorrelated.

Wu: spatially lagged term for original regression error,
ε: idiosyncratic error (i.e., the remaining error) and
λ: associated coefficient of spatially lagged term.

The SEM is also called Spatial Moving Average (SMA) model. The reduced form of the error in SEM is:

$$\mathbf{u} = (\mathrm{I} - \lambda\mathbf{W})^{-1}\varepsilon.$$

leading to the DGP of a SEM stated as:

$$\mathbf{y} = \mathbf{X}\beta + (\mathrm{I} - \lambda\mathbf{W})^{-1}\varepsilon. \tag{5.8a}$$

To understand a statistically significant presence of λ, we set null hypothesis, H_0: $\lambda = 0$ against H_1: $\lambda \neq 0$. The Lagrange Multiplier or Rao Score test statistic in this context is given by:

$$\mathrm{LM}_\lambda = \frac{d_\lambda^2}{T} \sim \chi^2(1), \text{ where } \mathrm{d}\lambda = \frac{e'\mathrm{W}e}{s_{ML}^2}.$$

e is vector of OLS residuals and We is its spatial lag, $s_{ML}^2 = \frac{e'e}{n}$ and $T = [\mathrm{tr}(\mathrm{WW} + \mathbf{W}'\mathrm{W})]$.

Robust Forms of the Lagrange Multiplier Tests

Anselin et al. (1996) identified that the LM_ρ statistic is also sensitive to the presence of spatial error correlation and thus, result of the test can be misleading in the sense that it may suggest a spatial lag model as the alternative, whereas the true alternative is a spatial error model. Similarly, the LM_λ statistic has power against a spatial lag alternative and can suggest a wrong model. To correct for this undesirable property, Anselin et al. (1996) introduced a robust form of the test statistics. The robust LM statistics are corrected for the covariance between $d_\rho and\ d_\lambda$ and they are given by:

$$LM_\rho^* = \frac{(d_\rho - d_\lambda)^2}{\{D - T\}} \sim \chi^2(1) \text{ for the spatial lag model and}$$

$$LM_\lambda^* = \frac{(d_\lambda - TD^{-1}d_\rho)^2}{[T\{1 - TD\}} \sim \chi^2(1) \text{ for the spatial error model.}$$

Spatial Autoregressive Moving Average (SARMA) Model

This is a combo model where spatial interaction is coming from two sources—spatial lag as well as spatial error term. Thus, we have the set of equations:

$$\mathbf{y} = \rho\mathbf{W}\mathbf{y} + \mathbf{X}\beta + \mathbf{u}.$$
$$\mathbf{u} = \lambda\mathbf{W}\mathbf{u} + \varepsilon.$$

The DGP of the combo model is expressed as:

$$\mathbf{y} = \rho\mathbf{W}\mathbf{y} + \mathbf{X}\beta + (\mathrm{I} - \lambda\mathbf{W})^{-1}\varepsilon. \tag{5.9}$$

Joint null hypothesis to check the significance of spatial lag as well as spatial error we consider,

$$\mathrm{H}_0 : \rho = \lambda = 0 \text{ against } \mathrm{H}_1 : \rho \neq 0 \text{ and } \lambda \neq 0.$$

The Lagrange Multiplier or Rao Score test statistic in this context is given by

$$\mathrm{LM}_{\rho\lambda} = \frac{d_\lambda^2}{T} + \frac{\left(d_\lambda - d_\rho\right)^2}{\{D - T\}} \sim \chi^2(2)$$

Extension of this idea to a joint test for the spatial parameters in SARMA (p, q) model is straightforward, but rarely used in practice. This formula can be further presented as:

$$\mathrm{LM}_{\rho\lambda} = \mathrm{LM}_\lambda + LM_\rho^* = \mathrm{LM}_\rho + LM_\lambda^*.$$

Specification Search

Sections 5.2.1 to 5.5.2.2 familiarises the reader with the diagnostics to assess the presence of spatial dependence which includes a collection of Lagrange Multiplier or Rao Score test statistics. Now, given the results of spatial diagnostics, which regression model should we choose? The decision rule has been presented in Fig. 5.12 following Anselin (2005).

If one of the LM_ρ and LM_λ is significant, and the other is not, then we proceed to the spatial regression model suggested by the significant statistics. If both the statistics are significant, then we proceed down the graph to the box with the robust diagnostics. In our example, as cited in Table 5.1, both LM_ρ and LM_λ are significant at 1% level of significance. Here in the next step, one needs to consider the robust LM statistics. In Table 5.1, LM^* is significant at 1% level of significance; however, LM_ρ^* is insignificant. Hence, we may conclude that the spatial error model is the

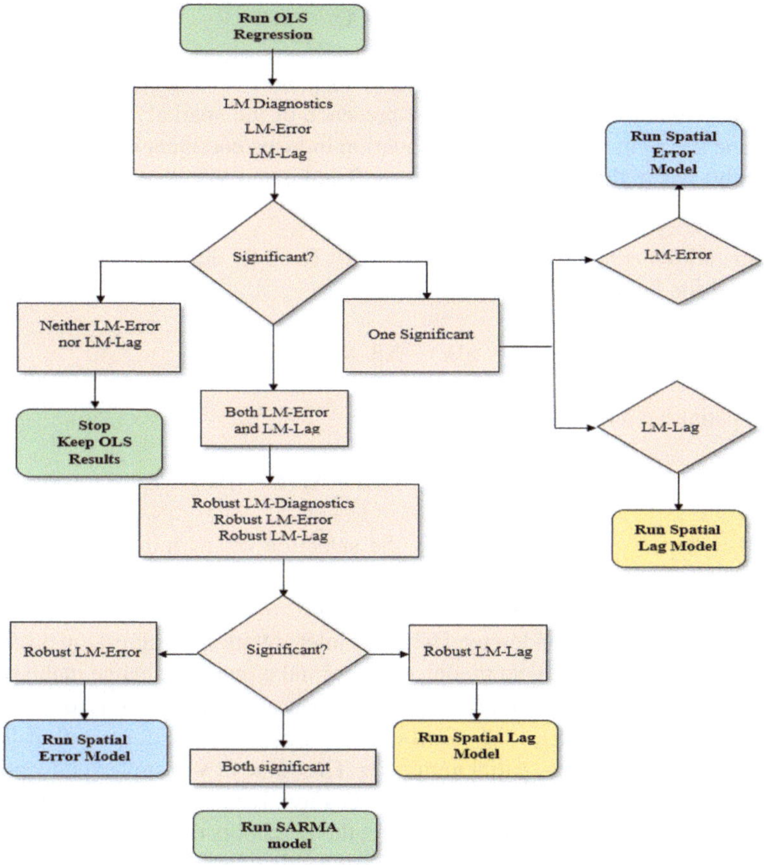

Fig. 5.12 Spatial specification search. Source: Anselin (2005)

appropriate one. Thus, malnutrition is not affected by proportion of child marriage or proportion of higher educated women in neighborhood; rather, it is affected by spillover effects of awareness and dietary practices, empowerment/agency of women.

In some cases, especially when the number of observations is large, both robust test statistics may still be significant. In those cases, the preferred alternative would be the one with the highest value for the statistic. In this context, one may use the SARMA model as well. However, decision rule needs to be applied with caution, since there may not be sufficient evidence to clearly select one alternative over another.

How to Estimate a Selected Model?

i. ***Spatial autoregressive model***

The spatial autoregressive model is given by:

$$\mathbf{y} = \rho\mathbf{W}\mathbf{y} + \mathbf{X}\beta + \mathbf{u},$$

where **Wy** is the spatially lagged dependent variable with associated autoregressive coefficient ρ as stated in Eq. (5.7). The presence of the spatially lagged dependent variable on the right hand side of the equation induces endogeneity or simultaneous equation bias as in this case Cov(u, **Wy**) ≠ 0 and u and **Wy** are associated.

Model can be estimated using two methods:

(a) Spatial two-stage least squares (S2SLS): To apply S2SLS model, we consider the DGP of the lag model given in (5.7a):

$$\mathbf{y} = (\mathbf{I} - \rho\mathbf{W})^{-1}\mathbf{X}\beta + (\mathbf{I} - \rho\mathbf{W})^{-1}\mathbf{u}.$$

The spatial lag term becomes:

$$\mathbf{W}\mathbf{y} = \mathbf{W}(\mathbf{I} - \rho\mathbf{W})^{-1}\mathbf{X}\beta + \mathbf{W}(\mathbf{I} - \rho\mathbf{W})^{-1}\mathbf{u}. \tag{5.10}$$

The key aspect of this approach is the selection of instruments to correct the endogeneity of **Wy**. Here several orders of spatially lagged explanatory variables, namely **WX**, $\mathbf{W}^2\mathbf{X}$, etc., are used as instruments.

Now, it is important to understand what the interpretation of regression coefficient, β, will be. In the standard linear model, the marginal effect of a change in the value of a continuous explanatory variable for a given observation corresponds to the value of that variable's coefficient $\beta_h = \frac{\delta y_i}{\delta x_{ih}}$. In a spatial lag model, this interpretation does not hold due to the presence of spatial multiplier. The value of y at any given location i is a function of the value not only of x at that location, but also of the x at neighboring locations (through the term ρ**WX** β), the x at the neighbors of the neighbors (through the term $\rho^2\mathbf{W}^2\mathbf{X}\,\beta$), etc., i.e., the value of the dependent variable at any given location depends on the explanatory variables at all locations in the system. The powers of the spatial autoregressive coefficients reflect the distance decay effect. In sum, here the total effect of a change in the explanatory variable on the dependent variable has two parts:

i. One due to the **direct effect** of x_h at each location, i.e., β_h, and
ii. The other is the **indirect effect** driven by the spatial multiplier, $\frac{\beta_h}{1-\rho}$.

(b) Maximum likelihood estimation (MLE): MLE is absed on stronger assumptions but provides estimates that have better properties (consistency and asymptotic efficiency). There is no need for instruments in this approach.

To understand the execution of a spatial lag or a SAR model in GeoDaSpace, let us once again consider our data set and the weight file. This time we consider a spillover effect of malnutrition introducing a new variable W_pwaste. This variable has been created by the software as a spatial lag of the variable waste. Here we intend to check whether percentage of wasted children in one district has any significant impact on percentage of wasted children in neighboring district along with other determinants as considered in Fig. 5.13 and Table 5.2.

Fig. 5.13 Execution of SAR model using GeoDaSpace

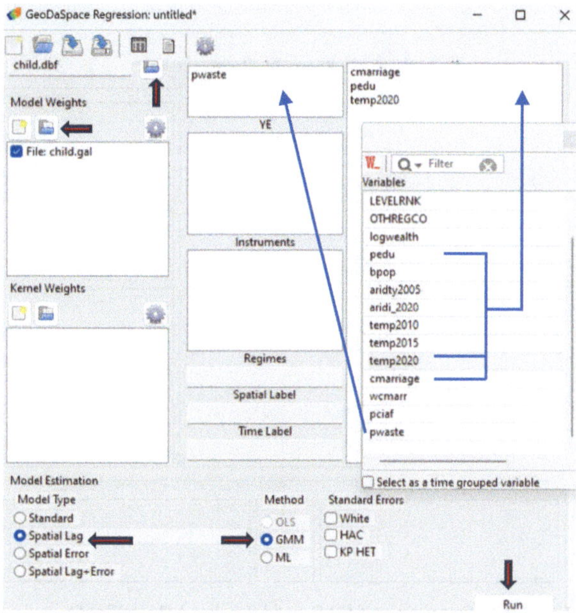

Table 5.2 Estimation of SAR model using S2SLS in GeoDaSpace

```
REGRESSION
----------
SUMMARY OF OUTPUT: SPATIAL TWO STAGE LEAST SQUARES
--------------------------------------------------
Data set            :    child.dbf
Weights matrix      :File: child.gal
Dependent Variable  :     pwaste                Number of Observations:       707
Mean dependent var  :    16.5360                Number of Variables   :         5
S.D. dependent var  :     5.7397                Degrees of Freedom    :       702
Pseudo R-squared    :     0.2403
Spatial Pseudo R-squared:  0.1154

------------------------------------------------------------------------------------
            Variable     Coefficient       Std.Error     z-Statistic     Probability
------------------------------------------------------------------------------------
            CONSTANT      11.4063577       2.6402152       4.3202379       0.0000156
            W_pwaste       0.1727662       0.1637094       1.0553224       0.2912779
           cmarriage       0.0627048       0.0220840       2.8393776       0.0045202
                pedu      -0.0415348       0.0169576      -2.4493308       0.0143122
            temp2020       0.1519561       0.0566915       2.6804037       0.0073533
------------------------------------------------------------------------------------
Instrumented: W_pwaste
Instruments: W_cmarriage, W_pedu, W_temp2020

DIAGNOSTICS FOR SPATIAL DEPENDENCE
TEST                           MI/DF       VALUE           PROB
Anselin-Kelejian Test            1         3.968          0.0464
================================ END OF REPORT =====================================
```

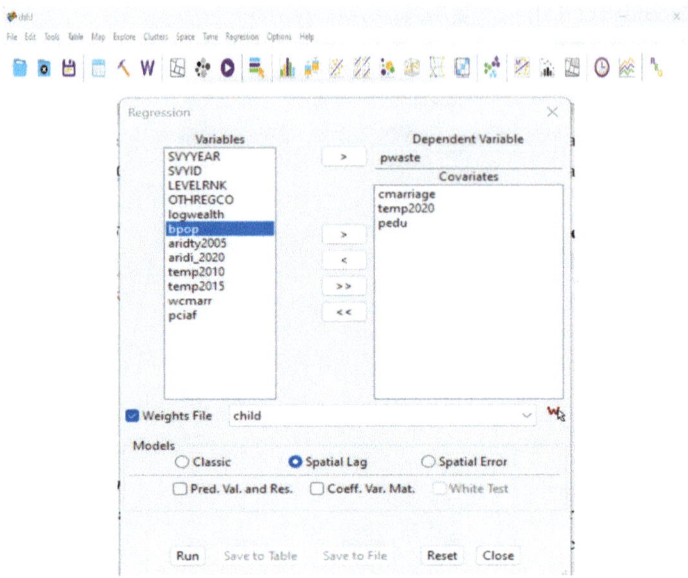

Fig. 5.14 Execution of SAR model using ML in GeoDa

The results show no significant presence of spatial lag for the dependent variable. Other factors are significant at either 1% or 5% level of significance.

The same model may be estimated using maximum likelihood estimation technique. The execution using GeoDa and the estimated model have been reported in Fig. 5.14 and Table 5.3.

Unlike 2SLS case, as we apply the ML, the spatial lag (W_pwaste) becomes significant at 1% level of significance as shown Table 5.3.

ii. ***Spatial error model***

To estimate a spatial error model, we limit the treatment of spatial error autocorrelation in a linear regression model where errors follow a spatial autoregressive process (SAR). A spatial error model is given by:

$$\mathbf{y} = \mathbf{X}\beta + \mathbf{u} \text{ and}$$

$$\mathbf{u} = \lambda \mathbf{W}\mathbf{u} + \varepsilon,$$

where $\boldsymbol{\varepsilon}$ can be heteroscedastic or homoscedastic; however, the errors are non-autocorrelated. The reduced form of the SAR component is:

$$\mathbf{u} = (\mathrm{I} - \lambda \mathbf{W})^{-1}\varepsilon. \tag{5.11}$$

Table 5.3 Results of SAR model using ML in GeoDa

```
>>07/25/24 19:23:39
REGRESSION
----------
SUMMARY OF OUTPUT: SPATIAL LAG MODEL - MAXIMUM LIKELIHOOD ESTIMATION
Data set              : child
Spatial Weight        : child
Dependent Variable    :      pwaste  Number of Observations:  707
Mean dependent var    :      16.536  Number of Variables   :    5
S.D. dependent var    :      5.7356  Degrees of Freedom    :  702
Lag coeff.   (Rho)    :    0.575284

R-squared             :    0.371703  Log likelihood        :    -2100.39
Sq. Correlation       : -            Akaike info criterion :     4210.79
Sigma-square          :     20.6691  Schwarz criterion     :     4233.59
S.E of regression     :     4.54633

-----------------------------------------------------------------------------
       Variable       Coefficient      Std.Error         z-value    Probability
-----------------------------------------------------------------------------
          W_pwaste       0.575284      0.0383302         15.0086        0.00000
          CONSTANT        6.69336        1.71897         3.89383        0.00010
         cmarriage      0.0270174      0.0151344         1.78516        0.07423
          temp2020       0.079296      0.0440207         1.80133        0.07165
              pedu     -0.0298502      0.0147374        -2.02547        0.04282
-----------------------------------------------------------------------------

REGRESSION DIAGNOSTICS
DIAGNOSTICS FOR HETEROSKEDASTICITY
RANDOM COEFFICIENTS
TEST                                      DF       VALUE            PROB
Breusch-Pagan test                        3         5.2029          0.15753

DIAGNOSTICS FOR SPATIAL DEPENDENCE
SPATIAL LAG DEPENDENCE FOR WEIGHT MATRIX : child
TEST                                      DF       VALUE            PROB
Likelihood Ratio Test                     1        188.7038         0.00000
----------------------------- END OF REPORT ------------------------------
```

The variance–covariance matrix for the errors u is not diagonal, as it takes the form:

$$E[uu'] = (I - \lambda \mathbf{W})^{-1} \Sigma (I - \lambda \mathbf{W}')^{-1},$$

with Σ being the variance–covariance matrix for the idiosyncratic errors. Assuming homoscedasticity, this can be simplified as:

$$E[uu'] = \sigma^2 [(I - \lambda \mathbf{W})'(\mathbf{I} - \lambda \mathbf{W})]^{-1}.$$

Substituting the reduced form in the original regression equation, we get the DGP of the SEM as expressed in equation (5.8a):

$$\mathbf{y} = \mathbf{X}\beta + (I - \lambda \mathbf{W})^{-1} \varepsilon.$$

Which further simplified to:

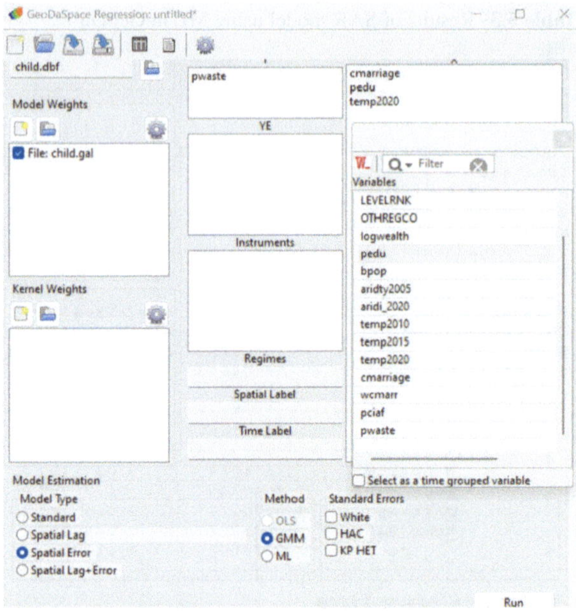

Fig. 5.15 Execution of spatial error model using GeoDaSpace

$$(\mathrm{I} - \lambda\mathbf{W})\mathbf{y} = (\mathrm{I} - \lambda\mathbf{W})\mathbf{X}\beta + \varepsilon. \tag{5.12}$$

This is referred to as spatial Cochrane-Orcutt transformation. The model is estimated using the

(a) Generalized method of moments (GMM) technique: The three main advantages of using GMM model are:

    i. The estimator is robust in presence of heteroscedasticity,
    ii. An asymptotic variance matrix is obtained for the parameter λ and
    iii. Joint inference is implemented for the spatial lag coefficient (ρ) and the spatial error coefficient (λ) in the so-called spatial combo model.

Let us illustrate the estimation of a spatial error model using our data set. This time we consider unobserved factors like women's agency of concerned district as well as neighboring districts may have impact on child malnutrition in the concerned district. As we select the option of spatial error model, in GeoDaSpace, a new variable lamda will be created which is the coefficient of the spatial error component. The steps in GeoDaSpace and the regression output have been reported in Fig. 5.15 and Table 5.4.

Spatial error coefficient, lamda (λ), is significant at 1% level of significance indicating factors like women's agency of concerned district as well as neighboring districts have significant impact on child malnutrition in the concerned district.

(b) Maximum likelihood estimation (MLE): Another possible estimation technique is MLE. As outlined by Anselin (1980), the ML estimate for β can be computed

Table 5.4 Output of spatial error model using weighted least square (assuming homoscedasticity) in GeoDaSpace

```
REGRESSION
----------
SUMMARY OF OUTPUT: SPATIALLY WEIGHTED LEAST SQUARES (HOM)
-----------------------------------------------------------
Data set            :    child.dbf
Weights matrix      :File: child.gal
Dependent Variable  :       pwaste              Number of Observations:       707
Mean dependent var  :      16.5360              Number of Variables   :         4
S.D. dependent var  :       5.7397              Degrees of Freedom    :       703
Pseudo R-squared    :       0.1121
N. of iterations    :            1

------------------------------------------------------------------------------------
            Variable     Coefficient       Std.Error     z-Statistic     Probability
------------------------------------------------------------------------------------
            CONSTANT      14.9603431       2.6911270       5.5591368       0.0000000
           cmarriage       0.0389815       0.0223354       1.7452771       0.0809366
                pedu      -0.0502204       0.0202955      -2.4744599       0.0133438
            temp2020       0.1790888       0.0795306       2.2518224       0.0243335
              lambda       0.5980401       0.0316037      18.9231285       0.0000000
------------------------------------------------------------------------------------
================================ END OF REPORT =====================================
```

as:

$$\hat{\beta}_{ML} = \hat{\beta}_0 - \hat{\rho}\hat{\beta}_1, \tag{5.13}$$

where $\hat{\beta}_0$ is the OLS estimate in a regression of $\mathbf{y}$ on $\mathbf{X}$:

$$\hat{\beta}_0 = \left[\mathbf{X}'\mathbf{X}\right]^{-1}\mathbf{X}'\mathbf{y},$$

and $\hat{\beta}_1$ is the OLS estimate in a regression of $\mathbf{Wy}$ on $\mathbf{X}$:

$$\hat{\beta}_1 = \left[\mathbf{X}'\mathbf{X}\right]^{-1}\mathbf{X}'\mathbf{Wy}.$$

In the next step, residual vectors are computed as:

$$e_0 = \mathbf{y} - \mathbf{X}\hat{\beta}_0,$$

$$\text{and } e_1 = \mathbf{Wy} - \mathbf{X}\hat{\beta}_1.$$

Next, we derive the value of ρ that maximizes the concentrated log-likelihood function (considering Joint normality of the error term $\mathbf{u}$):

$$\ln L = (n/2)\ln(1/n)(e_0 - \rho e_1)'(e_0 - \rho e_1).$$

Finally, substituting $\hat{\rho}$ in $\hat{\beta}_{ML} = \hat{\beta}_0 - \hat{\rho}\hat{\beta}_1$ we get the MLE of β. The steps of ML regression and the result have been presented in Fig. 5.16 and Table 5.5.

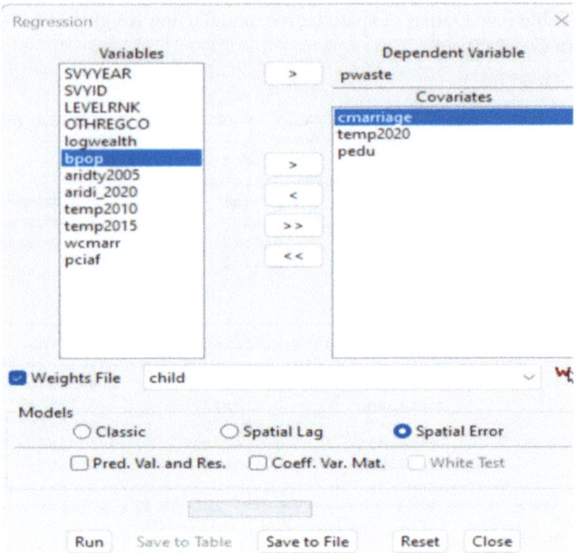

Fig. 5.16 Execution of spatial error model using ML estimation in GeoDa

Spatial error coefficient, lamda (λ), is significant at 1% level of significance like the weighted least square model. However, for other explanatory variables, there is some difference in the significance levels if we compare the coefficients of Tables 5.4 and 5.5.

iii. ***Spatial Autoregressive Moving Average (SARMA) model***

Models with autoregressive and error (Combo Model) as presented in Eq. (5.9) are given by:

$$\mathbf{y} = \rho\mathbf{Wy} + \mathbf{X}\beta + (\mathrm{I} - \lambda\mathbf{W})^{-1}\varepsilon.$$

We estimate the model using the GMM technique (Table 5.6).

The result presented in 16(b) shows that while the spatial lag of the dependent variable is insignificant, the coefficient related to the spatial error (lamda) is significant at 1% level of significance.

MLE can be used to estimate the SAR or the SMA models only—not the SARMA model.

Diagnostics for Spatial Autocorrelation

The regression output presented in Table 5.2 shows the regression diagnostics for Anselin-Kelejian test or AK test given by:

$AK = \frac{nI^2}{\emptyset^2} \sim \chi^2(1)$ where I is the Moran's I statistic computed from the S2LS residuals as:

Table 5.5 Results of spatial error model using ML estimation in GeoDa

```
>>07/25/24 19:36:01
REGRESSION
----------
SUMMARY OF OUTPUT: SPATIAL ERROR MODEL - MAXIMUM LIKELIHOOD ESTIMATION
Data set            : child
Spatial Weight      : child
Dependent Variable  :        pwaste  Number of Observations:  707
Mean dependent var  :     16.535960  Number of Variables   :    4
S.D. dependent var  :      5.735595  Degrees of Freedom    :  703
Lag coeff. (Lambda) :      0.622647

R-squared           :      0.395748  R-squared (BUSE)      : -
Sq. Correlation     : -              Log likelihood        :-2092.099072
Sigma-square        :       19.8781  Akaike info criterion :      4192.2
S.E of regression   :       4.45849  Schwarz criterion     :     4210.44

-----------------------------------------------------------------------------
       Variable       Coefficient      Std.Error        z-value    Probability
-----------------------------------------------------------------------------
          CONSTANT        15.164         2.75049          5.5132     0.00000
         cmarriage     0.0345011       0.0226265         1.52481     0.12731
          temp2020      0.175125       0.0817797         2.14142     0.03224
              pedu    -0.0499276       0.0204641        -2.43976     0.01470
            LAMBDA      0.622647       0.0372852         16.6996     0.00000
-----------------------------------------------------------------------------

REGRESSION DIAGNOSTICS
DIAGNOSTICS FOR HETEROSKEDASTICITY
RANDOM COEFFICIENTS
TEST                                    DF       VALUE          PROB
Breusch-Pagan test                      3          6.4729       0.09074

DIAGNOSTICS FOR SPATIAL DEPENDENCE
SPATIAL ERROR DEPENDENCE FOR WEIGHT MATRIX : child
TEST                                    DF       VALUE          PROB
Likelihood Ratio Test                   1        205.2922       0.00000
============================== END OF REPORT ========================
```

$$\mathrm{I} = \frac{e'We/S_0}{e'e/n} \text{when } \mathrm{S}_0 = \sum_i \sum_j w_{ij} \text{ is the sum of weights}$$

In our example, the H_0 of no spatial autocorrelation has been rejected at 5% level of significance. However, AK statistic is diffuse (and not focused) test—the rejection of H_0 does not point to either a lag or an error specification as the proper alternative. More robust statistics can be found which has not been discussed in this chapter.

All the ML results present measure of goodness of fit (AIC and SC) and regression diagnostics reports the LR test and Breusch-Pagan test. In the case of LR test, the rejection of H_0 of spatial randomness suggests the acceptance of the spatial lag or spatial error model, depending on which model we are estimating. Breusch-Pagan test on the other hand checks the presence of heteroscedasticity. Results presented in Tables 5.3 and 5.5 show the rejection of H_0 of spatial randomness at 1% level of

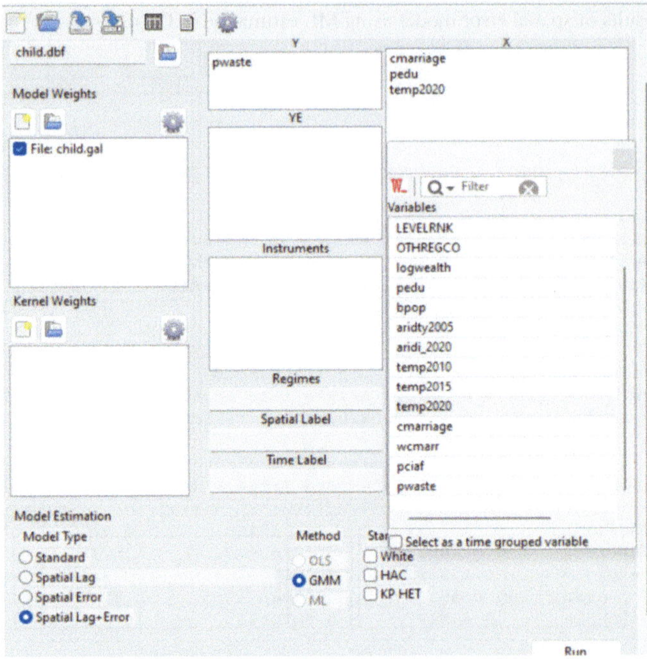

Fig. 5.17 Execution of SARMA model using GMM estimation in GeoDaSpace

Table 5.6 Results of SARMA model using GMM estimation in GeoDaSpace

```
REGRESSION
----------
SUMMARY OF OUTPUT: SPATIALLY WEIGHTED TWO STAGE LEAST SQUARES (HOM)
-------------------------------------------------------------------
Data set              :    child.dbf
Weights matrix        :File: child.gal
Dependent Variable    :       pwaste                Number of Observations:         707
Mean dependent var    :      16.5360                Number of Variables   :           5
S.D. dependent var    :       5.7397                Degrees of Freedom    :         702
Pseudo R-squared      :       0.2288
Spatial Pseudo R-squared:     0.1129
N. of iterations      :            1

------------------------------------------------------------------------------------
            Variable     Coefficient       Std.Error     z-Statistic     Probability
------------------------------------------------------------------------------------
            CONSTANT      12.3525745       3.3521396       3.6849821       0.0002287
            W_pwaste       0.1363713       0.1529083       0.8918499       0.3724734
           cmarriage       0.0427158       0.0236320       1.8075378       0.0706785
                pedu      -0.0434095       0.0210754      -2.0597279       0.0394246
            temp2020       0.1678097       0.0810245       2.0710979       0.0383497
              lambda       0.5481608       0.1029333       5.3253981       0.0000001
------------------------------------------------------------------------------------
Instrumented: W_pwaste
Instruments: W_cmarriage, W_pedu, W_temp2020
================================= END OF REPORT ====================================
```

significance and also report the acceptance of the H_0 of homoscedasticity at either 1% or 5% level of significance.

This regression analysis now can be extended to the context where we may incorporate spatial interactions not only for dependent variable (y), but also for the independent variables (**X**). This type of model is referred to as Durbin model.

5.5.3 Spatial Durbin Model

The spatial Durbin model (SDM) provides a general starting point for discussion of spatial regression model estimation since this model incorporates the SAR and the SEM. The SDM and its DGP have been summarized in equations (5.14) and (5.14a), respectively:

$$\mathbf{y} = \rho\mathbf{W}\mathbf{y} + \mathbf{X}\beta + \eta\mathbf{W}\mathbf{X} + \lambda\mathbf{W}\mathbf{u} + \varepsilon \tag{5.14}$$

$$\mathbf{y} = (\mathbf{I} - \rho\mathbf{W})^{-1}(\mathbf{X}\beta + \eta\mathbf{W}\mathbf{X} + \lambda\mathbf{W}\mathbf{u} + \varepsilon) \tag{5.14a}$$

This model can be written as a SAR model by defining $Z = [\mathbf{X}\ \mathbf{WX}]$ and $\delta = [\beta\ \eta]'$ resulting in:

$$\mathbf{y} = \rho\mathbf{W}\mathbf{y} + \mathbf{Z}\delta + \varepsilon \tag{5.15}$$

$$\mathbf{y} = (\mathbf{I} - \rho\mathbf{W})^{-1}\mathbf{Z}\delta + (\mathbf{I} - \rho\mathbf{W})^{-1}+ \tag{5.15a}$$

$$+ \sim N(0, \sigma^2 I_n) \tag{5.16}$$

The method of estimation is either the two-stage least square or the maximum likelihood estimation technique. Interpretation of regression coefficients and the regression diagnostics will be similar to SAR, apart from the fact that here we need to interpret the coefficients of the lagged explanatory variables.

Let us illustrate the estimation of a SDM using our data set. To estimate the SDM, we consider that the incidence of child marriage in one district gets affected by the incidence in the neighboring district as it is a practice which is related to the cultural norms of a broader region. Rest of the determinants are the same as discussed in the context of SAR. The steps to create a lagged explanatory variable (here lagged cmarriage) using GeoDa have been captured in Fig. 5.18. The execution of the regression is similar to that of a SAR model. The regression output of SDM using GeoDa has been reported in Table 5.7.

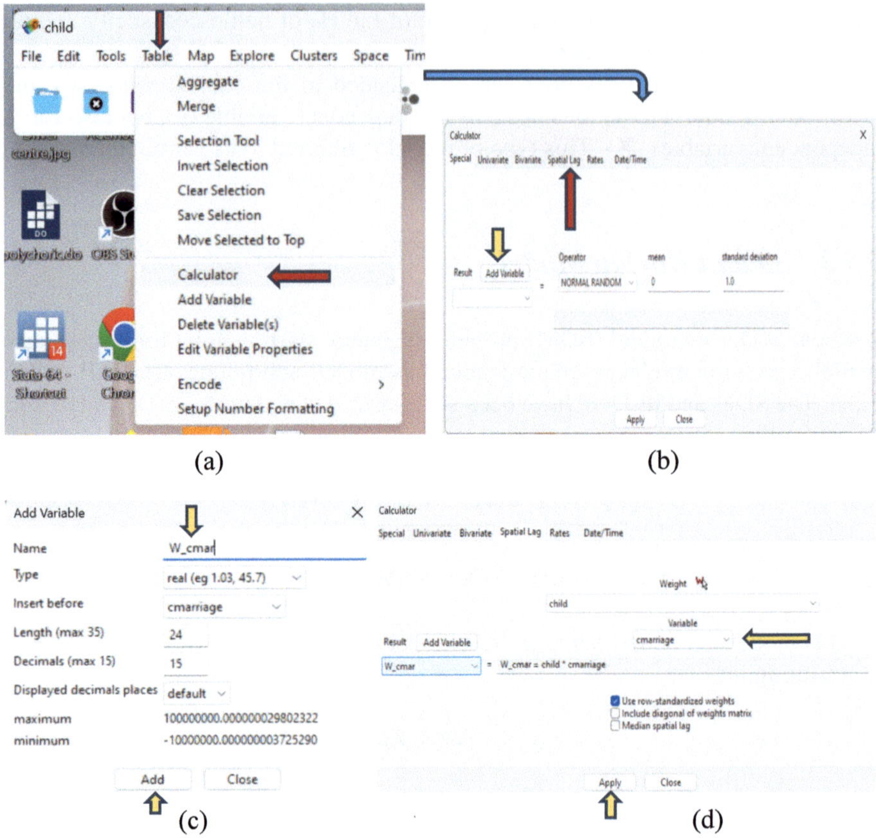

Fig. 5.18 Creating a lagged explanatory variable (lagged cmarriage) in GeoDa

Table 5.7 does not show any evidence of significant role of the lagged child marriage on the health outcome of a concerned district. Explanation of other regression coefficients and the analysis of regression dynamics will be similar to that of a SAR model.

Table 5.7 Results of SDM using ML estimation in GeoDa

```
REGRESSION
----------
SUMMARY OF OUTPUT: SPATIAL LAG MODEL - MAXIMUM LIKELIHOOD ESTIMATION
Data set            : child
Spatial Weight      : child
Dependent Variable  :      pstunt  Number of Observations:  703
Mean dependent var  :     30.7384  Number of Variables   :    6
S.D. dependent var  :     8.11621  Degrees of Freedom    :  697
Lag coeff.   (Rho)  :    0.481611

R-squared           :    0.474134  Log likelihood        :    -2261.29
Sq. Correlation     : -            Akaike info criterion :     4534.57
Sigma-square        :     34.6403  Schwarz criterion     :      4561.9
S.E of regression   :      5.8856

-----------------------------------------------------------------------------
       Variable       Coefficient      Std.Error       z-value    Probability
-----------------------------------------------------------------------------
          W_pstunt       0.481611      0.0427688       11.2608      0.00000
          CONSTANT        25.0462        2.59476       9.65261      0.00000
            w_cmar       0.160601        0.50363      0.318887      0.74981
         cmarriage      0.0667548       0.025405       2.62763      0.00860
          temp2020      0.0922464      0.0581526       1.58628      0.11268
              pedu      -0.163355      0.0200792      -8.13554      0.00000
-----------------------------------------------------------------------------

REGRESSION DIAGNOSTICS
DIAGNOSTICS FOR HETEROSKEDASTICITY
RANDOM COEFFICIENTS
TEST                                        DF       VALUE          PROB
Breusch-Pagan test                          4         17.5994       0.00148

DIAGNOSTICS FOR SPATIAL DEPENDENCE
SPATIAL LAG DEPENDENCE FOR WEIGHT MATRIX : child
TEST                                        DF       VALUE          PROB
Likelihood Ratio Test                       1        112.6487       0.00000
============================== END OF REPORT ================================
```

5.6 Conclusion

This chapter discusses the main components of spatial analysis. An excellent discussion is in Anselin and Rey's textbook (Anselin and Rey 2014). However, we have just covered the tip of the iceberg due to space constraints. The advances made in augmenting economic theories with spatial dimensions has been remarkable and there are several other methods of spatial data analysis that should be kept in mind by potential researchers in this field. For instance, spatial statistical methods are able to analyze qualitative data using both exploratory methods and in qualitative dependent variable regression models, we have not discussed them. Nor have we discussed spatio-temporal models. Interested readers may refer to advanced textbooks on spatial methods (LeSage et al. 2009).

The discussion in this chapter is based on the assumption of a homogenous space, with the strength of spatial dependence and interlinkages remaining constant over locations. But, if local conditions vary over space, Geographically Weighted Regression (GWR) models are more appropriate than the SARMA model discussed in this chapter (Fotheringham et al. 2002). It is represented as:

$$\eta_i = \Sigma_k \beta_k(u_i, v_i) x_{kj} = x_i^t \beta(u_i), \tag{5.5}$$

where x_{kj} and β_k are the kth explanatory variable and its coefficient, respectively. In this model, the coefficients vary with the geographical coordinates of the locality [$u_i = (u_i, v_i)$]. The GWR model will yield n-values of each coefficient, facilitating mapping and spatial analysis of the coefficients. A discussion of the GWR model is provided in a subsequent chapter.

Another important development in recent years has been the application of machine learning to spatial data analysis. Machine learning refers to the use of algorithms to imitate the process of learning by human beings and improve the accuracy of predictions based on previous results. An important application of machine learning in spatial analysis is in the spatial clustering of locations. Originating in the works of Wishart (1969) and Hartigan (1975), it comprises of both supervised learning and unsupervised learning to identify modal points in geographic space, or areas of high density surrounded by areas of low density. Examples of unsupervised learning include DBSCAN and its subsequent refinements, DBSCAN* and HDBSCAN (Kulldorff 1997; McInnes et al. 2017; Gan and Tao 2017). Among supervised clustering techniques are Automatic Zoning Procedure and Max-P method (Openshaw and Rao 1995; Duque et al. 2012).

An important application of machine learning techniques has been to the analysis of spatial big data. Big data refers to data characterized by large volume and fast speed of information generation in different forms and covering different dimensions. Such data poses challenges with regard to integration, storage and processing through advanced computation methods. About 80% of such data is geotagged (Farmer and Pozdnoukhov 2012), making them amenable for spatial analysis, severely exacerbated when big data problems concern geospatial data. This is because geospatial applications are intrinsically complex and involve large datasets; data are collected in real time through advanced geospatial data collection technologies that are popularly available in mobile devices and geospatial data are inherently multidimensional (Jiang and Shekhar 2017). Some of the techniques referred to in this section, and the earlier ones, are discussed along with R codes in Kopczewska (2021).

References

Abbott A (1997) Of time and space: the contemporary relevance of the Chicago school. Soc Forces 75:1150–1182

Anselin L (1980) Estimation methods for spatial autoregressive structures (No. 8; Regional Science Dissertation and Monograph Series)

Anselin L (1996) Interactive techniques and exploratory spatial data analysis

Anselin L (1995) Local Indicators of Spatial Association—LISA. Geogr Anal 27:93–115

Anselin L (2005) Exploring spatial data with GeoDa: a workbook. Urbana 51

Anselin L (2010) Local Indicators of Spatial Association—LISA. Geogr Anal 27:93–115. https://doi.org/10.1111/j.1538-4632.1995.tb00338.x

Anselin L (2019) A local indicator of multivariate spatial association: extending Geary's c. Geogr Anal 51:133–150. https://doi.org/10.1111/gean.12164

Anselin L (1988) Spatial econometrics: methods and models. Springer, Netherlands, Dordrecht

Anselin L, Amaral P (2024) Endogenous Spatial Regimes. J Geogr Syst 26:209–234. https://doi.org/10.1007/s10109-023-00411-2

Anselin L, Bera AK, Florax R, Yoon MJ (1996) Simple diagnostic tests for spatial dependence. Reg Sci Urban Econ 26:77–104

Anselin L, Getis A (1992) Spatial statistical analysis and geographic information systems. Ann Reg Sci 26:19–33. https://doi.org/10.1007/BF01581478

Anselin L, Rey SJ (2014) Modern spatial econometrics in practice: a guide to GeoDa, GeoDaSpace and PySAL. GeoDa Press LLC, Chicago, IL

Bao SD (2000) Determinants of human acid content in organic fertilisers. Beijing

Becker RA, Cleveland WS (1987) Brushing scatterplots. Technometrics 29:127–142. https://doi.org/10.1080/00401706.1987.10488204

Bivand RS (2010) Exploratory spatial data analysis. Handbook of applied spatial analysis. Springer, Berlin, Heidelberg, pp 219–254

Case AC, Rosen HS, Hines JR (1993) Budget spillovers and fiscal policy interdependence. J Public Econ 52:285–307. https://doi.org/10.1016/0047-2727(93)90036-S

Conley TG (1999) GMM estimation with cross sectional dependence. J Econom 92:1–45. https://doi.org/10.1016/S0304-4076(98)00084-0

Conley TG, Copa G (1999) Socio-economic distance and spatial patterns in unemployment. New York

Crespi A, Renner K, Zebisch M, Schauser I, Leps N, Walter A (2023) Analysing spatial patterns of climate change: climate clusters, hotspots and analogues to support climate risk assessment and communication in Germany. Clim Serv 30:100373. https://doi.org/10.1016/j.cliser.2023.100373

Davidson R, MacKinnon JG (2004) Econometric theory and methods. Oxford University Press, Oxford

Doreian P (1980) Linear models with spatially distributed data. Sociol Methods Res 9:29–60. https://doi.org/10.1177/004912418000900102

Douglas BR, Nora DV, Mei-Po K, Robert MK, Michael FG, Robert TC (2009) Spatial turn in health research. Science (80-) 30, 1740–1747. https://doi.org/10.1126/science.1232257.Spatial

Duncan EW, White NM, Mengersen K (2017) Spatial smoothing in Bayesian models: a comparison of weights matrix specifications and their impact on inference. Int J Health Geogr 16:47. https://doi.org/10.1186/s12942-017-0120-x

Duque JC, Anselin L, Rey SJ (2012) Themax-P regions problem*. J Reg Sci 52:397–419. https://doi.org/10.1111/j.1467-9787.2011.00743.x

Earnest A, Morgan G, Mengersen K, Ryan L, Summerhayes R, Beard J (2007) Evaluating the effect of neighbourhood weight matrices on smoothing properties of Conditional Autoregressive (CAR) models. Int J Health Geogr 6:54. https://doi.org/10.1186/1476-072X-6-54

Dormann C, McPherson JM, Araújo MB, Bivand R, Bolliger J, Carl G, Davies RG, Hirzel A, Jetz W, Daniel Kissling W, Kühn I, Ohlemüller R, Peres-Neto PR, Reineking B, Schröder B, Schurr FM, Wilson R (2007) Methods to account for spatial autocorrelation in the analysis of species distributional data: a review. Ecography (Cop) 30:609–628. https://doi.org/10.1111/j.2007.0906-7590.05171.x

Farmer C, Pozdnoukhov F (2012) Building streaming GIScience from context, theory, and intelligence. In: Janowicz K, Keßler C, Kauppinen T, Sreekumaran Nair N SS et al (eds) Proceedings of the Workshop on GIScience in the Big Data Age, pp 5–10

Fotheringham AS, Brunsdon C, Charlton M (2002) Geographically weighted regression: the analysis of spatially varying relationships. John Wiley, West Sussex, Chichester

Gan J, Tao Y (2017) On the hardness and approximation of Euclidean DBSCAN. ACM Trans Database Syst 42:1–45. https://doi.org/10.1145/3083897

Gelfand AE, Smith AFM (1990) Sampling-based approaches to calculating marginal densities. J Am Stat Assoc 85:398–409. https://doi.org/10.1080/01621459.1990.10476213

Getis A (1995) Cliff, A.D. and Ord, J.K. 1973: Spatial autocorrelation. London: Pion. Prog Hum Geogr 19:245–249. https://doi.org/10.1177/030913259501900205

Getis A, Ord JK (1992) The analysis of spatial association by use of distance statistics. Geogr Anal 24:189–206. https://doi.org/10.1111/j.1538-4632.1992.tb00261.x

Hartigan JA (1975) Clustering algorithms. John Wiley & Sons, New York

Hausman JA (1978) Specification tests in econometrics. Econometrica 46:1251. https://doi.org/10.2307/1913827

Jenks GF (1977) Optimal data classification for choropleth maps. Lawrence, Kansas

Jiang Z, Shekhar S (2017) Spatial big data science: classification techniques for earth observation imagery. Springer, Geneva

Kamel Boulos MN, Wilson JP (2023) Geospatial techniques for monitoring and mitigating climate change and its effects on human health. Int J Health Geogr 22:2. https://doi.org/10.1186/s12942-023-00324-9

Kelejian HH, Prucha I (2007) The relative efficiencies of various predictors in spatial econometric models containing spatial lags. Reg Sci Urban Econ 37(3):363–374

Kopczewska K (2021) Applied spatial statistics and econometrics: data analysis in R. Routledge, London

Kulldorff M (1997) A spatial scan statistic. Commun Stat Theory Methods 26:1481–1496. https://doi.org/10.1080/03610929708831995

Lee L-F (2004) Asymptotic distributions of quasi-maximum likelihood estimators for spatial autoregressive models. Econometrica 72:1899–1925. https://doi.org/10.1111/j.1468-0262.2004.00558.x

Leibenstein H (1950) Bandwagon, Snob, and Veblen effects in the theory of consumers' demand. Q J Econ 64:183. https://doi.org/10.2307/1882692

LeSage J, Pace RK, Pace LK (2009) Introduction to spatial econometrics. Chapman & Hall, CRC Press, Boca Raton

McInnes L, Healy J, Astels S (2017) hdbscan: Hierarchical density based clustering. J Open Source Softw 2:205. https://doi.org/10.21105/joss.00205

Monmonier M (2018) How to lie with maps, third. The University of Chicago Press, Chicago

Moran PAP (1948) The interpretation of statistical maps. J R Stat Soc Ser b-Methodological 10:243–251

Moreno JL (1932) Application of the group method to classification of labor. National Committee on Prisons and Prison, New York

Moscone F, Knapp M, Tosetti E (2007) Mental health expenditure in England: a spatial panel approach. J Health Econ 26:842–864

O'Sullivan D, Unwin DJ (2010) Geographic information analysis, 2nd edn. John Wiley & Sons, New Jersey

Openshaw S, Rao L (1995) Algorithms for reengineering 1991 census geography. Environ Plan A Econ Sp 27:425–446. https://doi.org/10.1068/a270425

Ord JKK, Getis A (1995) Local spatial autocorrelation statistics: distributional issues and an application. Geogr Anal 27:286–306. https://doi.org/10.1111/j.1538-4632.1995.tb00912.x

Thomas JJ, Cook KA (2006) A visual analytics agenda. IEEE Comput Graph Appl 26:10–13. https://doi.org/10.1109/MCG.2006.5

Tobler WR (1970) A computer movie simulating urban growth in the Detroit region. Econ Geogr 46:234–240. https://doi.org/10.2307/143141

Tukey JW (1963) Exploratory data analysis. Addison-Wesley, Reading, Massachusetts

Tulchinsky TH (2018) John Snow, Cholera, the broad street pump; waterborne diseases then and now. In: Case Studies in Public Health. Elsevier, pp 77–99

Wall PA, Devine OJ (2000) Interactive analysis of the spatial distribution of disease using a geographic information systems. J Geogr Syst 2:243–256. https://doi.org/10.1007/PL00011457
White HC (1963) An anatomy of kinship. Prentice-Hall, Englewood Cliffs, New Jersey
White MJ (1983) The measurement of spatial segregation. Am J Sociol 88(5):1008–1018. https://doi.org/10.1086/227768
Wishart D (1969) Mode analysis: a generalization of nearest neighbor which reduces chaining effects. In: Cole AJ (ed) Numerical taxonomy. Academic Press, New York, pp 282–311

Chapter 6
Geographically and Temporally Weighted Regression: Exploring Demographic Changes in Poland

Maria Kubara, Kateryna Zabarina, and Katarzyna Kopczewska

Abstract This chapter introduces Geographically and Temporally Weighted Regression (GTWR), an extension of Geographically Weighted Regression (GWR) that incorporates temporal variations into spatial analysis. Also referred to as Spatially and Temporally Geographically Weighted Regression (STGWR), this model enhances analytical insights by accounting for both spatial and temporal dependencies. The core element of GTWR is the spatio-temporal distance metric, which determines data inclusion in individual regressions. We estimate GTWR model in R software to examine demographic changes in Poland from 2018 to 2021, analyzing spatial and temporal patterns in fertility levels and the influence of population structure and economic factors. This chapter equips researchers with tools to uncover complex spatio-temporal relationships, improving their understanding of dynamic interplay between space and time.

6.1 Introduction

Geographically Weighted Regression is a separate class of econometric models, which deals with spatial heterogeneity understood as the diversity of relations over space. Even if seemingly similar to typical regressions, it has much less in common. Its dissidence is much more in conceptual and interpretational layers than in technical

Supplementary Information The online version contains supplementary material available at https://doi.org/10.1007/978-981-97-9060-9_6.

M. Kubara (✉) · K. Zabarina · K. Kopczewska
University of Warsaw, Warsaw, Poland
e-mail: maria.kubara@uw.edu.pl

K. Zabarina
e-mail: kzabarina@wne.uw.edu.pl

K. Kopczewska
e-mail: kkopczewska@wne.uw.edu.pl

F. Hamid and D. Mukherjee (eds.), *Data Science for Modeling Managerial and Socioeconomic Problems*, Contributions to Management Science,
https://doi.org/10.1007/978-981-97-9060-9_6

issues. GWR and GTWR give very different information than typical OLS (Ordinary Least Squares) or panel models.

Traditional econometric models generalize the relations between phenomena represented by variables. They estimate the coefficients of those relations and assume that if a model is right, all observations more or less follow this approximated relation. In other words, if the linear function which links dependent and explanatory variables is $y = 1 + 2x + \varepsilon$, then, on average, for each observation, the value of x multiplied by 2 and added 1 gives the value of y. This relation is disturbed by some nuisance or error e, which in a good model is neglectable. The simplicity and beauty of simple econometric models are useful, but the generalizations which it generates are under control. Traditional econometrics, based on p-value rules, operates on significance tests which check if observation-wise deviations from this general rule are not too big. A typical check of significance looks at the confidence interval built around the coefficient with the use of its standard deviation and analyses if it is small enough to be considered reliable. Typically, analysts decide on the impact of x on y based on those tests and consider variable x as having an insignificant impact on y when the standard deviation of the regression coefficient is too big. However, this reasoning has one more factor: the main driver of high standard deviation (high variance of relations) may be not their lack, but their diversity over space.

The major question that GWR asks is if in part of the territory, the relation is $y = 2 + 3x + \varepsilon$ and in another part of the territory it is $y = 5\text{-}1x + \varepsilon$, does it mean there is no relation? In light of typical global econometric models, one indeed assumes that x is not linked with y, as the impact can be positive or negative, so it is insignificant. This phenomenon of relation diversity is called spatial heterogeneity, and it is addressed with local models such as GWR. So, how does GWR deal with it? It estimates the local models which use the very neighboring observations only. Local models are estimated for each observation—i.e., in a dataset of 1000 observations one gets 1000 local models. Usually, the subsamples of data used for regression, which have just a few points from the nearby, are too small to make a reliable set for p-value analysis. Therefore, GWR is not focused on determining if a given variable is significant or not, as it assumes that the model is built in a reasonable way with important variables only. On the other hand, GWR looks for differences in coefficients among the models. If all models yield the same relation, there is no need for GWR. GWR's power appears only when local relations are diverse. Detection of those spatial instabilities is performed by mapping and clustering of coefficients (e.g., Kopczewska and Ćwiakowski 2021).

Geographically and Temporally Weighted Regression (GTWR), also referred to as Spatially and Temporally Geographically Weighted Regression (STGWR), extends classical GWR in the same way as panel models extend OLS models. It addresses the changes over time when phenomena are measured in some periods. In the case of traditional global panels, the estimated coefficients are fitted to reflect the relations in two dimensions—in cross-section of data and over time. This panel generalization can be supplemented by fixed or random effects for time, individuals or both (so-called two-way). However, those specific effects support the directional relation expressed by the coefficient with some vertical shift up or down. Spatial panels have an additional component of similarity in space which is addressed with spatial weights matrix and spatial lags of variables and modeled in a specific way (Kopczewska et al. 2017). Traditional panels do not distinguish between recent and old data—all periods are considered the same. In local models, the core issue is in selecting proper neighboring observations not only in space but also in time. This means that only the closest observations contribute to the regression—fresh in time and close in space. In consequence, for spatio-temporal data, one gets models for each observation, which already integrate changes in time into coefficients. Thus, again, as in GWR, one can compare spatial patterns of regression coefficients, while temporal changes are already included.

Thus, why is it worth estimating GWR and GTWR instead of OLS or panel models? It gives much deeper insight into details of relations between phenomena over time and space. It prevents deleting variables from models when insignificant in traditional global models—it inspects its specificity and enables finding clusters of similar relations, even if they are diverse. It follows the old rule which is not to average the multi-modal distributions but to treat the subsamples separately. Here those subsamples have mostly spatial dimensions, with control of temporal changes.

6.2 Geographically Weighted Regression—Definition and Main Features

6.2.1 General Form of the Model

Geographically Weighted Regression (or GWR) appeared first in Brunsdon et al. (1996) (and was widely described later in Fotheringham et al. (2002)) as a method to overcome the issue of spatial non-stationarity. Starting with an example of a theoretical OLS model explaining house price and analyzing the impact of possible regressors on a dependent variable, they claim that the use of so-called global model is not correct because of spatial heterogeneity: one factor may positively imply on price in location *A* and—at the same time—negatively in location *B*, which cannot be captured by a general model. They propose a solution in form of GWR, which looks as follows:

$$Y_i = \beta_0(u_i, v_i) + \sum_k \beta_k(u_i, v_i)X_{ik} + \varepsilon_i, i = 1, ..., n \qquad (6.1)$$

(Wu et al. 2014) where (u_i, v_i) are the spatial coordinates of location i, Y_i is a value of dependent variable, β_0 is the constant term, $\beta_k(u_i, v_i)$ represents a value of coefficient for regressor k ($k = 1, ..., K$), and ε_i is an error term in location i.

In simple terms, GWR involves estimation of local models for each point in the sample, using a small subset of the sample for each regression. These regressions are estimated locally, taking into account the spatial relations present in the given neighborhood. Instead of calculating a single global model, GWR provides results from n models (where n represents the number of observations in the sample), each based on the spatial division of the original sample. Later, the estimates from each local regression may be averaged to get the global trends or analyzed graphically to explore how the local relations are varying across different locations.

Main advantage of GWR is the consideration of the local heterogeneity and variability of regressors over the space. At the same time, the form of the model and interpretation of parameters remain simple. To account for spatial heterogeneity, GWR does not use spatial coordinates themselves, but distances between points. These distances are later used in the determination of spatial weights, a wider description of which is given below.

6.2.2 Weighting Coefficients and Kernel Functions

Calculation of fitted parameters in GWR reminds the calculation of $\hat{\beta}$ in OLS in a matrix form:

$$\beta(i) = (X'W(i)X)^{-1}X'W(i)y \tag{6.2}$$

(Brunsdon et al. 1996), where $\beta(i)$ is a vector of parameters and $W(i)$ is a diagonal matrix of weighting coefficients showing the importance of observations in the estimation of parameters for location i. In essence, each local regression within GWR selects a specific spatial subsample from the original dataset for parameter estimation. The influence of the neighboring observations included in each local estimation diminishes with distance from the central point using spatial weights. So, what is the spatial weight and how to calculate it?

The definition of spatial weight in GWR is impossible without the concept of bandwidth (here, only the definition is provided; a wider description is given in Sect. 6.2.3). In a very simple words, *bandwidth* is the radius of a circle around location i. Spatial weights for observations in the location i are calculated then as a function of a ratio between distances to neighboring points and bandwidth. Usually, the further a neighbor is, the lower the weight will be, and the smaller will be its impact on the results of each local estimation. Good visualization of $W(i)$ coefficients is given by de Bellefon and Floch (2018) (see Fig. 6.1).

Bandwidth parameter serves as an argument for *kernel function* used for spatial weights' calculation. According to Brunsdon et al. (1998), kernel function should possess the following features (all of them are clearly visible on a Fig. 6.1):

a. at the starting point (zero) its value is 1;
b. right-hand side limit at point d is equal to zero;
c. kernel function is "*monotone decreasing function for positive real numbers*" (Brunsdon et al. 1998, p. 434).

The most popular kernel functions are:

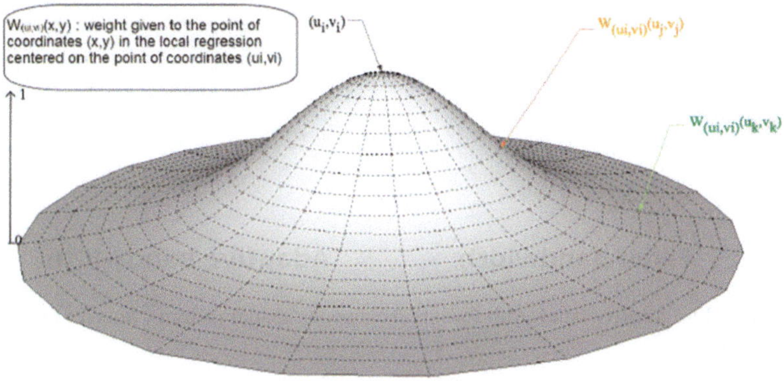

Fig. 6.1 Visual representation of $W(i)$ matrix. Source: de Bellefon and Floch (2018)

- Gaussian:

$$W_{ij} = exp\left(-d_{ij}^2/2h_i^2\right) \tag{6.3}$$

(Brunsdon et al. 1998),

- exponential:

$$W_{ij} = exp\left(-d_{ij}/h_i\right) \tag{6.4}$$

(Brunsdon et al. 1998),

- bisquare—used in the current study:

$$W_{ij} = (1 - \left(\frac{d_{ij}}{h_i})^2\right)^2, \ if\ d_{ij} < h_i\ and\ 0\ otherwise \tag{6.5}$$

(Huang et al. 2010),

where d_{ij} is the distance between locations i and j, and h_i is a bandwidth. By default, distance is Euclidean,[1] but can be any other—for example, Minkowski,[2] great circle or Manhattan distance (Lu et al. 2014). If bandwidth is *fixed* ($h_i = h$), then for each location the same radius is used, and the number of neighbors for each location i may be different. Keeping the bandwidth fixed may be problematic for

[1] Good overview of non-Euclidean metrics and its application to London house price data is given in Lu et al. (2014). Moreover, authors fairly claim that "*geographical space is usually too complicated to be measured simply by ED* [Euclidean distance] *metrics*" (Lu et al. 2014, p. 663), meaning that dependently on study topic and data different metrics may be applied (Cao et al. 2019).

[2] GWmodel:: package allows to calculate Minkowski distance for any p.

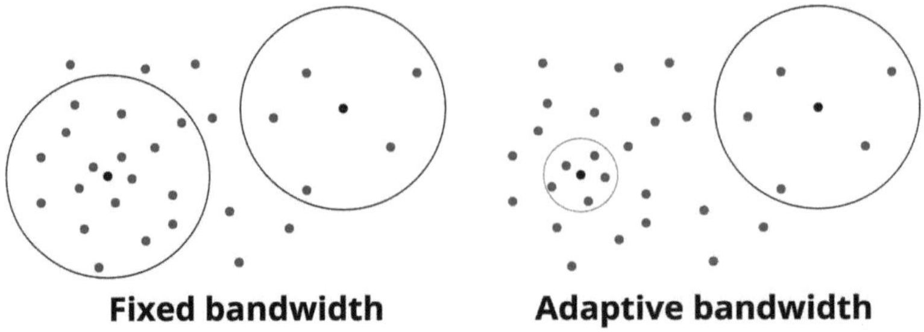

Fig. 6.2 Fixed vs adaptive bandwidth for inhomogeneous spatial patterns. Source: Own work in Canva software

inhomogeneous spatial patterns, i.e., where we observe areas with diverse density of spatial entities (Fig. 6.2). For such patterns, fixed bandwidth in one place may include only a few neighbors in the local regression, while in the denser areas the same bandwidth distance may yield many observations included in the local regression. In such case, it may be better to use *adaptive* bandwidth—which utilizes various radii, but a fixed number of neighbors for each location. Visualization of this problem is presented in Fig. 6.2 which shows the same point pattern with varying density of point locations with the utilization of fixed and adaptive bandwidth. For the same two local centers (black dots) in the fixed bandwidth case, local regressions would include 17 and 5 neighboring observations, while in the adaptive bandwidth case both local regressions would include just 5 neighbors—this was ensured by the adjusted radius size for areas with higher and lower density.

One can divide kernel functions on *continuous* and functions with a *compact support*—according to Heine-Borel theorem, "*compactness of a set is equivalent to the set's being closed and bounded.*"[3]. Kernel functions are different in terms of the values, which they can take at the edges of the selected bandwidth. From the example kernels listed above, exponential and Gaussian kernels are continuous, since their supports are open sets (from $-\infty$ to $+\infty$), and bisquare kernel is a kernel with a compact support—values for d_{ij} are bounded from above with a bandwidth value.

Very good graphical representation of different kernel functions (both continuous and compact support) is given in Fig. 6.3. Figure 6.3a compares uniform kernel (which is basically an OLS case—all locations have the same importance) to Gaussian and exponential, and Fig. 6.3b represents kernel functions with compact support. The comparison of kernels illustrates the variation of spatial weights in each local regression, relative to the distance from the central point. In GWR, the goal is to progressively assign lower weights to observations that are further from the central point. The choice of the kernel defines how these weights will change as the distance increases.

[3] Speaking more formally, compactness, closedness, and boundedness are slightly different definitions, but in this chapter, we will consider Heine-Borel theorem and Euclidean space.

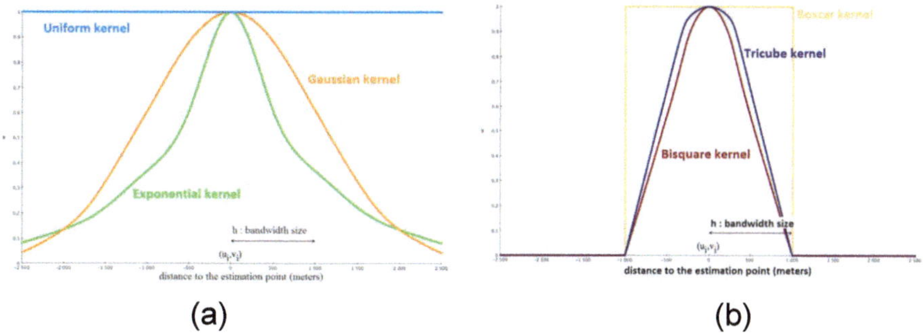

Fig. 6.3 Difference between kernels: (a) continuous kernels, (b) compact support kernels. Source: de Bellefon and Floch (2018)

6.2.3 Bandwidth Choice

As it was mentioned before, bandwidth determines how many neighboring points around the location i are taken into account in each local regression and—as a parameter of kernel function—what is a weight of each observation in a final equation. Thus, the choice of bandwidth is crucial. The higher the bandwidth, the more observations are included into estimation of each local regression. Depending on whether the bandwidth is adaptive or fixed, the choice will be made based on distance or the number of neighbors included in each local regression. With a smaller bandwidth, the estimated effects are more localized, and with a larger bandwidth the analysis captures spatial effects over a larger area. At the extreme, when the (adaptive) bandwidth equals the total number of observations in the sample, GWR will produce results of a single global model. Figure 6.4 shows how kernel function may change with a change of bandwidth.

According to existing literature, optimal bandwidth is chosen using one of the three[4] approaches—arbitral choice based on previous studies/knowledge of a research problem (rather a rare option), leave-one-out cross-validation (LOOCV), or value of AIC (or also corrected AIC, denoted as *AICc*, which is used more often). Idea of *leave-one-out cross-validation* method lies in a minimization of the squared difference between the observed value of the dependent variable and its fitted value in location i estimated by omission of this location from data:

$$CV = \sum_i (y_i - y_{\neq i}(h))^2 \tag{6.6}$$

(Guo et al. 2008), where h is a bandwidth.

---

[4] Among other approaches, one can mention Information Complexity Criteria by Koç (2022), various algorithmic approaches by da Silva and Mendes (2018), combination of LOOCV with spatial error LM statistic by Cho et al. (2010).

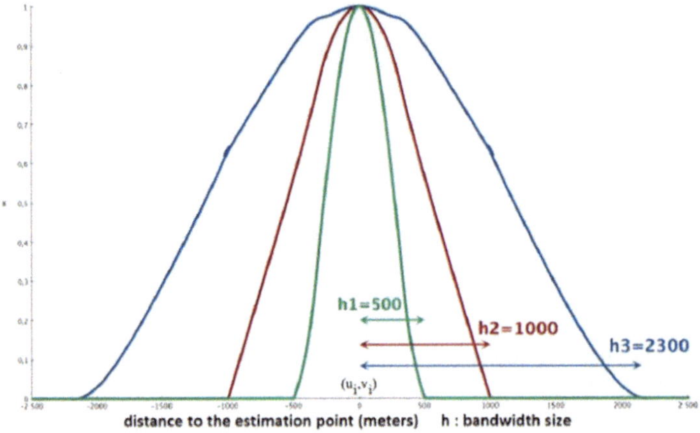

Fig. 6.4 Impact of bandwidth. Source: de Bellefon and Floch (2018)

AICc optimization works similarly—the algorithm looks through all the different bandwidths and chooses such that minimizes AICc value for the estimated model. Corrected AIC can be found as follows:

$$AICc = 2nlog(\hat{\sigma}) + nlog(2\pi) + n\left\{\frac{n + tr(S)}{n - 2 - tr(S)}\right\} \tag{6.7}$$

(Koç 2022), where n is a sample size, $\hat{\sigma}$ is the standard deviation of residuals, and $tr(S)$ is a trace of $\hat{y}$ matrix.

There is a discussion in the literature about which of these two methods is better for bandwidth determination. Whereas CV "*only accounts for model prediction accuracy,*" AIC "*accounts for ... trade-off between prediction accuracy and complexity*" (Lu et al. 2014, p. 662). Guo et al. (2008, p. 2528) also point out that AIC "*has the advantage of taking into account the fact that the degrees of freedom may vary among the models centered on different locations.*" On the other hand, in some cases bandwidths produced by CV and AIC are almost the same or CV performs even better (da Silva and Mendes 2018).

6.2.4 Interpretation of Coefficients in GWR

GWR coefficients can be interpreted at two levels. Firstly, we can examine local models to understand the specific influences of each variable at diverse locations. Alternatively, global model results, which average local estimations, offer a broader perspective comparable to a single OLS estimation. Whether from local or global models, GWR coefficients are interpreted similarly to traditional OLS estimates, but they reflect results from varying spatial samples.

Commonly, interpreting GWR results involves examining the distribution of coefficients from local regressions to identify if a variable's impact varies across locations. This process is usually accompanied with the graphical analysis. Visualization of coefficient values received from local regressions on a map provides a clear visual representation of how the impact of variables changes across the sample, highlighting areas with the highest or lowest influences. In the spatio-temporal version of GWR, which is the Geographically and Temporally Weighted Regression, the same rules for interpreting model coefficients will apply. Example of such analysis with interpretation is provided in Sect. 6.5.

6.3 Geographically and Temporally Weighted Regression as an Augmented Form of GWR

Huang et al. (2010) developed a GTWR (Geographically and Temporally Weighted Regression) as an extension of GWR with a third dimension—temporal, to account for variability both in space and in time. They—as well as Brunsdon et al. (1996)—considered a classic case of housing prices and argued that such prices not only vary spatially, but also temporally.

The general form of GTWR model is very similar to GWR—the only difference is that we now have a third dimension:

$$Y_i = \beta_0(u_i, v_i, t_i) + \sum_k \beta_k(u_i, v_i, t_i)X_{ik} + \varepsilon_i, i = 1, ..., n \tag{6.8}$$

(Huang et al. 2010).

In the same way, calculation of betas changes:

$$\hat{\beta}(u_i, v_i, t_i) = (X'W(u_i, v_i, t_i)X)^{-1}X'W(u_i, v_i, t_i)y \tag{6.9}$$

(Huang et al. 2010).

But despite the fact that Eqs. (6.8) and (6.9) are very similar to GWR case, $W(i)$ matrix is constructed differently (though it is also a diagonal matrix). In GTWR, local regressions include neighboring observations based not just on spatial proximity, but also on temporal closeness. Bandwidth here will be specified on spatio-temporal distance measures. Kernel functions in GTWR enable spatio-temporal filtering for the observations in each local regression. As Fotheringham et al. (2015, p. 435) state, "*there is no reason to assume* a priori *a constant spatial bandwidth over time*" and provide with an equation allowing to find spatio-temporal weights for various bandwidths over time:

$$w^t_{ijS,T} = k_S\left(d_{s_{ij}}, b_{St}\right) \times k_T\left(d_{t_{ij}}, b_T\right) \tag{6.10}$$

(Fotheringham et al. 2015).

Here, k_S and k_T are respectively spatial and temporal kernel functions, $d_{s_{ij}}$ and $d_{t_{ij}}$ are respectively spatial and temporal distances between points i and j, b_T is a temporal bandwidth, and finally, b_{St} is a "*spatial bandwidth specific to time t*" (as above, p. 436).

In GWR, closeness between points is determined usually by simple Euclidean distance. In GTWR, we have a new dimension, and therefore, we need to consider both temporal and spatial distances. Huang et al. (2010) introduced an ellipsoidal model (see Fig. 6.5a) of space and time relationship—axes U and V determine spatial location and axis T shows moments in time, the dark gray point is the location i and other white points are its neighbors. The shape of the relationship is determined by the fact that distances in space and time are not equal. Authors propose a simple formula for distance calculation:

$$d^{ST} = d^S \otimes d^T \tag{6.11}$$

(Huang et al. 2010), where $\otimes$ is an operator determining the relationship between spatial distance d^S and temporal distance d^T. Huang et al. (2010) consider a simple case of addition sign, which gives a following formula:

$$d^{ST} = d^S + d^T = \sqrt{\lambda[(u_i - u_j)^2 + (v_i - v_j)^2] + \mu(t_i - t_j)^2} \tag{6.12}$$

(Huang et al. 2010), where (u_i, v_i) is a location of point i, (u_j, v_j) is a location of point j and t_i and t_j are times observed at locations i and j. Simply speaking, such spatio-temporal distance is a combination of Euclidean distance in space and distance in time. Authors claim that weightening with λ and μ is important—otherwise, in the case when one of the distances is larger, it may weaken the impact of another.

Although Huang et al. (2010) apply λ and μ weights in order to have a balanced impact of space and time, Wu et al. (2014, p. 1190) fairly argue that in such "*orthogonal spatiotemporal coordination system ... the distance measured in space dimensions has no effect on temporal distance and is inadequate for modeling their interactions*" and propose a new way of spatio-temporal distance calculation (illustration of a new approach is given in Fig. 6.5b):

$$d_{ij}^{ST} = d_{ij}^S \otimes d_{ij}^T = \lambda d_{ij}^S + \mu d_{ij}^T + 2cos(\xi)\sqrt{\lambda\mu d_{ij}^S d_{ij}^T}, \; if \; t_j < t_i \, and \; if \; t_i < t_j \tag{6.13}$$

(Wu et al. 2014), where t_i and t_j are times observed at locations i and j, d_{ij}^S and d_{ij}^T are spatial and temporal distance, respectively, and finally, λ, μ, and $\xi \in [0, \pi]$ are adjustment parameters. Similarly to GWR, optimization of λ, μ, and ξ in GTWR can be done with LOOCV or AIC minimization. As in the case of GWR, one can apply Gaussian, bisquare, or exponential kernels with fixed or adaptive bandwidth.

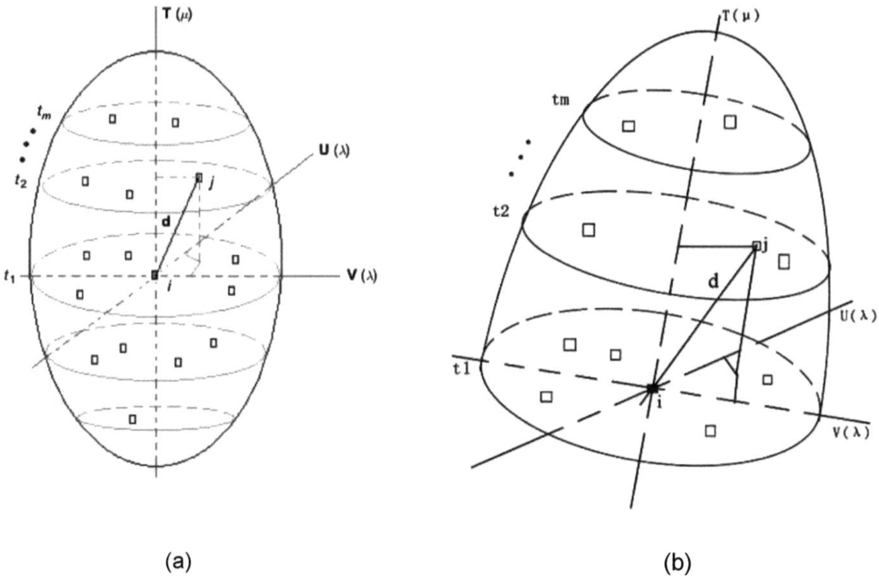

Fig. 6.5 Distance across time and space. Source: (a) Huang et al. (2010), (b) Wu et al. (2014)

6.4 Applications of GWR and GTWR in Practice

Practical applications of GTWR reach very diverse topics but are often relying on data availability.

Early applications of GTWR were in economics and housing markets and exploited the examples analyzed before with GWR without temporal dimension. One of the first papers using GTWR was by Huang et al. (2010) who studies housing data in Calgary, Canada. The dataset included 5000 points—transactions in the years 2002–2004. The authors underline the value added of GTWR, especially in the reduction of absolute errors and yielding better goodness of fit. Similarly, Fotheringham et al. (2015) in the analysis of housing data in London found that GTWR compared to GWR was better at addressing the bias-variance trade-off. However, their study used data for almost 20 years (1980–1998)—geo-located points with information on a year of transaction. In this study, the authors underline that the selection of bandwidth and kernel functions for spatial and temporal dimensions matters for the outcome, which makes GTWR highly advanced in proper usage.

The literature presents also social studies using geographically and temporally weighted regressions. Nugroho and Sumarminingsih (2021) analyzed the percentage of the poor population in East Java Province using annual regional data from 38 territorial units and 6 years. They show that GTWR by capturing jointly spatial and temporal variability gives robust results of estimation. Among other social studies are those linked to mobility. Xu et al. (2020) analyzed geo-located and time-marked ca. 1600 pedestrian-related injuries in Hong Kong including their severity. The main conclusion is that adding a temporal dimension over a spatial one improves the goodness of fit. Another study that deals with mobility is on transit over the city of Beijing and the traffic it generates. Ma et al. (2018) show for 1027 city neighborhoods and hourly ridership data that modeling time and space jointly significantly improves the fit of the model.

Another huge stream of spatio-temporal research is the environment. Recent studies analyze precipitation hydrogen isotopes in Northeastern United States (Que et al. 2020) and show that GTWR improves predictions by using the most recent observations. Also, air quality studies prove the usefulness of GTWR. Wei et al. (2019) study air pollution with $PM_{2.5}$ in China and examine the relationship between $PM_{2.5}$ and 5 criteria air pollutants (SO_2, NO_2, PM_{10}, CO, and O_3). For weekly data for 13 regions, they confirm the better quality of fit but also remark that the data require a wider spatial dimension than 13 units to reveal the potential of spatial data. GTWR may use not only territorially aggregated data, but oppositely, data of low granulation. Bai et al. (2016) show for a satellite-derived 500 m resolution grid that a ground-level $PM_{2.5}$ can be better explained with spatio-temporal local modeling. For satellite data on water quality (Chu et al. 2018), modeling with GTWR also yields a relatively high goodness of fit, improves the explanation of variance, and shows a better pace–time explanatory power.

Climate data are also being linked to epidemiological data. Liu et al. (2018) run a study for 22 countries using the weekly status of influenza surveillance and explaining it with weather conditions: temperature, precipitation, vapor pressure, and relative humidity. In their logistic GTWR, they discover new spatial and temporal varying patterns—spatial clustering and weather dependence. This study, beyond an interesting topic, also developed a binary-choice version of GTWR. Similarly, Wu et al. (2020) proposed neural network GTWR for environmental modeling—they achieved the best fit and prediction with this model and also exactly quantified spatio-temporal non-stationary relationships.

To conclude, all studies using the temporal version of geographically weighted regression claim it improves the estimation, supports better fit, and reduces prediction errors. What is important, the studies can be on point data, but also regions represented with centroids. The other common factor is that datasets are of moderate size. This method is resource-demanding and for now, it is not suitable for big data.

6.5 Case Study for GTWR Application

6.5.1 Research Problem and Data

Traditionally, fertility levels were considered outcomes of purely cultural factors. However, recognizing the strong correlation between demographic trends and labor market stability, economists have delved into this topic. Becker's influential theory (Becker 1992) established a new perspective, paving the way for policies utilizing economic nudges to impact fertility levels globally. Here, we adapt Becker's theory to regional data on fertility in Poland (NUTS4).

Our model explains fertility in a region by combining demographic and economic factors (Table 6.1). We begin with the proportion of women in the reproductive age group (20 to 45) in a given region, representing the general fertility potential. We consider economic variables, like the number of preschools per 1000 inhabitants, which impact fertility rates by facilitating quicker return to the labor market after childbirth (Lee and Lee 2014). We also consider the unemployment rate among men in the area, presuming that stable employment increases the likelihood of procreation. Moreover, we include the average salary in the region, which should have a positive impact on fertility, though this effect may be nuanced due to potential income loss for mothers after childbirth (Angelov et al. 2016). In non-panel models, we introduce a categorical variable "year" to account for time effects to some extent.

The investigated phenomenon presents considerable levels of spatial heterogeneity. Fertility is influenced by various endogenous factors intricately tied to specific cultural and regional mindsets. Moreover, economic factors exhibit diversity across the diverse landscape of Poland. Temporally, we expect further diversity, given the research period of 2018–2021. In July 2019, a new social transfer for every child was introduced, incentivizing childless couples to extend their families. Soon after, the pandemic began, leading to increased time spent together during lockdowns, associated with higher fertility rates (Anser et al. 2020). However, pandemic-related labor market challenges might have discouraged people from having children (Anser et al. 2020; Vignoli et al. 2020). Notably, these effects were intertwined with regional heterogeneity, adding to the spatio-temporal diversity that makes the GTWR method an excellent testing case (Fig. 6.6).

Table 6.1 Statistical distribution of variables included in the analysis

Variable	Statistical description			
	min	mean	s.d	max
Birth Live births per 1000 inhabitants	5.650	9.098	1.434	16.520
k_fertil1000 Women aged 20–45 per 1000 inhabitants	137.032	170.354	9.335	213.248
preschool1000 Preschools per 1000 inhabitants	0.028	0.312	0.099	0.649
salary Average salary in the region	3183.340	4642.660	713.424	10,076.640

Source: Own work. Data retrieved from GUS (2023)

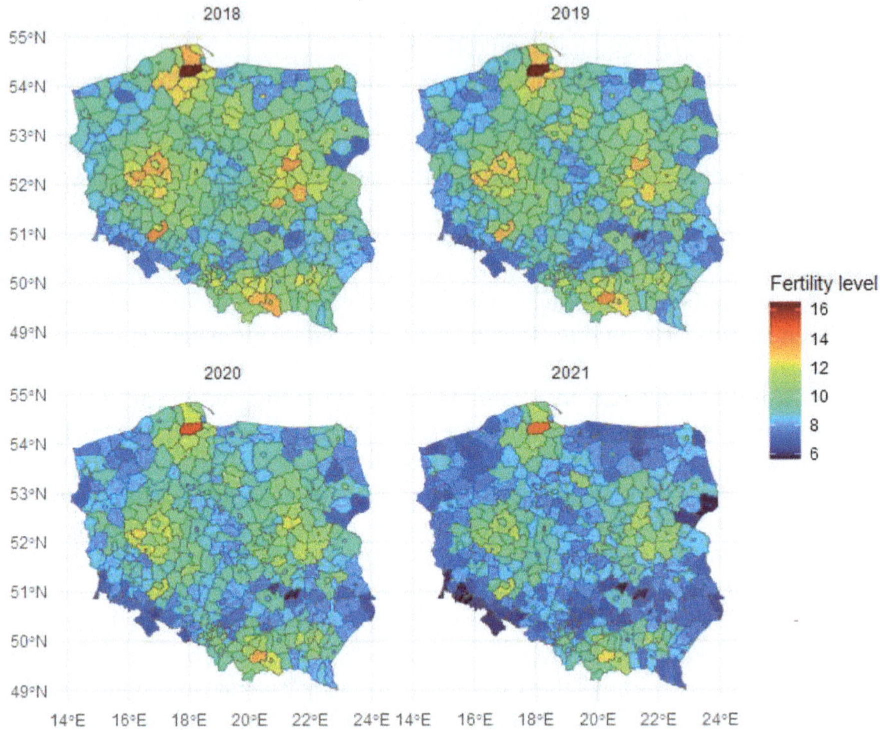

Fig. 6.6 Life births per 1000 inhabitants at NUTS4 level in Poland in 2018–2021. Source: Own work in R with ggplot2:: package

6.5.2 Computation of GWR and GTWR in R: GWmodel:: Package

GWmodel:: package in R provides well-developed functions for estimating various extensions of the GWR framework, such as GTWR (Gollini et al. 2015). It operates on spatial objects based on the sp:: package (Pebesma and Bivand 2005). In this chapter, we will discuss the essential functions from GWmodel:: required for estimating both GWR and GTWR models.

Estimating GWR and GTWR models begins by determining an appropriate bandwidth, which is considered a hyperparameter and optimized through simulations to achieve the best fit to the data. The bandwidth defines the neighborhood range used in each local regression and can be expressed in terms of a distance (fixed spatial kernel) or the number of observations included in the neighborhood (adaptive spatial kernel). The adaptive kernel is preferred for areas with varying density, ensuring that all local models are computed on the same sample size (Ćwiakowski 2020; Kopczewska 2020).

In the GWmodel:: package, the bandwidth for GWR and GTWR models can be calculated using the bw.gwr() and bw.gtwr() functions (Table 6.2). These functions share a set of parameters, requiring the model formula and a spatial dataset (in appropriate sp:: class). Users can modify the optimization method, as well as the form of the kernel function. Additionally, users can specify whether the bandwidth kernel should be adaptive or fixed (default is *adaptive* = FALSE). The parameter *p* determines the power of the Minkowski distance. The *theta* parameter allows for the rotation angle of the coordinate system in radians. If the dataset is reported in the *SpatialPointsDataFrame* class, the *longlat* parameter enables the bandwidth to be reported in kilometers instead of degrees. Parameters *dMat* and *st.dMat* allow for pre-specification of a distance matrix, which can speed up bandwidth computation, especially in spatio-temporal cases. This matrix can be calculated with the st.dist() function, which takes the dataset in the point data only (users need to provide the coordinates of the centroids for regional data). The *parallel.method* and *parallel.arg* parameters enable the utilization of multithreading computing, further expediting the calculation of the optimal bandwidth. In the bw.gtwr() function, additional parameters *lambda* and *ksi* allow for spatio-temporal distance modification, following the approach proposed by Wu et al. (2014). Furthermore, the *t.units* parameter permits the specification of the time unit if needed. Ultimately, both functions provide a single value, the calculated bandwidth, which can then be utilized in the model estimation process (Table 6.2).

Once the bandwidth is determined, the GWR model estimation process can begin. For the classic GWR setting, as proposed by Brunsdon et al. (1996), the function gwr.basic() is used. Some parameters in this function share similarities with the bw.gwr() function and should be kept consistent to maintain accuracy. The *bw* parameter is set to the previously estimated bandwidth value. In case specific geographical points need evaluation, the *regression.points* parameter allows for their specification; otherwise, the model is evaluated automatically. Parameters *cv* and *F123.test* can be modified to return additional model evaluation statistics. The *W.vect parameter* allows for specifying an additional weighting vector to scale the spatial weight matrix.

Table 6.2 Bandwidth optimization functions for GWR and GTWR models

```
bw.gwr(formula,data,approach = "CV",
       kernel = "bisquare",adaptive = FALSE,
       p = 2,theta = 0,longlat = F,
        dMat,parallel.method = F,parallel.arg = NULL)
```

```
bw.gtwr(formula,data,obs.tv,approach = "CV",
       kernel = "bisquare",adaptive = FALSE,
       p = 2,theta = 0,longlat = F,
       lamda = 0.05,t.units = "auto",ksi = 0,
       st.dMat,verbose = T)
```

Source: documentation of GWmodel:: package

Table 6.3 Functions estimating GWR and GTWR models

```
gwr.basic(formula,data,regression.points,
        bw, kernel = "bisquare",
        adaptive = FALSE,p = 2,theta = 0,longlat = F,dMat,
        F123.test = F,cv = F,W.vect = NULL,
        parallel.method = FALSE,parallel.arg = NULL)
```

```
gtwr(formula,data,regression.points,
        obs.tv,reg.tv,
        st.bw,kernel = "bisquare",
        adaptive = FALSE,p = 2,theta = 0,longlat = F,
        lamda = 0.05,t.units = "auto",ksi = 0,st.dMat)
```

Source: documentation of GWmodel:: package

The GWmodel:: package implements the GTWR model based on the concept by Huang et al. (2010). Estimating the GTWR model involves new parameters in the gtwr() function call. The *obs.tv* parameter requires the vector of time dimension identifiers (usually extracted from the column with time tickers). If regression points are provided, a similar time tag must be provided in the *reg.tv* parameter. The spatio-temporal bandwidth is specified through the *st.bw* parameter, and the spatio-temporal weighting matrix can be passed through the *st.dMat* field (Table 6.3).

Both functions, gwr.basic() and gtwr(), produce a single object of class *gwrm*, which contains various slots with details of the model estimation. The first element, *GW.arguments*, stores the parameters used in the GWR model call. The *GW.diagnostic* element holds essential statistics describing the model's fit, such as AIC or residual sum of squares. The next element, *lm*, presents the outcomes of the global regression—the general, non-spatial model computed globally for the entire dataset. It matches the results of a (pooled) OLS model on a given dataset. A crucial element in the resulting object is the *SDF* slot, which contains a spatial data frame with the local regression details, like local coefficients, their standard errors, and more. The information stored in this element is vital for visualizing the distribution of GWR coefficients in space.

The GWR and GTWR model results are displayed in the R console when the resulting object is called. The print() function for *gwrm* class objects provides an overview of the estimated model, including the global model with its statistics, GWR estimation parameters, and the statistical distribution of coefficients from local regressions.

6.5.3 Linear Model of the Investigated Phenomenon

We start the analysis with the basic OLS model, employing the formula discussed in Sect. 6.5.1 (see Table 6.1 for variable details). We estimate two model structures, one with the year categorical variable and another without it, to match the global models of GWR and GTWR. The codes used to initiate the estimation are provided below.

```
#Loading necessary packages
library(GWmodel)
library(sp)

#specifying the formula for OLS and GWR
formula.gwr<-birth~k_fertil1000+preschool1000+salary+
    men_unempl+as.factor(year)
ols <- lm(formula.gwr,data=demoData) #OLS model

#specifying the formula for pooled OLS and GTWR
formula.gtwr<-birth~k_fertil1000+preschool1000+salary+men_
unempl
ols2<-lm(formula.gtwr,data=demoData)
```

The results from the linear model show that the selected variables significantly influence fertility levels in Poland (Table 6.4). Both models exhibit relatively strong predictive power, explaining approximately 44% of the phenomenon's variance. The OLS model with year effects demonstrates a slightly better fit compared to the alternative specification, as indicated by the AIC and R-squared.

The observed impact directions align with theoretical expectations. Fertility is expected to increase with higher shares of women in the reproductive age group and a greater number of available preschools. Conversely, increased men's unemployment negatively affects fertility, indicating the relationship between stable employment status for fathers and their inclination to start a family. The salary coefficient result is somewhat less intuitive, revealing that higher income levels are associated with

Table 6.4 Results from the linear regression models

Model	OLS	OLS2
Constant	–6.040***(–0.649)	–5.882***(–0.59)
k_fertil1000	0.091***(–0.004)	0.091***(–0.003)
preschool1000	2.392***(–0.297)	2.419***(–0.295)
salary	–0.0002***(–0.00005)	–0.0003***(–0.00004)
men_unempl	–0.0001***(–0.00003)	–0.0001***(–0.00003)
year_2019	–0.109(–0.08)	
year_2020	0.163*(–0.093)	
year_2021	–0.187*(–0.107)	
Observations	1520	1520
R2	0.447	0.439
AIC	4526.074	4542.001

Note *p < 0.1; ** p < 0.05; *** p < 0.01
Standard deviations are reported in parentheses
Source: Own work in R

decreased fertility rates in the region. This may be linked to less favorable trade-offs between a mother's income and the time she spends on maternity leave, thereby foregoing income during that period.

Despite the OLS model providing an initial overview of the modeled phenomenon, it lacks consideration of crucial spatial effects necessary to explain the high levels of heterogeneity across space. Additionally, the OLS setting poorly reflects the interplay of spatial and temporal heterogeneity in the explained variable. These aspects will be better addressed by the models presented in the upcoming sections.

6.5.4 GWR Estimation

In R, the implementation of the GWR model begins with data preparation. In addition to variable values, we need to provide the spatial distribution of the observations using spatial objects from the sp:: package. The GWmodel:: functions accept either point (*SpatialPointsDataFrame*) or polygon data (*SpatialPolygonsDataFrame*). If polygons are used, the functions automatically approximate the spatial polygons by their centroids for all necessary calculations. For our analysis, we will utilize polygon data linked to the shapefile information pertaining to NUTS4 regions in Poland.

To estimate the model, we start with setting an appropriate bandwidth with bw.gwr() function. We specify the bandwidth to be adaptive, while the rest of the parameters are kept at their default values. The function will return the number of closest neighbors which will be incorporated in each local regression. Then, we will estimate the GWR model, once again changing the *adaptive* parameter to TRUE to match the bandwidth calculation.

```
bw.poly.adapt<-bw.gwr(formula.gwr,data=poviat_data_sp,
     adaptive=TRUE)

gwr.model.adapt<-gwr.basic(formula.gwr,data=poviat_data_sp,
     bw=bw.poly.adapt,kernel="bisquare",
     adaptive=TRUE)
```

As discussed previously, the results of the global model align with the OLS estimation presented earlier (the first OLS specification with the year categorical variable). The main advantage of the GWR is its ability to fit local models, which include the closest neighbors of given observations (according to the bandwidth value). This approach effectively captures the spatial heterogeneity of the modeled phenomenon. In this case study, the local estimations of the coefficients are not stable across the sample (Table 6.5), which confirms that the fertility levels are characterized by high levels of spatial heterogeneity. The GWR model is better fitted to the data, which is further evidenced by lowered AIC scores than for OLS. However, the high R-squared value suggests that the GWR model might be overfitting the data, likely due to the inclusion of a relatively small set of observations in each local model (bandwidth set to 34).

Table 6.5 GWR results—statistical distribution of GWR coefficients

Statistics	Min	1st Qu	Median	3rd Qu	Max
Constant	–74.394	–12.210	–2.028	7.134	40.618
k_fertil1000	–0.149	0.017	0.067	0.122	0.415
preschool1000	–25.611	–1.968	1.027	5.052	26.846
salary	–0.008	–0.001	0.000	0.001	0.004
men_unempl	–0.008	–0.001	0.000	0.000	0.003
year_2019	–2.166	–0.503	–0.177	0.081	2.276
year_2020	–5.064	–0.828	0.013	0.656	6.217
year_2021	–8.012	–1.508	–0.396	0.538	8.911
Bandwidth	Adaptive, 34				
R2	0.9518				
AIC	1568.083				

Source: Own work in R with GWmodel:: package

For a more detailed understanding of the diverse impacts of the modeled variables in different local contexts, we turn to the graphical analysis of the coefficients. To accomplish this, we need to retrieve the local coefficient values from the SDF slot of the GWR model and connect them to a spatial dataset for easy plotting over the map.

```
# extracting local GWR coefficients for preschool1000
gwr.model.adapt$SDF$preschool1000
```

The analysis of the local distribution of coefficients for the "preschool1000" variable reveals diversified impacts across regions (Fig. 6.7). While most regions exhibit a positive relationship between the number of preschools and fertility levels, some show an opposite influence. It is important to note that the results obtained from GWR are not smooth across space, indicating rapid changes in influence between neighboring regions. This unexpected outcome suggests that the GWR model struggles to appropriately capture the spatial heterogeneity of the modeled phenomenon. We would anticipate a more gradual change of coefficients, reflecting the local heterogeneity of impacts connected to unobservable factors, such as culture or local value specificities. The GWR model's main flaw in this setting lies in its inability to capture the spatio-temporal interplay of the studied effects. The coefficients of local GWR models remain constant across time (Fig. 6.7), which is a significant drawback when modeling spatio-temporal phenomena. This limitation will be addressed in the GTWR model.

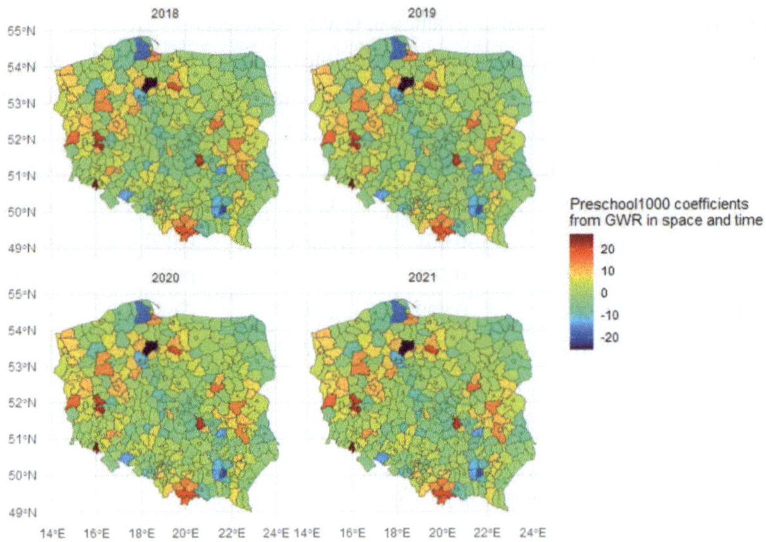

Fig. 6.7 Local coefficients of the "preschool1000" variable calculated with GWR. Source: Own work in R with ggplot2:: package

6.5.5 GTWR—Capturing Spatio-Temporal Heterogeneity

GTWR estimation is conducted analogously to the GWR model. The main difference between these two approaches is the spatio-temporal depiction of neighborhoods. Consequently, we estimate the spatio-temporal bandwidth for GTWR, which defines the number of spatio-temporal neighbors which will be included in each estimation. In this context, neighbors will be distanced not only across geography but across the time dimension (previous observation from the same region is as distant as its spatial neighbors in a current period sharing a common border).

To reduce the computational burden of the GTWR bandwidth estimation, we first create a spatio-temporal distance matrix with st.dist().

```
#distance matrix
x <- st.dist(coordinates(poviat_data_sp), #centroids
     obs.tv=poviat_data_sp@data$year)


bw.gtwr.adapt.value <- bw.gtwr(formula.gtwr,
     data=poviat_data_sp,obs.tv=poviat_data_sp@data$year,
     adaptive=TRUE,st.dMat=x)


gtwr.adapt.result <- gtwr(formula.gtwr, data=poviat_data_sp,
     obs.tv=poviat_data_sp@data$year,
     st.bw=bw.gtwr.adapt.value,adaptive=TRUE)
```

In the GTWR model estimation again, we utilize the adaptive bandwidth, which will determine the number of spatio-temporal neighbors included in the local estimations. We use the model formula without the year variable, as the temporal dimension is handled within GTWR itself. Once more, the results of the global model correspond to the corresponding OLS model fit (the second OLS model from Sect. 6.5.3). The GTWR model fit and the distribution of coefficients are presented in Table 6.6.

At first glance, it may appear that the GTWR model performs worse than the GWR specification. However, upon closer examination, it becomes evident that the GTWR model provides more reliable and consistent estimates compared to the corresponding OLS specification and GWR model. The model fit of 65% indicates a good yet reasonable level of explained fertility variance, signifying the better alignment of the GTWR model with the data than the OLS specification. Moreover, the GTWR model does not suffer from the overfitting problem, as evidenced by more coherent coefficients from local regressions (less coefficient variance across local models). This conclusion becomes even more apparent when observing the spatio-temporal representation of the coefficients on the map (Figs. 6.8 and 6.9).

Results from the GTWR are smooth across time and space and seem to effectively capture the intricacies of the spatio-temporal heterogeneity patterns in the data (Figs. 6.8 and 6.9). Thanks to this trait, we may draw meaningful conclusions from the spatio-temporal distribution of coefficients from the GTWR model.

For example, we observe a diminishing impact of preschools on fertility levels over time (Fig. 6.8). This diminishing return of available preschool facilities is most evident in the central part of the country. The strongest effects of preschools are visible in the northwestern part of the country, and yet their influence is still diminishing in time.

A similar analysis can be conducted for the men's unemployment variable (Fig. 6.9). By analyzing the spatio-temporal distribution of GTWR coefficients, we can conclude that the employment stability of fathers is most crucial in the western side of the country. This effect strengthens over time and extends to the neighboring regions. This may be an exhibit of a gradual cultural shift within the population living

Table 6.6 GTWR results—statistical distribution of GTWR coefficients

Statistics	Min	1st Qu	Median	3rd Qu	Max
Constant	–24.677	–13.176	–7.945	–4.909	4.781
k_fertil1000	0.025	0.085	0.106	0.130	0.190
preschool1000	–3.487	0.748	2.402	3.594	8.415
salary	–0.001	0.000	0.000	0.000	0.001
men_unempl	–0.001	0.000	0.000	0.000	0.000
Bandwidth	Adaptive, 123				
R2	0.6537				
AIC	3938.685				

Source: Own work in R with GWmodel:: package

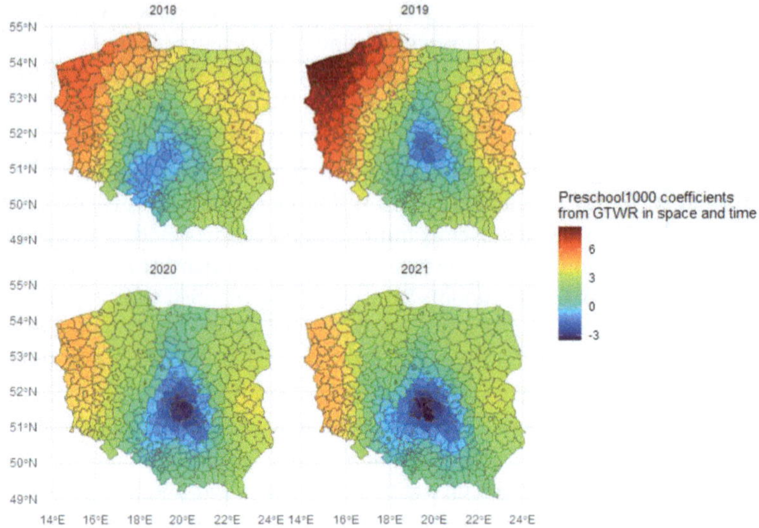

Fig. 6.8 Local coefficients of the "preschool1000" variable calculated with GTWR. Source: Own work in R with ggplot2:: package

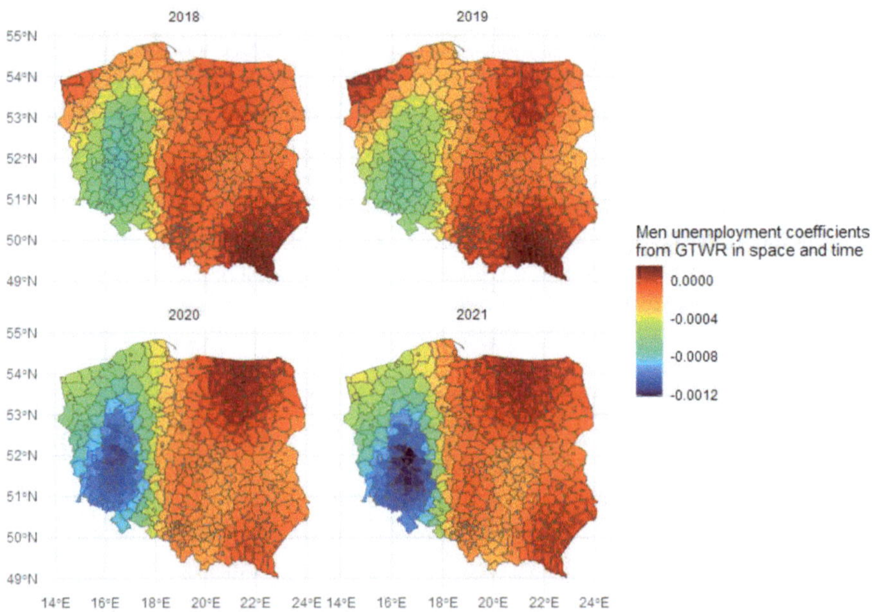

Fig. 6.9 Local coefficients of the "men_unempl" variable calculated with GTWR over space and time. Source: Own work in R with ggplot2:: package

in that area. It seems that, in this region, the economic situation of the fathers over time becomes a more influential factor in determining local fertility levels.

6.5.6 *Discussion of the Results*

The case study presented in this chapter highlights the unique opportunities provided by GTWR model estimation for measuring phenomena characterized by high spatio-temporal variability. The capability to capture both dimensions of the effects enables drawing meaningful conclusions regarding spatio-temporal trends in variable influences. Moreover, it became apparent that applying the GWR model for panel data may result in severe overfitting issues, which can be resolved by incorporating the temporal dimension when measuring neighborhoods in the modeled phenomena.

The discussed examples and provided code guidance will enable readers to familiarize themselves with the method and apply it to their own study designs. Presenting examples using the R language and GWmodel:: packages facilitates widespread utilization of these techniques, thanks to the open-source nature of the software used.

6.6 Concluding Remarks

This chapter shows that temporal dimension in spatial data may easily improve the quality of model fit and prediction. Spatial–Temporal Geographically Weighted Regression gains popularity as a stable and robust solution to spatio-temporal data. The core of GTWR modeling is a proper definition of distance and kernel function in both dimensions. It prejudges the quality of the final model.

The methodological solutions described in this chapter constitute a standard approach to this form of modeling. Implementations in GWmodel:: package in R software were based on the best-developed theoretical approaches in the statistical literature and can be considered as a safe toolbox for this kind of analysis. The chapter presents the major R codes and formulas; however, the package itself includes a few more for more specific cases.

An important aspect of GTWR is data. As the literature review and empirical examples show, this method is quite flexible in terms of granulation. One kind of data is geo-localized point data which include also time labels. The other type of data is regional data—they are adjusted by adding geo-coordinates of the centroid of the region. Also, novel types of data such as satellite images are suitable as input for GTWR. The size of both dimensions matters for the outcomes. As the method relies on selecting subsamples by time and space to include the most recent and closest observations, it requires a number of observations that can reasonably create subsamples for each observation. With few regions (or points) or a few points in time, the potential of GTWR cannot be developed. Kernel function and bandwidth determine what will be the spatial range of subsample included in local regression and what will be the strength of decay. Therefore, one can reasonably construct the dataset to meet the requirements of the method. One important feature is a limitation for big data operations.

An interesting methodological development of GWR and GTWR is substituting OLS regression (which is the regression method for spatio-temporally selected subsamples) with machine learning algorithms. This builds more complex models, but as evidenced by the majority of ML studies, is more efficient in terms of computation time and fit quality. This trend appears in whole spatial modeling and should be considered an important methodological approach to modeling (Kopczewska 2022).

There are wide possibilities for the implementation of the GTWR method in many research areas. Empirical examples show that GTWR can be successfully conducted on environmental, epidemiological, pollution, social, and economic data. In general, all data that include spatial and temporal dimensions can be used in this modeling. The value added of the method lies in tracking the local heterogeneities in data—looking for specific spatial or temporal patterns. Therefore, the method is highly suitable for discovering irregularities, structural changes, and local differences, which cannot be found with methods that generalize the patterns.

References

Angelov N, Johansson P, Lindahl E (2016) Parenthood and the gender gap in pay. J Law Econ 34(3):545–579. https://doi.org/10.1086/684851

Anser MK, Yousaf Z, Khan MA, Voo XH, Nassani AA, Alotaibi SM, Abro MMQ, Zaman K (2020) The impacts of COVID-19 measures on global environment and fertility rate: double coincidence. Air Qual Atmos Health 13(9):1083–1092. https://doi.org/10.1007/S11869-020-00865-Z/TABLES/3

Bai Y, Wu L, Qin K, Zhang Y, Shen Y, Zhou Y (2016) A geographically and temporally weighted regression model for ground-level PM2. 5 estimation from satellite-derived 500 m resolution AOD. Remote Sens 8(3):262

Becker GS (1992) Fertility and the economy. J Popul Econ 5(3):185–201. https://doi.org/10.1007/BF00172092

Britannica, T. Editors of Encyclopaedia (2024) compactness. Encyclopedia Britannica, 5 Feb. https://www.britannica.com/science/compactness

Brunsdon C, Fotheringham AS, Charlton ME (1996) Geographically weighted regression: a method for exploring spatial nonstationarity. Geogr Anal 28(4):281–298. https://doi.org/10.1111/J.1538-4632.1996.TB00936.X

Brunsdon C, Fotheringham S, Charlton M (1998) Geographically weighted regression—Modelling spatial non-stationarity. J R Stat Soc Ser D: Stat 47(3):431–443. https://doi.org/10.1111/1467-9884.00145

Cao K, Diao M, Wu B (2019) A big data–based geographically weighted regression model for public housing prices: a case study in Singapore. Ann Am Assoc Geogr 109(1):173–186

Cho SH, Lambert DM, Chen Z (2010) Geographically weighted regression bandwidth selection and spatial autocorrelation: an empirical example using Chinese agriculture data. Appl Econ Lett 17(8):767–772

Chu HJ, Kong SJ, Chang CH (2018) Spatio-temporal water quality mapping from satellite images using geographically and temporally weighted regression. Int J Appl Earth Obs Geoinf 65:1–11

Ćwiakowski P (2020) Geographically weighted regression – modelling spatial heterogeneity. In: Kopczewska K (ed) Applied spatial statistics and econometrics, Routledge, pp 289–322. https://doi.org/10.4324/9781003033219-6

da Silva AR, Mendes FF (2018) On comparing some algorithms for finding the optimal bandwidth in geographically weighted regression. Appl Soft Comput 73:943–957

de Bellefon MP, Floch JM (2018) Geographically weighted regression. Handbook of spatial analysis, Insee Methodes. Insee, Eurostat

Fotheringham AS, Brunsdon C, Charlton M (2002) Geographically weighted regression: the analysis of spatially varying relationships. Wiley, Chichester

Fotheringham AS, Crespo R, Yao J (2015) Geographical and temporal weighted regression (GTWR). Geogr Anal 47(4):431–452

Gollini I, Lu B, Charlton M, Brunsdon C, Harris P (2015) GWmodel: An R package for exploring spatial heterogeneity using geographically weighted models. J Stat Softw 63(17): 1–50. https://doi.org/10.18637/JSS.V063.I17

Guo L, Ma Z, Zhang L (2008) Comparison of bandwidth selection in application of geographically weighted regression: a case study. Can J for Res 38(9):2526–2534

GUS (2023) GUS—Bank Danych Lokalnych. https://bdl.stat.gov.pl/bdl/start

Huang B, Wu B, Barry M (2010) Geographically and temporally weighted regression for modeling spatio-temporal variation in house prices. Int J Geogr Inf Sci 24(3):383–401. https://doi.org/10.1080/13658810802672469

Koç T (2022) Bandwidth selection in geographically weighted regression models via information complexity criteria. J Math 2022(1):1527407

Kopczewska K (2020) Applied spatial statistics and econometrics: data analysis in R. Routledge

Kopczewska K (2022) Spatial machine learning: new opportunities for regional science. Ann Reg Sci 68(3):713–755

Kopczewska K, Ćwiakowski P (2021) Spatio-temporal stability of housing submarkets. Tracking spatial location of clusters of geographically weighted regression estimates of price determinants. Land Use Policy 103: 105292

Kopczewska K, Kudła J, Walczyk K (2017) Strategy of spatial panel estimation: Spatial spillovers between taxation and economic growth. Appl Spat Anal Policy 10:77–102

Lee GHY, Lee SP (2014) Childcare availability, fertility and female labor force participation in Japan. J Jpn Int Econ 32:71–85. https://doi.org/10.1016/J.JJIE.2014.01.002

Liu Y, Lam KF, Wu JT, Lam TTY (2018) Geographically weighted temporally correlated logistic regression model. Sci Rep 8(1):1417

Lu B, Charlton M, Harris P, Fotheringham AS (2014) Geographically weighted regression with a non-Euclidean distance metric: a case study using hedonic house price data. Int J Geogr Inf Sci 28(4):660–681

Ma X, Zhang J, Ding C, Wang Y (2018) A geographically and temporally weighted regression model to explore the spatiotemporal influence of built environment on transit ridership. Comput Environ Urban Syst 70:113–124

Nugroho WH, Sumarminingsih E (2021 March) Geographically and temporally weighted regression model with Gaussian Kernel weighted function and Bisquare Kernel weighted function. In IOP conference series: materials science and engineering 1115(1): 012063. March 2021. IOP Publishing

Pebesma E, Bivand RS (2005) S classes and methods for spatial data: the sp package. R News 5(2):9–13

Que X, Ma X, Ma C, Chen Q (2020) A spatiotemporal weighted regression model (STWR v1. 0) for analyzing local nonstationarity in space and time. Geosci Model Dev 13(12): 6149–6164

Vignoli D, Mencarini L, Alderotti G (2020) Is the effect of job uncertainty on fertility intentions channeled by subjective well-being? Advances in Life Course Research 46:100343. https://doi.org/10.1016/J.ALCR.2020.100343

Wei Q, Zhang L, Duan W, Zhen Z (2019) Global and geographically and temporally weighted regression models for modeling PM2. 5 in Heilongjiang, China from 2015 to 2018. Int J Environ Res Public Health 16(24), 5107

Wu B, Li R, Huang B (2014) A geographically and temporally weighted autoregressive model with application to housing prices. Int J Geogr Inf Sci 28(5):1186–1204. https://doi.org/10.1080/13658816.2013.878463

Wu S, Wang Z, Du Z, Huang B, Zhang F, Liu R (2021) Geographically and temporally neural network weighted regression for modeling spatiotemporal non-stationary relationships. Int J Geogr Inf Sci 35(3):582–608

Xu X, Luo X, Ma C, Xiao D (2020) Spatial-temporal analysis of pedestrian injury severity with geographically and temporally weighted regression model in Hong Kong. Transport Res f: Traffic Psychol Behav 69:286–300

Chapter 7
Econometric Modeling of Market Risk: Application to Indian Stock Exchange

Suparna Biswas and Rituparna Sen

Abstract This chapter explores econometric modeling techniques for assessing market risk in equity-based investments, which are inherently subject to price fluctuations. Investors and researchers alike seek to evaluate and compare market risk while computing risk-adjusted returns for index funds. The chapter introduces key risk measures, including Value at Risk (VaR), Expected Shortfall, Median Shortfall, Sharpe ratio, and Treynor ratio. It further discusses estimation approaches for these measures using historical asset price data, covering both parametric and nonparametric techniques, with a particular focus on Extreme Value Theory (EVT). Additionally, the chapter examines time series modeling methods, specifically the GARCH and MSGARCH models. Finally, it presents back-testing procedures to evaluate the performance of these models in an out-of-sample context. The methodologies are applied to real data from the Indian equity market, with all empirical analyses conducted using the R programming language.

7.1 Introduction

In many economic applications including finance and insurance, the quantity of interest is random. In finance, the investor is interested in the returns of portfolios. In insurance, the company is interested in the amount of claims. Both of these quantities are random. In such situations, the typical size of the random quantity is of interest. This is usually obtained from the mean or average of the distribution of the random quantity. In addition, it is also important to know how bad the situation can be, that is

This chapter draws in part, with necessary permissions, from the authors' previously published work: Sen and Das (2023) and Biswas and Dutta (2015).

Supplementary Information The online version contains supplementary material available at https://doi.org/10.1007/978-981-97-9060-9_7.

S. Biswas · R. Sen (✉)
Indian Statistical Institute, Bangalore, India
e-mail: rsen@isibang.ac.in

F. Hamid and D. Mukherjee (eds.), *Data Science for Modeling Managerial and Socioeconomic Problems*, Contributions to Management Science,
https://doi.org/10.1007/978-981-97-9060-9_7

the amount of risk involved. In particular, we will focus on market risk which is the risk to the portfolio posed by a drop in the market price the portfolio's components.

Riskiness enables us to compare portfolios with similar levels of mean returns. In general, for the same level of mean return, investors prefer portfolios that have lower risk. This concept is formalized in the theory of portfolio optimization in Markowitz (1952), where variance is used as a measure of risk. Although variance gives us some idea of risk, it does not exactly capture how bad the situation can be. This is because variance takes into account all the deviations from the mean, higher as well as lower. Moreover, variance is based on squared deviations and hence captures the riskiness only for Gaussian distribution where the density involves squared deviations from the mean. The distributions of financial returns are observed to have heavier tails than normal and are skewed. This is another reason that variance is unsuitable as a measure of risk for financial returns. For more discussion on this, see Michaud (1989). In addition to variance, other quantities that measure the spread of the distribution, like range, inter-quartile range, mean absolute deviation, etc., can also be used as measures of risk. They suffer from the same drawback as variance that, to estimate the spread, the full distribution of returns is taken into account. However, when considering risk, only lower return values are informative.

In this paper, we shall present some common measures of risk and their properties. We also present methods to estimate these measures and compare the performance of these estimation methods. We shall demonstrate all the results with real data from the Indian market. We demonstrate the methods with daily prices of Nifty 50 and S&P BSE Sensex. The plot of daily log returns of Nifty 50 is presented in Fig. 7.1. Since the Sensex's daily log returns plot exhibits a similar pattern to the Nifty 50's, it is not included here. We also compare 21 index funds, which track the Sensex or the Nifty index. We have gathered information on the daily NAV (or closing price for an exchange-traded fund) and the closing price of the underlying index for each fund from January 1, 2012, to December 31, 2019. The data presented here are secondary, as they have been sourced from the websites of AMFI, NSE, and BSE. The data set includes daily negative log returns based on the daily NAV and the closing price of the underlying index.

7.2 Measures of Risk

As discussed in the introduction, it is desirable for measures of risk to concentrate on the last tail of the distribution of returns, in other words, the extreme losses. This idea is known as safety first and was introduced by Roy (1952). A related concept is downside risk which is the risk associated with a decline in the value of an asset as a result of changes in market conditions.

Since the 1990 financial crisis, risk management and financial institutions regulation have been at the center of public discourse and policymaking. There has been a revolution in risk management since then. Value at Risk (VaR), a novel technique for assessing financial market risk that was created in reaction to the early 1990s

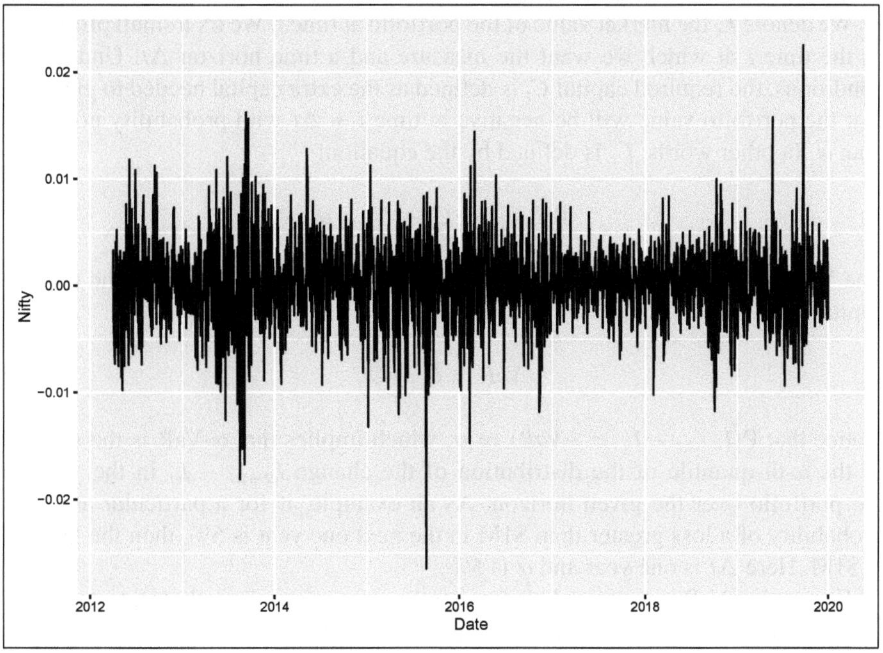

Fig. 7.1 Daily returns of Nifty 50 from January 1, 2012, to December 31, 2019.

financial disasters, kicked off this process. There are several well-publicized losses such as Orange County, Procter and Gamble, Metallgesellschaft, Long-Term Capital Management, and so many others that stunned the market. The common lesson of these disasters is that billion of dollars can be lost because of poor supervision and management of financial risks. In this section, we introduce some common measures of downside risk.

7.2.1 Value at Risk

The concept of Value at Risk is very simple. We fix a small probability α. We are looking for the largest amount a, so that the loss incurred by our portfolio at a fixed time in the future is greater than that a with probability less than α. In layman terms, this means that at a fixed time horizon losses are less than the VaR with a high probability $(1 - \alpha)$. It is clear that this is an important quantity for investors and insurers. A minor use of VaR is also in testing proposed models of financial returns. In the early 1990s, VaR-based risk management had already become a prevalent market practice. Regulators selected VaR as the regulatory reporting instrument in the 1996 Market Risk Amendment of the original Basel Accord. See Philippe (2001) for a thorough review of VaR.

We denote L_t the market value of the portfolio at time t. We fix a small probability α, the time t at which we want the measure and a time horizon Δt. Under these conditions, the required capital C_t is defined as the extra capital needed to guarantee that the portfolio value will be negative at time $t + \Delta t$ with probability no greater than α. In other words, C_t is defined by the equation:

$$\mathrm{P}(L_{t+\Delta t} + C_t < 0) = \alpha. \tag{7.1}$$

The VaR at time t is defined as the sum of the current endowment and the required capital

$$VaR = L_t + C_t. \tag{7.2}$$

Notice that $\mathrm{P}(L_{t+\Delta t} - L_t < -VaR) = \alpha$, which implies that α-VaR is the negative of the α-th quantile of the distribution of the change $L_{t+\Delta t} - L_t$ in the value of the portfolio over the given horizon. As an example, if for a particular asset, the probability of a loss greater than \$1M in the next one year is 5%, then the 5% VaR is \$1M. Here Δt is one year and α is 5%.

Expressing VaR in units of L_t is frequently more convenient, that is, to set $VaR = \widetilde{VaR} * L_t$. Expressed in units of L_t, the negative of the VaR is the α-quantile of the distribution of the raw returns over the period in question. It is worth noting that for tiny relative changes, raw returns and log returns are extremely close and we have:

$$r_t = \frac{L_{t+\Delta t} - L_t}{L_t} \sim \log\left(\frac{L_{t+\Delta t}}{L_t}\right) \tag{7.3}$$

which justifies the fact that one often defines α-VaR at the horizon Δt, as the negative of the α-quantile of the distribution of the log returns over a period of length Δt.

7.2.2 Expected Shortfall

VaR's clear advantage is that it is simple and straightforward to use. But VaR not only ignores any loss beyond the VaR level, it also do not satisfy one of the axioms of coherence, i.e., the sub-additivity property, hence VaR is not a coherent risk measure (see Artzner et al. 1999). This means that the risk of a portfolio, as assessed by VaR, can thus be greater than the sum of the individual risks of its components. Hence, managing risk by VaR may fail to stimulate diversification.

To address these issues, Expected Shortfall (ES) was introduced which is a coherent risk measure. ES is thought to be an excellent complement to VaR and is closely related to it. The average amount of money that can be lost in states that are above the VaR level is what the ES measures. It is also known as Conditional Value at Risk and is defined as the expected value of the portfolio loss given that a

VaR exceedance has occurred. That is

$$ES = E(r_t | r_t < -VaR). \tag{7.4}$$

The main advantage of ES is that it encourages diversification because of its coherence nature and also it takes into account the actual sizes of the losses. However, estimation of ES from observed data becomes problematic as we are trying to capture the mean of extreme observations and it involves integration on an unbounded support, which is a problem for numerical integration.

Although the ES is a reliable risk measure and superior than VaR in theory, the standard errors of ES estimates are significantly higher than those of VaR estimations. Since empirical quantiles are resilient to extreme outliers, unlike traditional estimators of ES, back-testing becomes simpler even though disregarding the severity of losses at the tail of the distribution for VaR results in a theoretically flawed risk measure. Again given that the ES takes into account the entire right tail of the loss distribution when it encounters a fat-tailed loss distribution, estimates are more influenced by the occurrence of large and rare losses. Large and infrequent losses, however, have less of an impact on VaR since they do not account for losses over the VaR level. Therefore, when the probability of rare and large losses is high, the ES estimate becomes more volatile and has a larger estimation error than the VaR estimate.

In Tables 7.1 and 7.2, we present the VaR and ES of 21 Indian index funds. We find that the VaR for the majority of Indian index funds is very similar to the corresponding benchmark indices. Compared to the benchmark indices, only two of the twenty-one index funds appear to have higher absolute values. This implies that these two index funds' extreme daily losses may be marginally worse than those of the benchmark index. Such index funds are ICICI Prudential Nifty Next 50 Index Fund-Growth Plan and HDFC Index Fund-Sensex Plan-Growth Plan. We also observe from Tables 7.1 and 7.2 that none of the Indian index funds have higher ES values compared to their equivalent benchmark indices.

According to Tables 7.1 and 7.2, the absolute values of the VaR for 12 of the 21 Indian index funds are lower than those of the corresponding market indices they replicate. These funds are the Aditya Birla Sun Life Nifty 50 Index Fund-Growth Plan, Franklin India Index Fund-Growth Plan, IDBI Nifty Index Fund-Growth Plan, IDFC Nifty ETF, IDFC Nifty Fund-Growth Plan, LIC MF Index Fund-Nifty Plan-Growth Plan, Nippon India Index Fund-Nifty Plan-Growth Plan, Tata Index Fund-Nifty Plan-Growth Plan, IDFC Sensex ETF, LIC MF Index Fund-Sensex-Growth Plan, Nippon India Index Fund-Sensex-Growth Plan, and Tata Index Fund-Sensex-Growth Plan. We also observe that for all the Indian index funds, the absolute values of the ES are less than the same for the respective market indices, which they imitate except for one index fund, i.e., UTI Nifty ETF. According to this, the daily losses in these funds should be less severe than those in the corresponding benchmark indices.

Table 7.1 VaR and ES of Nifty Index Funds at 5%

Index	VaR(Historical)	ES(Historical)
Nifty 50	**0.006656**	**0.010716**
Aditya Birla Sun Life Nifty 50 Index Fund-Growth Plan	0.006646	0.010553
Aditya Birla Sun Life Nifty ETF	0.006657	0.010691
Franklin India Index Fund-Growth Plan	0.006555	0.010462
HDFC Index Fund-Growth Plan	0.006657	0.010696
ICICI Prudential Nifty ETF	0.006662	0.010702
ICICI Prudential Nifty Next 50 Index Fund-Growth Plan	0.006975	0.010024
IDBI Nifty Index Fund-Growth Plan	0.006641	0.010677
IDFC Nifty ETF	0.006592	0.010562
IDFC Nifty Fund-Growth Plan	0.006602	0.010598
LIC MF Index Fund-Growth Plan	0.006599	0.010622
Nippon India Index Fund-Growth Plan	0.006644	0.010584
SBI Nifty Index Fund-Growth Plan	0.006693	0.010672
Tata Index Fund-Growth Plan	0.006604	0.010482
Taurus Nifty Index Fund-Growth Plan	0.006660	0.010421
UTI Nifty ETF	0.006676	0.010728
UTI Nifty Index Fund-Growth Plan	0.006678	0.010713

Source: The entries in the table have been computed based on data collected from AMFI and NSE websites (https://www1.nseindia.com/products/content/equities/indices/historical_index_data.htm; https://www.amfiindia.com/net-asset-value/nav-history)

Table 7.2 VaR and ES of Sensex Index Funds at 5%

Index	VaR(Historical)	ES(Historical)
S & P BSE Sensex	**0.006737**	**0.010956**
HDFC Index Fund-Sensex Plan-Growth Plan	0.006762	0.010955
IDFC Sensex ETF	0.006654	0.010805
LIC MF Index Fund-Growth Plan	0.006673	0.010859
Nippon India Index Fund-Growth Plan	0.006703	0.010785
Tata Index Fund-Growth Plan	0.006639	0.010714

Source: The entries in the table have been computed based on data collected from AMFI and BSE websites (https://www.bseindia.com/indices/IndexArchiveData.html; https://www.amfiindia.com/net-asset-value/nav-history)

7.2.3 Other Measures of Risk

Median Shortfall (MS) is simply the median loss when VaR is exceeded and it is defined as (see So and Wong 2012)

$$MS = \text{Median}(r_t | r_t < -VaR).$$

Under the assumption that the distribution function is continuous and strictly increasing, it is easy to verify that

$$MS = VaR_{0.5\alpha}.$$

We can see that both ES and MS provide information about the loss beyond VaR. ES is always coherent, while MS exhibits coherence under the condition that asset losses or returns adhere to elliptically contoured distributions, a scenario applicable to numerous financial markets.

The **mean absolute deviation** (MAD) is calculated by summing the absolute values of the differences between the returns and their mean, then dividing by the total number of returns. This measure is two-sided, similar to variance, yet exhibits greater robustness to extreme values.

$$\frac{\sum_{i=1}^{n} | r_i - \bar{r} |}{n}. \tag{7.5}$$

The subsequent measures are derived from the Capital Asset Pricing Model (CAPM) and are consequently associated with a benchmark security. According to Bacon (2008), the CAPM beta multiplied by market risk is the definition of **systematic risk**. **Market risk** is the benchmark's standard deviation. In the CAPM regression equation, the standard deviation of the error term represents the **specific risk**.

The following measures are based on drawdown. A drawdown is a peak-to-trough fall for an investment, trading account, or fund over a defined time period. A drawdown is usually quoted as the percentage between the peak and the subsequent trough. Drawdowns are crucial for determining the historical risk of various assets, evaluating fund performance, or keeping track of one's own trading results. The average of the drawdowns across the whole analysis period is the **pain index**. In a visual representation of a drawdown chart, the pain index is the area enclosed by the horizontal line at zero percent and the drawdown line.

The following measures are based on a minimum acceptable return (MAR). **Downside variance** is defined as

$$\frac{\sum_{i=1}^{n} \min(R_t - MAR, 0)^2}{n} \tag{7.6}$$

The square-root of this value represents the Downside Deviation, while the sum excluding the square denotes the Downside Potential.

7.3 Estimation of Risk Measures from Data

In this section, we review various methods for the estimation of VaR. We discuss both parametric and the nonparametric estimation methods of VaR.

7.3.1 *Gaussian Estimation*

The basic VaR calculation uses the α-quantile of a parametric model of the return distribution to determine the VaR. The predominant models include the unconditional normal model, which utilizes a frequently updated variance estimate. That is, we consider $r_t \sim \mathcal{N}(\mu, \sigma^2)$. The α-VaR is $-\mu - \sigma\Phi^{-1}(\alpha)$. Maximum likelihood is used to compute the parameters with the usual estimators: $\bar{M}$ for the mean and $\hat{\sigma^2}$ for the variance. The parametric VaR is expressed by

$$VaR = -\bar{M} - \hat{\sigma} * z_{\alpha} \tag{7.7}$$

where z_{α} is the α-th quantile of the standard normal distribution.

The next thing that we are going to do is to examine the time evolution of VaR for these two indices, the Nifty and the Sensex. For the purpose of estimating the VaR, we use a rolling window of the preceding four years for each day. For each day, we use a rolling window of the previous 4 years to estimate the VaR. Our data set consists of daily negative log returns considering the end-of-day prices of the two indices, Nifty and Sensex, between January 1, 2012, and December 31, 2019. The results are presented in Fig. 7.2. Notice that the risk goes down for most of the time period with a sharp decline in late 2017. The risk of the two indices move very closely.

A generic parametric VaR model includes an assumed return distribution whose parameters θ are unknown, that is $r_t \sim F_{\theta}$ and parameters are estimated by maximum likelihood technique. The VaR is then $-F_{\hat{\theta}}^{-1}(\alpha)$, the negative of the α-quantile of the estimated distribution. Non-normal parametric VaR estimation utilizes a different distribution for the losses to more accurately take into consideration the potential fat-tailed character of downside risk. This will be further discussed in Sect. 7.3.4.

7.3.2 *Modified and Robust Estimation*

The traditional VaR method is no longer an accurate way to quantify risk if the return on a portfolio is not Gaussian distributed. To accommodate non-Gaussian distributions, one may employ the Cornish-Fisher approach, which is accurate when returns are close to the Gaussian distribution. Using the Cornish-Fisher expansion, Zangari (1996) offers a modified VaR computation considering the upper moments of

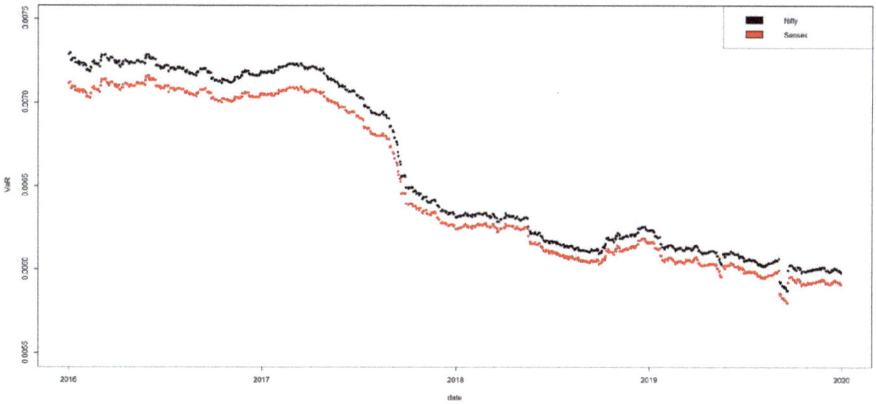

Fig. 7.2 Time evolution of Gaussian VaR of *Nifty 50* and *Sensex* at α=5% using daily returns for previous four years

non-normal distributions. The Cornish-Fisher VaR will provide a higher loss estimate than the typical VaR when the returns exhibit negative skewness or fat tails (that is platykurtic). The loss estimation, however, will be less than classical VaR when returns have positive skewness or are leptokurtic. When returns follow a Gaussian distribution, this approach aligns with the standard parametric VaR. A common name for this measurement is modified VaR, and it is now frequently quoted and utilized in the literature. The modified VaR is defined in the following manner:

$$\begin{aligned} z_{cf} &= z_c + [(z_c^2 - 1)S]/6 + [(z_c^3 - 3z_c)K]/24 - [(2z_c^3 - 5z_c)S^2]/36 \\ VaR &= -\bar{R} - \hat{\sigma} * z_{cf} \end{aligned} \tag{7.8}$$

where S is the skewness and K is the excess kurtosis of R.

It is crucial that the risk measure's estimation is robust to outliers (see Cont et al. 2010). This is especially the case for ES and other moment-based methods. By default, the sample counterparts of the portfolio moments are used to estimate them. However, we should clean the data first if we believe that the data contains outliers and we use the method mentioned by Boudt et al. (2008) to remove the outliers. Here the $1-p$ most extreme observations are winsorized if they are detected as being outliers, for a suitably chosen p. It might be more prudent to decrease distortion brought on by autocorrelation if the data includes returns for extremely illiquid assets. The proper return series in this circumstance is the Geltner autocorrelation adjusted return series, which is the return series adjusted for first-order autocorrelation.

7.3.3 Historical Simulation

If a sound assumption is established on the return distribution, the parametric model-based estimates are effective. However, it can be challenging or impossible to find a specification that inevitably covers the genuine distribution. At the other end of the spectrum is a pure nonparametric estimate of the VaR using the α-quantile of the empirical distribution of the returns. A popular method for VaR assessment is historical simulation (HS). In order to predict future returns, historical returns are used rather than distributional assumptions.

Let r_t be the return at time t has been observed for $t = 1, \cdots, T$. The HS procedure is defined as follows:

$$\hat{\text{VaR}}_t^{HS} = -R_{(\omega)},$$

where $R(\omega)$ is the ω-th order statistic of returns, with $\omega = [T\alpha]$. The method most frequently used by commercial banks to forecast VaR is by far HS. Pérignon and Smith (2010) indicate that almost 75% of banks who reveal their VaR methodology utilize the HS approach.

These estimations are approximate, and a single new data point could result in very different VaR values. Smoothing the estimated quantiles typically lowers the variance of the estimate in comparison with those computed straight from the sorted returns. If the sample size is low, this becomes even more apparent. Žiković and Aktan (2011) suggested to forecast the VaR using weighted historical simulation, adding an exponentially weighted moving average to the forecast to give the most recent observations more weight, along with the right smoothing parameter.

Non-parametric estimates of VaR have the advantages of being easily computed and being generally consistent under very weak conditions. But the main disadvantage is that very large samples are required for estimating the quantiles particularly for small values of α.

Next, using the two indices, we compare the various estimation techniques previously discussed. Robustness issues discussed in Sect. 7.3.2 are incorporated. We compare the three approaches, which are the Gaussian, modified, and historical ones. It follows that for every index, there are nine VaR estimates, derived from three different methodologies and three different types of robustness. The results are reported in Table 7.3. There is a consistent pattern where the modified estimates are at the bottom and the historical estimates are at the top. Also, the corrected estimates are lower in most cases.

7.3.4 EVT-Based Estimation

The area of probability theory known as Extreme Value Theory (EVT) deals with the extreme high and low values of a random variable. It specifically provides findings

Table 7.3 Estimated VaR for Nifty 50 and S&P BSE Sensex at α=5% for one day horizon at December 2019 using daily log returns starting January 2012. The values are scaled by dividing by the VaR obtain using the Gaussian method with no robustness correction. These VaR values are 0.006246487 and 0.006117069 respectively for Nifty 50 and S&P BSE Sensex

	Nifty 50			S&P BSE Sensex		
	gaussian	modified	historical	gaussian	modified	historical
none	1.0000	0.9694	1.0720	1.0000	0.9693	1.0679
boudt	0.9928	0.9694	1.0619	0.9907	0.9690	1.0561
geltner	0.9717	0.9717	1.0586	0.9827	0.9827	1.0578

regarding the distribution of maxima (minima) of a sequence of random variables and approximations to the distribution of a random variable over a particular threshold. These findings suggest that a Generalized Pareto Distribution (GPD) provides a good approximation of the return distribution below a certain threshold u_0. Hence we can say that EVT framework is related to Pickands-Balkema-de-Haan theorem. So basically we first find the threshold, then estimate the parameters of the GPD, and finally, use the quantiles of this distribution to estimate the Value at Risk.

Finding relatively stable areas of the Hill plot is a frequent technique for determining the threshold. The procedure is described below. Keep in mind that the distribution of returns above a threshold resembles a GPD. For $\xi > 0$, the cdf of a (GPD) is given by

$$G_{\xi,\beta}(x) = 1 - \left(1 + \frac{\xi x}{\beta}\right)^{-1/\xi} \tag{7.9}$$

The Hill tail-index estimator of ξ is given by

$$\hat{\xi}(k) = \left(\frac{1}{k} \sum_{i=n-k+1}^{n} \ln(X_{(i)}) - \ln(X_{(n-i+1)})\right)^{-1} \tag{7.10}$$

where $X_{(i)}$ is the i-th order statistic of $X_1, \ldots, X_n$ and the estimator is based on the highest k order statistics. The plot of $\hat{\xi}(k)$ on k is called the Hill plot. A good fit to the GPD is indicated by a range of k values for which the estimator does not fluctuate excessively.

The Hill plot for Nifty 50 is shown in the left panel of Fig. 7.3. From the Hill plot, k=200 seems to be an appropriate threshold. Using this, we fit a GPD by maximum likelihood. The goodness of the fit is assessed in the right panel of Fig. 7.3. Finally, estimate of the VaR is obtained as the corresponding quantile of the fitted GPD. We obtain the VaR of Nifty 50 by this method as 0.00614827.

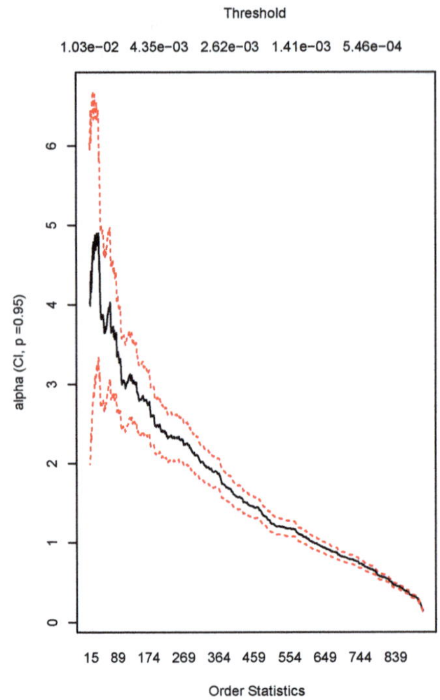

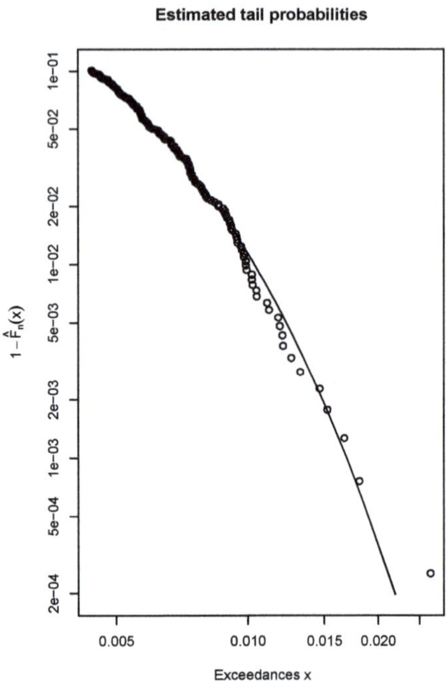

Fig. 7.3 Left: Hill plot of daily returns index *Nifty 50* from January 2012 to December 2019. Right: The estimated tail probabilities according to the GPD fitted model (line) and the corresponding empirical probabilities (dots)

7.4 Estimation of Time Series Data

In this section, we will introduce generalized autoregressive conditional heteroskedasticity (GARCH) and Markov Switching model and the estimation of VaR using these two models.

7.4.1 GARCH Model

Financial time series with time-varying volatility and volatility clustering are frequently modeled using ARCH models. ARCH models were introduced by Engle (1982). Here the conditional variance process is given an autoregressive structure and the returns are modeled as a white noise multiplied by the volatility. We say that a time series $\{u_t\}_{t=1}^{T}$ follows an ARCH(p) process if:

$$u_t = \sigma_t \varepsilon_t,$$

$$\sigma_t^2 = a + \sum_{i=1}^{p} \alpha_i u_{t-i}^2,$$

where ε_t (the "innovations") are i.i.d. with expectation 0 and variance 1 and are assumed independent from σ_k for all $k \leq t$. The lag length $p \geq 0$ is integral to the model specification and can be ascertained by the Box-Pierce test or other tests for the significance of autocorrelation, with the scenario $p = 0$ representing a white noise process. To ensure that σ_t^2 remains positive, $a, \alpha_i \geq 0 \ \forall \ i$ is required.

According to Bollerslev (1986), the ARCH model was expanded such that it could accommodate the presence of an additional autoregressive structure within σ_t^2. The GARCH(p,q) (generalized ARCH) model is given by

$$u_t = \sigma_t \varepsilon_t,$$
$$\sigma_t^2 = a + \sum_{i=1}^{p} \alpha_i u_{t-i}^2 + \sum_{i=1}^{q} \beta_i \sigma_{t-i}^2, \quad (7.11)$$

where the innovations $\varepsilon_t \sim$ i.i.d.$\mathcal{N}(0, 1)$ are distributed according to a standard Gaussian. GARCH processes (Bollerslev 1986), particularly the GARCH(1,1) process with normal innovations, have been the most successful models for accounting for the time-dependent volatility in financial time series. Common volatilities can be well predicted by GARCH models with normal innovations, but it is difficult to anticipate extreme observations, or spikes, in returns using this method. Furthermore, whereas the GARCH-normal model has thick tails unconditionally, they are thin conditionally. A GARCH model with non-normal innovations can be easily estimated, with the Student-t distribution being the most popular specification. Student-t innovations have the advantage of being thick-tailed, which will generally result in superior predictive densities. It should be noted that the Student-t contains Gaussian errors as a particular case. The drawbacks of non-normal innovations for the VaR exercise are numerous. For example, estimating multivariate versions of these models is sometimes difficult, and since the GARCH process is not self additive, calculating recursive forecasts for multi-step forward VaR levels is problematic. The widely used RiskMetrics method employs the often updated normal model, which is asymptotically identical to an IGARCH model.

7.4.1.1 Estimation of VaR Using GARCH Model

The goal of this investigation is to provide a procedure for accurately estimating VaR in the presence of time-varying volatility[1]. The data set includes daily negative log

[1] The estimation technique used here is a standard technique which is used to estimate VaR using GARCH model available in RStudio. https://rstudio-pubs-static.s3.amazonaws.com/534670_78d421a4edc94d27a69d49a1c966c9ff.html.

returns considering the closing prices of the Nifty index between January 1, 2012, and December 31, 2019.

1. First, we use the Augmented Dickey-Fuller test to confirm the stationarity of the returns, where null hypothesis indicates non-stationary time series.
   Augmented Dickey-Fuller Test
   Dickey-Fuller = -13.115, Lag order = 13, p-value = 0.01
   alternative hypothesis: stationary
   A small p-value ($<$ 0.01) indicates enough evidence to reject the null hypothesis, implying that time series are stationary.
2. To obtain the best fit of a time series model that represents the stochastic process that generates time series, the Box-Jenkins approach uses autoregressive integrated moving average (ARIMA) models. Three stages of modeling are used in this method: identification, estimation, and diagnostic checking.
   Identification: The time series must be stationary in order to apply the Box-Jenkins approach. Here, we utilize the returns that we previously examined for stationarity in the preceding section. Moreover, based on the Autocorrelation Function (ACF) and Partial Autocorrelation Function (PACF), it is possible to determine p order of the autoregressive part, d number of nonseasonal differences needed for stationarity, and q order of the moving average part order of the ARIMA model. The model can also be identified using the Akaike Information Criterion (AIC). AIC deals with both the risk of over fitting and under fitting. The model is chosen based on lowest AIC. We use the auto.arima function from forecast package to find the best model.
   Best model: ARIMA(0,0,1) with non-zero mean
   From the calculated AIC scores for a number of ARIMA models (refer to Appendix), we conclude that 1-order Moving Average (MA(1)) is the best model.
   Estimation: Maximum Likelihood technique is used to estimate the coefficients of the parameters. Using ARIMA(0, 0, 1) as selected model, the coefficients obtained are as follows:
   MA1: Coefficient: 0.0833, S.E: 0.0234
   Mean: Coefficient: -0.0002, S.E: 0.0001
   $\sigma^2 = 0.00001491$: log likelihood = 7925.06
   AIC=-15844.12 AICc=-15844.1 BIC=-15827.44
   Diagnostic Checking: The process entails looking at the residual plot, its ACF and PACF diagram, and checking the Ljung-Box test outcome. The chosen model is suitable if ACF and PACF of the model residuals reveal no significant lags (see Fig. 7.4).
   Next we use the Ljung-Box test to confirm that the residuals are uncorrelated.

$$R_{LB} = T(T+2)\sum_{i=1}^{m}\frac{\hat{\rho}_i^2}{T-i},$$

where $\hat{\rho}_i$ is the sample autocorrelation at lag i, T is the sample size, and m is the number of lags tested. The R_{LB} follows χ^2 distribution with $m-p-q$, where p and q are the number of parameters in the ARIMA(p,q) model. The null hypothesis refers to $H_0 : \hat{\rho}_1 = \hat{\rho}_2 = \cdots = \hat{\rho}_m = 0$

Box-Ljung test

X-squared = 14.505, df = 13, p-value = 0.3393

We cannot reject the null hypothesis; therefore, the residual process behaves like white noise and gives no sign of a pattern that may be modeled.

3. **GARCH Implementation**: The time series plot of the residuals exhibits some cluster volatility despite the absence of significant lags in the ACF and PACF of the residuals. In order to model volatility, we use the GARCH model. Let the time series r_t be divided into the predictable and unpredictable component,

$$r_t = E(r_t|I_{t-1}) + e_t,$$

where I_{t-1} is the information set at time $t - 1$ and $E(r_t|I_{t-1}) = -2 \times 10^{-4}$ and e_t is the unpredictable part. e_t can be expressed as GARCH process defined in Eq. (7.11). The next stage is to define the conditional variance, σ^2, which is the second component of the error term decomposition. For such a task, we use the most popular GARCH(1,1) model expressed as,

$$\sigma_t^2 = a + \alpha_1 e_{t-1}^2 + \beta_1 \sigma_{t-1}^2.$$

The squared residuals must be correlated in order for the GARCH process to be viable. ACF and PACF plots clearly indicate significant correlation in Fig. 7.5.

Significance testing on α_1 and β_1 is a different method for determining whether the squared residuals are heteroscedastic.

Since α_1 and β_1 are significantly different from zero, we assume that the volatility of the residuals will fluctuate over time.

4. **Distributional properties**: To calculate VaR, we must first correctly define the corresponding quantile of the assumed distribution. For a normal distribution, the quantile at $\alpha = 5\%$ is 1.645. According to empirical data, making the assumption of normality frequently results in poor outcomes. The Jarque-Bera (JB) test helps determine whether the distribution of returns is normal. The JB score for a normally distributed sample would be 0. The low p-value suggests that stock returns do not follow a normal distribution.

Jarque-Bera Test

X-squared = 679.43, df = 2, p-value < 0.00000000000000022

Other assumptions about the distribution of stock returns can be used to accurately model the thickness of tails. Like the normal distribution, the t-distribution is symmetric and shaped like a bell, but it has heavier tails, which means it is more likely to have values that deviate greatly from the mean. To obtain the fitting parameters of t-distribution, we utilize the rugarch package's fitdist function. The parameters obtained are $\mu = -0.0002297204$, $\sigma = 0.0039012899$, and shape = 5.1074269941.

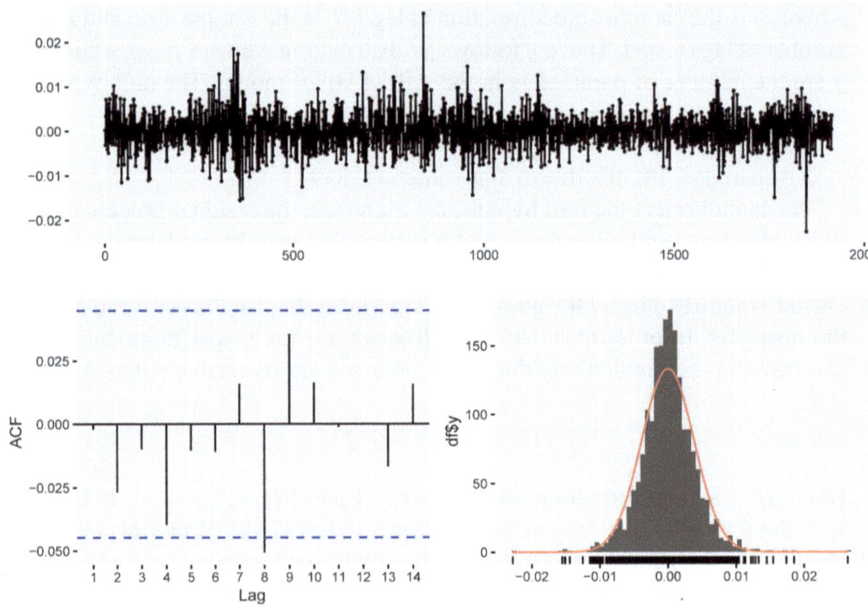

Fig. 7.4 Plot shows ACF and PACF of the residuals of ARIMA(0,0,1) model. The histogram of the residuals in comparison with a normal distribution N(0, σ^2) may be seen in the lower right corner of the plot

5. **GARCH-VaR**: We have already noted that returns exhibit volatility that varies with time. Therefore, we employ the conditional variance provided by the GARCH(1,1) model to estimate VaR. We utilize the Student's t-distribution to determine the distribution features of the underlined asset's. Here, the VaR is defined as

$$VaR = \mu + \sigma_{t|t-1} F^{-1}(\alpha),$$

where $\sigma_{t|t-1}$ is the conditional standard deviation given the information at $t - 1$ and F is the distribution function of Student's t-distribution.

7.4.2 *Markov Switching Model*

Recent studies show that volatility predictions by GARCH-type models may fail to capture the true variation in volatility in the case of regime changes in the volatility dynamics (see Lamoureux and Lastrapes 1990; Bauwens et al. (2014). This issue can be resolved by allowing the GARCH model's parameters to change over time in accordance with a latent discrete Markov process. This method, referred to as

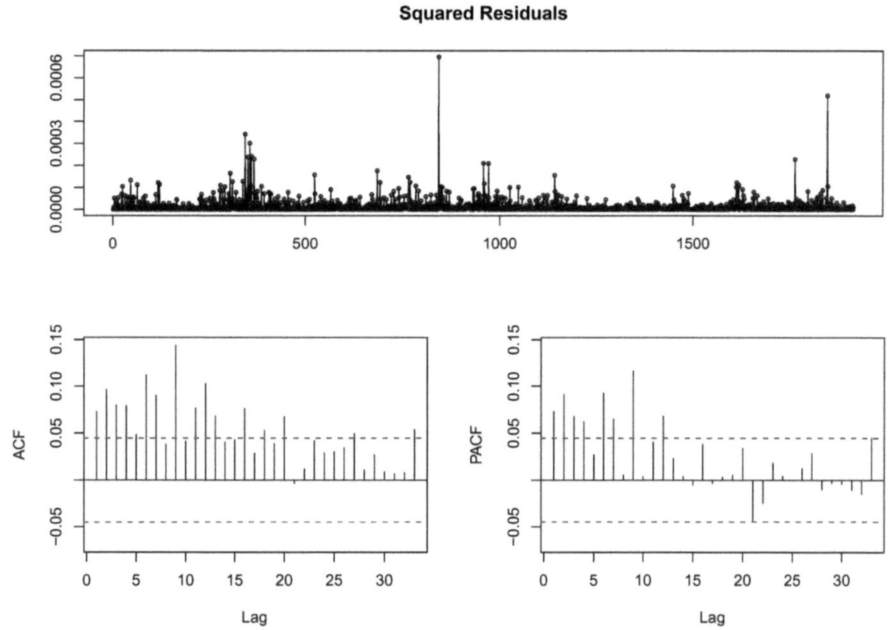

Fig. 7.5 Plot shows ACF and PACF of the squared residuals of ARIMA(0,0,1) model

Table 7.4 Significance testing on α_1 and β_1

Coef.	Estimate	Std. Error	t value	Pr(> $\|t\|$)
μ	−0.0000045090949	0.000082363782	−0.05474609	0.95634075
ω	0.0000003322784	0.000001079598	0.30777989	0.75824982
α_1	0.0623043615376	0.034248935445	1.81916199	0.06888672
β_1	0.9171409915231	0.033975492845	26.99419242	0.00000000

the Markov-switching GARCH (MSGARCH) model, produces volatility forecasts that can swiftly adjust to changes in the unconditional volatility level. MSGARCH models are mainly used in finance. Other fields of applications include the analysis of business cycles in economics (see, e.g., Kim and Nelson 1999) and the forecasting of windpower (see, e.g., Zhang et al. 2014). The major reasons for their limited utilization in industrial applications are their laborious implementation and difficult estimation. Utilizing the R package MSGARCH, which applies the specification of Haas et al. (2004), we have the benefit that the estimation does not encounter the path-dependency problem, in contrast to the MSGARCH model in Gray (1996) seminal work.

Let the variable of interest at time t be y_t. We assume that y_t has zero mean and is not serially correlated, that is, the following moment conditions are assumed: $E[y_t] = 0$ and $E[y_t y_{t-l}] = 0$ for $l \neq 0$ and for all $t > 0$. This assumption is realistic for high frequency returns for which the (conditional) mean is often assumed to be zero; see, for instance, McNeil et al. (2015). The general MSGARCH specification can be expressed as:[2]

$$y_t | (s_t = k,\ I_{t-1}) \sim \mathcal{D}(0, h_{k,t}, \xi_k), \tag{7.12}$$

where I_{t-1} is the information set observed up to time $t-1$, that is, $I_{t-1} \equiv \{y_{t-i}, i > 0\}$, $\mathcal{D}(0, h_{k,t}, \xi_k)$ is a continuous distribution with zero mean, time-varying variance $h_{k,t}$, and additional shape parameters gathered in the vector ξ_k. The integer-valued stochastic variable s_t, defined on the discrete space $\{1, \ldots, K\}$, characterizes the MSGARCH model. The standardized innovations are defined as

$$\eta_{k,t} \equiv y_t / h_{k,t}^{1/2} \overset{i.i.d}{\sim} \mathcal{D}(0, 1, \xi_k).$$

Further we assume that s_t evolves according to an unobserved first-order ergodic homogeneous Markov process with $K \times K$ transition probability matrix $\mathbf{P}$:

$$\mathbf{P} = \begin{bmatrix} p_{1,1} & p_{1,2} & p_{1,3} & \cdots & p_{1,K} \\ p_{2,1} & p_{2,2} & p_{2,3} & \cdots & p_{2,K} \\ \cdots & \cdots & \cdots & \cdots & \cdots \\ p_{K,1} & p_{K,2} & p_{K,3} & \cdots & p_{K,K} \end{bmatrix},$$

where $p_{i,j} \equiv P[s_t = j | s_{t-1} = i]$ is the probability of a transition from state $s_{t-1} = i$ to state $s_t = j$. Also, the following constraints hold: $0 < p_{i,j} < 1\, \forall\, i, j \in \{1, \ldots, K\}$, and $\sum_{j=1}^{K} p_{i,j} = 1$, $\forall\, i \in \{1, \ldots, K\}$. Given the parametrization of $\mathcal{D}(\cdot)$, we have $E[y_t^2 | (s_t = k, I_{t-1})] = h_{k,t}$, that is, $h_{k,t}$ is the variance of y_t conditional on the realization of $s_t = k$.

7.4.2.1 Estimation of VaR Using MSGARCH Model

An important application of MSGARCH models is in quantitative finance where we observe that investors invest their money in various risky investment options. Here our observations are usually the log returns of financial assets and we try to estimate the quantile of their future distribution at a given level $\alpha \in (0, 1)$ as well as their expected value below this level. Formally, the VaR forecast in $T+1$ at level α (given the information set up to at time T) is defined as:

$$VaR_{T+1}^{\alpha} \equiv \inf\{y_{T+1} \in \mathbb{R} | F(y_{T+1} | I_T) = \alpha\}, \tag{7.13}$$

[2] The definition is taken from the article: Ardia et al. (2019) and this is an open access work.

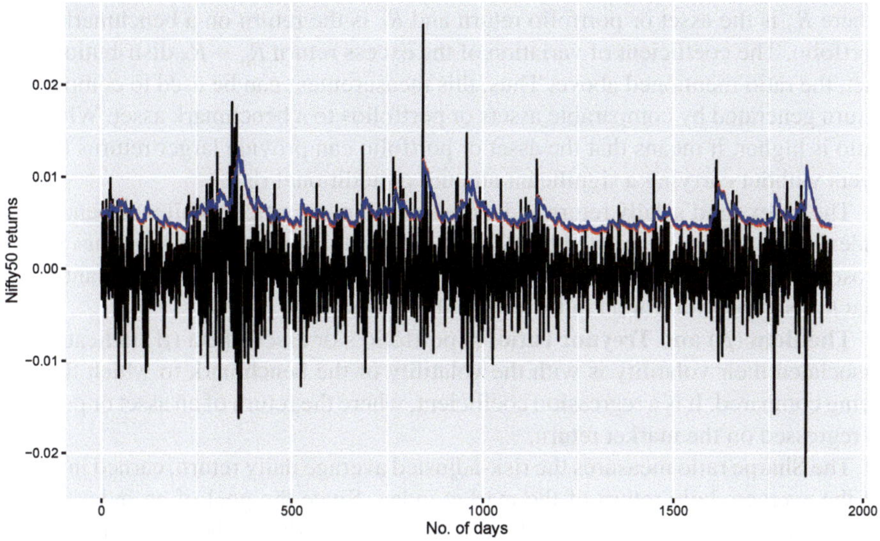

Fig. 7.6 VaR comparison of Nifty 50 at 5% (Red is GARCH-VaR and Blue is MSGARCH-VaR)

where $F(y|I_T)$ is the one-step ahead cumulative distribution function evaluated in y. The ES is obtained as:

$$ES^{\alpha}_{T+1} \equiv E[y_{T+1}|y_{T+1} \leq VaR^{\alpha}_{T+1}, I_T]. \tag{7.14}$$

Next we show the time evolution of MSGARCH-VaR of Nifty 50 at 5% using the daily returns. From the previous observation, we know that the returns are not normally distributed. So, for the underlined asset's distribution properties we use the student's t-distribution. The parameters are estimated using maximum likelihood technique.

7.5 Ratios to Compare Portfolios

In the previous sections, we have reviewed the various risk measures and how to assess them. Now it is time to look at how to combine the various risk measures with returns to create an ideal portfolio.

The excess return per unit of the deviation in an asset or portfolio is measured by the **Sharpe ratio**, which is also referred to as the reward-to-variability ratio. Sharpe (1994) defined this ratio as follows:

$$S = \frac{\mathrm{E}[R_a - R_b]}{\sqrt{\mathrm{var}[R_a - R_b]}},$$

where R_a is the asset or portfolio return and R_b is the return on a benchmark asset/portfolio. The coefficient of variation of the excess return $R_a - R_b$ distribution is, in fact, the ratio mentioned above. Thus, this measurement can be used to compare the return generated by comparable assets or portfolios to a benchmark asset. When this ratio is higher, it means that the asset or portfolio can provide larger returns than its peers without carrying a significant amount of additional risk.

The index fund's daily return is denoted by R_a in our article, while the benchmark index's daily return is denoted by R_b. It is assumed that the R_a and R_b values will be close because an index fund seeks to replicate a benchmark index. Thus, we anticipate that the Sharpe ratio for an index fund will be near zero.

The Beta (β) and Treynor ratio: A portfolio's or asset's beta (β) indicates how associated their volatility is with the volatility of the benchmark to which they are being compared. It is a regression coefficient, where the return of an asset or portfolio is regressed on the market return.

The Sharpe ratio measures the risk-adjusted average daily return, earned in excess of the average daily return of the market index. Since the goal of an index fund is to replicate a market index, the fund's Sharpe ratio should be near zero because the average return of the fund and the index should be similar. We observe that the Sharpe ratio of most of the index funds is much close to zero (see Table 7.3 and 7.4). But the Sharpe ratio of Aditya Birla Sun Life Nifty Exchange Traded Fund and ICICI Prudential Nifty Next 50 Index Fund-Growth Plan is substantially different from zero. Among the Nifty-based funds, the Sharpe ratio of IDBI Nifty Index Fund-Growth Plan is much closer to zero than the same for the other index funds. Among the Sensex-based funds, the Sharpe ratio of LIC MF Index Fund-Growth Plan is much closer to zero than the same for the other index funds.

The Treynor and Sharpe ratios are similar. While the denominator is different—that is, the β coefficient of the asset or portfolio takes the place of σ—the numerator stays the same. Thus, the excess return per unit of systematic risk is measured by the Treynor ratio. The Treynor ratio is defined as follows:

$$T = \frac{\mathrm{E}[R_a - R_b]}{\beta},$$

where

$$\beta = \frac{\mathrm{cov}(R_a, R_b)}{\mathrm{var}(R_b)}.$$

Given that the ideal index fund aims to replicate a wide market index rather than beating it, it is reasonable to anticipate that the index fund's Treynor ratio will be around zero.

As previously stated, the Treynor ratio of an index fund should be near zero because it calculates the excess return per unit of beta, a measure of systematic risk. From Tables 7.3 and 7.4, the Treynor ratio of most of the index funds is much close to zero. But the Treynor ratio of Aditya Birla Sun Life Nifty Exchange Traded Fund (ETF) is substantially different from zero. Among the Nifty-based funds, the Treynor ratio of IDBI Nifty Index Fund-Growth Plan is much closer to zero than the same for the other index funds. Among the Sensex-based funds, the Treynor ratio of Tata Index Fund-Growth Plan is much closer to zero than the same for the other index funds.

Among the above-mentioned funds, the Sharpe ratio and Treynor ratio for two funds, namely the Aditya Birla Sun Life Nifty ETF and ICICI Prudential Nifty Next 50 Index Fund-Growth Plan, are substantially different from zero in comparison with the same for the other index funds (see Table 7.3 and 7.4). Therefore, the funds, Aditya Birla Sun Life Nifty 50 Index Fund-Growth Plan, Franklin India Index Fund-Growth Plan, IDBI Nifty Index Fund-Growth Plan, IDFC Nifty ETF, IDFC Nifty Fund-Growth Plan, LIC MF Index Fund-Nifty Plan-Growth Plan, Nippon India Index Fund-Nifty Plan-Growth Plan, Tata Index Fund-Nifty Plan-Growth Plan, IDFC Sensex ETF, LIC MF Index Fund-Sensex-Growth Plan, Nippon India Index Fund-Sensex-Growth Plan, and Tata Index Fund-Sensex-Growth Plan, seem to be preferable (among Indian index funds) in terms of all the four criteria.

We can observe from Tables 7.3 and 7.4 that the beta coefficients of all Indian index funds have absolute values below one. This suggests that these funds are appropriate for risk-averse investors and are low-risk investments.

7.6 Back-Testing VaR Models

Back-testing is a technique that evaluates precision and potency using past data. So we wish to determine the validity of a certain VaR model. Back-testing VaR is used to compare the forecast/predicted losses from the actual calculated losses realized at the end of a fixed time horizon. The outcomes of the back-testing give us details on specific instances when VaR is underestimated or where the losses exceed the initial expected VaR value. If the back-testing values are inaccurate, these VaR values can then be recalculated, lowering the risk of unexpected losses for researchers and institutions.

The plot of the one-step ahead VaR and actual observed returns for the Sensex, utilizing the Geltner adjusted modified method with data from the last six years for 490 time points from January 2018 to December 2019, is presented in Fig. 7.7. Observe that there are 19 incidences where the black dot is above the red line indicating a VaR exceedance. We have also shown the one-step ahead VaR using the MSGARCH method for the same data set in Fig. 7.7 and observe that there are 21 incidences where the black dot is above the blue line indicating a VaR exceedance.

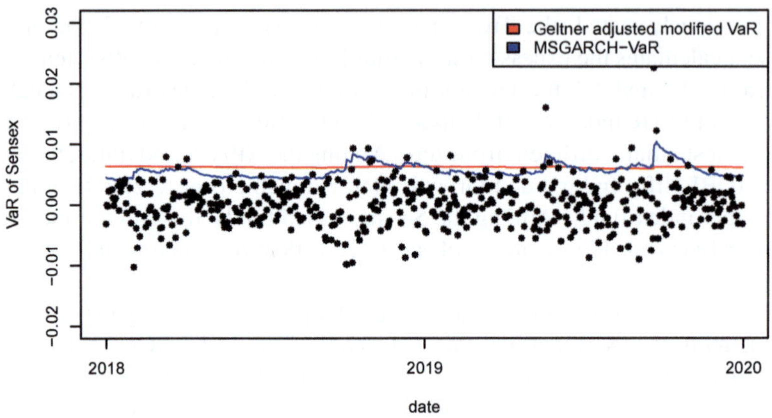

Fig. 7.7 Backtest: One-step ahead VaR using Geltner adjusted modified and MSGARCH method and actual observed returns for Sensex

The expected number of exceedances is 5% of 490, that is 24.5. So we see that the procedure exhibits less than expected number of exceedances. To verify that the exceedance probability is actually 5%, a formal test must now be conducted. The most basic of these tests is a binomial likelihood ratio test known as the unconditional coverage (UC) test, which was suggested by Kupiec (1995). We denote $I_t(\alpha)$ the hit variable associated with the ex-post observation of an αVaR exception at time t:

$$I_t(\alpha) = \begin{cases} 1 \text{ if } & r_t < -VaR_{t|t-1}(\alpha) \\ 0 \text{ otherwise.} \end{cases} \tag{7.15}$$

Note that if we assume that the VaR exceptions on different days are independent, then $B = \sum_{t=1}^{n} I_t$ follows a binomial (n, α) distribution. We are interested in testing the null hypothesis that α =5% vs the alternative $\alpha \neq 5\%$. The likelihood ratio statistic can be derived as

$$LR_{UC} = \frac{(1-\alpha)^{n-B}\alpha^{B}}{(1-B/n)^{n-B}(B/n)^{B}} \tag{7.16}$$

and from general results $-2\ln LR_{UC}$ has an asymptotic χ_1^2 distribution.

For the example, the one-sided p-value for the count 19 is 0.099. So we do not reject the null hypothesis and conclude that the VaR is estimated correctly by the Geltner adjusted modified method.

VaR exceedances are assumed to be independent in the analysis above, despite the fact that actual evidence shows that these exceedances occur in clusters. The conditional coverage test proposed by Christoffersen (1998) offers an alternate by assuming that the violation process may be described as a Markov chain with two states and the transition matrix

$$\Pi = \begin{pmatrix} 1 - \pi_{01} & \pi_{01} \\ 1 - \pi_{11} & \pi_{11} \end{pmatrix} \tag{7.17}$$

where $\pi_{ij} = \mathrm{P}(I_t(\alpha) = j \mid I_{t-1}(\alpha) = i)$. The likelihood ratio statistic, in this case, can be derived as

$$LR_{CC} = \frac{(1-\alpha)^{n-B}\alpha^{B}}{(1-\hat{\pi_{01}})^{n_{00}}\hat{\pi_{01}}^{n_{01}}(1-\hat{\pi_{11}})^{n_{10}}\hat{\pi_{11}}^{n_{11}}} \tag{7.18}$$

where n_{ij} are the observed transition counts. In this case, $-2\ln LR_{CC}$ has an asymptotic χ_2^2 distribution.

For the example, the p-value for the count 21 is 0.1801. So we do not reject the null hypothesis and conclude that the VaR is estimated correctly by the MSGARCH method.

7.7 Concluding Remarks

It is impossible to say which fund is consistently the best across the four measures. According to VaR and ES, the majority of Indian index funds appear to offer similar market risk as the respective benchmark indices. However, we have identified twelve Aditya Birla Sun Life Nifty 50 Index Fund-Growth Plan, Franklin India Index Fund-Growth Plan, IDBI Nifty Index Fund-Growth Plan, IDFC Nifty ETF, IDFC Nifty Fund-Growth Plan, LIC MF Index Fund-Nifty Plan-Growth Plan, Nippon India Index Fund-Nifty Plan-Growth Plan, Tata Index Fund-Nifty Plan-Growth Plan, IDFC Sensex ETF, LIC MF Index Fund-Sensex-Growth Plan, Nippon India Index Fund-Sensex-Growth Plan, and Tata Index Fund-Sensex-Growth Plan, which, according to all four criteria, appear to be better than the other Indian index funds. These funds' risk-adjusted daily returns are comparable to the average daily returns of the Nifty or Sensex indices, whichever they are meant to replicate, and they appear to show somewhat less extreme market risk than the corresponding benchmark indices.

Table 7.5 Sharpe ratio and Treynor ratio of Nifty Index Funds

Index	Sharpe	Treynor	Beta
Aditya Birla Sun Life Nifty 50 Index Fund-Growth Plan	−0.001170	−0.000189	0.043833
Aditya Birla Sun Life Nifty ETF	0.026902	0.005265	0.149262
Franklin India Index Fund-Growth Plan	−0.000567	-8.9135×10^{-5}	0.045075
HDFC Index Fund-Growth Plan	−0.002351	−0.000373	0.045018
ICICI Prudential Nifty ETF	−0.003314	−0.000525	0.045088
ICICI Prudential Nifty Next 50 Index Fund-Growth Plan	0.006987	0.000574	0.085174
IDBI Nifty Index Fund-Growth Plan	**0.000115**	$\mathbf{1.8334} \times 10^{-5}$	0.044886
IDFC Nifty ETF	−0.003042	−0.000494	0.043709
IDFC Nifty Fund-Growth Plan	−0.002669	−0.000430	0.044067
LIC MF Index Fund-Growth Plan	0.000354	-5.5499×10^{-5}	0.045429
Nippon India Index Fund-Growth Plan	−0.000602	-9.0181×10^{-5}	0.047536
SBI Nifty Index Fund-Growth Plan	−0.001419	−0.000224	0.045298
Tata Index Fund-Growth Plan	−0.001715	−0.000283	0.042869
Taurus Nifty Index Fund-Growth Plan	−0.001612	0.000181	−0.066386
UTI Nifty ETF	−0.002095	−0.000328	0.045739
UTI Nifty Index Fund-Growth Plan	−0.002291	−0.000367	0.044757

Source: The entries in the table have been computed based on data collected from AMFI and NSE websites (https://www1.nseindia.com/products/content/equities/indices/historical_index_data.htm; https://www.amfiindia.com/net-asset-value/nav-history)

Table 7.6 Sharpe ratio and Treynor ratio of Sensex Index Funds at 5%

Index	Sharpe Ratio	Treynor Ratio	Beta
HDFC Index Fund-Sensex Plan-Growth Plan	−0.002236	−0.000292	0.055189
IDFC Sensex ETF	−0.002890	−0.000381	0.054189
LIC MF Index Fund-Growth Plan	**-0.000517**	-6.82046×10^{-5}	0.054313
Nippon India Index Fund-Growth Plan	−0.001456	−0.000186	0.055732
Tata Index Fund-Growth Plan	−0.001402	$\mathbf{-5.6752} \times 10^{-5}$	0.165025

Source: The entries in the table have been computed based on data collected from AMFI and BSE websites (https://www.bseindia.com/indices/IndexArchiveData.html; https://www.amfiindia.com/net-asset-value/nav-history)

R Codes

Codes for Figure 7.1

```
install.packages("tidyverse")
library(tidyverse)
```

```
Returns = read.csv('E:/Returns.csv' , header = T)
e=Returns[,1]
E=as.Date(e)
qplot(E , -Returns[,2] ,xlab='Date',ylab='Nifty', geom = 'line')#Returns[,2] refers to Nifty50 returns
```

Codes for Figure 7.2

```
install.packages("PerformanceAnalytics")
library(PerformanceAnalytics)
md = c("gaussian", "modified", "historical")
d = matrix(0,1000)
d1=matrix(0,1000)
for(i in 1:1000)
{
j = i+915
d[i,] = VaR(Returns[,2][i:j], p = 0.95, method = md[1], invert=FALSE)
d1[i,]=VaR(Returns[,3][i:j], p = 0.95, method = md[1], invert=FALSE)#Returns[,3] refers to Sensex returns
}
e=x[,1][916:1915]
E=as.Date(e)
plot(E,d[,1],xlab='date',ylab='VaR',pch=20,type='l')
lines(E,d1[,1],col="red",pch=20)
legend("topright",c("Nifty","Sensex"),fill=c("black","red"))
```

Codes for Table 7.3

```
install.packages("robustbase")
library(robustbase)
cn = c("none","boudt","geltner")
VaRk = matrix(0,3,3)
for(i in 1:3)
{
for(j in 1:3)
{
Rc = Return.clean(ts(Returns[,3]), method = cn[j], alpha = .01)
Rc1=Return.clean(ts(Returns[,2]), method = cn[j], alpha = .01)
VaRk[i,j]=VaR(Rc, p = 0.95, method = md[i], invert=FALSE)
}
}
```

Codes for Figure 7.3

```
install.packages("QRM")
library(QRM)
Rc = Return.clean(ts(Returns[,2]), method = cn[2], alpha = .01)
hillPlot(-Rc)
out = fit.GPD(-Rc,nextremes=25)
r = RiskMeasures(out,.95)[2]
```

Codes for ADF Test

```
install.packages("tseries")
library(tseries)
adf.test(Returns[,2])
```

Codes for Model Identification

```
install.packages("forecast")
library(forecast)
model.arima=auto.arima(Returns[,2], max.order=c(3,0,3), stationary=TRUE,
trace=T, ic='aicc')
```

output:

```
Fitting models using approximations to speed things up..
ARIMA(2,0,2) with non-zero mean : -15838.24
ARIMA(0,0,0) with non-zero mean : -15833.55
ARIMA(1,0,0) with non-zero mean : -15842.35
ARIMA(0,0,1) with non-zero mean : -15844.11
ARIMA(0,0,0) with zero mean : -15831.06
ARIMA(1,0,1) with non-zero mean : -15842.56
ARIMA(0,0,2) with non-zero mean : -15843.55
ARIMA(1,0,2) with non-zero mean : -15844.07
ARIMA(0,0,1) with zero mean : -15842.27
Now re-fitting the best model(s) without approximations...
ARIMA(0,0,1) with non-zero mean : -15844.1
```

Codes for Figure 7.4

```
install.packages("ggthemes")
library(ggthemes)
model.arima$residuals %>% ggtsdisplay(plot.type = 'hist' , lag.max = 14)
```

Codes for Ljung-Box Test

```
ar.res = model.arima$residuals
Box.test(ar.res, lag = 14 , fitdf =1 , type = 'Ljung-Box')
```

Codes for Figure 7.5

tsdisplay($ar.res^2$, main = 'Squared Residuals')

Codes for testing heteroscedasticity

```
install.packages("rugarch")
library(rugarch)
model.spec=ugarchspec(variance.model=list(model='sGARCH',
garchOrder=c(1 , 1)),mean.model= list(armaOrder=c(0, 0)))
model.fit = ugarchfit(spec = model.spec , data = ar.res , solver = 'solnp')
options(scipen = 999)
model.fit@fit$matcoef
```

Codes for Jarque-Bera test

```
jarque.bera.test(Returns[,2])
```

Codes for obtaining parameters of *t*-distribution

```
fitdist(distribution = 'std' , x = Returns[,2])$pars
```

Codes for quantile

```
qdist(distribution = 'std' , shape = 5.1074269941 , p = 0.95)
```

Codes for GARCH-VaR

```
qplot(y=Returns[,2],x=1:length(Returns[,2]),geom='line')+geom_line(aes(y=model.fit@fit$sigma*(1.564482), x=1:length(Returns[,2])), colour='red')
```

Codes for MSGARCH-VaR

```
install.packages("MSGARCH")
library(MSGARCH)
spec=CreateSpec(variance.spec=list(model=c("sGARCH")),distribution.spec = list(distribution=c("std")), switch.spec=list(do.mix = FALSE))
fit=FitML(spec = spec, data = Returns[,2])
risk=Risk(object = fit,alpha=0.95,do.its=TRUE)
qplot(y = Returns[,2] , x = 1:length(Returns[,2]) , geom = 'line', xlab='No. of days',ylab='Nifty50 returns')+geom_line(aes(y = risk$VaR , x = 1:length(Returns[,2])) , colour = 'red')
```

Codes for Backtesting

```
Rc = Return.clean(ts(Returns[,3]), method = cn[3], alpha = .01)
d = matrix(0,491)
for(i in 1:491)
{
j = i+1424
d[i,] = VaR(Rc[i:j], p = 0.95, method = md[2], invert=FALSE)
}
risk = matrix(0,491)
for(i in 1:491)
{
j = i+1424
spec=CreateSpec(variance.spec = list(model = c("sGARCH")), distribution.spec=list(distribution= c("std")),switch.spec=list(do.mix = FALSE))
fit=FitML(spec = spec, data = Returns[,3][i:j])
risk[i]=Risk(object = fit,alpha=0.95,do.its=FALSE,do.es=FALSE)
}
e=Returns[,1][1425:1915]
E=as.Date(e)
plot(E,d[,1],ylim=c(-0.020,0.03),xlab='date',ylab='VaR of Sensex',type='l', col='red')
lines(E,risk,col='blue')
points(E,Returns[,3][1425:1915],pch=8,cex=0.5)
legend("topright",c("Geltner adjusted modified VaR","MSGARCH-VaR"),fill=c("red","blue"))
```

Codes for Coveragetest

```
install.packages("quarks")
library(quarks)
cvgtest(obj=list(loss=Returns[,3][1425:1915],VaR=d[,1],p=0.95))
cvgtest(obj=list(loss=Returns[,3][1425:1915],VaR=risk,p=0.95))
```

References

Ardia D, Bluteau K, Boudt K, Catania L, Trottier D-A (2019) Markov-switching garch models in r: the msgarch package. J Stat Softw 91(4)
Artzner P, Delbaen F, Eber J-M, Heath D (1999) Coherent measures of risk. Mathe Fin 9(3):203–228
Bacon CR (2008) Practical portfolio performance measurement and attribution, with CD-ROM. Wiley
Bauwens L, De Backer B, Dufays A (2014) A Bayesian method of change-point estimation with recurrent regimes: application to garch models. J Emp Fin 29:207–229
Biswas S, Dutta S (2015) Assessing market risk of indian index funds. Glob Bus Rev 16(3):511–523
Bollerslev T (1986) Generalized autoregressive conditional heteroskedasticity. J Economet 31(3):307–327
Boudt K, Peterson BG, Croux C (2008) Estimation and decomposition of downside risk for portfolios with non-normal returns. J Risk 11(2):79–103
Christoffersen PF (1998) Evaluating interval forecasts. Int Econ Rev:841–862
Cont R, Deguest R, Scandolo G (2010) Robustness and sensitivity analysis of risk measurement procedures. Quan Fin 10(6):593–606
Engle RF (1982) Autoregressive conditional heteroscedasticity with estimates of the variance of united kingdom inflation. Econ J Economet Soc:987–1007
Gray SF (1996) Modeling the conditional distribution of interest rates as a regime-switching process. J Fin Econ 42(1):27–62
Haas M, Mittnik S, Paolella MS (2004) A new approach to markov-switching garch models. J Fin Economet 2(4):493–530
Kim C-J, Nelson CR (1999) Has the us economy become more stable? A Bayesian approach based on a markov-switching model of the business cycle. Rev Econ Stat 81(4):608–616
Kupiec P (1995) Techniques for verifying the accuracy of risk measurement models. J Deriv 3(2)
Lamoureux CG, Lastrapes WD (1990) Persistence in variance, structural change, and the garch model. J Bus Econ Stat 8(2):225–234
Markowitz H (1952) Portfolio selection. J Fin 7(1):77–91
McNeil AJ, Frey R, Embrechts P (2015) Quantitative risk management: concepts, techniques and tools-revised edition. Princeton University Press
Michaud RO (1989) The Markowitz optimization enigma: Is 'optimized'optimal? Fin Anal J 45(1):31–42
Pérignon C, Smith DR (2010) The level and quality of value-at-risk disclosure by commercial banks. J Bank Fin 34(2):362–377
Philippe J (2001) Value at risk: the new benchmark for managing financial risk. McGraw-Hill Professional, NY
Roy AD (1952) Safety first and the holding of assets. Econ J Econ Soc:431–449
Sen R, Das S (2023) Statistical risk analysis. In: Computational finance with r. Springer, pp 213–228
Sharpe WF (1994) The sharpe ratio. J Portf Manage 21(1):49–58
So MKP, Wong C-M (2012) Estimation of multiple period expected shortfall and median shortfall for risk management. Quan Fin 12(5):739–754
Zangari P (1996) A var methodology for portfolios that include options. RiskMet Mon 1:4–12
Zhang Y, Wang J, Wang X (2014) Review on probabilistic forecasting of wind power generation. Renew Sustain Ener Rev 32:255–270
Žiković S, Aktan B (2011) Decay factor optimisation in time weighted simulation-evaluating var performance. Int J Forecast 27(4):1147–1159

Chapter 8
Causal Inference with Matching: Evaluation of Programs and Policies

Vikram Dayal and M. Rahul

Abstract A significant focus of modern empirical economics is causal inference. This chapter begins with a gentle introduction to the Neyman-Rubin potential outcomes causal model, which defines the causal effect of a treatment, and outlines the necessary assumptions for valid causal inference. We then explore matching techniques, a method applicable to observational data, in certain contexts. Since observational studies require careful adjustment for relevant regressors, matching helps achieve balance in the distribution of regressors between treated and control groups. Once balance is attained, we analyze the differences in outcomes between these groups. The chapter covers propensity score matching and genetic matching methods, demonstrating their application using data from a training program evaluation. Additionally, we illustrate the implementation of these methods using the Matching package in R programming language.

8.1 Introduction

Recently, economists have been greatly interested in causal inference—estimating the impact of some intervention or program (the cause) (Abadie and Cattaneo 2018; Imbens and Wooldridge 2009). They often use the Neyman-Rubin potential outcomes causal model (henceforth the NRPO model).[1] The NRPO model clarifies the target of estimation, and the assumptions required. In other words, it helps clarify when a

[1] The Nobel laureate in economics, Guido Imbens, worked closely with Rubin, teaching with him, co-authoring papers (for example, Angrist et al. (1996) and the book Imbens and Rubin (2015)).

Supplementary Information The online version contains supplementary material available at https://doi.org/10.1007/978-981-97-9060-9_8.

V. Dayal (✉) · M. Rahul
Institute of Economic Growth, University of Delhi, Delhi, India
e-mail: vikday@iegindia.org

F. Hamid and D. Mukherjee (eds.), *Data Science for Modeling Managerial and Socioeconomic Problems*, Contributions to Management Science,
https://doi.org/10.1007/978-981-97-9060-9_8

statistical analysis results in sound causal inference. The NRPO causal model focuses on the treatment assignment mechanism.

We can use the matching method with observational data in some contexts. We can match persons who received treatment to similar persons who did not receive treatment. So, for example, we match a person who is 20 years old and received treatment to a person who is 20 years old and did not receive treatment. The treatment could be a program or a policy.[2] Treatment assignment concerns the probability of a unit receiving treatment. The propensity score is the probability of a unit with certain regressors receiving treatment. When analyzing observational data, we can use propensity score matching to achieve balance in regressors between treatment and control groups. Propensity score matching has been a very popular method for causal inference. Dehejia and Wahba (1999) demonstrated the usefulness of propensity score matching. They examined observational data related to the evaluation of a labor training program.

However, in a recent paper, King and Nielsen (2019) argued that propensity scores should not be used for matching. They wrote that using estimated propensity scores for matching can give poor results, increasing imbalance rather than balance, and along with this, leading to bias. Earlier, Diamond and Sekhon (2013) had pointed out that while using propensity score matching, the researcher had to iteratively check balance and improve it. They also noted that applied researchers did not follow the required iterative approach. Moreover, Sekhon (2011), using the dataset used by Dehejia and Wahba (1999), demonstrated that propensity score matching can worsen balance. Matching made the balance in a specific regressor, earnings in 1974, worse. Therefore, we need to check balance and iteratively improve it while using propensity score matching.

Given these difficulties with using propensity scores for matching, we can use the 'Matching' package in R (Sekhon 2011). This package uses a powerful algorithm to achieve better balance between treatment and control groups. The analyst does not have to correctly specify the propensity score.

This chapter presents the NRPO causal model, reviews matching, and then provides an empirical example using R so that the reader of the chapter can implement the method, while also understanding its key features.

8.2 NRPO Causal Model

The NRPO causal model is based on the potential outcomes experienced by a unit with and without an intervention or treatment. We use a simple example to explain some key ideas in the model.

[2] Rubin himself has worked on matching over a long time span, from the seventies when few other researchers were working on it (Rubin 2006). Subsequently, more recent decades have seen a large number of conceptual and applied papers on matching. Imbens also contributed to work on matching (for example, Abadie and Imbens 2006).

Suppose we observe a person, the first of five people, who takes a medicine to reduce pain. Taking a medicine is the treatment, that we denote by T. The person can either take the medicine ($T = 1$) or not ($T = 0$). Pain is an outcome, and we denote it by Y, and assume that it is a continuous variable that we are able to measure an hour later. We use i to denote each person. Let $Y_i(1)$ denote the outcome for person i when taking the medicine, and $Y_i(0)$ denote the outcome for person i when not taking the medicine. The subscript denotes the person and the number inside the parentheses denotes whether the medicine was taken or not: $Y_i(1)$ corresponds to $Outcome_{\text{unit i}}(Treated)$.

If person 1 ($i = 1$) takes the medicine ($T = 1$), the level of pain is 3, so $Y_1(1) = 3$. If person 1 does not takes the medicine, the level of pain is 7, so $Y_1(0) = 7$. So the potential outcomes for person 1 when taking the medicine or not are 3 and 7. The effect of the medicine is to lower the level of pain by 4.

The causal effect of medicine on pain for person 1 is the difference in the potential outcome had the person taken the medicine and the potential outcome had the person not taken the medicine. In symbols, $\tau_1 = Y_1(1) - Y_1(0) = 3 - 7 = -4$, where τ is the causal effect. However, in reality, we have to confront a basic issue in causal inference. Person 1 takes the medicine, so we will observe person 1's pain when taking the medicine, $Y_1(1)$, but we do not person 1's pain when not taking the medicine, $Y_1(0)$. We say that the pain when not taking the medicine, $Y_1(0)$, is counterfactual. Person 1 takes the medicine so for this person $T = 1$, and an hour later the level of pain for person 1 is recorded. For person 1, $T = 1$, and so for person 1, $Y_1(1) = Y_1^{OBS}$, where Y_1^{OBS} is observed Y.

Suppose we observe four more persons, who take a medicine (T) to reduce pain (Y). We see, for each person, the potential outcomes, treatment received, observed outcome, and effect of medicines in Table 8.1.

At first glance, using symbols for potential outcomes may seem unremarkable. But the use of symbols to differentiate between the potential outcomes and the observed outcomes is critical. This helps to clarify key issues in causal inference. The toy example above was used to introduce the notation and basic concepts of the NRPO causal model. We now move to a more general, formal treatment.

We define $Y_i(1)$ to be the potential outcome for the unit i under treatment and $Y_i(0)$ to be the potential outcome for unit i if it were in the control group. We cannot observe both potential outcomes. The observed outcome is given by

Table 8.1 Hypothetical example

Person i	$Y_i(1)$	$Y_i(0)$	T_i	Y_i^{OBS}	*Effect*$_i$
1	3	7	1	3	–4
2	1	6	1	1	–5
3	5	11	0	11	–6
4	3	7	0	7	–4
5	1	6	0	6	–5

$$Y_i^{OBS} = T_i Y_i(1) + (1 - T_i) Y_i(0) \tag{8.1}$$

where $T_i = 1$ if the unit received treatment and $T_i = 0$ if the unit did not receive treatment.

The causal (treatment) effect for a unit i is then given by the difference between these two potential outcomes. That is,

$$\text{Causal (treatment) effect for unit i} = Y_i(1) - Y_i(0) \tag{8.2}$$

However, the causal effect for an individual unit cannot be estimated by using observed data since only one of the potential outcomes is observed. One cannot observe both states simultaneously for the same unit. This is the 'fundamental problem of causal inference' (Holland 1986).

Depending on the data, and corresponding plausible assumptions, we can estimate an average treatment effect, τ:

$$\tau = E[Y_i(1) - Y_i(0)] \tag{8.3}$$

With observational data, assumptions are required to identify the average treatment effect (Rosenbaum and Rubin 1983; Imbens 2004; Sekhon 2008):

Assumption 1: The treatment assignment is independent of the potential outcomes, conditional on relevant regressors. That is,

$$(Y_1, Y_0) \perp T|X \tag{8.4}$$

This is a crucial assumption and should not be made lightly.

Assumption 2: The probability of any unit receiving treatment lies between 0 and 1, conditional on the relevant regressors.

$$0 < Pr(T = 1|X) < 1 \tag{8.5}$$

Assumption 3: The potential outcome of any particular unit is not affected by the treatment assignment of other units. That is, there is no interference or spillover between the units under study.

These three assumptions allow us to overcome the counterfactual nature of causal inference. Consider $E[Y_i(0)|X_i, T_i = 1]$—because $T_i = 1$ here, $Y_i(0)$ is not observable. However, because of our assumptions, $E[Y_i(0)|X_i, T_i = 1] = E[Y_i(0)|X_i, T_i = 0]$, and $E[Y_i(0)|X_i, T_i = 0] = E[Y_i^{OBS}|X_i, T_i = 0]$.

As we noted above, Assumption 1 is crucial. We need to choose which variables to match on keeping in mind Assumption 1. A basic principle is to match on 'pre-treatment' variables and not match on variables affected by the treatment.

Fig. 8.1 Causal graph.
Source: Created by authors

Table 8.2 Hypothetical example

Person i	$Y_i(1)$	$Y_i(0)$	T_i	Y_i^{OBS}	Effect$_i$	X_i
1	3	7	1	3	–4	High
2	1	6	1	1	-5	Low
3	5	11	0	11	–6	Very high
4	3	7	0	7	–4	High
5	1	6	0	6	-5	Low

We can use causal graphs to represent our causal assumptions and decide which regressors to match on.[3] Each node in the causal graph represents a variable and the direction of the arrow in the graph goes from the cause to the affected variable. For example, if our causal assumption is that T affects Y, it can be represented as $T \rightarrow Y$.

Typically, we may have several regressors denoted by X that affect both T and Y. Then, the causal graph would be as shown in Fig. 8.1. X introduces an indirect path from T to Y. X confounds the effect of T on Y. The indirect path going from T to Y through X ($T \leftarrow X \rightarrow Y$) is called a 'backdoor path.' Any estimation of causal effect of T on Y without adjusting for X would leave the backdoor path open and lead to a biased estimate. Adjusting for X, through matching or other methods, closes this backdoor path. Or, in this simplified example, Assumption 1 holds once we match on X.

8.3 Matching

We had seen earlier that we cannot estimate causal effects at the level of the individual unit. We do not observe both potential outcomes for an individual unit. Let us go back to our earlier hypothetical example and assume that we also observe a regressor X, proneness to headaches (Table 8.2). Since person 1, who has a high proneness, is treated, we only observe $Y_1(1)$, and not $Y_1(0)$. However, person 4 also has a high proneness to headaches, and we can impute $Y_1(0)$ by using $Y_4(0)$ which we can observe, since person 4 did not take the medicine. Similarly, we can match person 2 with person 5, since both have a low proneness to headaches. Once we match each unit in the treatment group to a unit in the control group, we can estimate an average effect of treatment for the treated.

[3] Causal graphs are often called Causal Directed Acyclic Graphs and have an underlying mathematics.

Person 3 did not take the medicine, but has a very high proneness to headaches. No person with a very high proneness to headaches takes the medicine, so we should avoid using the data for person 3 while estimating the average causal effect on the treated.

If we match units in the treated group to units in the control group with very similar regressors, and then take the difference between the average outcome for these two groups, we end up with an average effect of treatment on the treated (ATT).

8.3.1 Matching Reduces Model Dependence

Generally, social scientists use regressions to estimate causal effects. The researcher is unlikely to know the true model and has to choose among different functional forms, which lead to varying estimates, some of which may be biased. This is called model dependence.

We show an example of model dependence using synthetic data. y is the outcome variable, and x is the regressor that affects y. t is a binary variable that denotes treatment if 1 and control if 0; the true effect of t on y is –0.03. The data is created such that there is overlap between the x values of the units of the treatment and control group only for values of x between 5 and 20. For values of x below 5 and above 20, we only observe units in the control group and do not observe units in the treatment group.

We plot the data and the linear and quadratic fits (Fig. 8.2). We use blue to show the units in the control group and orange for the units in the treated group. The vertical gap between the fitted lines for the two groups represents the estimate of the treatment effect. In the linear case, the orange line is above the blue line indicating a positive effect. In the quadratic fit, the orange curve lies below the blue one. To see what is happening, note that by construction, in our data, there are no treated units for $x < 5$ and $x > 20$. Yet, the fitted lines for the treated units extend into these regions. The fitted lines are based on extrapolation using the control units. Because of extrapolation, even small changes in the specification of the model can change the estimates greatly (Ho et al. 2007, King and Zeng 2007).

We can estimate the treatment effect on the whole data. We use regression with two specifications:

a linear specification: $y = \alpha + \delta t \beta x + \epsilon$, and
a quadratic one: $y = \alpha + \delta t + \beta x + \gamma x^2 + \epsilon$

The linear model gives us an estimate of the treatment effect equal to 0.06. On the other hand, the treatment effect estimated from the quadratic model is -0.02. Here the linear regression is way off the mark, while the quadratic specification is closer, but we have model dependence.

We now restrict ourselves to the observations with x values that are common for both the treated and control groups (Fig. 8.3). The control units that have x values less than 5 or x values more than 20 are at a distance from the treated observations,

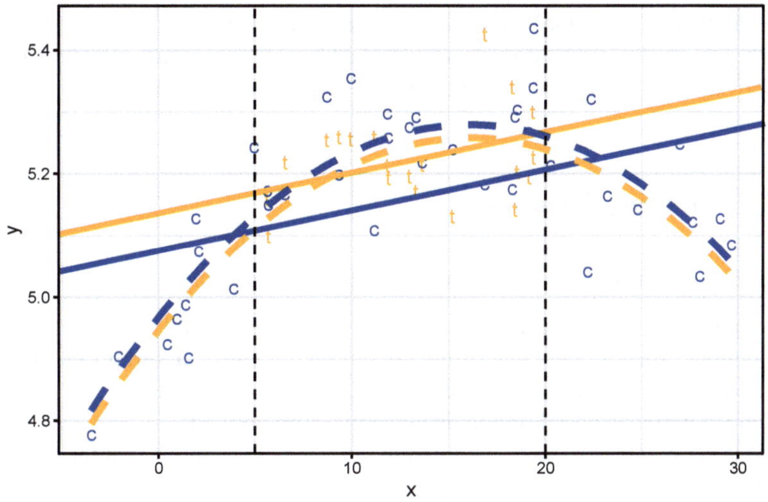

Fig. 8.2 Entire span of data is used even though there are no treatment group observations for $x < 5$ and $x > 20$. This leads to model dependence. Source: Authors' estimation based on simulated data

which all have x values between 5 and 20. Essentially, by restricting the observations in this way, we are matching treatment units to control units that are close to them.

For both the linear and quadratic model specifications, we get a treatment effect of –0.03, which is the true effect by construction. We can again plot and see what has changed now.

Estimating using treated and control groups only in the region of overlap has reduced model dependence and bias (Fig. 8.3). In our example, we had just one regressor. In practice, we have a large number of regressors. In such cases, matching

Fig. 8.3 Data is used only where there is overlap in the control and treatment observations. So we do not have model dependence, and both models provide good estimates of the true effect. Source: Authors' estimation based on simulated data

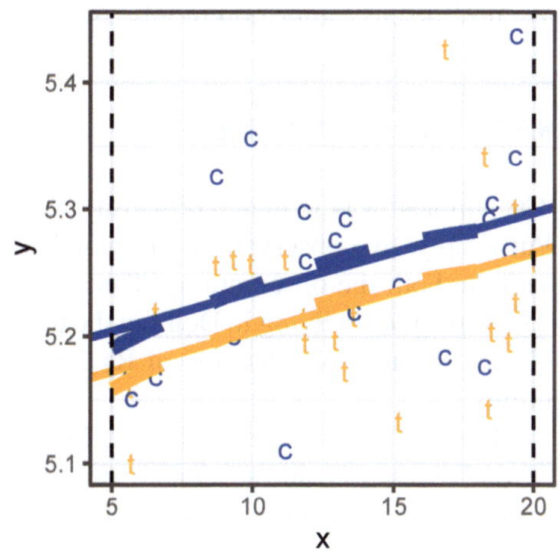

can be employed to 'prune' our data and obtain similar treated and control observations for our analysis. Once we have matched our data, we can either compare the means of the two groups or use a parametric model to estimate the causal effects.

8.3.2 *Matching Methods*

Stuart (2010) points out that there are four steps in matching:

1. Define a distance measure to be used, once the variables to be used in the matching procedure are determined based on our causal assumptions. The lower the 'distance' between two observations, the better the match.
2. Identify matches based on the distance measure defined.
3. See if the matched sample obtained is satisfactory. If not, repeat the above steps till a satisfactorily matched sample is achieved.
4. Estimate the treatment effect using the matched sample obtained from the above steps.

In the matching workflow, we first aim to achieve good balance using relevant regressors. Until we achieve good balance, we do not work with the outcome. Once we have achieved balance, we then proceed to estimation. So even if regression is used, because it is based on balance already achieved, the level of model dependence will be low. This is in contrast to a typical exercise that uses regression from the start. The researcher tries out several specifications. The researcher may report only a subset of the specifications, often based on the results in terms of the coefficient and statistical significance. This means there is researcher discretion and model dependence.

Matching methods differ from each other based on how the distance measure is defined and how the particular distance measure is used to arrive at a set of matches. We now discuss some commonly used matching methods.

8.3.2.1 Exact Matching

The simplest way to match observations is to match exactly based on the regressors. Put simply, two observations are matched only if the value of each regressor is the same in both the observations. Thus, a 30-year-old male with a past income of Rs 25000 per month living in Delhi in the treatment group will get matched to a 30-year-old male with a past income of Rs 25000 per month living in Delhi in the control group.

However, in practice, exact matching is difficult to implement because we have several regressors, and several are continuous. This implies that exact matching may result in few matches or none at all.

8.3.2.2 Propensity Score Matching

A commonly used distance measure is an estimate of the propensity score (Rosenbaum and Rubin 1983). The propensity score is the conditional probability of assignment to treatment given the regressors. That is, propensity score for an observation i can be denoted by

$$e_i(X_i) = Pr(T_i = 1|X_i) \tag{8.6}$$

Propensity score is a balancing score. Basically, instead of trying to match on several regressors, we can match on the propensity score. In practice, with observational data, propensity scores have to be estimated. Usually, a logistic regression is used to estimate the propensity score.

King and Nielsen (2019) contend that propensity score matching may, instead of reducing regressor balance, increase it, leading to bias and greater model dependence. We need to make sure that regressor balance improves after propensity score matching before going ahead with estimation of effects.

8.3.2.3 Genetic Matching

The researcher needs to check whether matching on the estimated propensity scores leads to good balance, and revise the model accordingly. It is not easy for a researcher to keep improving regressor balance. Genetic matching uses an evolutionary search algorithm which iteratively checks and improves regressor balance (Diamond and Sekhon 2013). Genetic matching (GenMatch) maximizes the balance of observed regressors between treated and control groups by searching among a set of distance metrics. The following generalized version of Mahalanobis distance is[4]:

$$GMD(X_i, X_j, W) = \sqrt{(X_i - X_j)'S^{-\frac{1}{2}}WS^{-\frac{1}{2}}(X_i - X_j)} \tag{8.8}$$

where W denotes a weight matrix and S denotes the variance-covariance matrix of X[5].

---

[4] Mahalanobis distance builds on the Euclidean distance. Euclidean distance is a measure of distance between two vectors in a Euclidean space. The Euclidean distance between two vectors is given by

$$D_{ij} = \sqrt{\Sigma_i^n (X_i - X_j)^2} \tag{8.7}$$

The Euclidean distance is sensitive to scale and units of the variables involved. The Mahalanobis distance builds on the Euclidean distance and is defined as follows:

$$D_{ij} = \{(X_i - X_j)'S^{-1}(X_i - X_j)\}^{1/2} \tag{8.9}$$

where S is the sample covariance matrix of X (Sekhon 2009).

[5] GenMatch tries to achieve balance by minimizing the largest individual discrepancy, based on p-values from Kolmogorov-Smirnov tests and paired t-tests for all matched variables (Diamond and

Having discussed some popular matching methods, we now move on to illustrate propensity score and genetic matching using R with data used by Dehejia and Wahba (1999)[6]. The following section provides some details on this dataset.

8.4 Empirical Example

8.4.1 The National Supported Work Program Data

A well-conducted randomized experiment can provide valid estimates of causal effects. But can we get valid estimates from observational data? LaLonde (1986) discussed issues in the econometric evaluation of training programs with observational data. The author assessed how well different econometric methods estimated causal effects with observational data by comparing them to the estimates from a randomized experiment.

LaLonde (1986) used data from the National Supported Work (NSW) program. The government-funded NSW program provided job training and work experience to disadvantaged individuals. The assignment of training under the NSW was random. The individuals in the treatment group received training while those in the control group received no benefits. LaLonde (1986) used the experimental controls to first estimate the effect of training on earnings for males and females separately.

Having estimated the experimental benchmark, LaLonde (1986) created an observational dataset. The author sets aside the experimental control group and used a non-experimental control group. The control groups were drawn from the Panel Study of Income Dynamics (PSID) and Current Population Survey-Social Security Administration File (CPS). Regression-based methods were used with the observational dataset to adjust for observed characteristics of treatment and control groups. LaLonde (1986) found that econometric estimation did not always replicate the experimental estimates.

Dehejia and Wahba (1999) used a male sub-sample of the dataset used by LaLonde (1986). They used propensity score matching to estimate the treatment effect of the training program. They found that the propensity score-based estimates were close to the experimental estimates of LaLonde (1986). We use the male sub-sample of the CPS dataset used by Dehejia and Wahba (1999). The dataset contains pre-treatment variables such as pre-intervention earnings, education, age, ethnicity, and marital status, along with the outcome and treatment.

Sekhon 2013). The algorithm starts with the generation of a batch of weights (*Ws*) of a fixed size (pop.size). With each *W* corresponding to a different distance metric as per equation (8.9), we get a matched sample for each of these weights in the batch. Each of the matched samples is evaluated to identify the lowest value of the loss function. The number of generations is determined by the choice of limits on the maximum number of generations as well as the number of generations after which if there is no improvement in the objective function, optimization will stop (Sekhon 2011).

[6] which in turn relied on the data from LaLonde (1986)

8.4.2 *Using R*

We illustrate the use of matching to estimate the treatment effect of the training program. As mentioned earlier, the dataset used is the same as Dehejia and Wahba (1999). First, the NSW data is used to establish the experimental estimate of the treatment effect. We then use propensity score matching and genetic matching to estimate the treatment effects with non-experimental data.

8.4.2.1 Experimental Benchmark Estimates

We use data available on a website.[7] We first read the stata file and then write it to a csv file. We do this once and subsequently work with the csv file stored on the computer. Note that we use `typewriter` font for R code. Text after # is a comment.

```
# load packages
library(haven)
library(tidyverse)
# read file from website
data <- read_dta("http://www.nber.org/~rdehejia/data/nsw_dw.dta")
# write file as csv
write_csv(data, "data_nsw.csv")
```

The dataframe `data_nsw` is data from the experimental evaluation of a labor training program, the NSW Demonstration. This program provided work experience for six to eighteen months to individuals. The participants were randomly selected.

```
# read file from computer
library(readr)
data_nsw <- read_csv("data_nsw.csv")
```

We look at the dimensions of the data and the first four rows and columns.

[7] https://users.nber.org/~rdehejia/.

```
# dimensions
dim(data_nsw)
## [1] 445  11
# first four rows and columns
data_nsw[1:4, 1:4]
## # A tibble: 4 x 4
##   data_id                treat   age education
##   <chr>                  <dbl> <dbl>     <dbl>
## 1 Dehejia-Wahba Sample      1    37        11
## 2 Dehejia-Wahba Sample      1    22         9
## 3 Dehejia-Wahba Sample      1    30        12
## 4 Dehejia-Wahba Sample      1    27        11
```

We have loaded data from the experimental evaluation. We select a few variables, grouping by treatment, and calculate means. The data is balanced in the distribution of regressors between the treatment and control group (Table 8.3).

```
# Get means of variables in data_nsw by treat
data_nsw %>%
  # select variables
  dplyr::select(treat, black, hispanic,
          married, nodegree, age, re74) %>%
  # group by treatment
  group_by(treat) %>%
  # means by treatment with summarise
  summarise(across(,mean, na.rm = TRUE)) %>%
  # make table with kable
  kable(, booktabs = T, digits = T,
  caption = "Means of variables by
  Treatment in the experimental data") %>%
  footnote("Source: Calculated by the authors using the NSW data",
  general_title=" ")
```

We regress the earnings in 1978 on the treatment to get an estimate of the effect of the program on earnings. This provides us with a benchmark average treatment effect estimate of 1794 dollars (Table 8.4).

Table 8.3 Means of variables by treatment in the experimental data

Treat	Black	Hispanic	Married	Nodegree	Age	re74
0	0.8	0.1	0.2	0.8	25.1	2107.0
1	0.8	0.1	0.2	0.7	25.8	2095.6

Source: Calculated by the authors using the NSW data

Table 8.4 Regression for dependent variable earnings in 1978 with experimental data

	Model 1
(Intercept)	4554.80
	(408.05)
treat	1794.34
	(632.85)
R^2	0.02
Num. obs.	445

Standard errors in parentheses. Source: Authors' estimation based on NSW data

```
# regress outcome on treatment
modexp1 <- lm(re78 ~ treat, data = data_nsw)
# load texreg pacakge
library(texreg)
library(tibble)
# create regression table using estimated model
texreg(list(modexp1),
  caption = "Regression for dependent variable
    earnings in 1978 with experimental data.",
       caption.above = TRUE,
       stars = numeric(0),
       include.ci = FALSE,
       include.adjrs = FALSE,
  custom.note = "Standard errors in parentheses.
  Source: Authors' estimation based on NSW data",
       label = "tab:tab_reg1")
```

8.4.2.2 Propensity Score Matching with Observational Data

We now use the observational data from the Current Population Survey-Social Security Administration File (CPS), used by Lalonde for the control group, for our illustration.

```
#Non-experimental control groups
cps <- read_dta(
  "http://www.nber.org/~rdehejia/data/cps_controls.dta")
```

We look at the dimensions of the CPS data and the first four rows and columns.

```
dim(cps)
## [1] 15992    11
cps[1:4, 1:4]
## # A tibble: 4 x 4
##   data_id treat   age education
##   <chr>   <dbl> <dbl>     <dbl>
## 1 CPS1        0    45        11
## 2 CPS1        0    21        14
## 3 CPS1        0    38        12
## 4 CPS1        0    48         6
```

We now remove the control group from the experimental data, we use the CPS data observations as a control group, creating the observational dataset.

```
data_cps <- data_nsw %>%
  filter(treat==1) %>% # remove control group
  rbind(cps)  # add cps data as control group
```

We now have our observational data frame, `data_cps`. We examine balance between the treatment and control group with respect to the regressors. We can see considerable imbalance in the variables black, married, and nodegree in the data (Table 8.5).

```
data_cps %>%
  dplyr::select(treat, black, hispanic, married, nodegree, age, re74) %>%
  group_by(treat) %>%
  summarise(across(, mean, na.rm = TRUE)) %>%
  kable(, booktabs = T, digits = T,
        caption =
          "Means of variables by Treatment
        before matching in CPS based data") %>%
  footnote("Source: Calculated by the authors using the CPS data",
  general_title=" ")
```

We will now use the Matching package to balance the treatment and control groups. First, we load the required package. The Matching package requires inputs of the outcome and treatment variables, which we create as separate variables.

Table 8.5 Means of variables by treatment before matching in CPS based data

Treat	Black	Hispanic	Married	Nodegree	Age	re74
0	0.1	0.1	0.7	0.3	33.2	14016.8
1	0.8	0.1	0.2	0.7	25.8	2095.6

Source: Calculated by the authors using the CPS data

```
library(Matching)
set.seed(112) # for reproducibility
# Outcome Y
Y <- data_cps$re78
# Treatment Tr
Tr <- data_cps$treat
```

For matching using propensity scores, we first need to estimate the propensity score.

```
# propensity score model (logit)
psm <- glm(Tr ~ age + education + black +
              hispanic + married + nodegree +
              re74 + re75,
family = binomial(link = "logit"), data = data_cps)
# get fitted values
ps <- psm$fitted.values
```

We can now feed the estimated propensity score `psm$fitted` into the 'Match' function for executing propensity score matching.

```
# match with estimated propensity score
match_psm <- Match(Y = Y, Tr = Tr, X = ps)
```

The Matching package has the `MatchBalance` function that provides detailed and comprehensive output related to the balance metrics before and after matching. To conserve space, the results to the command have been suppressed, but we extract and present some of the output from `MatchBalance`. Table 8.6 presents the means of select regressors for the treated and control groups after the propensity score matching exercise.

```
MatchBalance(Tr ~ age + education + black +
                 hispanic + married + nodegree +
                  re74 + re75,
                data=data_cps, digits = 1,
                match.out = match_psm)
```

Comparing the means, it seems that balance has generally improved after propensity score matching except for the `hispanic` variable for which the balance has

Table 8.6 Means of variables by Treatment after propensity score matching

treat	black	hispanic	married	nodegree	re74
0	0.8	0.1	0.2	0.7	2273
1	0.8	0.0	0.2	0.7	2096

Source: Calculated by the authors using the CPS data

slightly reduced. For continuous regressors, we can see how the distributional balance has improved between the treatment and control groups. We use the Quantile-Quantile (Q-Q) plots of the regressors. Q-Q plots help us compare the distributions of particular regressors between the treatment and control groups. A Q-Q plot is a scatter plot of the quantiles of two distributions plotted against each other. The closer the points are to the 45 degree line, the better is the balance between the two distributions. Here we focus on one of the regressors, `age`. Figure 8.4 shows the Q-Q plots for the raw data (left) as well as the matched data (right) based on the propensity score. Propensity score matching does not seem to have done a great job at balancing the `age` regressor across the distribution with the points being considerably far away from the 45 degree line in the plot even after matching.

```
# plot side by side
par(mfrow = c(1, 2))
# First Q-Q plot (raw data)
qqplot(data_cps$age[data_cps$treat==1],
       data_cps$age[data_cps$treat==0],
       xlab = "age of control group",
        ylab = "age of treatment group",
       main="Raw data")
# add line of equality
abline(coef = c(0,1), col = "brown", lty = 3, lwd=3)
# Second Q-Q plot (psm matched data)
qqplot(data_cps$age[match_psm$index.treated],
       data_cps$age[match_psm$index.control],
       xlab = "age of control group",
        ylab = "age of treatment group",
       main="Propensity Score matched data")
abline(coef = c(0,1), col = "brown", lty = 3, lwd=3)
```

Balance can be improved, in principle, but requires further refinement of the propensity score model and redoing the matching in an iterative process till better balance is achieved.

The summary command can be called to access the treatment effect estimates. We have suppressed the output.

```
summary(match_psm)
```

The estimated treatment effect of the program comes out to be around 1440 dollars with an Abadie-Imbens standard error of about 820. This implies that the average impact of the program on the treated was to increase the earnings by 1440 dollars compared to those who did not receive the treatment.

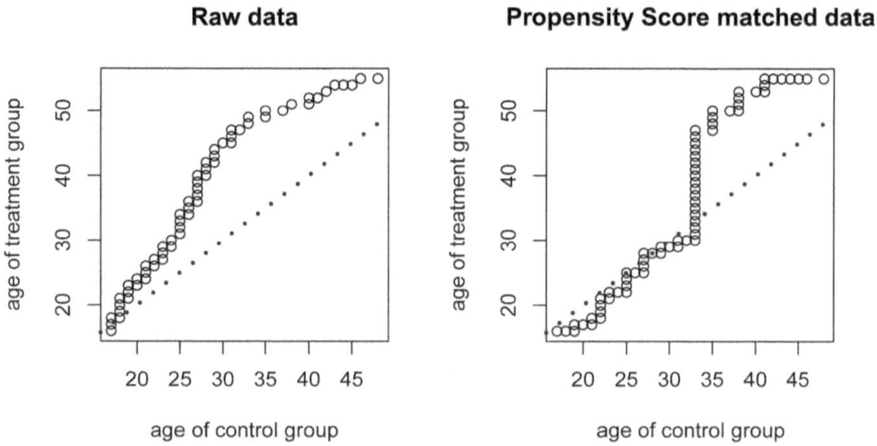

Fig. 8.4 Graphical check of balance of age between treatment and control groups. Quantile-quantile plots for the raw data (left) and the matched data (right). Although balance has improved, imbalance remains after matching. Source: Authors' calculation based on CPS data

8.4.2.3 Genetic Matching with Observational Data

As the iterative process of propensity score matching is cumbersome and time consuming, an alternative is to automate the process of improving balance as in the case of genetic matching. In this section, we demonstrate how genetic matching can be used for estimating the treatment effect in our observational data. For convenience, we use the attach function. Once the dataset is attached, we can refer to a variable directly without even providing the name of the dataset.

```
attach(data_cps)
# regressors
X <- cbind(age, education, black, hispanic,
           married, nodegree, re74, re75)
```

We now use the main function in the package, `GenMatch`, providing the treatment indicator, and the regressors to match on.

```
gen <- GenMatch(Tr = Tr, X = X, pop.size = 150)
```

We follow this up using the `Match` function.

```
match_gen <- Match(Y = Y, Tr = Tr, X = X,
                   Weight.matrix = gen)
```

We again use the `MatchBalance` function to assess our balance after genetic matching, though the output has been suppressed. We present the means of the regressors after genetic matching in the treatment and control groups in Table 8.7.

There has been a considerable improvement in the balance. Compared to propensity score matching, Table 8.7 shows that genetic matching has been able to do a better job of balancing in terms of the means of the treatment and control groups, for the variables `hispanic` and `re74`.

```
MatchBalance(Tr ~ age + education + black +
               hispanic + married + nodegree +
              re74 + re75,
             data = data_cps, digits = 1,
             match.out = match_gen)
```

Further, we can plot the Q-Q plot to see the balance across the distribution of a particular variable. Figure 8.5 presents the Q-Q plot for the `age` variable after genetic matching. It can be seen that the points are now much closer to the 45 degree line of balance compared to the raw data as well as the propensity score matched data shown earlier in Figure 8.4.

```
par(mfrow=c(1,1))
# Q-Q plot for genetic matching
qqplot(data_cps$age[match_gen$index.treated],
       data_cps$age[match_gen$index.control],
       xlab = "age of control group",
        ylab = "age of treatment group",)
abline(coef = c(0,1), col = "brown", lty = 3, lwd=3)
```

The summary command provides an estimate of the causal effect along with Abadie-Imbens standard errors.

```
summary(match_gen)
```

The estimate of the average treatment effect on the treated is 1970.5 dollars, with an Abadie-Imbens standard error equal to about 938. This estimate of the average treatment effect is higher than the estimate obtained through propensity score matching.

Table 8.7 Means of variables by Treatment after genetic matching

Treat	Black	Hispanic	Married	Nodegree	re74
0	0.8	0.1	0.2	0.7	2054
1	0.8	0.1	0.2	0.7	2096

Source: Calculated by the authors using the CPS data

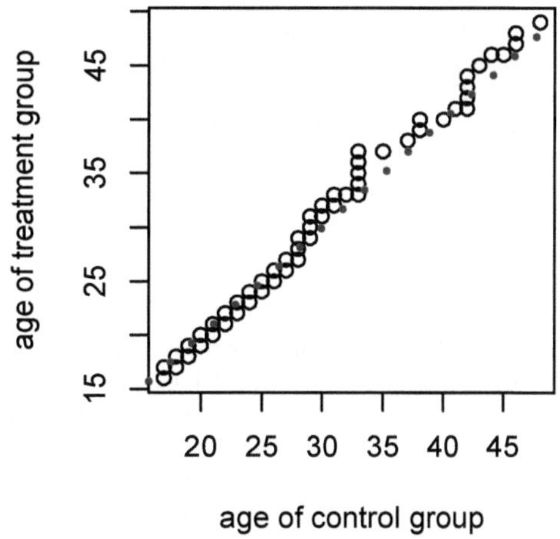

Fig. 8.5 Graphical check of balance of age between treatment and control groups. Quantile-quantile plot after genetic matching. Balance has improved compared to propensity score matching. Source: Calculated by the authors using the CPS data

8.5 Concluding Remarks

This chapter placed the spotlight on the NRPO causal model and matching methods. The chapter is a practical introduction to using the Matching package in the R software. Causal inference has some nuances, which can be appreciated by reading some of the literature cited in the chapter. Imbens and Rubin (2015) provides a detailed treatment of the NRPO model and causal inference. Sekhon and co-authors discuss the implementation (Sekhon 2011) and rationale (Diamond and Sekhon 2013) of genetic matching, and the links with the NRPO causal model (Sekhon 2009). King and Nielsen (2019) examine issues with propensity score matching. Matching will provide valid inferences if we match on the required set of regressors. VanderWeele (2019) explains this issue drawing on recent research in causal inference.

References

Abadie A, Cattaneo MD (2018) Econometric methods for program evaluation. Annu Rev Econ 10(1):465–503

Abadie A, Imbens GW (2006) Large sample properties of matching estimators for average treatment effects. Econometrica 74(1):235–267

Angrist JD, Imbens GW, Rubin DB (1996) Identification of causal effects using instrumental variables. J Am Stat Assoc 91(434):444–455

Dehejia RH, Wahba S (1999) Causal effects in nonexperimental studies: reevaluating the evaluation of training programs. J Am Stat Assoc 94(448):1053–1062

Diamond A, Sekhon JS (2013) Genetic matching for estimating causal effects: a general multivariate matching method for achieving balance in observational studies. Rev Econ Stat 95(3):932–945

Ho DE, Imai K, King G, Stuart EA (2007) Matching as nonparametric preprocessing for reducing model dependence in parametric causal inference. Polit Anal 15(3):199–236

Holland PW (1986) Statistics and causal inference. J Am Stat Assoc 81(396):945–960

Imbens GW (2004) Nonparametric estimation of average treatment effects under exogeneity: a review. Rev Econ Stat 86(1):4–29
Imbens GW, Rubin DB (2015) Causal inference for statistics, social, and biomedical sciences: an introduction. Cambridge University Press, Cambridge
Imbens GW, Wooldridge JM (2009) Recent developments in the econometrics of program evaluation. J Econ Lit 47(1):5–86
King G, Nielsen R (2019) Why propensity scores should not be used for matching. Polit Anal 27(4):435–454
King G, Zeng L (2007) When can history be our guide? The pitfalls of counterfactual inference. Int Stud Q 51(1):183–210
LaLonde RJ (1986) Evaluating the econometric evaluations of training programs with experimental data. Am Econ Rev 76(4):604–620
Rosenbaum PR, Rubin DB (1983) The central role of the propensity score in observational studies for causal effects. Biometrika 70(1):41–55
Rubin DB (2004) Teaching statistical inference for causal effects in experiments and observational studies. J Educ Behav Stat 29(3):343–367
Rubin DB (2006) Matched sampling for causal effects. Cambridge University Press, Cambridge
Sekhon JS (2008) The Neyman-Rubin model of causal inference and estimation via matching methods. The Oxford Handbook of Political Methodology 2:1–32
Sekhon JS (2009) Opiates for the matches: matching methods for causal inference. Annu Rev Polit Sci 12:487–508
Sekhon JS (2011) Multivariate and propensity score matching software with automated balance optimization: the matching package for R. Stat Softw 42(7)
Stuart EA (2010) Matching methods for causal inference: a review and a look forward. Stat Sci 25(1):1–21
VanderWeele TJ (2019) Principles of confounder selection. Eur J Epidemiol 34(3):211–219

Chapter 9
LASSO Regression and K-means Clustering: Application to Indian Labor Market

Nandana Sengupta

Abstract This chapter introduces fundamental machine learning (ML) concepts, including the bias-variance trade-off, regularization and cross-validation, contrasting them with traditional econometric approaches and illustrating their application in social sciences. It explores one technique from each of the supervised and unsupervised learning paradigms—LASSO regression and *k*-means clustering. The chapter also provides an overview of the basic principles of supervised and unsupervised learning principles while demystification commonly used terms in ML parlance such as deep learning, neural networks, random forests, recommendation systems and matrix factorization. Following the theoretical discussion, an empirical application is presented using R, based on primary data on job market outcomes and skills from 784 respondents in Bangalore, India. This application examines (i) the relationship between skill claims and job outcomes in informal work, and (ii) the clustering of skills among informal workers. The chapter concludes with a summary and outlines potential research directions for applied researchers interested in working at the intersection of ML and econometrics.

9.1 Glossary of Useful Terms

- *Dataset or sample:* It refers to tabular data where rows correspond to different entities (for example: individual respondents, documents, firms, etc.) and columns correspond to characteristics of these entities (for example: age, number of words, revenue, etc.). In the discussion within this paper, the entries in the dataset are assumed to be numerical valued. A specific row is typically denoted by the subscript i and a specific column is typically denoted by the subscript j.

Supplementary Information The online version contains supplementary material available at https://doi.org/10.1007/978-981-97-9060-9_9.

N. Sengupta (✉)
School of Public Policy, Indian Institute of Technology Delhi, Delhi, India
e-mail: nandana.sengupta@sopp.iitd.ac.in

F. Hamid and D. Mukherjee (eds.), *Data Science for Modeling Managerial and Socioeconomic Problems*, Contributions to Management Science,
https://doi.org/10.1007/978-981-97-9060-9_9

- *Outcome of interest:* It refers to a specific characteristic that the analyst is interested in studying or predicting. In ML parlance, this is often referred to as *'target' or 'label'*. The outcome of interest is typically denoted by y and y_i typically refers to the value of the outcome of interest for a specific row i.
- *Covariates:* It refers to the other characteristics that the analyst utilizes to study or predict the target. In ML parlance, these are often referred to as *'features'*. Covariates are typically denoted by x, and x_j refers to the jth covariate and x_{ij} refers to the jth covariate corresponding to the ith entity.
- *Model:* refers to the set of mathematical rules or functions that define the system being studied. For instance a model where the target variable is some function of the features and some parameters can be represented as follows:

$$y = f(x_1, x_2, \cdots x_j; \beta).$$

- *Parameters:* It refers to the constant values that, along with the functional form, define the model. Parameters often have a specific interpretation in econometric models and obtaining the best estimate of the key model parameters is usually the main aim of econometrics. In this chapter, the model parameters will be denoted by β and estimated values by $\hat{\beta}$.
- *Model fitting or estimation:* Given the data and a model, the optimal model parameters are obtained via a process known as model fitting or estimation. This involves optimizing a loss function defined for the specific model. For instance, in the case of a simple linear regression, the model is defined as $y_i = \beta_0 + \beta_1 x_i + \varepsilon_i$ where y is the outcome of interest, x is a covariate, ε captures random error, and β_0, β_1 are the model parameters. The model is fit by minimizing the least squared errors loss function:

$$\underset{\beta_0, \beta_1}{\text{Min}} \sum_{i=1}^{N} (y_i - (\beta_0 + \beta_1 x_i))^2$$

where N is the total number of observation. The β_0, β_1 values that minimize the loss function given the data are referred to as the parameter estimates, denoted by $\hat{\beta}_0, \hat{\beta}_1$ and the fitted model is denoted by $\hat{y} = \hat{\beta}_0 + \hat{\beta}_1 x$. Note that since they depend on the data used for model fitting, estimated parameter values are not fixed and vary with the data.
- *Bias of estimated parameter:* Assuming the model used for estimation is correct (i.e. it is the 'true' model) and the corresponding true value of the parameter is β, the bias of the estimated parameter $\hat{\beta}$ is defined as:

$$b(\hat{\beta}) = E(\hat{\beta}) - \beta.$$

The bias is a measure of how far the estimate is from the 'true' parameter value. A bias of zero corresponds to an unbiased estimate with an expected value that coincides with the true parameter value.

- *Variance of estimated parameter:* The variance of the estimated parameter $\hat{\beta}$ is defined as:

$$Var(\hat{\beta}) = E\big(\hat{\beta} - E(\hat{\beta})\big)^2.$$

  The variance is a measure of the stability of the estimate. The lower the variance, the higher is the estimate stability.
- *Out-of-sample prediction:* During model fitting, the aim is to optimize the loss function on the given data, but often analysts are interested in measuring how well the estimated model fits new data, which is not part of the original sample. This is known as out of sample prediction. If a model fits the sample data very well but has poor out of sample prediction, the model is said to suffer from the problem of 'overfitting'.

9.2 Introduction

Quantitative public policy and empirical social sciences have long been dominated by applied econometric methods. As ML techniques become more popular among data scientists and as computers become capable of storing and processing large quantities of data, there have been some recent efforts to incorporate them in the analysis of social and economic phenomena. This chapter aims to provide readers with an introduction to fundamental ML concepts as well as an understanding of both the opportunities and limitations of using them in social science applications.

The chapter starts by asking the following key questions. (1) How is ML different from econometrics? (2) Can ML ideas be meaningfully incorporated into social science? (3) What are some limitations of using ML techniques in social science applications? Let's address each of these questions separately.

How are ML techniques different from econometrics?

The major differences between econometric and ML models can be understood by considering the differences in the underlying models, the properties of the estimated parameters and the typical datasets on which these models are run.

- *Underlying model:* The main focus in econometrics is to provide an explanation of various observed outcomes. The estimation process is based on conditions implied by economic theory and thus the parameters in the model have an economic or social interpretation. On the other hand, the main focus in ML is to provide accurate predictions of the outcomes of interest. While these techniques are extremely powerful for forecasting, it can be very hard to interpret the underlying structure implied by them and the models used to predict outcomes are often a black box.
- *Properties of parameter estimates:* Econometricians are primarily interested in obtaining *unbiased* estimates of the parameters of interest and then choosing

those estimation techniques which lead to lowest variance among the set of unbiased estimators (i.e. $b(\hat{\beta}) = 0$). Since ML models are typically a black box, the parameters associated with these models do not typically have a meaning in the real-world context. ML researchers are primarily interested in obtaining more stable and accurate estimates of the outcome of interest. They are willing to add in some bias to the estimation model, if it leads to a dramatic reduction in the variance. This process is known as 'regularization' (Hastie et al. 2009).

- *Datasets:* Econometricians tend to use datasets which are comparatively smaller and are usually fully observed. In case of missing entries in the datasets, most econometricians either restrict their analysis to fully observed subsets of the data or obtain imputed values of the missing data *before* running their models on the data. In contrast, ML models can handle larger, more messy datasets which are often sparse, i.e. contain a large proportion of missing values.

Can ML techniques be meaningfully incorporated in social science?

The preceding discussion on the differences between ML and econometric techniques provides some pointers on answering this question. ML models are *not* a substitute for econometrics, given that the underlying models are often black boxes. Standard econometric models are more relevant for social scientists interested in learning about the processes and explanations of social and economic phenomena. Having said that, ML models are powerful at predictions and thus can be very useful tools for social scientists interested in forecasting future outcomes and obtaining accurate imputations of out of sample and missing data.

The article 'AI vs MD' (Mukherjee 2017) provides an intuitive framework for understanding the kinds of questions ML models are suited for within the context of medical science. The author argues that knowledge can be divided into three broad categories of questions: knowing *what*, knowing *how* and knowing *why*. Existing mainstream ML methods are extremely powerful tools for answering the first category of questions: *knowing what*[1] and a number of interdisciplinary research studies have successfully utilized ML models for predictions in social science applications. Shmueli (2010) provides an in-depth discussion on the important distinction between explanation and prediction. Some examples of such studies include poverty mapping at scale using novel sources of data like remote sensing (Xie et al. 2016) and mobile phone metadata (Blumenstock et al. 2015). An area where ML models have been used is in the prediction of economic behaviour in specific contexts like predicting teacher effectiveness or predicting police misconduct (Chalfin et al. 2016). Another application area is the analysis of text data to infer underlying categories or clusters, see, for example, Bond et al. (2012); Rutherford et al. (2018). This list is by no means exhaustive and is intended to give readers a flavour of some applied papers utilizing ML methods in social science settings. While these are all examples of applied papers, there is also an emerging literature on incorporating ML techniques

[1] 'Explainable AI' is a fast evolving area of research which is pushing the boundaries on the last 2 categories of questions.

for methodological advances for social scientists including work Wager and Athey (2018), Taddy (2013), Sengupta and Sowell (2020) and others.

What are some limitations of using ML techniques in social science applications?

While ML models may prove very useful in obtaining predictions and forecasts, they have limited usefulness when it comes to providing explanations or uncovering underlying mechanism in social science contexts. Thus, attempting to draw causal interpretations from 'off the shelf' ML models is not recommended. Further, whereas the modelling of the data generating process to account for issues like selection bias has a rich history of research within econometrics, the identification and mitigation of such biases is a nascent area of research within machine learning.[2]

With this preliminary understanding of how ML models fit into the social science discourse, the chapter now switches gears to more technical details of ML methods. The rest of the chapter is organized as follows. Section 9.3 elaborates on some basics of ML methods. Sections 9.4 and 9.5 provide a detailed discussion on Supervised and Unsupervised ML methods. A practical demonstration of introductory ML of LASSO and k-means clustering in `R` is presented in Sect. 9.6 using primary data from India on job market outcomes and skills. Section 9.7 concludes the discussion with a summary and a guide to references on advanced topics at the intersection of ML and econometrics.

9.3 Basics of Machine Learning

A few basic technical concepts used in ML models are elaborated in this section and these are contrasted with standard practice in econometric models.[3]

9.3.1 Mean Squared Error

The objective of a typical ML model is to provide better forecasts on out of sample data. This objective is served by minimizing what is known as the mean squared error or MSE of parameter estimates. Technically, MSE is defined as:

$$MSE(\hat{\beta}) = E[(\hat{\beta} - \beta)^2]$$

[2] This includes questions of fairness, accountability and transparency in data collection and model prediction. Research in this area is nascent and outside of the purview of this chapter.

[3] Readers are referred to the glossary of terms at the start of the chapter for detailed notations and definitions of terms that will be used in Sects. 9.3–9.5.

where β refers to the true value of the parameter and $\hat{\beta}$ refers to the estimated value. Note that the 'true' value of the parameter is a theoretical concept and can never be known in any practical setting.[4]

In contrast, the objective of a typical econometric model is to provide the best 'fit' on the given sample data. This objective is served within the regression framework by minimizing as the least squares error or LSE which is defined as:

$$LSE(\hat{\beta}) = \sum_{i=1}^{N} [(y_i - \hat{y}_i(\hat{\beta}))^2].$$

9.3.2 Bias-Variance Trade-off

MSE can be shown to be mathematically equal to the sum of variance and bias-squared of the parameter estimates[5]:

$$MSE(\hat{\beta}) = Var(\hat{\beta}) + b(\hat{\beta})^2.$$

This implies that biased parameter estimates can lead to *lower* MSE if the associated variance is drastically reduced in comparison with an unbiased parameter estimate. This property of the MSE is known as the 'bias variance trade-off'. In fact, most ML models contain a term that biases the estimate towards a pre-set prior value. In adding such a 'prior', these models lead to a reduction in variance and thus provide more stable estimates.

In contrast, econometric models aim to obtain unbiased and efficient estimates of the parameters of interest. Efficient estimates are those that have the lowest variance among the set of unbiased estimators.

---

[4] One can, however, assume true parameter values in simulation settings and use such simulations to evaluate various models. Simulation studies on MSE are beyond the scope of this discussion.

[5] MSE equals sum of variance and bias-squared:

$$\begin{aligned}
MSE &= E[(\hat{\beta} - \beta)^2] \\
&= E[\hat{\beta} - E(\hat{\beta}) + E(\hat{\beta}) - \beta]^2 \\
&= E[\hat{\beta} - E(\hat{\beta}) + b(\hat{\beta})]^2 \\
&= E[\beta - E(\hat{\beta})]^2 + [b(\hat{\beta})]^2 + 2b(\hat{\beta})E[\hat{\beta} - E(\hat{\beta})] \\
&= Var(\hat{\beta}) + b(\hat{\beta})^2
\end{aligned}$$

9.3.3 Regularization and Tuning Parameters

In practice, the introduction of a bias term is done by augmenting the standard loss function with a 'regularization' term which penalizes estimates which are far from the prior. The following is an example of a loss function to be minimized which is augmented by a regularization term.

$$\mathcal{L}(\beta) = \sum_{i=1}^{N}\left(y_i - \hat{y}_i(\hat{\beta})\right)^2 + \lambda\hat{\beta}^2$$

where the first term on the right hand side is a standard least squared error loss function. The second term is a regularization term which prevents overfitting by penalizing values of $\hat{\beta}$ with larger magnitude (since the second term is minimized when $\hat{\beta} = 0$). λ is referred to as a 'tuning parameter' which controls the relative weight between the original loss function and the regularization penalty.

In contrast, the practice of regularization is not followed in standard econometric models, since the aim is to obtain unbiased estimates of the parameters of interest.

9.3.4 Training and Validation Sets

Given that the loss functions corresponding to ML models contain a regularization term, the estimated parameters of interest are a function of the tuning parameter λ. In practice, the tuning parameter is typically selected via a data-driven procedure known as validation.

The basic idea of validation is to leave out some of the data when fitting the model. The data on which the model is fit is known as the 'training' data and the data that is left out is known as the 'validation' set. The value of the tuning parameter λ that minimizes the error on the validation set is picked as the optimal tuning parameter value. This procedure provides a data-driven way to obtain models with superior performance on out of sample data. Many ML models use an extension of this basic validation process by averaging over k random splits of training and validation sets. This process is known as k fold cross-validation.

In contrast, standard econometric models typically use all the data in the sample, leaving no data out to measure predictive ability, since their aim is to find a model that fits the given data well, rather than a model which predicts out of sample data well.

9.3.5 *Supervised and Unsupervised Models*

Traditional ML models are broadly divided into two categories[6]: supervised learning models and unsupervised learning models.

Supervised Learning Models: aim to predict the outcome of interest using some combination of covariates and parameters, along with a regularization penalty to prevent overfitting. A majority of econometric models that are commonly used in practice such as linear regression and logistic regression are analogous to supervised learning models. In this chapter, the supervised learning models of LASSO regression will be discussed in detail whereas other related methods will be introduced.

Unsupervised Learning Models: aim to infer structure in datasets where there is no outcome of interest or target variable and all variables are considered to be covariates or features. Techniques such as principal component analysis and singular value decomposition are analogous to unsupervised learning models since they are used to infer latent dimensions within a dataset. In this chapter, the unsupervised learning technique of k-means clustering will be discussed in detail whereas other related methods will be introduced.

9.4 Supervised Machine Learning

Supervised learning methods learn the relationship between covariates and the outcome of interest. The basic form of a supervised learning model is the following:

$$y = f(x_1, \cdots, x_p; \beta)$$

and the associated loss function to be minimized is of the form:

$$\mathcal{L}\big(y, f(x_1, \cdots, x_p; \beta)\big) + \lambda \cdot g(\beta)$$

where β is a vector of parameters, $f(.)$ is the model to be fit, $\mathcal{L}(.)$ is an appropriate loss function, $g(.)$ is a regularization penalty and λ is the tuning parameter. The main objective of supervised learning models is to predict out of sample values of the outcome of interest y given new values of the covariates $x_1, \cdots, x_p$.

9.4.1 *LASSO Regression*

A popular and ubiquitous supervised learning method that has found a number of applications in econometrics is the LASSO (short for Least Absolute Shrinkage

[6] There are new categories of models emerging in recent years including active learning, semi-supervised learning and reinforcement learning which are outside the purview of this chapter.

and Selection Operator) regression, which allows for simultaneous model selection and estimation. LASSO techniques have been utilized by economists for estimating stable variance covariance matrices for asset returns (Welsch and Zhou 2007), for the construction and estimation of financial networks (Hallac et al. 2017) and selecting a subset of instruments in an instrumental variable estimation (Windmeijer et al. 2019). The LASSO regression is an extension of the standard econometric Ordinary Least Squares or OLS regression model. Consider the OLS model with outcome of interest y and covariates $x_1, \cdots, x_p$:

$$y_i = \beta_0 + \beta_1 x_{1i} + \beta_2 x_{2i} + \cdots + \beta_p x_{pi} + \varepsilon_i$$

where $i = \{1, 2, \cdots, N\}$ refers to each observation in the sample. The corresponding OLS minimization loss function is:

$$\underset{\beta_0,\cdots,\beta_p}{\text{Min}} \sum_{i=1}^{N} (y_i - (\beta_0 + \cdots + \beta_p x_{pi}))^2.$$

In matrix notation, the OLS model and the associated minimization can be written as:

$$Y = X\beta + \epsilon; \ \underset{\beta}{\text{Min}} \{(Y - X\beta)'(Y - X\beta)\}$$

where $Y = \{y_1, y_2, \cdots y_N\}'$, $\beta = \{\beta_0, \beta_2, \cdots \beta_p\}'$ and $X = \{1, X_1, X_2, \cdots X_p\}$ where $X_j = \{x_{1j}, x_{2j}, \cdots x_{Nj}\}' \quad \forall j = 1, \cdots, p$. The solution of the minimization problem leads to the following closed form solution for the parameter estimates:

$$\hat{\beta} = (X'X)^{-1}X'Y.$$

This solution leads to valid estimates only when $N > p$, i.e. the number of observations is strictly more than the number of covariates p. Further, the problem of *multicollinearity* occurs if a subset of covariates is highly correlated with each other. Multicollinearity leads to a situation of unstable parameter estimates associated with high variances.

The LASSO regression augments the standard OLS loss function with a regularization penalty term as follows:

$$\underset{\beta_0,\cdots,\beta_p}{\text{Min}} \sum_{i=1}^{N} (y_i - (\beta_0 + \cdots + \beta_p x_{pi}))^2 + \lambda \sum_{j=1}^{p} |\beta_j|$$

where the first term is the standard OLS loss function and the second term is the regularization penalty for LASSO. Note that any value of $\beta_j \neq 0$ (i.e. $|\beta_j| > 0$) increases the value of the minimization function, and thus the second term penalizes solutions which are far from 0.

The recommended estimation procedure for LASSO takes place in two steps. In the first step, the optimal value $\hat{\lambda}$ is typically selected via cross-validation. In the second step, the final parameter estimate $\hat{\beta}$ is obtained by minimizing the LASSO loss function after plugging in the value of $\hat{\lambda}$ from the first step.[7] In practice, both the steps above are in-built into a number of statistical programming toolkits, as we will see below.

LASSO regression is preferred over OLS by analysts when the number of covariates or features $\{X_1, \cdots X_p\}$ is large and there is not sufficient theory to pre-select a subset of covariates. In fact, unlike OLS regression, LASSO regression produces a result even in the case that $p > N$, i.e. the number of covariates is *higher* than the number of observations. The LASSO regression is therefore described as a supervised learning model that does simultaneous estimation and covariate or feature selection. This property also makes LASSO attractive in situations where OLS may suffer from multicollinearity issues.

It should be noted, however, that LASSO estimates are inherently biased towards 0 as long as $\hat{\lambda} > 0$ since the regularization term penalizes movement away from the origin. Further, variances associated with LASSO parameter estimates do not have a closed form solution and are non-standard (Osborne et al. 2000). Finally, the covariates selected by LASSO should not be viewed as 'causally' related to the outcome of interest, since the usual caveat of *'correlation does not imply causation* still holds.

9.4.1.1 Geometry of LASSO (Advanced)

For advanced readers interested in the mechanism behind LASSO regression's covariate or feature selection property, it helps to visualize the geometry of the LASSO minimization problem. Consider Fig. 9.1 which represents the geometry of LASSO when there are 2 covariates X_1 and X_2. The 2 axes represent possible values that parameters β_1 and β_2 can take. The red ellipses represent the contour plots of the OLS loss function, which takes its minimum at the point $\hat{\beta}_{ols}$. The blue region depicts the constraint sets associated with a given value of λ. The intersection of the contour plots and the constraint set represents the LASSO estimate $\hat{\beta}_{lasso}$ for the given value of λ. The nature of the constraint set leads to some LASSO estimates being set to 0. Note that the LASSO estimate $\hat{\beta}_{1,lasso} = 0$ and $\hat{\beta}_{2,lasso} < \hat{\beta}_{2,ols}$ leading to the selection of covariate X_2 in the final model.

9.4.2 *Other Related Models*

While the LASSO regression discussed above is a supervised learning method which has found a number of applications in social sciences, there are a number of related

[7] For a more detailed discussion please refer to Hastie et al. (2009).

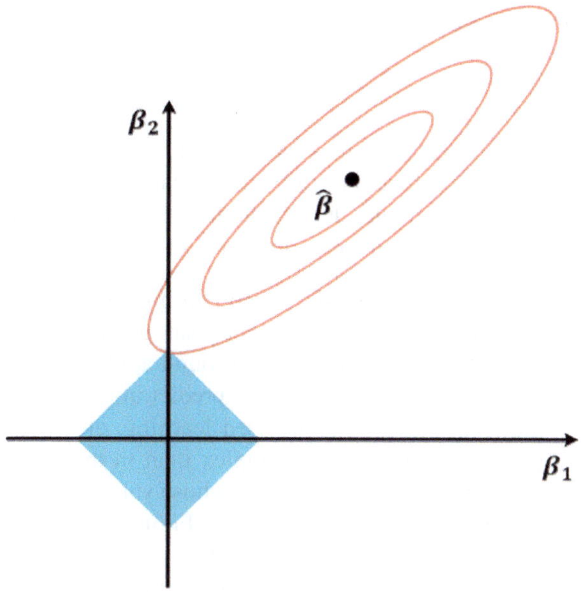

Fig. 9.1 Geometry of LASSO regression

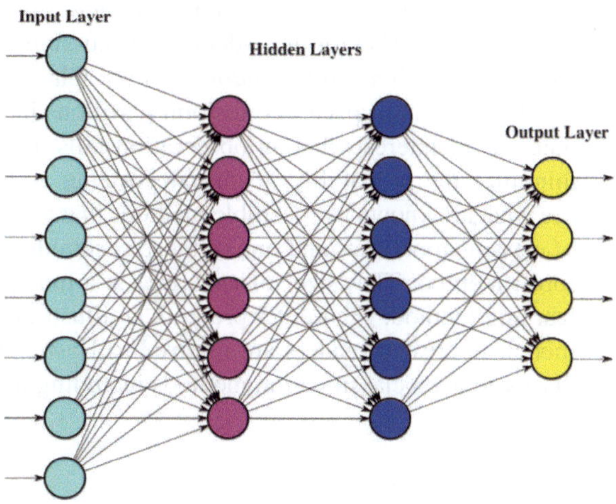

Fig. 9.2 Illustrative neural network model

supervised learning techniques which are becoming increasingly popular especially within the computational social science literature. These techniques include techniques related to LASSO such as ridge regression (Marquardt and Snee 1975) and elastic net regression (Zou and Hastie 2005) as well as some more involved techniques such as artificial neural networks (Hassoun 1995) and support vector machines (Kecman 2005), among others. A detailed discussion on each of these is beyond the purview of this chapter; however, we introduced readers to two popular techniques whose use has become ubiquitous in recent years across multiple applications. Recall that the loss function to be minimized for supervised learning methods is of the form:

$$\mathcal{L}\big(y, f(x_1, \cdots, x_p; \beta)\big) + \lambda \cdot g(\beta).$$

The following techniques broadly fit this objective, albeit with more complex loss and regularization functions, as well as a larger number of tuning parameters. Solutions from both techniques depend on the values of the tuning parameters, which are often set using validation techniques, like in the case of the LASSO regression.

- *Deep Neural Networks:* Deep Learning and Neural Network models have gained a lot of prominence in recent years given their powerful predictive abilities. In particular, these models have found application in a number of computer vision problems where the aim is to predict image labels using image inputs. Here too, the idea is that of supervised learning, albeit the associated loss function and model are much more complex than in the LASSO regression and there are a number of tuning parameters whose values are either manually set by the analyst or selected via cross-validation. Figure 9.2 presents an illustrative example of a neural network with inputs (or covariates) on the left hand side, a network of linkages and hidden layers corresponding to the underlying model in the middle and the output layer (or outcome of interest) on the right. Each layer is typically associated with at least one tuning parameter. Interested readers are referred to Samek et al. (2021) for more detailed discussions on deep neural networks.
- *Classification Trees and Random Forests:* Classification or decision trees partition the covariate space into disjoint partitions in order to predict the 'class' (or outcome of interest) that an observation falls into. The objective is to typically minimize the probability of miscategorizing an observation. Most statisticians use what are known as ensemble methods that average predictions from a number of classification trees and one of these ensemble methods is known as a random forest method. The 'size' of each tree is a tuning parameter of this supervised learning method. Figure 9.5 presents an illustrative example of a random forest model. Interested readers are referred to Hastie et al. (2009) for more detailed discussions on classification methods including trees and random forests.

9.5 Unsupervised Machine Learning

Unsupervised learning models infer structure in datasets where there is no 'target' variable and all variables are considered to be features. Such models are useful for exploratory data analysis and 'making sense' of large unstructured datasets by projecting observations from a high-dimensional feature space $\{X_1, X_2, \cdots X_p\}$ (where p is large) onto a lower k-dimensional subspace ($k << p$) based on latent similarities.

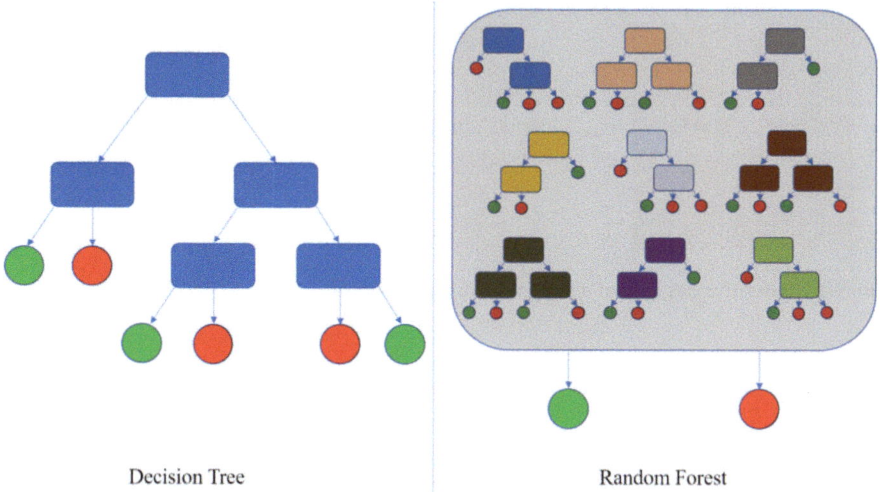

Fig. 9.3 Illustrative classification tree and random forest models

The method of dividing the data points into similar groups based on their various features is referred to as 'clustering'. As the data points are unlabeled, clustering relies on unsupervised machine learning. The simplest and most intuitive form of clustering, and therefore a good starting point is the k-means clustering algorithm (Hartigan and Wong 1979) which is described in more detail next.

9.5.1 k-means Clustering

The k-means clustering algorithm partitions all observations into k clusters $(C_1, C_2, \cdots C_k)$ around $k << p$ central points $(\mu_1, \mu_2, \cdots, \mu_k)$ based on the value of features $\{X_1, X_2, \cdots X_p\}$. Social scientists have applied this technique in contexts ranging from clustering online shopping consumers (Kim and Ahn 2008) to clustering types of behaviour in addiction research (Mak et al. 2019) and online opinion mining (Riaz et al. 2019).

The k-means clustering algorithm is an intuitive algorithm which aims to find k clusters such that the sum of the squares of the euclidean distance between each observation vector $x^{(i)}$ and the cluster centre it is assigned to, $\mu^{(i)}$, is minimized. Note that $(\mu_1, \mu_2, \cdots, \mu_k)$ refer to the list of final cluster centres that the algorithm produces whereas $\mu^{(i)}$ refers to the interim cluster centre that an observation i is assigned to at a given iteration.

The algorithm then consists of the following two repeated steps:

Step 1: Cluster assignment. Starting with a set of randomly initialized cluster centres, each observation is assigned to a single cluster such that the following cluster assignment loss function is minimized.

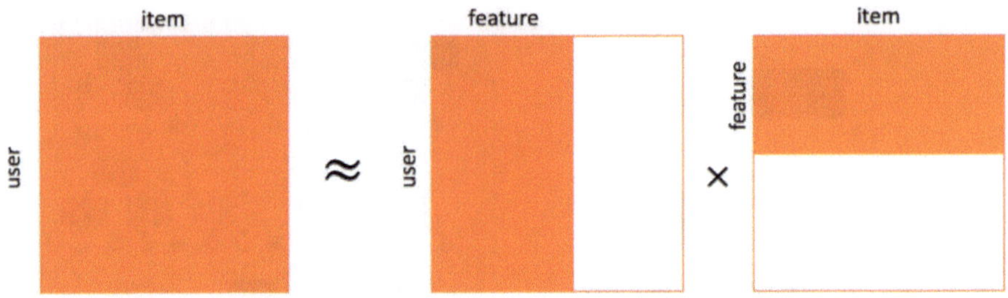

Fig. 9.4 Illustrative example of a matrix factorization model

$$\min_{\mu_1,\cdots\mu_k} \frac{1}{N} \sum_{i=1}^{N} ||x^{(i)} - \mu^{(i)}||^2 \quad \text{where} \quad \mu^{(i)} \in (\mu_1, \cdots, \mu_k).$$

Step 2: Updating cluster centres. In the next step, the cluster centres are updated by taking the mean of the covariate vectors of all observations in a given cluster.

$$\mu_j = \frac{1}{|C_j|} \sum_{x^{(i)} \in C_j} x^{(i)} \qquad \forall \; j \in \{1, 2, \cdots k\}.$$

Steps 1 and 2 are repeated until convergence[8] or until a maximum number of iterations are reached. Apart from the dataset, the primary input to the k-means clustering algorithm is the total number of appropriate clusters, k. This number itself can be estimated using the 'gap-statistic' which compares changes in within-cluster dispersion. Further details of this heuristic are beyond the scope of the discussion here. Interested readers are referred to Tibshirani et al. (2001) for a deeper dive into the topic.

9.5.2 Other Related Methods

k-means clustering is a simple and useful technique for finding patterns and for exploratory analysis. However, the underlying idea of taking a set of observations in a high-dimensional space and grouping them into a smaller list is related to a key idea in unsupervised ML models: project high-dimension objects to lower-dimensional spaces based on similarities. This idea becomes especially meaningful in the context of the 'digital data economy' where a large number of consumer characteristics are often collected by tech firms, and these need to be condensed into easier to understand groupings. While a detailed discussion on these topics is beyond the scope of this

[8] Convergence here implies that after subsequent iterations the cluster centres become stable, i.e. each update leads to a set of cluster centres which are within some small distance ϵ of the previous set of cluster centres.

chapter, readers are provided with a preliminary overview of some of these methods next.

- *Matrix Factorization:* As the name suggests, matrix factorization techniques factorize tabular data of interest, say $X \in \mathcal{R}^{N \times P}$ into two lower-dimensional component matrices $U \in \mathcal{R}^{N \times K}$, $V \in \mathcal{R}^{K \times P}$ such that the product UV recovers the original matrix, as illustrated in Fig. 9.4 where the rows represent a set of users and columns represent items over which user ratings are elicited. Matrix factorization models simultaneously take into account similarities across observations and similarities across covariates or features. Note the analogy to k-means clustering where the idea is to find k clusters to which each observation is assigned based on similarities in only the feature space. Matrix factorization can be thought of as a two-dimensional extension of k-means clustering which simultaneously finds observations which are similar in the feature space as well as finding features that are similar in the space of observations. In fact, it has been shown that k-means clustering can be expressed as a special case of matrix factorization (Ding et al. 2005).[9] Matrix factorization methods are not new to social science. Principal component analysis (PCA) and singular value decomposition (SVD) are types of matrix factorization techniques which have long been in use by econometricians. However, the ML literature has augmented the standard matrix factorization models to include regularization and tuning parameters, making these models applicable to the study of large matrices with multiple missing values, also known as *sparse matrices*. Such models have proved very effective in varied settings such as recommendation systems (Nagarnaik and Thomas 2015), signal processing (Weng and Wang 2012) and document clustering (Xu et al. 2003).
- *Recommendation Systems:* Matrix factorization models are the underlying models used in recommendation systems which consist of very large and sparse matrices of content ratings from millions of users on thousands of items (for instance on online platforms). These (mostly empty) matrices are projected into lower-dimensional subspaces using matrix factorization methods and the reconstructed matrix is used to 'impute' missing ratings for providing recommendations. Interested readers are referred to Koren et al. (2009) for a review of recommendation systems literature and Sengupta et al. (2023) for extensions of the techniques to social science.
- *Topic Modelling:* Another application of unsupervised learning and matrix factorization is for text analysis. Consider a large set of documents (for example, CVs, reviews, legal texts, etc.). Each word appearing in the corpus of documents can be thought of as a feature X_j. Each document i either contains the word ($x_{ij} = 1$) or does not ($x_{ij} = 0$). This can be used to construct a very sparse matrix of words in documents. Matrix factorization methods can be utilized to project these large sparse matrices to a lower-dimensional subspace for a variety of applications, such as the popular method of 'Topic Modeling' which learns a set of abstract 'topics' in a set of documents. Interested readers are referred to Vayansky and Kumar

[9] See also Bauckhage (2015).

(2020) for a review of recommendation systems literature and Mützel (2015) for extensions of the techniques to social science.

9.6 Practical Demonstration in R

Practical demonstrations of the two algorithms discussed in detail in this chapter, LASSO regression and *k*-means clustering, are presented next. The data used in this demonstration come from a representative survey of informal workers residing in Bangalore slums collected in 2017 (Sengupta et al. 2021).

By most estimates informal workers, i.e. persons who are employed without formal contracts and typically without social security benefits, comprise over 90% of the Indian labour force (National Commission for Enterprises in the Unorganised Sector 2007). Such workers often have no formal degrees or skill certifications, although several state-sponsored skilling programmes have been implemented in recent years. In fact, it has been found that less than 5% of the Indian workforce have any formal vocational training (Mehrotra et al. 2013). Traditional econometric approaches to study worker skills rely on an enumeration of a small set of skills sourced from experts in a top-down manner. On the other hand, ethnographic research on skills in the informal sector produces a richer set of more diverse skills (see for instance Darrah (2013), Okada (2004), Garrido et al. (2010)).

While quantitative techniques produce distributions over a small set of skill attributes which are representative of larger populations, qualitative techniques provide complex representations over a large set of skill attributes but are not typically suitable for producing distributional results. By using the two ML techniques described in this chapter, we will demonstrate how such techniques can bridge this gap between traditional quantitative and qualitative approaches.

In the dataset used for the demonstration in this chapter, informally employed workers (across a wide range of sectors) provide information on their weekly wages along with self-assessments on a set of 106 skills. These 106 skills were sourced from other workers using a previous qualitative questionnaire. We use this dataset to i) find skills associated with higher weekly wages using LASSO regression and ii) map skills that co-occur using *k*-means clustering. This unique dataset is well suited for these ML applications given the information on a large number of skills assessments for each worker—this allows us to select a smaller subset that is most predictive of wages and also to group workers based on their skill profile.

9.6.1 *LASSO Regression*

The dataset for running the LASSO regression includes responses from 727 individuals on their weekly wages and their self-assessments on over 106 skills. The aim of the exercise is the identification of skill claims (as measured by the respondents' self-assessments) that are associated with higher wages. The outcome of interest is the log of weekly wages[10] while the remaining variables form the feature matrix. Note that LASSO is the appropriate tool in this context since the number of covariates or features is large (over 100) and the objective is the selection of relevant skills. The code for running this application in R is presented next.

First, start by clearing out the coding environment, loading the appropriate packages[11] and setting the working directory to the one where the data is stored.

```
#### SETTING UP THE ENVIRONMENT

# clean environment
rm(list = ls())

# read in the relevant package
library(glmnet)
library(corrplot)

#Change Working Directory
setwd("/Users/Desktop/ML Tutorial")
```

The data used for this exercise is stored in the file `DataSupervisedML.csv` which is read in next. It is always a good practice to get a sense of the data by seeing the header row, listing out the column names and summarizing the columns of the data.

```
# read in the data
dataSuper = read.csv("DataSupervisedML.csv")

# print out the column names
colnames(dataSuper)
```

From the output reproduced below, note that the dataset consists of weekly wages and the list of 106 skill self-assessments.

[10] It is standard practice in econometrics to consider the log of wages when studying labour market outcomes based on the widely used Mincerian wage equation specification Lemieux (2006).

[11] The command for installing a new package is `install.packages("package name")`, in this case `install.packages("glmnet")` and `install.packages("corrplot")`.

```
> colnames(dataSuper)
  [1] "WKLY.WAGES"                 "Bank"                      "Handling.cash"              "Document.handling"
  [5] "Measurement"                "Simple.maths"              "Mobile.phone.usage"         "Internet.usage"
  [9] "Safety.awareness"           "General.knowledge"         "Geographic.knowledge"       "Tracking.prices"
 [13] "Traffic.rules"              "English.listening"         "English.speaking"           "English.reading"
 [17] "English.writing"            "Hindi.listening"           "Hindi.speaking"             "Hindi.reading"
 [21] "Hindi.writing"              "Kannada.listening"         "Kannada.speaking"           "Kannada.reading"
 [25] "Kannada.writing"            "Beauty.services"           "Designing.skills"           "Stitching"
 [29] "Embroidery"                 "Color.knowledge"           "Working.with.flowers"       "Working.with.cloth"
 [33] "Working.with.wood"          "Handling.small.machines"   "Handling.heavy.machines"    "Cutting.meat"
 [37] "Cutting.metal"              "Working.with.tools"        "Lifting.heavy.items"        "Lifting.stones"
 [41] "Cement.work"                "Construction"              "Work.at.heights"            "Work.in.bathrooms"
 [45] "Work.in.kitchens"           "Work.in.unclean.area"      "Work.outdoors"              "Cleaning.house"
 [49] "Cleaning.roads"             "Washing.clothes"           "Ironing"                    "Cleaning.vessels"
 [53] "Cooking"                    "Housekeeping"              "Wedding.Work"               "Packing"
 [57] "Bike.riding"                "Knowledge.of.bike.parts"   "Fixing.a.bike"              "Driving"
 [61] "Knowledge.of.car.parts"     "Car.mechanic"              "Delivery"                   "Electric.work"
 [65] "Applying.putty"             "Painting.walls"            "Computer.handling"          "Data.entry"
 [69] "Quality.Verification"       "Inventory"                 "Management"                 "Customer.service"
 [73] "Sales"                      "Following.instructions"    "Self.defense"               "Caring"
 [77] "Supervising"                "Smiling..Body.language."   "Communication"              "Confidence"
 [81] "Friendliness"               "Helpful"                   "Neatness"                   "Patience"
 [85] "Physical.health"            "Physical.strength"         "Hardwork"                   "Positive.attitude"
 [89] "Punctuality"                "Politeness"                "Honesty"                    "Respectful"
 [93] "Discipline"                 "Adjusting.attitude"        "Strong.and.loud.voice"      "Bold.attitude"
 [97] "Active.personality"         "Creativity"                "Dressing.sense"             "Relationship.with.people"
[101] "Relationship.with.owners"   "Teamwork"                  "Speed.of.work"              "Interest.in.work"
[105] "Stress.management"          "Time.management"           "Work.ethic"
```

Continue the initial sanity checks on the data by looking at the first six rows (or the ‘head’) of the data and the summary statistics.

```
# see the top 6 rows of the dataset
head(dataSuper)

# obtain summary statistics of all the columns
summary(dataSuper)
```

Next specify the outcome of interest variable `y.out` to the log of the column corresponding to weekly wages. Following this, construct the matrix of covariates or features using the remaining variables in the dataset.

```
y.out = log(dataSuper[,1]) # log of weekly wages
x.out = dataSuper[, -1]  # self-assessments on all skills
```

Before running the LASSO regression, it is always useful to run an OLS regression as a baseline.

```
## correlation plot
corrplot(cor(x.out),  tl.pos='n' )

## Simple regression
ols.reg = lm(y.out~as.matrix(x.out))
summary(ols.reg)
```

The correlation plot and regression output are not reproduced here, but readers can confirm from the correlation plot that a number of features are highly correlated with each other (as evident from the patches of dark blue in the plot) which can lead to issues of multicollinearity in standard OLS settings. There is further evidence of this in the OLS regression summary which has an R^2 value of close to 50% but only a handful of significant covariates. Multicollinearity issues lead to unreliable estimates and therefore a LASSO regression model is appropriate in this setting.

Readers are now ready to run the LASSO regression model for which the R package `glmnet` will be used. The simplest version of running a LASSO regression is to plug in a value of the tuning parameter λ into the model fitting command, for instance the following command plugs in the value of $\lambda = 0.5$.[12]

```
# Plugging in a tuning parameter value of 0.5 (without doing cross-validation)
fit.lasso = glmnet(as.matrix(x.out), y.out, alpha=1, lambda=0.5)
```

However as discussed above, the choice of tuning parameter is critical in any supervised ML model and the preferred method of selection is the use of cross-validation. The next set of codes first conducts cross-validation for the LASSO regression model and then picks the optimal tuning parameter from the cross-validation exercise and finally fits the optimal LASSO regression. Note that the optimal tuning parameter here, λ_{1se}, corresponds to the value of λ that gives the most regularized model such that the cross-validated error is within one standard error of the minimum.[13]

```
set.seed(1234)

# crossvalidation for tuning parameter selection
CV = cv.glmnet(as.matrix(x.out),y.out, alpha=1)

# optimal tuning parameter
CV$lambda.1se

# fitting the data using the optimal tuning parameter
fit.lasso.optimal = glmnet(as.matrix(x.out), y.out, alpha=1, lambda=CV$lambda.1se)
```

As a final step, the covariates selected by the optimal LASSO regression, i.e. the covariates corresponding to non-zero estimated coefficient estimates, are reported next.

```
# parameter estimates
coefs.lasso = fit.lasso.optimal$beta[,1]

# which covariates get selected (have non-zero values)
selected.coefs = coefs.lasso[coefs.lasso !=0]

# print out the list of selected covariates
selected.coefs
```

The output of the previous command, reproduced below, lists out the selected covariates along with their coefficient estimates.

[12] While the remaining arguments of the command are self-explanatory, the argument $\alpha = 1$ here corresponds to the LASSO model. The package can be used to also run ridge regression and elastic net regression which are related to LASSO and correspond to different values of α.

[13] Interested readers can re-run the model here using the value of λ_{min}, the resulting model has a substantially larger number of selected covariates, making the interpretation more involved. They can also explore how results vary using different random seed generators using the set.seed() function.

```
> selected.coefs
       General.knowledge    Geographic.knowledge
             0.008381718             0.125792076
           Traffic.rules       English.listening
             0.012641667             0.002331366
          Beauty.services           Cutting.metal
             0.006319417             0.014636913
          Washing.clothes                 Cooking
            -0.067347354            -0.044246783
             Housekeeping             Bike.riding
            -0.018255019             0.089034417
Knowledge.of.bike.parts                  Driving
             0.056610706             0.024273787
 Knowledge.of.car.parts        Computer.handling
             0.006000092             0.032728631
               Data.entry             Supervising
             0.021549340             0.052703120
                 Teamwork
             0.037868880
```

From the output, it can be noted that out of the 106 skill claims, only 17 get selected by the LASSO regression model. Of these skill claims, a majority of five ('Geographic knowledge', '(Knowledge of) Traffic rules', 'Bike riding', 'Knowledge of bike parts' and 'Knowledge of car parts') correspond specifically to the high growth sector of delivery services and ride-sharing services in the city of Bangalore. Interestingly, three skill claims corresponding to domestic work ('Washing clothes', 'Cooking' and 'Housekeeping') have a *negative* association with weekly wages. This is likely because workers with these skills correspond to domestic help workers who have very low bargaining abilities. More general skills like 'Computer handling', 'Data entry' and 'English listening (i.e. understanding spoken English)' give a clue about the importance of digital literacy and English language skills even in informal work contexts in India. Finally, the skill claims of 'Supervising' and 'Teamwork' point to the role of soft skills in predicting wage levels.

Why is the use of LASSO so critical in this application? First, the results point to specific skills that informal workers can be trained in such as English and digital literacy. Second, the results uncover the extent of disparities in wages associated with specific job types—for instance jobs with skills corresponding to the IT services sector and ride-sharing sectors are associated with higher wages, whereas jobs with skills corresponding to domestic work are associated with low wages. None of these insights could be gleaned from the OLS results above, which would be unreliable given the multicollinearity concerns discussed. Given the collection of large datasets in the digital economy, LASSO regression models will be useful in a number of settings where the selection of a subset of features or covariates can provide important insights into economic and social variables of interest.

9.6.2 k-means Clustering

One of the aims of the investigation in Sengupta et al. (2021) is the identification of skill claims that are clustered together. k-means clustering is the appropriate tool in this context since there is no specific outcome of interest Y; rather, the objective is to find skills that co-occur. For this application, the transpose of the matrix of skill claims is used so that the covariates for clustering, $x_1, \cdots, x_p$ correspond to the self-assessment on each skill by every respondent. A snapshot of the dataset is presented in the following figure corresponding to a set of skills assessments by 24 respondents across seven skills.

Note that if we were to run the k-means algorithm on this dataset as is then the output would consist of clusters of *respondents*. Since our interest is in clustering *skills* instead, we need to first transpose the matrix so that the rows correspond to the skills whereas the columns correspond to respondents. This implies that skills where respondents rate themselves similarly will be clustered together. The code for running the algorithm is presented next.

As always, start by clearing out the coding environment, loading the appropriate packages and setting the working directory to the one where the data is stored.

The data used for this exercise is stored in the file `DataUnsupervisedML.csv`. First, the input matrix is transposed and a sanity check is conducted by printing out the row names and the dimensions of the

	Bank	Handling.cash	Document.handling	Measurement	Simple.maths	Mobile.phone.usage	Internet.usage
1	0	1	1	0	1	1	0
2	0	1	1	1	1	2	0
3	2	2	2	2	2	2	2
4	0	1	1	0	1	1	0
5	0	1	1	0	1	1	0
6	0	1	0	0	1	1	0
7	2	2	2	1	2	2	0
8	2	2	1	1	1	2	1
9	1	1	0	0	1	2	0
10	2	1	2	1	2	1	0
11	1	1	2	1	1	2	1
12	1	1	1	1	1	2	2
13	2	2	2	0	2	2	2
14	1	1	1	0	1	1	0
15	0	1	1	0	1	1	0
16	0	1	1	0	1	1	0
17	0	0	1	1	1	2	0
18	0	1	2	0	1	2	0
19	0	1	2	1	1	2	0
20	1	1	2	0	1	2	1
21	1	2	2	2	2	2	0
22	1	2	1	0	1	2	0
23	1	1	2	2	2	2	0
24	2	2	2	2	1	2	2

Fig. 9.5 Snapshot of skills assessment data

matrix. The output confirms that the row names correspond to the different skills and that the columns correspond to self-assessments across all individuals. The entries of the matrix take the values 0 (not comfortable at all), 1 (somewhat comfortable) and 2 (very comfortable). There are a total of 106 skills and 784 clustering covariates corresponding to the 784 respondents. An additional sanity check is to take a look at the data using the command `View(.)` which displays the data in a new tab within the `RStudio` environment.

```
#### SETTING UP THE ENVIRONMENT

# clean environment
rm(list = ls())

# read in the relevant package
library("factoextra")

#Change Working Directory
setwd("/Users/Desktop/ML Tutorial")

# reading data
dataUnsuper = read.csv("DataUnsupervisedML.csv")

# transpose so that skills are rows and self assessments are columns
dataskills = t(dataUnsuper)

rownames(dataskills)

dim(dataskills)

View(dataskills)
```

Recall that the only input required for k-means clustering is the number of clusters k. The appropriate number of clusters can be obtained by looking at the *gapstat*. The R package `factoextra` has a nice implementation of this which produces a plot indicating the number of clusters, as shown next. Note that in this demonstration the maximum number of clusters `k.max` is set to 25, the number of random trials `nboot` is set to 50 and a specific seed has been set for the generation of random initializations. Interested readers are encouraged to consider different settings to check the robustness of the resulting number of optimal clusters. It is always a good practice to scale the input matrix as we have done. We implement the algorithm in R using the command `kmean()` for clustering and the package `factoextra` for estimating the appropriate number of clusters.

```
# seed for random number generation
set.seed(123)

# plot to find optimal number of clusters
fviz_nbclust(dataskills, kmeans, method = "gap_stat", k.max = 25,  nboot  = 50)

# appropriate no. clusters = 10
```

From the resulting plot of the *gapstat* reproduced in Fig. 9.6, we note that the appropriate number of clusters $k = 10$ based on the dotted line. Recall that clustering algorithm starts with k randomly chosen centroids $(\mu_1, \mu_2, \cdots \mu_k)$. The

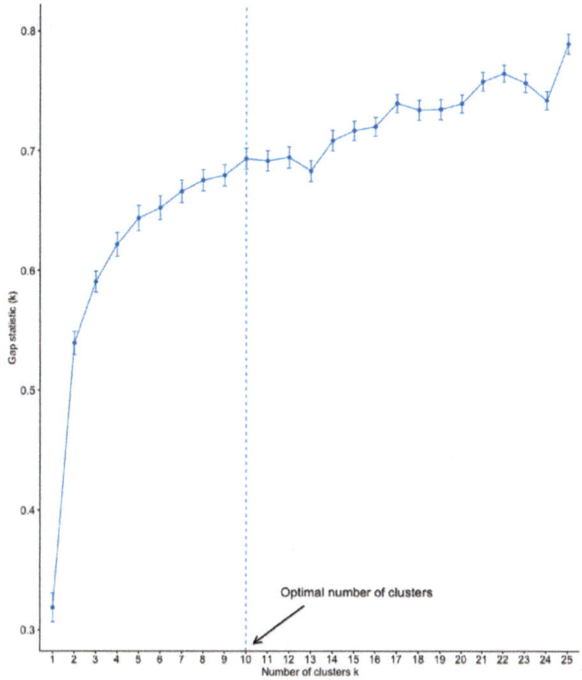

Fig. 9.6 Plot to find appropriate number of clusters using *gapstat*

`nstart` option attempts multiple random initial configurations of which the best one is reported. Interested readers are encouraged to try out different initializations by changing both `set.seed()` and `nstart` inputs. The `kmeans` command produces a number of result attributes. The attribute we are interested in is `"cluster"` which reports the cluster each row unit is assigned to.

```
# kmeans clustering
k = 10
cluster_summary <- kmeans(dataskills, k, nstart = 50)

# output of k means clustering
cluster_summary$cluster
```

While the clustering is now complete, reporting each cluster with all members is more easy to interpret in this application. The following set of commands helps achieve this by creating *k* lists of skills corresponding to the *k* clusters.

```
# output cluster members
skill_cluster_membership = vector("list", k)

for (i in 1: k){
skill_cluster_membership[[i]] = which(cluster_summary$cluster == i)}

# clustered skills
skill_cluster_membership
```

Next, the clusters are ordered by number of members in each cluster.

```
# sorting clusters by number of members
clusters_sorted_by_size = sort(cluster_summary$size, index.return = TRUE)$ix

# output cluster members arranged in number of members
skill_cluster_membership_bysize = vector("list", K)

for (i in 1: N){
  skill_cluster_membership_bysize[[i]] =
  skill_cluster_membership[[clusters_sorted_by_size[i]]]}

####################################################
# clustered skills by number of members

skill_cluster_membership_bysize
```

Table 9.1 presents the 10 optimal clusters obtained by the *k*-means clustering algorithm, in increasing order of number of skills per cluster. An informative way of

Table 9.1 10 clusters obtained by the k-means algorithm, in increasing order of number of skills per cluster

Cluster no.	List of skills
1	Work in bathrooms, Cleaning roads
2	Handling small machines, Cutting metal, Cement work, Work in unclean area
3	Packing, Management, Customer service, Sales, Creativity
4	Simple maths, Mobile phone usage, Color knowledge, Following instructions, Confidence, Adjusting attitude, Active personality
5	Cutting meat, Work in kitchens, Cleaning house, Washing clothes, Ironing, Cleaning vessels, Cooking, Housekeeping
6	Measurement, General knowledge, Hindi listening, Hindi speaking, Beauty services, Working with tools, Bike riding, Knowledge of bike parts, Driving
7	Bank, Internet usage, English listening, English reading, English writing, Kannada reading, Kannada writing, Computer handling, Data entry
8	Handling cash, Geographic knowledge, Tracking prices, Traffic rules, Quality Verification, Inventory, Supervising, Dressing sense, Teamwork
9	English speaking, Hindi reading, Hindi writing, Designing skills, Stitching, Embroidery, Working with flowers, Working with cloth, Working with wood, Handling heavy machines, Lifting heavy items, Lifting stones,Construction, Work at heights, Wedding Work, Fixing a bike, Knowledge of car parts, Car mechanic, Delivery, Electric work, Applying putty, Painting walls
10	Document handling, Safety awareness, Kannada listening, Kannada speaking, Work outdoors, Self defence, Caring, Smiling Body language , Communication, Friendliness, Helpful, Neatness, Patience, Physical health, Physical strength, Hardwork, Positive attitude, Punctuality, Politeness, Honesty, Respectful, Discipline, Strong and loud voice, Bold attitude, Relationship with people, Relationship with owners, Speed of work, Interest in work, Stress management, Time management, Work ethic

analysing the clusters is to see which skills co-exist within a cluster. Let us consider the skills co-located in each cluster one by one.

- Cluster 1 consists of skills associated with jobs often undertaken by workers belonging to the most disadvantaged communities, particularly women. These skills are not associated with any other general language or task skills in the same cluster.
- Cluster 2 likely refers to skills associated with working in small informal workshops, again not associated with other general task and language skills.
- Cluster 3 consists of skills associated with sales and inventory work. The cluster also contains the soft skills of 'management', 'customer service' and 'creativity'.
- Cluster 4 consists of skills which do not lend themselves immediately to an easy occupational categorization, unlike the other clusters. Most of the skills in the cluster consists of general task skills—showing evidence that it is likely that individuals who (claim to) possess a general task skill, likely (claim to) possess a whole range of other general task skills as well.
- Cluster 5 consists of skills associated with domestic work, also a predominantly female vocation. Again, there are no general language or task skills within the same cluster.
- Cluster 6 consists of skills associated with the beauty industry and with the ride services industry. Both of these often interface with the large migrant Hindi-speaking population in Bangalore—and thus it is not surprising to see that both speaking and understanding Hindi are also co-located in this cluster.
- Cluster 7 consists of skills associated with the burgeoning IT services sector within Bangalore as is apparent from the skills of 'internet usage', 'computer handling' and 'data entry'. This cluster also contains the skills of 'bank work', English language reading, writing and listening. Interestingly, the cluster also contains skills associated with the native residents of Bangalore as demonstrated by the skills of writing and reading in Kannada (the native language of the region).
- Cluster 8 is a cluster corresponding to sales and delivery work, another rapidly growing sector in the city. Here skills like 'handling cash', 'tracking prices', 'inventory' and 'geographic knowledge' are co-located with softer skills like 'supervising', 'dressing sense' and 'teamwork'.
- Cluster 9 consists of reading and writing skills in Hindi, speaking skills in English, along with a number of specialized skills (from 'driving' and 'working with tools' to 'stitching' and 'cement work'). This cluster likely corresponds to the population of respondents who can perform a large number of odd jobs and interact regularly with Bangalore's large migrant non-Kannada-speaking population of corporate employees.
- Cluster 10 consists of a range of personal and social skills. It is likely that most respondents rate themselves very highly on these, being less objective when assessing these softer values.

Why is *k*-means clustering useful in this setting? Because it provides a numerical technique to club different skills together, in the process providing us insights on types of occupations that are associated with particular sub-populations. This goes beyond simple pairwise correlations between skills that are more standard in applied econometric settings.

9.7 Discussion

ML models can be very useful within social science context when the primary focus of the analysis is (1) predictions and (2) efficient exploration of large-scale data. However, when it comes to questions of explanations and associations, ML models are generally a black box. Thus, while ML models can be a useful tool for social scientists and policy makers to answer questions which fall under the category of 'knowing what', they cannot replace standard econometrics especially when the questions fall under the categories of 'knowing how' and 'knowing why'. Having said that, as ML becomes more ubiquitous, social scientists will ignore advances in these techniques at their own risk.

In this chapter, the basic ingredients of ML models are described in detail, including regularization, tuning parameters, training and validation sets. An overview of the primary sub-categories of ML models, i.e. supervised and unsupervised learning, is then provided, along with a deeper dive into LASSO regression and *k*-means clustering. A number of other related techniques including neural networks, random forests, matrix factorization, etc., are then introduced. These newer techniques are now making an entry into the economics and social science literature. For an overview paper on these topics, readers are referred to Athey and Imbens (2019). For a deeper dive into the fundamentals on more general ML models, readers are referred to Hastie et al. (2009).

The chapter ends with the following question: what can econometricians and social scientists bring to the table for ML research and applications that intersect with society? Three directions of ongoing interdisciplinary engagement are highlighted in order to answer this question. First, the incorporation of ML ideas into structural econometric models has already become an active area of research (Wager and Athey 2018; Windmeijer et al. 2019; Sengupta and Sowell 2020). Second, with their in-depth knowledge of local contexts, social scientists are beginning to have a key role in contributing to recent academic debates on the implementation of ML systems and processes in the real world (Jordan and Mitchell 2015; Ipsos 2017; van den Broek et al. 2021). Third, the measurement of potential algorithmic bias is an active area of research in computer science (Kirkpatrick 2016; Cowgill and Tucker 2019) which can be informed by the field of economics which has a long history of methods for identifying, modelling and mitigating sources of bias in the data generating process.

References

Athey S, Imbens GW (2019) Machine learning methods that economists should know about. Ann Rev Econ 11:685–725

Bauckhage C (2015) K-means clustering is matrix factorization. arXiv preprint arXiv:1512.07548

Blumenstock J, Cadamuro G, On R (2015) Predicting poverty and wealth from mobile phone metadata. Science 350(6264):1073–1076

Bond RM, Fariss CJ, Jones JJ, Kramer AD, Marlow C, Settle JE, Fowler JH (2012) A 61-million-person experiment in social influence and political mobilization. Nature 489(7415):295–298

van den Broek E, Sergeeva A, Huysman M (2021) When the machine meets the expert: an ethnography of developing ai for hiring. MIS Quart 45(3)

Chalfin A, Danieli O, Hillis A, Jelveh Z, Luca M, Ludwig J, Mullainathan S (2016) Productivity and selection of human capital with machine learning. Am Econ Rev 106(5):124–27

Cowgill B, Tucker CE (2019) Economics, fairness and algorithmic bias. preparation for: J Econ Persp

Darrah CN (2013) Learning and work: an exploration in industrial ethnography. Routledge

Ding C, He X, Simon HD (2005) On the equivalence of nonnegative matrix factorization and spectral clustering. In: Proceedings of the 2005 SIAM international conference on data mining, SIAM, pp 606–610

Garrido M, Sullivan J, Gordon A (2010) Understanding the links between ict skills training and employability: an analytical framework. In: Proceedings of the 4th ACM/IEEE international conference on information and communication technologies and development, pp 1–10

Hallac D, Park Y, Boyd S, Leskovec J (2017) Network inference via the time-varying graphical lasso. In: Proceedings of the 23rd ACM SIGKDD International Conference on Knowledge Discovery and Data Mining, pp 205–213

Hartigan JA, Wong MA (1979) Algorithm as 136: A k-means clustering algorithm. J Royal Stat Soc Series C (Appl Stat) 28(1):100–108

Hassoun MH (1995) Fundamentals of artificial neural networks. MIT press

Hastie T, Tibshirani R, Friedman JH, Friedman JH (2009) The elements of statistical learning: data mining, inference, and prediction, vol 2. Springer

Ipsos M (2017) Public views of machine learning. Report title 92

Jordan MI, Mitchell TM (2015) Machine learning: trends, perspectives, and prospects. Science 349(6245):255–260

Kecman V (2005) Support vector machines–an introduction. In: Support vector machines: theory and applications, Springer, pp 1–47

Kj Kim, Ahn H (2008) A recommender system using ga k-means clustering in an online shopping market. Exp Syst Appl 34(2):1200–1209

Kirkpatrick K (2016) Battling algorithmic bias: how do we ensure algorithms treat us fairly? Commun ACM 59(10):16–17

Koren Y, Bell R, Volinsky C (2009) Matrix factorization techniques for recommender systems. Computer 42(8):30–37

Lemieux T (2006) The "mincer equation" thirty years after schooling, experience, and earnings. In: Jacob Mincer a pioneer of modern labor economics, Springer, pp 127–145

Mak KK, Lee K, Park C (2019) Applications of machine learning in addiction studies: a systematic review. Psyc Res 275:53–60

Marquardt DW, Snee RD (1975) Ridge regression in practice. Am Stat 29(1):3–20

Mehrotra S, Gandhi A, Sahoo BK (2013) Estimating india's skill gap on a realistic basis for 2022. Econ Pol Weekly 48(13):102–111

Mukherjee S (2017) URL https://www.newyorker.com/magazine/2017/04/03/ai-versus-md, [Online; posted 03-April-2017]

Mützel S (2015) Facing big data: making sociology relevant. Big Data Soc 2(2):2053951715599179

Nagarnaik P, Thomas A (2015) Survey on recommendation system methods. In: 2015 2nd International Conference on Electronics and Communication Systems (ICECS), IEEE, pp 1603–1608

Okada A (2004) Skills development and interfirm learning linkages under globalization: lessons from the indian automobile industry. World Devel 32(7):1265–1288

Osborne MR, Presnell B, Turlach BA (2000) On the lasso and its dual. J Comput Graph Stat 9(2):319–337

Riaz S, Fatima M, Kamran M, Nisar MW (2019) Opinion mining on large scale data using sentiment analysis and k-means clustering. Cluster Comput 22(3):7149–7164

Rutherford A, Lupu Y, Cebrian M, Rahwan I, LeVeck BL, Garcia-Herranz M (2018) Inferring mechanisms for global constitutional progress. Nat Human Beh 2(8):592–599

Samek W, Montavon G, Lapuschkin S, Anders CJ, Müller KR (2021) Explaining deep neural networks and beyond: a review of methods and applications. Proc IEEE 109(3):247–278

Sengupta N, Sowell F (2020) On the asymptotic distribution of ridge regression estimators using training and test samples. Econometrics 8(4):39

Sengupta N, Gaurav S, Evans J (2021) The skills space in informal work: insights from bangalore slums. J Devel Stud 57(10):1662–1689

Sengupta N, Udell M, Srebro N, Evans J (2023) Sparse data reconstruction, missing value and multiple imputation through matrix factorization. Sociol Methodol 53(1):72–114

Shmueli G (2010) To explain or to predict?

Taddy M (2013) Multinomial inverse regression for text analysis. J Am Stat Assoc 108(503):755–770

Tibshirani R, Walther G, Hastie T (2001) Estimating the number of clusters in a data set via the gap statistic. J Royal Stat Soc Series B (Stat Methodol) 63(2):411–423

National Commission for Enterprises in the Unorganised Sector GoI (2007) Report on conditions of work and promotion of livelihoods in the unorganised sector. Tech. rep

Vayansky I, Kumar SA (2020) A review of topic modeling methods. Inform Syst 94:101582

Wager S, Athey S (2018) Estimation and inference of heterogeneous treatment effects using random forests. J Am Stat Assoc 113(523):1228–1242

Welsch RE, Zhou X (2007) Application of robust statistics to asset allocation models. REVSTAT-Stat J 5(1):97–114

Weng Z, Wang X (2012) Low-rank matrix completion for array signal processing. 2012 IEEE International Conference on Acoustics. IEEE, Speech and Signal Processing (ICASSP), pp 2697–2700

Windmeijer F, Farbmacher H, Davies N, Davey Smith G (2019) On the use of the lasso for instrumental variables estimation with some invalid instruments. J Am Stat Assoc 114(527):1339–1350

Xie M, Jean N, Burke M, Lobell D, Ermon S (2016) Transfer learning from deep features for remote sensing and poverty mapping. In: Thirtieth AAAI Conference on Artificial Intelligence

Xu W, Liu X, Gong Y (2003) Document clustering based on non-negative matrix factorization. In: Proceedings of the 26th annual international ACM SIGIR conference on Research and development in informaion retrieval, pp 267–273

Zou H, Hastie T (2005) Regularization and variable selection via the elastic net. J Royal Stat Soc Series B: Stat Methodol 67(2):301–320

Chapter 10
Dimensionality Reduction Techniques: Application to Indian Bank Customer Survey Data

Maharnab Naha, Kanish Debnath, and Karan Singh Khati

Abstract Data dimension reduction techniques enable researchers to reduce the number of variables (dimensions) in a dataset while preserving essential information. In this chapter, we demonstrate the application of two widely used linear methods, Exploratory Factor Analysis (EFA) and Principal Component Analysis (PCA), for data dimensionality reduction. Both EFA and PCA aim to simplify the structure of observed data and to identify patterns of association among variables, making them particularly valuable for assessing questionnaire scale dimensionality and exploring item relationships. This chapter applies these methods to identify key service quality factors in the banking sector, with a specific focus on public sector banks in Kanpur City, Uttar Pradesh, India. A survey of 506 participants was conducted in 27 bank branches using an 18 item service quality questionnaire. Our methods and results offer actionable insights for deriving latent customer service quality constructs for Indian public sector banks. The analysis in this chapter mainly aims to provide practical guidance for implementing EFA and PCA in both R and Python, including the use of polychoric correlations to appropriately handle Likert-scale responses. More broadly, the chapter illustrates the applicability of EFA and PCA, which can be applied across domains such as public policy, social science and management research.

Supplementary Information The online version contains supplementary material available at https://doi.org/10.1007/978-981-97-9060-9_10.

M. Naha
Department of Economics, University of Houston, Houston, TX, USA
e-mail: mnaha2@uh.edu

K. Debnath (✉)
MYRA School of Business, Mysuru, India
e-mail: kanish.debnath@myra.ac.in

K. S. Khati
New Delhi Institute of Management, Delhi, India
e-mail: karan.khati@ndimdelhi.org

F. Hamid and D. Mukherjee (eds.), *Data Science for Modeling Managerial and Socioeconomic Problems*, Contributions to Management Science,
https://doi.org/10.1007/978-981-97-9060-9_10

10.1 Introduction

Often data can have a large number of dimensions or variables. If a dataset has many variables that are almost equal to the number of observations, conducting a statistical analysis on such data becomes challenging or infeasible. Techniques for reducing data dimensions enable minimization of variables in datasets while retaining crucial information. Data dimension reduction can be achieved primarily in two ways: variable selection and feature extraction. While the first approach selects a subset of relevant variables that may lead to a possible loss of information, the second approach transforms existing data into a new lower-dimensional dataset. Data dimensionality reduction using the second approach can be accomplished through both linear and nonlinear methodologies. Linear methods that include principal component analysis (PCA), linear discriminant analysis (LDA), and factor analyses—exploratory factor analysis (EFA) and confirmatory factor analysis (CFA)—are popular choices. Nonlinear methods include t-Distributed Stochastic Neighbor Embedding (t-SNE) and Kernel PCA.

This chapter concentrates on two widely used linear dimensionality reduction techniques, EFA and PCA. High correlation coefficients among certain variables may represent different aspects of a common underlying dimension, called factors or latent variables. Whereas EFA models this continuum explicitly as one or more latent factors, PCA treats the same pattern of associations as an opportunity to express the data in a reduced set of orthogonal linear combinations called principal components. It is crucial to acknowledge that the constructs revealed by either technique are not inherently meaningful; their interpretation is a subjective endeavor guided by the researcher's theoretical framework and domain knowledge. In EFA, the extracted latent factors gain significance only when they are aligned with clearly defined constructs. Similarly, in PCA, the orthogonal principal components require careful interpretation before they can contribute to substantive conclusions (Henson and Roberts 2006; Abdi and Williams 2010).

Regardless of whether the objective is latent variable modeling using EFA or descriptive summarization using PCA, the methodological analysis mostly begins with an appropriate structure of the correlation matrix (Fabrigar et al. 1999; Field 2009). When items are measured on ordinal Likert-type scales, Pearson correlations can understate the true relationships, and therefore it is standard practice to substitute them with a polychoric correlation matrix, which estimates the correlations among latent continuous variables presumed to generate the observed categorical responses (Olsson 1979; Drasgow 2006; Flora and Curran 2004; Holgado-Tello et al. 2010). Using this matrix, both EFA and PCA strive for parsimony by capturing as much shared variance as possible with the fewest factors or components required for meaningful interpretation (Kolenikov and Angeles 2009; Fabrigar et al. 1999).

In our study, we analyze a primary survey of Indian Public Sector Banks' (PSBs) customers' preferences toward bank service quality. The survey data was collected over a three-month period in 2021 from Kanpur city in Uttar Pradesh, India. The dataset contains 501 responses—436 complete and the rest with less than 10%

missing values were imputed assuming that these values were missing completely at random (MCAR). The survey had 18 five-point Likert scale items (see Appendix 10.6 A) that provides a realistic context for demonstrating how dimensional reduction techniques, such as EFA and PCA, can identify the key dimensions of banking service quality as latent constructs within the data. Consequently, the value of any retained factor or component ultimately depends on a well-reasoned interpretation anchored in domain knowledge.

The rest of the article is organized as follows. Section 10.2 discusses the theoretical frameworks of exploratory factor analysis and principal component analysis. Section 10.3 discusses a guided application, demonstrating the practical application of EFA and PCA in R and detailing each step of the methodology used to identify latent factors in customer satisfaction of banking services. These steps are also available in Python as a Python notebook file in the supplementary files. One can easily get started with Python using Google Colab online as demonstrated in Appendix 10.6 B. Finally, Section 10.4 concludes with a discussion of the study findings, its limitations, and future directions.

10.2 Statistical Tools for Data Dimension Reduction

10.2.1 Exploratory Factor Analysis

Factor analysis is a multivariate statistical model that is used to reduce a large number of variables into a smaller set of underlying factors. This method works by explaining the variance among a large number of observed variables through a set of smaller latent (unobserved) factors and their associated weights. There are two main types of factor analysis models:

- EFA: Used when the underlying structure of the items is unknown, exploring observed variables to create a model explaining latent dimensions, and
- CFA: Used to test a hypothesized structure, given that the latent variables for a set of observed variables is already understood.

Factor analysis finds wide application in various fields dealing with multivariate data, including social sciences, management, psychology, and marketing. For example, consider a company seeking to understand employee productivity based on a comprehensive survey covering various aspects of the work-life of the employees in the company, including: Company Workload, Salary Satisfaction, Relationship with the Manager, Recognition for Work, Career Growth Opportunities, Job Autonomy, etc. A method like factor analysis examines the correlations between these response variables to identify underlying latent structures that explain most of the data's variance. In this context, the observed variables might be clustered into three main latent factors: Job Satisfaction (e.g., salary, recognition), Work Environment (e.g., office conditions, company culture), and Professional Development (e.g., career growth, job autonomy). Given this structure, each latent factor would have different weights, or "factor loadings", for each observed variable, indicating how strongly that variable contributes to the factor. Identifying this latent factor through this process reduces the complexity from numerous individual variables to a few key factors, revealing what drives employee productivity and guiding targeted action.

In factor analysis, for a dataset that has p variables with m factors, typically $m < p$, we assume that the observed variables are linear combinations of latent factors plus error terms, that is, the variables are linear in common factors (Khosla 2004). The factor analysis model, for each variable X_i and its mean μ_i, is denoted by:

$$\begin{aligned} X_1 - \mu_1 &= \lambda_{11} f_1 + \lambda_{12} f_2 + \ldots + \lambda_{1m} f_m + \epsilon_1 \\ X_2 - \mu_2 &= \lambda_{21} f_1 + \lambda_{22} f_2 + \ldots + \lambda_{2m} f_m + \epsilon_2 \\ &\vdots \\ X_p - \mu_p &= \lambda_{p1} f_1 + \lambda_{p2} f_2 + \ldots + \lambda_{pm} f_m + \epsilon_p \end{aligned} \tag{10.1}$$

In matrix form, this can be expressed as:

$$(X - \mu)_{p\times 1} = \Lambda_{p\times m} F_{m\times 1} + \varepsilon_{p\times 1} \tag{10.2}$$

The interpretability of factor analysis comes from Λ, the coefficient weights or "factor loadings". λ_{ij} represents how much the ith variable is influenced by the jth factor. When some λ_{ik} is larger for certain variables i, it indicates that the kth factor strongly influences these i variables, distinguishing them from others with lower loadings.

Note that the vectors F and ε are non-observable random vectors (that is $E(F) = 0$, $Cov(F) = I$(identity matrix), $Cov(f_i, f_j) = 0$) and the factors are assumed to be independent ($E(\varepsilon) = 0$, $Var(\varepsilon) = \Psi$(diagonal matrix)). It is also assumed that F and ε are independent, allowing different error variances for each variable ($Cov(\varepsilon, F) = 0$).

Based on these assumptions, the covariance structure can be defined as:

$$\begin{aligned} Cov(X) &= E(X - \mu)(X - \mu)' \\ &= \Lambda E(FF')\Lambda' + E(\varepsilon F')\Lambda' + \Lambda E(F\varepsilon') + E(\varepsilon\varepsilon') \\ &= \Lambda\Lambda' + \Psi \end{aligned} \tag{10.3}$$

In addition, the covariance between the variables and the factors is Λ, defined as:

$$\begin{aligned} Cov(X, F) &= E(X - \mu)F' \\ &= \Lambda E(FF') + E(\varepsilon F') \\ &= \Lambda \end{aligned} \tag{10.4}$$

So, in standardizing the variables, λ_{ij} represents the correlation between factor j and variable i.

The variance of the ith variable can be divided into two components: The common variance (or ith communality, denoted as h_i^2) represents the shared variability between the common factors, and the specific variance (or residual variance, denoted as Ψ_i) accounts to unique variability associated with the ith factor. The ith communality, which represents the portion of variance explained by the common factors, is calculated as the sum of squares of the loadings of the ith variable on these factors, thus providing a measure of how well the variable is accounted for by the underlying factor structure.

$$\begin{aligned} Var(X_i) &= (\lambda_{i1}^2 + \lambda_{i2}^2 + \ldots + \lambda_{ip}^2) + \psi_{ii} \\ &= h_i^2 + \Psi_i \qquad\qquad i = 1, 2 \ldots, p \end{aligned} \tag{10.5}$$

Factor analysis approximates the original data's covariance structure using a simplified model using $Cov(X) = \Lambda\Lambda' + \Psi$. The quality of this approximation depends on the number of factors used in the model. When the estimated factor analysis structure does not fit well, it indicates that the model may be inadequate and suggests that more factors might be needed.

On the other hand, there is a limit to how many factors we can include while still maintaining a simpler model than the original data. With p variables, there are $p + p(p - 1)/2 = p(p + 1)/2$ unique elements in the variance-covariance matrix. When the number of factors m equals the number of variables p, the covariance matrix equals $\Lambda\Lambda'$ and $\Psi = 0$. However, factor analysis is most effective as a dimensionality reduction tool when using fewer factors than variables, offering a parsimonious explanation of covariation with substantially fewer parameters than the original $p(p + 1)/2$. In practice, several methods and rules exist for choosing the appropriate number of factors, including Kaiser's eigenvalue criterion, scree plot analysis, parallel analysis, and also examining the proportion of variance explained by the factors.

For a factor analysis model, the factor loadings are not unique. Consider an orthogonal matrix T of dimensions $m \times m$, such that $TT' = T'T = I$, where I is the identity matrix; we have:

$$\begin{aligned} X - \mu &= \Lambda F + \varepsilon \\ &= \Lambda TT'F + \varepsilon \\ &= \Lambda^* F^* + \varepsilon \quad \text{where } \Lambda^* = \Lambda T,\ \ F^* = T'F \end{aligned} \tag{10.6}$$

The transformed factors F^* maintain the same statistical properties as the original factors F (that is $E(F^*) = 0$, $Cov(F^*) = I$). Although factor loadings differ between F and F^*, the overall covariance structure remains unchanged. So, both the common variance and specific variance components are preserved in this transformation. This is where factor rotation comes into place to address the inherent ambiguity in the initial factor solution. To improve interpretability, the loading matrix is subjected to a rotation process, which involves multiplying it by an orthogonal or oblique matrix, chosen based on specific criteria, to redistribute the variance explained by each factor.

Factor loadings can be estimated through various methods. Common approaches include the Principal Component Method, the Principal Factor Method, the Maximum Likelihood Method, and the Iterated Principal Factor Method. Each method offers unique advantages and is suitable for different data structures and research objectives. For a comprehensive theoretical framework of these factor extraction techniques, Sects. 13.3.1–13.3.4 in Rencher and Christensen (2002) provide detailed instructions.

Given that factors are unobservable constructs, it is necessary to estimate them based on the observed variables and their corresponding factor loadings. Such an estimation process gives us "factor scores", which represent approximations of the underlying latent factor for each data point. Using these factor scores, we can explore how specific data points relate to these uncovered factors and serve them as variables for further multivariate analysis. There are several approaches to finding factor scores, and the appropriate method depends on the context of the analysis. DiStefano et al. (2019) provides a comprehensive discussion on different types of factor scores.

Although the general factor analysis model provides the theoretical foundation, its practical application often takes the form of EFA when the underlying structure is unknown. EFA is a systematic process that builds upon the concepts discussed earlier and typically involves several key steps: First, relevant indicators or items are selected and the sample size and characteristics are evaluated to ensure the suitability of factor analysis. To start with the analysis, a crucial step is deciding on the number of factors to retain. Next, an extraction method is chosen to estimate the factor model considering the characteristics of the data. To enhance interpretability, a rotation technique is applied to factor loadings. Finally, a method for computing factor scores is selected to represent the underlying factors for further statistical analysis.

For more detailed instruction and to delve into the topic in greater depth, the following resources are recommended. The theoretical framework of the factor analysis model and its underlying principles can be explored in the works of Rencher and Christensen (2002), Habing (2003), Khosla (2004), Yong et al. (2013), and Kline (2014). For greater understanding of its practical applications, valuable guidance can be found in Streiner (1994), Taherdoost et al. (2014), Watkins (2018), Costello and Osborne (2019), and Shrestha (2021).

10.2.2 Principal Component Analysis

Principal Component Analysis (PCA) is another widely-used technique for dimensionality reduction that takes a fundamentally different approach from factor analysis. Rather than modeling latent constructs with measurement error, PCA transforms observed variables into a smaller set of orthogonal, uncorrelated components that capture the maximum variance in the data (Jolliffe and Cadima 2016). This method works by creating linear combinations of the original variables, with each of the orthogonal and uncorrelated variables called the principal components (PCs) explaining successively less variance than the previous one. This transformation is purely descriptive, it reorganizes existing information without making assumptions about underlying factors or error structures. The result is a more compact representation of the data that preserves most of the original variation while reducing complexity.

Formally, consider a data matrix X with n observations (rows) and p variables (columns),

$$X = \begin{bmatrix} x_{11} & x_{12} & \dots & x_{1p} \\ x_{21} & x_{22} & \dots & x_{2p} \\ \vdots & \vdots & \ddots & \vdots \\ x_{n1} & x_{n2} & \dots & x_{np} \end{bmatrix}_{n \times p} \tag{10.7}$$

Firstly, the data must be centred by subtracting the mean of each variable as $X_c = X - \mathbf{1}_n \bar{x}^T$, where $\bar{x}$ is the vector of column means, and $\mathbf{1}_n$ is an $n \times 1$ vector of ones. From the centered data, we can compute the sample covariance matrix C as

$$C = \frac{1}{n-1} X_c^T X_c \tag{10.8}$$

When the variables have different units or scales, it is common practice to also standardize each variable to have unit variance, which effectively means that we can substitute this by performing on the correlation matrix rather than the covariance matrix. In this case, all variables contribute equally regardless of their original scale.

PCA proceeds via the eigen-decomposition of C. Because C is symmetric and positive semidefinite, it admits an eigen decomposition,

$$Cv_j = \lambda_j v_j, \quad j = 1, \dots, p, \quad \lambda_1 \geq \dots \geq \lambda_p \geq 0 \tag{10.9}$$

where v_j are the eigenvectors (principal directions) and λ_j are their associated eigenvalues, sorted in descending order ($\lambda_1 \geq \lambda_2 \geq \dots \geq \lambda_p \geq 0$). We collect the eigenvectors into a matrix $V = [v_1, \dots, v_p]$, which gives the principal component score matrix $Z = [z_1, \dots, z_p]$ as a linear combination of the original variables:

$$Z = X_c V \tag{10.10}$$

Here, $z_i \in Z$, defined as the PCA score, provides the coordinates of the observation in the space defined by the principal components. The PCA scores Z can be visualized (e.g., scatter plots of the first two PCs) to explore patterns such as clusters of similar observations or trends in the data. Importantly, extreme observations (outliers) often emerge as points far from the origin in the PC score plots. Such observations may reflect unique patterns or noise and should be carefully examined in the context of the data domain.

The matrix V, whose columns are the principal component directions v_j, is often referred to as the loading matrix. Each entry v_{ij} describes how strongly the original variable i contributes to the principal component j. Squared loadings v_{ij}^2 quantify the proportion of variance of variable i explained by component j. Variables with large loadings (in magnitude) on the same component are often grouped together, suggesting they capture similar patterns of variation in the data.

To reduce the dimension from p to m, where $m < p$, only the first m principal components are retained, to give the lower rank approximation to the original data as

$$X_{approx} = Z_m V_m^T + \mathbf{1}_n \bar{x}^T \tag{10.11}$$

where $V_m = [v_1, \ldots, v_m] \in \mathbb{R}^{p\times m}, \quad Z_m = X_c V_m \in \mathbb{R}^{n\times m}$.

This completes the dimensional reduction methodology, and the cumulative proportion of variance explained (PVE) is given by

$$\text{PVE}(m) = \frac{\sum_{j=1}^{m} \lambda_j}{\sum_{j=1}^{p} \lambda_j} \tag{10.12}$$

Having obtained the eigenvalues and eigenvectors and assessed their cumulative variance via the PVE curve, we can reinterpret these quantities as orthogonal building blocks of the covariance matrix. The eigen-expansion rewrites the covariance structure as a weighted sum of rank-one outer products:

$$\text{Cov}(X) = C = \sum_{j=1}^{p} \lambda_j v_j v_j^{\top} \tag{10.13}$$

Because the terms are mutually orthogonal, discarding the smallest eigenvalues yields an optimal Frobenius-norm approximation. The Frobenius norm of a matrix is a measure of the "size" of the matrix, calculated as the square root of the sum of the squares of all its elements. The goal of Frobenius-norm approximation is to find a low-rank matrix that closely resembles the original matrix in terms of its Frobenius norm.

$$\text{Cov}(X) \approx \sum_{j=1}^{m} \lambda_j v_j v_j^{\top}, \qquad m \ll p \tag{10.14}$$

This makes the retained Principal Components preserve the PVE(m) of total variance while filtering directions dominated by noise.

In cases where the observed variables are ordinal (such as Likert scales), the Pearson-based covariance or correlation matrix may not be appropriate for PCA.Instead, a polychoric correlation matrix can be computed, which estimates the correlations between the latent continuous variables assumed to underlie the observed ordinal data. Performing PCA on the polychoric correlation matrix typically yields more accurate components that better reflect the underlying structure of the ordinal data.

Beyond the standard eigen-decomposition approach, PCA can also be efficiently computed via singular value decomposition (SVD), which offers greater numerical stability and flexibility. SVD is especially preferred in high-dimensional settings where the number of variables exceeds the number of observations (p $\gg$ n), or when dealing with sparse or incomplete data. Additionally, SVD-based methods are often computationally faster, particularly when only a subset of the principal components is required, as they avoid explicitly forming the full covariance matrix. The thin SVD of the centered data matrix X_c is given by

$$X_c = U\Sigma V^\top \tag{10.15}$$

where U and V are orthogonal matrices, and Σ is a diagonal matrix of singular values. In this decomposition, the columns of V correspond to the principal directions (analogous to the eigenvectors of the covariance matrix), and the singular values σ_j relate to the variance explained via $\lambda_j = \sigma_j^2/(n-1)$.

For a more thorough review in the theory and recent methodological coverage of PCA,see Jolliffe (2002), Abdi and Williams (2010), Shlens (2014), Jolliffe and Cadima (2016), Hasan and Abdulazeez (2021) and Greenacre et al. (2022). Applied work is provided by Kolenikov and Angeles (2009), Finch et al. (2017), and Salem and Hussein (2019).

10.2.3 Choosing Between PCA and EFA

Both PCA and EFA start from the same correlation or covariance structure, yet their objectives diverge. PCA offers a descriptive low-rank approximation of the data by constructing orthogonal linear combinations of principal components that together explain as much total variance as possible. In contrast, EFA is model-based; it hypothesizes that the observed covariation among observed variables is driven by a smaller set of latent factors, with each variable's unique (error) variance modeled separately. This distinction can be easily represented and distinguished through path diagrams.

Figure 10.1 shows a two-factor ($m = 2$) and three-variable ($p = 3$) models of PCA and EFA in a path diagram representation. The arrows in the path diagrams (Figure 10.1) indicate the dependencies between the variables and the components or factors. One can observe that the directions of the arrows linked to the variables are not the same. For PCA, the components are a weighted average of the independent variables. In contrast, for EFA the causal arrows are reversed, the variables are a weighted average of the latent factors plus an error term ε_i. This is highlighted through the set of equations for PCA, Eq. (10.16) and EFA Eq. (10.17).

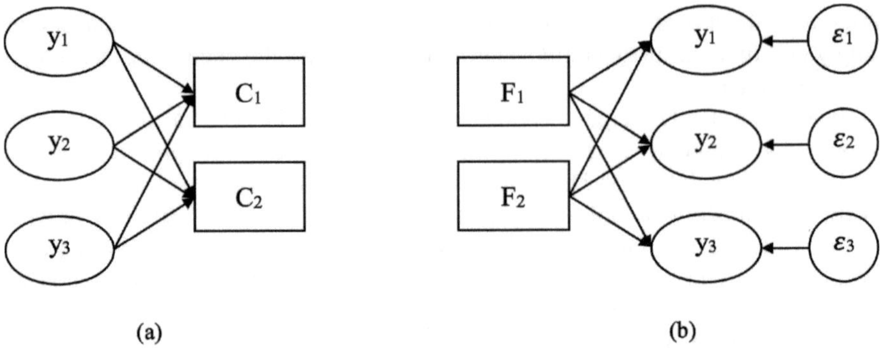

Fig. 10.1 Path Diagrams for (a) PCA and (b) EFA (Source: Authors' representation)

$$\begin{aligned} C_1 &= \gamma_{11} \cdot y_1 + \gamma_{12} \cdot y_2 + \gamma_{13} \cdot y_3 \\ C_2 &= \gamma_{21} \cdot y_1 + \gamma_{22} \cdot y_2 + \gamma_{23} \cdot y_3 \end{aligned} \tag{10.16}$$

$$\begin{aligned} y_1 &= \lambda_{11} \cdot F_1 + \lambda_{21} \cdot F_2 + \varepsilon_1 \\ y_2 &= \lambda_{12} \cdot F_1 + \lambda_{22} \cdot F_2 + \varepsilon_2 \\ y_3 &= \lambda_{13} \cdot F_1 + \lambda_{23} \cdot F_2 + \varepsilon_3 \end{aligned} \tag{10.17}$$

Notice that in PCA the weights are chosen to maximise successively the remaining variance. In EFA the linear combination of latent factors plus an error term means that only the shared (common) variance is attributed to the factors. However, Bentler and Kano (1990) show algebraically that, when certain model constraints are met (e.g., full-rank factor loadings and specific error structures), a common-factor model can be re-parameterised so that the resulting loadings and reproduced covariances are the same as those obtained from a principal-components solution. Stevens (2009) identified that the methods likely have similar results when variables exceed 30 and communalities scores are not less than 0.70. When commonalities scores are low and variables are less than 20, slight variations in results were noticeable (Fabrigar and Wegener 2012).

The general guideline is as follows: employ PCA when a concise set of composite predictors is required for purposes such as regression, clustering, or visualization, and when concerns regarding measurement error or latent meaning are not paramount. Employ EFA when your theoretical framework suggests the presence of latent constructs, and it is necessary to distinguish between common and unique variance prior to proceeding with confirmatory factor analysis or structural equation modeling. Regardless of the chosen methodology, it is essential to report the proportion of variance accounted for (total variance for PCA, common variance for EFA) and provide a rationale for the number of dimensions retained.

10.3 A Guided Application of EFA and PCA

10.3.1 Motivation for the Study

Banks play a pivotal role in driving modern economies through their financial services that stimulate investment, job creation, income generation, and, most importantly, act as financial intermediaries. However, in today's competitive landscape, maintaining strong customer relationships is paramount for banks to gain a competitive edge. To do so, understanding the dimensions of bank services that influence customer satisfaction becomes crucial in enhancing customer experiences, fostering loyalty, and nurturing long-term relationships. Satisfied customers are more likely to increase their deposits, investments, and credit utilization with banks, which collectively boosts economic growth. Therefore, by prioritizing customer satisfaction and building robust relationships, banks can not only strengthen their market position but also play a role in national economic growth.

Customer satisfaction in the banking sector has been extensively studied, often using the SERVQUAL model (Parasuraman et al. 1988). For example, Lenka et al. (2009) explored whether the perception of service quality affects customer satisfaction and loyalty by surveying 350 respondents in Odisha. They found that better human, technical, and other tangible aspects of services significantly increased satisfaction, which was directly correlated with loyalty. Similarly, Kaura et al. (2015) examined the determinants of customer satisfaction using data from 445 respondents in Rajasthan. Through factor and regression analyses, they discovered that customer satisfaction mediates loyalty. In another study, Mittal et al. (2015) investigated the relationship between service quality and satisfaction among 350 respondents in Delhi and the National Capital Region (NCR), finding that tangible factors, especially the physical environment, are crucial for satisfaction. Garg et al. (2014) developed a psychometric scale to measure customer experience in Indian banks, examining 14 factors and their impact on satisfaction, providing a validated tool for assessing customer experiences. Notwithstanding these insights, significant gaps persist in the comprehension of the precise mechanisms that connect service quality dimensions to customer satisfaction and loyalty across India's diverse banking contexts.

Recent technological advancements and increased competition have transformed the banking sector. In this rapidly evolving landscape, customer expectations have also evolved. To meet these changing expectations, banks must leverage digital services and offer personalized solutions tailored to individual needs. Researchers have examined customer adoption and barriers to new technologies. For example, Sivathanu (2019) studied the adoption of digital payments post-demonetization with 766 respondents in Pune, finding that behavioral intention and resistance to innovation influence adoption. They recommend addressing security concerns and usability to facilitate the transition from cash to digital payments. Furthermore, Ligon et al. (2019) investigated the low adoption of digital payment services among 1003 small merchants in Jaipur, attributing it to customer disinterest and tax liability concerns. Sankaran and Chakraborty (2021) analyzed mobile banking usage factors with 457 respondents, and noted that the monetary, emotional, and quality value components, along with trust, were significant, while the performance and social value were not. Saxena et al. (2022) found similar results in a study with 536 respondents in Delhi NCR. Chawla and Joshi (2023) examined mobile wallet adoption among 744 respondents, using structural equation modeling to show that perceived usefulness, trust, and attitude mediate the adoption of technology in nonlinear ways.

Our study shows the application of EFA and PCA to condense the dimensions of banking service quality into smaller thematic constructs. This study therefore aims to address the following research question: What are the key factors of customer service quality in the banking sector?

In the subsequent sections, we will use EFA to primarily assess the factors that affect the satisfaction level of customers of PSBs in India. In the initial stage, before data collection, based on literature reviews and discussions with bank officials and some customers, a questionnaire was developed. The questionnaire collected information using a five-point Likert scale (see Table 10.2 in Appendix A) on eighteen items related to bank service qualities, customer satisfaction, and adoption of mobile banking services and wallet usage, as shown in Table 10.1 (Appendix A). In addition, data was collected on demographic information such as age, sex, location, education level, income level, and banking preferences for each respondent. It should also be noted that to ensure the correctness of the questions, the questionnaire was pilot tested on fifty respondents and modified based on their feedback.

10.3.2 Data and Data Cleaning

We collected data from Kanpur, a large city in Uttar Pradesh, India, over a three-month period in 2019. As people are usually skeptical of anyone who asks questions related to bank services, we approached the officials at the bank regional offices to help us conduct the survey in the bank branches. From the various PSBs approached, only five banks allowed the survey and guided us to their specific bank branches to administer the questionnaire.

Consequently, data collection took place in 27 different branches of these banks using a systematic sampling procedure. Every third person entering a branch was asked to participate in the survey. If a person declined, the next individual was approached. This method resulted in a total of 506 responses. Of these, 436 responses were complete without missing data points, 65 responses had fewer than 10% missing data points, and only 5 responses had more than 10% missing data points. In general, all data variables had less than 10% missing values. In the accompanying files to this chapter, we provide the missing value imputed data to the reader to follow through the guided application of EFA.

```
# Navigate to the data is stored and make it working directory
setwd("path where the data file is stored")
# Load the data
ch11_data <- read.csv("ch11_survey_data.csv")
```

```
> summary(ch11_data)
 Q01      Q02      Q03        Q04          Q05        Q06        Q07          Q08          Q09
 1: 55    1: 42    1: 43    1    : 22   1    : 47   1: 24    1    : 59   1    : 58   1    :101
 2: 24    2: 75    2: 54    2    : 34   2    : 44   2: 41    2    : 65   2    : 36   2    : 52
 3: 10    3: 69    3: 12    3    : 20   3    : 42   3: 35    3    : 50   3    : 45   3    : 72
 4:100    4:118    4:112    4    : 76   4    :134   4: 78    4    :112   4    :137   4    :105
 5:317    5:202    5:285    5    :353   5    :235   5:328    5    :216   5    :225   5    :158
                            NA's:  1    NA's:  4             NA's:  4   NA's:  5   NA's: 18

   Q10        Q11       Q12         Q13       Q14          Q15        Q16       Q17          Q18
 1    : 20   1: 32   1    :  4   1: 42   1    : 52   1    : 24   1: 58   1    : 11   1    : 15
 2    : 24   2: 42   2    : 24   2: 59   2    : 13   2    : 52   2: 50   2    :  6   2    : 25
 3    : 46   3: 18   3    : 23   3: 76   3    : 11   3    : 33   3: 19   3    :  3   3    : 18
 4    :108   4:128   4    : 51   4:122   4    : 27   4    : 47   4: 82   4    : 41   4    :104
 5    :293   5:286   5    :387   5:207   5    :402   5    :348   5:297   5    :443   5    :316
 NA's: 15            NA's: 17            NA's:  1   NA's:  2            NA's:  2   NA's: 28

                               C01                    E01                    E02
 _Completely Disagree        : 43   _No          :331   _No          :337
 _Neither Agree nor Disagree:192   _Yes         :118   _Yes         : 84
 _Completely Agree           :271   _AlreadyUsing: 56   _AlreadyUsing: 82
                                    NA's         :  1   NA's         :  3
```

Little's (1988) test statistic was used to assess whether the missing data followed a completely random pattern, with the null hypothesis assuming that values were missing completely at random (MCAR) (Li 2013). The test did not yield a statistically significant result, indicating that the missing data pattern is likely random. Specifically, with a p-value of 0.421, we do not have sufficient evidence to conclude that the missingness is related to the data itself. This is advantageous for handling missing data, as it allows for the use of methods like data imputation without the risk of biased estimates due to non-random missing data.

```
> # Check if the data is missing completely at random (MCAR)
> library(naniar)
> mcar_test_result <- mcar_test(ch11_data)
> print(mcar_test_result)
# A tibble: 1 × 4
  statistic    df p.value missing.patterns
      <dbl> <dbl>   <dbl>            <int>
1      932.   924   0.421               33
```

To enhance the robustness of our analysis, we retained only individuals with less than 10% missing data points, resulting in 501 responses. We then performed data imputation for the remaining individuals with missing data using the Multivariate Imputation by Chained Equations (MICE) method, assuming that the missing data adhered to the missing at random (MAR) mechanism (Van Buuren and Oudshoorn 2000). The mice() function generates multiple imputations for missing values in the data frame df using different techniques: predictive mean matching (pmm) for numeric data, logistic regression imputation (logreg) for binary data with two-factor levels, polytomous regression imputation (polyreg) for unordered categorical data with more than two-factor levels, and the proportional odds model (polr) for ordered categorical data with more than two-factor levels.

```
> # Filter out rows with more than 10% missing data
> calculate_missing_percentage <- function(x) {
+   sum(is.na(x)) / length(x) * 100
+ }
> missing_percentage_rows <- apply(df, 1, calculate_missing_percentage)
> ch11_data <- ch11_data[missing_percentage_rows <= 10, ]
```

```
> # Impute missing values
> library(mice)
> imp <- mice(ch11_data, defaultMethod = c("pmm", "logreg", "polyreg", "polr"), seed = 1)
> ch11_data <- complete(imp)
```

10.3.3 Exploratory Factor Analysis

We employ EFA to examine the dataset and explore the relationships among the service quality items in the questionnaire. These items, labeled Q01 to Q18, are measured on a 5-point Likert scale. To prepare the data for analysis, we select the first 18 columns of the data frame (Q01 to Q18), which correspond to the service quality items. We convert these categorical items from class factor to class numeric, enabling the application of the necessary functions for factor analysis.

```
> # Subsetting the data
> EFA_data <- ch11_data[,c(1:18)]
```

```
> # Converting the factor columns (if any) to numeric columns
> EFA_data[] <- lapply(EFA_data, as.character)
> EFA_data[] <- lapply(EFA_data, as.numeric)
```

Our approach involves using polychoric correlations, particularly suitable for our ordinal construction from the Likert scale of the questionnaires (Holgado-Tello et al. 2010). Polychoric correlations have received extensive consideration because they assume that an unobservable normally distributed continuous variable underlies each observed categorical variable and estimates the Pearson correlation between those underlying hypothetical variables (Watkins 2018). Figure 10.2 illustrates the polychoric correlation between the ordinal responses in service quality responses.

```
> # Calculate the polychoric correlation matrix
> polychoric_result = polychoric(EFA_data)
>
> # Extract the correlation matrix
> correlation_matrix = polychoric_result$rho
>
> # Plot the polychoric correlation matrix
> cor.plot(correlation_matrix,
+          numbers = TRUE,
+          upper = FALSE,
+          main = "Polychoric Correlation",
+          show.legend = FALSE)
```

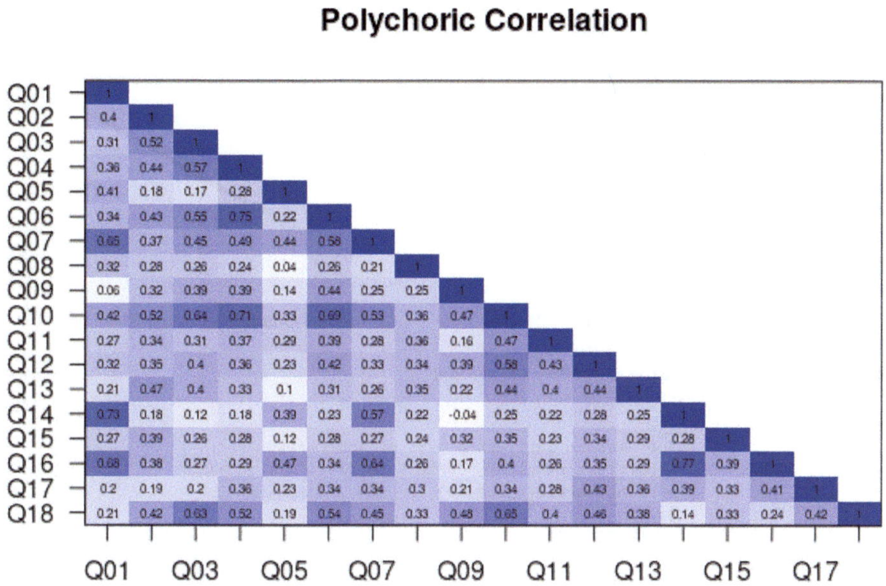

Fig. 10.2 Polychoric Correlation Table

10.3.4 Checking if Data is Valid for a Factor Analysis

To assess the suitability of the dataset for factor analysis. The Kaiser-Meyer-Olkin measure (Kaiser 1974) of sampling adequacy yielded an overall value of 0.88 for the complete model. This surpassed the recommended threshold of 0.7, indicating that the data is highly suitable for factor analysis (Dabbagh et al. 2023). Moreover, individual variables exceeded the minimum adequacy criterion, with all values surpassing 0.5.

```
> # Perform Kaiser-Meyer-Olkin (KMO) measure of sampling adequacy
> library(psych)
> kmo_result <- KMO(EFA_data)
> print(kmo_result)
Kaiser-Meyer-Olkin factor adequacy
Call: KMO(r = EFA_data)
Overall MSA =  0.88
MSA for each item =
 Q01  Q02  Q03  Q04  Q05  Q06  Q07  Q08  Q09  Q10  Q11  Q12  Q13  Q14  Q15  Q16  Q17  Q18
0.81 0.90 0.93 0.88 0.85 0.88 0.89 0.88 0.90 0.92 0.89 0.92 0.87 0.79 0.90 0.86 0.86 0.91
```

To complement this, Bartlett's test of sphericity (Bartlett 1954) was employed to examine the null hypothesis that the correlation matrix is an identity matrix (Shrestha 2021). A significant result (low p-value) indicates that the correlation matrix significantly deviates from an identity matrix, suggesting that factor analysis is appropriate.

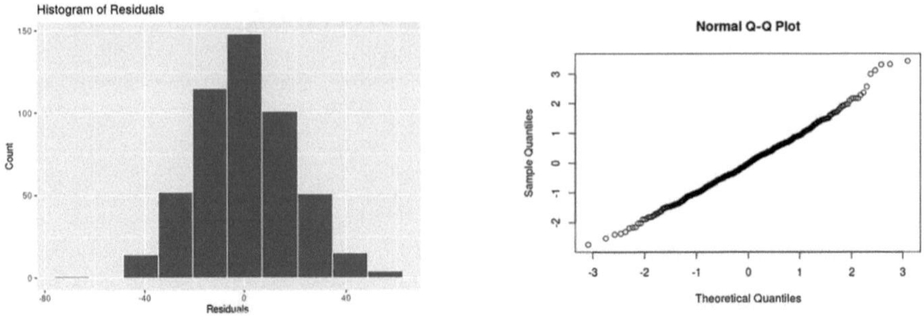

(a) Histogram of Residuals (b) Quantile-Quantile Plot

Fig. 10.3 Plots to check the normality of residuals

```
> # Perform Bartlett's test of sphericity
> bartlett_result <- cortest.bartlett(EFA_data)
> print(bartlett_result)
$chisq
[1] 2896.612

$p.value
[1] 0

$df
[1] 153
```

Additionally, we ensured the normality of residuals by generating random data and fitting a linear model, where the sample quantile corresponded closely to the theoretical quantile. Figures 10.3a and 10.3b present residuals' histogram and QQ plot, indicating that the residuals are approximately normally distributed (Pérez and Medrano 2014).

```
> # Checking for normality
> set.seed(123) # Setting seed for reproducibility
> random <- rchisq(nrow(EFA_data), 190)
> fake <- lm(random ~ ., data = data.frame(EFA_data))
> standardized <- rstudent(fake)
> fitted <- scale(fake$fitted.values)
> ggplot2::ggplot(data = data.frame(Residuals = standardized), aes(x = Residuals)) +
+   geom_histogram(bins = 10, fill = "grey", color = "black") +
+   ggplot2::labs(title = "Residuals Histogram") +
+   ggplot2::theme_minimal()
>
> # Checking for linearity
> qqnorm(standardized)
```

10.3.5 Factor Extraction

Several factor extraction methods are available, each with unique characteristics and advantages, including unweighted least squares, generalized least squares, maximum likelihood, alpha factoring, image factoring, and principal component analysis. In R, the choice of factor extraction method can be specified using the "fm" parameter. The fm = "pa" method uses the principal factor solution, and fm = "ml" performs a maximum likelihood factor analysis.

Choosing the best extraction method often requires empirical testing, considering factors such as the distribution of the data, the size of the dataset, and the number of variables involved. The study by Mabel and Olayemi (2020) and Coughlin (2013) provides a comparison of different extraction techniques. For our subsequent factor analysis, we will use principal axis factoring (PAF) as a reliable method to extract factors from a large set of variables. PAF has the advantage of being able to analyze both correlations and covariances, similar to PCA. Instead of using the full correlation matrix among variables, PAF uses a reduced correlation matrix.

10.3.6 Determining Number of Factors

Various approaches can be used to determine the number of factors to be extracted. It is widely recommended in numerous studies to adopt a comprehensive strategy, combining multiple criteria and theoretical considerations to arrive at the optimal number of factors to retain in the model (Yong et al. 2013; Ruscio and Roche 2012). Kaiser's eigenvalue criterion (Kaiser 1960) suggests retaining any factors with an eigenvalue above one. The scree test (Cattell 1966) helps determine the number of initial unrotated factors by identifying the distinct point of break in the slope of the scree plot. Additionally, parallel analysis (Horn 1965) generates a random dataset to simulate eigenvalues, allowing us to compare them with the eigenvalues obtained from the actual data. From Figure 10.4, the eigenvalue test recommended retaining two factors, while both the scree test and parallel analysis suggested three factors as a reasonable number of latent factors to extract. After careful consideration, we proceeded with a three-factor model.

Fig. 10.4 Determining the Number of Factors

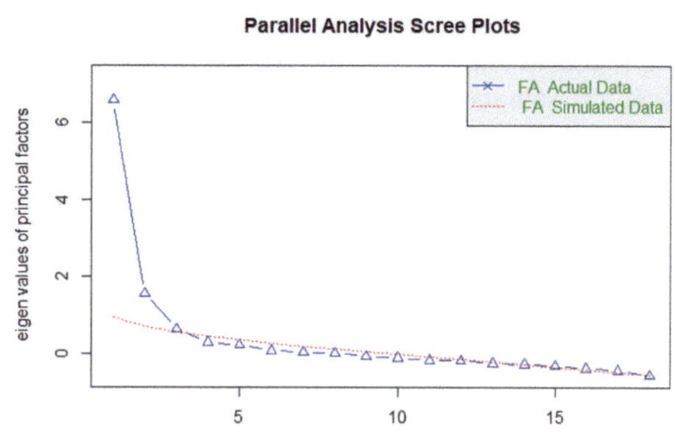

```
> # Perform parallel analysis, using factor extraction method as principal axis factoring
> parallel_test <- fa.parallel(correlation_matrix, fm="pa", fa="fa")
> print(parallel_test)

> # Display the eigenvalues
> print(parallel_test$fa.values)

 [1]  6.572649907  1.548819773  0.629660581  0.286182229  0.213310888  0.073996006
 [7]  0.037518552 -0.002956986 -0.069719840 -0.121090441 -0.167386763 -0.188952433
[13] -0.251081118 -0.281773733 -0.323323921 -0.385474897 -0.436518271 -0.560890936
```

10.3.7 Factor Rotation

Factor rotation is a technique used in factor analysis to improve the interpretability of the results. While it does not enhance the fundamental aspects of factor analysis, it increases the utility of the identified factors, making the findings more meaningful and easier to understand. Choosing the appropriate rotation method is crucial. Orthogonal rotations, such as varimax, quartimax, and equamax, produce uncorrelated factors. In contrast, oblique rotations, including direct oblimin, quartimin, and promax, allow the factors to correlate. We are utilizing the varimax rotation (Kaiser 1958), which attempts to load each variable highly on one factor and minimally on others, which helps to achieve a clearer and more interpretable factor structure. This method is particularly useful when the goal is to obtain orthogonal factors, meaning the factors remain independent and do not correlate with each other.

10.3.8 Factor Scores

Factor scores are composite variables representing an individual's placement on the factor(s) identified in EFA. They are used to summarize individual responses on a set of observed variables, allowing for further analysis such as hypothesis testing or regression. Factor scores can be computed using refined methods (like regression, Bartlett, and Anderson-Rubin scores) or non-refined methods (like sum scores or weighted sum scores). The choice of method affects the properties of the factor scores, such as their correlation with the estimated factor and other factors and whether they are unbiased estimates. We must consider the assumptions of the method and the quality of the data when using factor scores. DiStefano et al. (2019) provides a comprehensive overview on the use of factor scores following EFA, highlighting methods to create them and their applications.

10.3.9 Implementing EFA on the Bank Customer Survey Data

Three factors were extracted using exploratory factor analysis with polychoric correlations and employing principal axis factoring with varimax rotation. Varimax rotation is justified for achieving an orthogonal structure in factor analysis, as it maximizes the sum of variances of squared loadings, resulting in uncorrelated rotated factors and maintaining the ability to reproduce the original correlation matrix (Kline 2014). This approach successfully explained 52% of the variance present in the original dataset. To ensure factor adequacy, the factor loadings are of at least 0.3 to be deemed practically significant in assessing the influence and relevance of variables within the factor structure (Streiner 1994). Figure 10.5 highlights the highest factor loading of at least 0.45 for each variable on the respective factor component. Given the ordinal nature of the data, we applied an ordinal standardized item alpha coefficient to assess reliability (Zumbo et al. 2007). The results indicate strong reliability across all three factors. The raw alpha coefficient for factor one was 0.83, while for factor two, it was 0.81, and for factor three, it stood at 0.62. These coefficients consistently exceeded the acceptable threshold for reliability (Cronbach 1951).

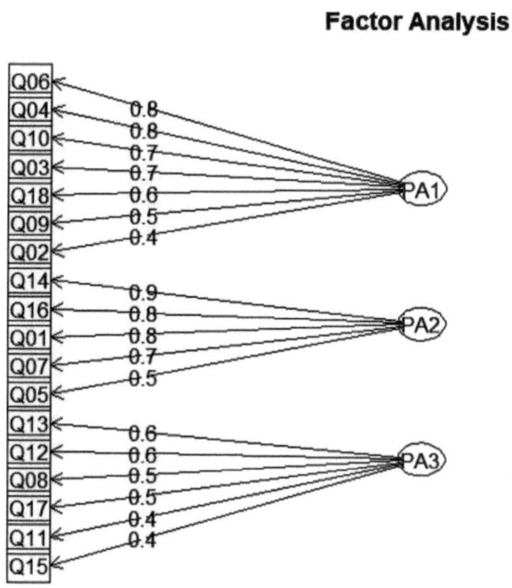

Fig. 10.5 Visual Representation of the Exploratory Factor Analysis

```
> # Perform factor analysis
> EFA_fit <- fa(EFA_data, nfactor = 3, rotate = "varimax", fm = "pa", cor = "poly",
scores = "Anderson")
> summary(EFA_fit)

Factor analysis with Call: fa(r = EFA_data, nfactors = 3, rotate = "varimax",
scores = "Anderson", fm = "pa", cor = "poly")

Test of the hypothesis that 3 factors are sufficient.
The degrees of freedom for the model is 102  and the objective function was  1.34
The number of observations was  501  with Chi Square =  657.98  with prob <  3.7e-82

The root mean square of the residuals (RMSA) is  0.04
The df corrected root mean square of the residuals is  0.05

Tucker Lewis Index of factoring reliability =  0.828
RMSEA index =  0.104  and the 10 % confidence intervals are  0.097 0.112
BIC =  23.89
>
> # Print the factor analysis results with
> # 'cut = 0.45': suppress factor loadings less than 0.45 for clarity
> # 'digits = 2': display loadings with 2 decimal places
> # 'sort = TRUE': sort the loadings within factors
> print(EFA_fit, cut = 0.45, digits = 2, sort = TRUE)
Factor Analysis using method =  pa
Call: fa(r = data, nfactors = 3, rotate = "varimax", scores = "Anderson",
    fm = "pa", cor = "poly")
Standardized loadings (pattern matrix) based upon correlation matrix
    item  PA1  PA2  PA3   h2   u2 com
Q06    6 0.75            0.68 0.32 1.4
Q04    4 0.75            0.66 0.34 1.3
Q10   10 0.74            0.77 0.23 1.8
Q03    3 0.67            0.55 0.45 1.5
Q18   18 0.62            0.59 0.41 1.9
Q09    9 0.49            0.33 0.67 1.7
Q02    2                 0.39 0.61 2.4
Q14   14      0.87       0.85 0.15 1.2
Q16   16      0.80       0.76 0.24 1.4
Q01    1      0.78       0.68 0.32 1.2
Q07    7 0.48 0.68       0.71 0.29 1.9
Q05    5      0.48       0.27 0.73 1.4
Q13   13           0.61 0.44 0.56 1.3
Q12   12           0.59 0.48 0.52 1.8
Q08    8           0.50 0.29 0.71 1.3
Q17   17           0.47 0.32 0.68 1.9
Q11   11                0.32 0.68 2.1
Q15   15                0.27 0.73 1.9

                       PA1  PA2  PA3
SS loadings           3.59 3.15 2.61
Proportion Var        0.20 0.17 0.15
Cumulative Var        0.20 0.37 0.52
Proportion Explained  0.38 0.34 0.28
Cumulative Proportion 0.38 0.72 1.00
```

```
Mean item complexity =  1.6
Test of the hypothesis that 3 factors are sufficient.

df null model =  153  with the objective function =  10.16 with Chi Square =  5010.66
df of  the model are 102  and the objective function was  1.34

The root mean square of the residuals (RMSR) is  0.04
The df corrected root mean square of the residuals is  0.05

The harmonic n.obs is  501 with the empirical chi square  288.27  with prob <  1.1e-19
The total n.obs was  501  with Likelihood Chi Square =  657.98  with prob <  3.7e-82

Tucker Lewis Index of factoring reliability =  0.828
RMSEA index =  0.104  and the 90 % confidence intervals are  0.097 0.112
BIC =  23.89
Fit based upon off diagonal values = 0.99
Measures of factor score adequacy
                                                  PA1  PA2  PA3
Correlation of (regression) scores with factors   0.92 0.95 0.83
Multiple R square of scores with factors          0.86 0.90 0.70
Minimum correlation of possible factor scores     0.71 0.79 0.39
>
> # Visual interpretation of the factor analysis
> fa.diagram(EFA_fit)
>
> # Define the items associated with each factor
> factor1 <- c("Q06", "Q04", "Q10", "Q03", "Q18", "Q09", "Q02")
> factor2 <- c("Q14", "Q16", "Q01", "Q07", "Q05")
> factor3 <- c("Q13", "Q12", "Q08", "Q17", "Q11", "Q15")
>
> # Check the normal internal reliability for each factor
> alpha_factor1 <- psych::alpha(data[, factor1], check.keys = TRUE)
> alpha_factor2 <- psych::alpha(data[, factor2], check.keys = TRUE)
> alpha_factor3 <- psych::alpha(data[, factor3], check.keys = TRUE)
>
> print(alpha_factor1)

Reliability analysis
Call: psych::alpha(x = data[, factor1], check.keys = TRUE)

  raw_alpha std.alpha G6(smc) average_r S/N   ase mean   sd median_r
       0.83      0.84    0.83      0.43 5.2 0.012  4.1 0.87     0.41

    95% confidence boundaries
         lower alpha upper
Feldt      0.8  0.83  0.85
Duhachek   0.8  0.83  0.85
```

```
 Reliability if an item is dropped:
    raw_alpha std.alpha G6(smc) average_r S/N alpha se  var.r med.r
Q06      0.80      0.81    0.79      0.41 4.2    0.014 0.0091  0.39
Q04      0.80      0.81    0.79      0.42 4.4    0.014 0.0079  0.41
Q10      0.78      0.79    0.78      0.39 3.9    0.015 0.0079  0.37
Q03      0.80      0.81    0.80      0.42 4.3    0.014 0.0123  0.37
Q18      0.80      0.82    0.80      0.43 4.4    0.014 0.0124  0.43
Q09      0.83      0.84    0.82      0.46 5.2    0.012 0.0089  0.44
Q02      0.82      0.83    0.82      0.45 4.9    0.013 0.0109  0.44

 Item statistics
      n raw.r std.r r.cor r.drop mean  sd
Q06 501  0.74  0.75  0.71   0.63  4.3 1.2
Q04 501  0.71  0.73  0.68   0.60  4.4 1.1
Q10 501  0.80  0.82  0.79   0.72  4.3 1.1
Q03 501  0.75  0.74  0.68   0.62  4.1 1.3
Q18 501  0.70  0.71  0.65   0.59  4.4 1.0
Q09 501  0.64  0.60  0.49   0.45  3.3 1.5
Q02 501  0.65  0.64  0.53   0.49  3.7 1.3

Non missing response frequency for each item
       1    2    3    4    5 miss
Q06 0.05 0.08 0.07 0.16 0.64    0
Q04 0.04 0.07 0.04 0.15 0.70    0
Q10 0.04 0.05 0.10 0.22 0.59    0
Q03 0.09 0.11 0.02 0.22 0.57    0
Q18 0.03 0.06 0.05 0.22 0.64    0
Q09 0.21 0.11 0.15 0.22 0.32    0
Q02 0.08 0.15 0.14 0.23 0.40    0
> print(alpha_factor2)

Reliability analysis
Call: psych::alpha(x = data[, factor2], check.keys = TRUE)

  raw_alpha std.alpha G6(smc) average_r S/N   ase mean sd median_r
       0.81      0.81    0.79      0.46 4.3 0.013    4  1     0.46

    95% confidence boundaries
         lower alpha upper
Feldt     0.78  0.81  0.84
Duhachek  0.79  0.81  0.84

 Reliability if an item is dropped:
    raw_alpha std.alpha G6(smc) average_r S/N alpha se  var.r med.r
Q14      0.77      0.77    0.72      0.45 3.3    0.017 0.0076  0.45
Q16      0.75      0.75    0.71      0.43 3.0    0.018 0.0109  0.40
Q01      0.76      0.76    0.72      0.44 3.1    0.017 0.0111  0.40
Q07      0.78      0.78    0.74      0.46 3.4    0.016 0.0163  0.47
Q05      0.82      0.82    0.78      0.53 4.5    0.013 0.0045  0.54
```

```
 Item statistics
      n raw.r std.r r.cor r.drop mean  sd
Q14 501  0.76  0.77  0.70   0.62  4.4 1.3
Q16 501  0.82  0.81  0.76   0.68  4.0 1.4
Q01 501  0.79  0.79  0.73   0.66  4.2 1.3
Q07 501  0.76  0.75  0.66   0.60  3.7 1.4
Q05 501  0.64  0.65  0.49   0.45  3.9 1.3

Non missing response frequency for each item
       1    2    3    4    5 miss
Q14 0.10 0.03 0.02 0.05 0.79    0
Q16 0.12 0.10 0.04 0.16 0.58    0
Q01 0.11 0.05 0.02 0.20 0.62    0
Q07 0.12 0.13 0.10 0.22 0.43    0
Q05 0.09 0.09 0.09 0.27 0.47    0
> print(alpha_factor3)

Reliability analysis
Call: psych::alpha(x = data[, factor3], check.keys = TRUE)

  raw_alpha std.alpha G6(smc) average_r S/N   ase mean   sd median_r
      0.62      0.62    0.59      0.22 1.7 0.026  4.2 0.67     0.21

    95% confidence boundaries
         lower alpha upper
Feldt     0.56  0.62  0.67
Duhachek  0.57  0.62  0.67

 Reliability if an item is dropped:
    raw_alpha std.alpha G6(smc) average_r S/N alpha se  var.r med.r
Q13      0.54      0.55    0.50      0.20 1.2    0.031 0.0038  0.20
Q12      0.56      0.55    0.51      0.20 1.2    0.030 0.0045  0.18
Q08      0.57      0.58    0.54      0.22 1.4    0.029 0.0050  0.21
Q17      0.61      0.62    0.57      0.25 1.6    0.027 0.0032  0.24
Q11      0.55      0.56    0.51      0.20 1.3    0.031 0.0025  0.21
Q15      0.60      0.60    0.56      0.23 1.5    0.027 0.0052  0.24

 Item statistics
      n raw.r std.r r.cor r.drop mean   sd
Q13 501  0.67  0.64  0.54   0.42  3.8 1.31
Q12 501  0.59  0.64  0.53   0.42  4.6 0.88
Q08 501  0.63  0.58  0.44   0.36  3.9 1.36
Q17 501  0.41  0.51  0.32   0.25  4.8 0.72
Q11 501  0.64  0.62  0.51   0.41  4.2 1.22
Q15 501  0.56  0.54  0.37   0.29  4.3 1.23

Non missing response frequency for each item
       1    2    3    4    5 miss
Q13 0.08 0.12 0.15 0.24 0.41    0
Q12 0.01 0.05 0.05 0.11 0.78    0
Q08 0.12 0.07 0.09 0.27 0.45    0
Q17 0.02 0.01 0.01 0.08 0.88    0
Q11 0.06 0.08 0.04 0.25 0.57    0
```

```
Q15 0.05 0.10 0.07 0.09 0.69    0
>
> # Extracting the factor scores
> factor_score <- as.data.frame(EFA_fit$scores)
> names(factor_score)<- c("ande_pa1","ande_pa2","ande_pa3")
```

10.3.10 Results from EFA

The exploratory factor analysis established a three-factor structure. This method helps uncover latent constructs within the dataset, allowing us to conceptualize these constructs theoretically (Bandalos and Finney 2018). Within this structure of 18 items of customer service quality, we identified the following factor groupings that are presented below, ordered in decreasing value of factor loading.

Factor 1, which can be labeled as "Customer-Employee Interaction", consists of the following items (see Table 10.1 in Appendix A): Q06 (*Bank employees always listen attentively to my problems*), Q04 (*Employees of this bank are consistently courteous to me*), Q10 (*Behavior of employees of this bank instills confidence in me toward the bank*), Q03 (*I am able to complete my transactions quickly*), Q18 (*Transaction processes are easy which saves a lot of labor*), Q09 (*Employees of this bank give me personal attention*), Q02 (*The bank provides complete information about different terms and conditions*). Each item within this factor reflects different aspects of this interaction, focusing on elements such as attentiveness, courtesy, confidence building, efficiency, personal attention, and the provision of comprehensive information.

Factor 2, which can be denoted as "Convenience and Efficiency", comprises these questionnaire items (see Table 10.1 in Appendix A): 14 (*Forms needed for various services in the bank are readily available*), 16 (*The bank officials are always available when I need them*), 1 (*The hours of operation of the bank are very convenient*), 7 (*Bank always takes very little time in solving my problem*), 5 (*The bank uses new and modern equipment/machines*). Factor 2 includes elements related to the ease of accessing banking services and the efficiency with which these services are provided.

Factor 3, which can be characterized as "Aesthetics, Comfort and Transparency", includes the following items (see Table 10.1 in Appendix A): 13 (*Various information boards, pamphlets, advertisement in the bank premises are visually appealing and informative*), 12 (*The bank employees always have neat and attractive appearance*), 8 (*Bank premises have good security provisions*), 17 (*Location of the bank branch is easy to access*), 11 (*The bank's facilities like furniture, etc., are comfortable and put me at ease*), 15 (*There are no hidden charges in the services offered by the bank*). This factor revolves around and mostly directs toward aspects related to the physical environment and security provisions, the aesthetics of the bank, as well as the level of comfort and transparency it provides to customers.

10.3.11 Implementing PCA on the Bank Cutomer Survey data

The analysis for PCA were performed with the principal() and fa.parallel()routines of the **psych** package. Following the analytical framework presented earlier (from Sect. 10.3.2 to Sect. 10.3.8), we conducted a polychoric principal component analysis (PCA) on the set of 18 ordinal SERVQUAL items. After converting the Likert-type responses into ordered factors, we estimated the latent correlations among items with the polychoric() function, resulting in an 18×18 polychoric correlation matrix. To determine the appropriate number of components, a parallel analysis (fa.parallel() with fa = "pc" and 100 iterations) was performed. This procedure indicated that three principal components had eigenvalues exceeding their corresponding random benchmarks, thus supporting a three-component PCA solution.

```
> # Read in the data
> ch10_data <- read.csv("ch10_survey_data.csv")
>
> # Subselect and change the class
> PCA_data <- ch10_data[, 1:18]
> PCA_data[] <- lapply(PCA_data, function(x) as.numeric(as.character(x)))
>
> # Find the polychoric correlation
> PCA_ord <- as.data.frame(lapply(PCA_data, ordered))
> poly_out <- polychoric(PCA_ord)
Converted non-numeric input to numeric
> Rho      <- poly_out$rho    # extract the correlation matrix
>
> # Run a Parallel Analysis to decide how many components to retain
> fa.parallel(
+   Rho,
+   n.obs  = nrow(PCA_data),
+   fa     = "pc",
+   n.iter = 100,
+   main   = "Parallel Analysis Scree Plots (PCA)"
+ )
Parallel analysis suggests that the number of factors =  NA
and the number of components =  3
>
> # Perform PCA on the polychoric correlation matrix
> # nfactors = parallel analysis suggests 3 components
> pc_poly <- principal(
+   r         = Rho,
+   nfactors = 3,
+   rotate    = "varimax",
+   scores    = FALSE
+ )
>
> # Summary of the analysis
> pc_poly
```

```
Principal Components Analysis
Call: principal(r = Rho, nfactors = 3, rotate = "varimax", scores = FALSE)
Standardized loadings (pattern matrix) based upon correlation matrix
     RC1   RC2   RC3   h2   u2 com
1   0.16  0.82  0.20 0.74 0.26 1.2
2   0.50  0.20  0.39 0.44 0.56 2.2
3   0.76  0.12  0.21 0.63 0.37 1.2
4   0.79  0.24  0.14 0.70 0.30 1.3
5   0.22  0.63 -0.03 0.44 0.56 1.2
6   0.78  0.26  0.16 0.71 0.29 1.3
7   0.47  0.72  0.10 0.75 0.25 1.8
8   0.12  0.07  0.67 0.47 0.53 1.1
9   0.61 -0.09  0.25 0.44 0.56 1.4
10  0.77  0.26  0.35 0.78 0.22 1.6
11  0.32  0.16  0.50 0.38 0.62 1.9
12  0.37  0.16  0.61 0.53 0.47 1.8
13  0.25  0.05  0.69 0.55 0.45 1.3
14 -0.11  0.86  0.31 0.84 0.16 1.3
15  0.19  0.19  0.53 0.36 0.64 1.5
16  0.10  0.82  0.34 0.80 0.20 1.4
17  0.15  0.26  0.59 0.43 0.57 1.5
18  0.69  0.06  0.39 0.64 0.36 1.6

                       RC1  RC2  RC3
SS loadings           4.20 3.41 3.03
Proportion Var        0.23 0.19 0.17
Cumulative Var        0.23 0.42 0.59
Proportion Explained  0.39 0.32 0.28
Cumulative Proportion 0.39 0.72 1.00


Mean item complexity =  1.5
Test of the hypothesis that 3 components are sufficient.

The root mean square of the residuals (RMSR) is  0.06

Fit based upon off diagonal values = 0.97>
> # Find the factor scores, using regression method score
> fs <- factor.scores(
+   x      = PCA_data,
+   f      = pc_poly,
+   method = "regression"
+ )
>
> # Inject those scores back into your psych-PCA object
> pc_poly$scores <- fs$scores
>
> # Extract the first three columns of the scores matrix
> PC12 <- fs$scores[, 1:3]
>
> # Extracting the principal components as a dataset
```

```
> Z <- as.data.frame(PC12)
> colnames(Z) <- c("PC1_score", "PC2_score" , "PC3_score")
> dim(Z)
[1] 501    3
>
> # Create the Biplot using 2 factors
> pc_poly2 <- principal(r = Rho, nfactors = 2, rotate    = "varimax", scores    = TRUE)
> fs2 <- factor.scores( x = PCA_data, f = pc_poly2, method = "regression")
> pc_poly2$scores <- fs2$scores
> biplot(pc_poly2, main = " ", cex = c(0.6, 0.8))
```

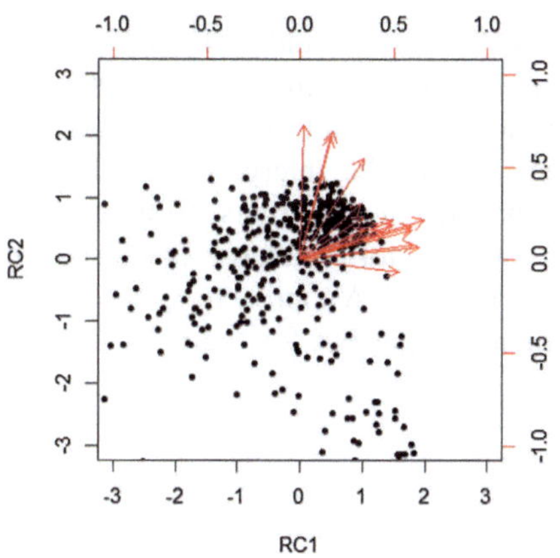

(a) Biplot of PC1 and PC2 loadings

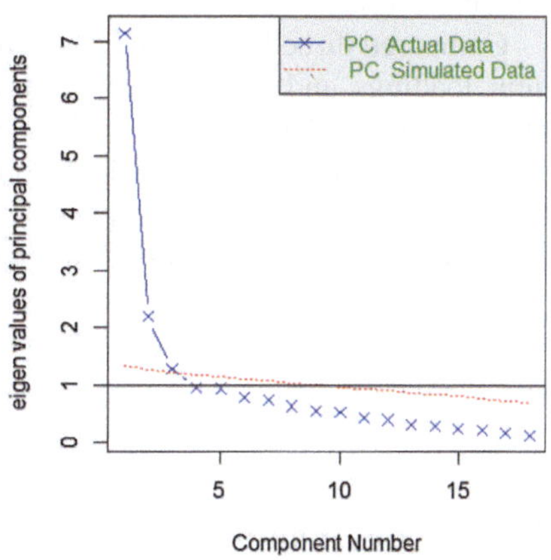

(b) Screeplot

Fig. 10.6 PCA results

Figure 10.6a illustrates a Biplot chart that represents the two most significant principal components, identified from the sequence of components arranged in descending order of significance. Based on Fig. 10.6b, three components were subsequently extracted using the principal() function and subjected to orthogonal rotation employing the varimax criterion. The varimax rotation was applied to enhance interpretability by maximizing the variance of squared loadings across components. This orthogonal rotation approach maintains component independence, distinguishing PCA's total variance-oriented nature from exploratory factor analysis (EFA), which primarily isolates common variance.

The resulting three-component PCA model accounted for approximately 59% of the total observed variance: 23% attributed to Component 1, 19% to Component 2, and 17% to Component 3. This compares favourably to the previously reported three-factor EFA solution, which explained 52% of the common variance. The observed increase in explained variance under PCA reflects its broader objective of maximising the reconstruction of total variance, as opposed to isolating only shared variance as in EFA.

Inspection of rotated loadings (thresholded at |0.45| for interpretability) revealed distinct patterns consistent with, yet different from, those obtained in the EFA analysis.The first principal component clearly emphasized customer-employee interaction,combining items related to employee attentiveness, courtesy, and transaction efficiency (items Q03, Q04, Q06, Q09, Q10, and Q18). The second component captured aspects of convenience and accessibility, including items on branch hours, availability of assistance,and modern equipment (Q01, Q05, Q07, Q14, and Q16). The third component grouped items focused on ambience, transparency of charges, and branch location (Q08, Q11,Q12, Q13, Q15, and Q17), forming a coherent dimension related to physical environment and transparency.

Compared directly to the EFA results, the PCA solution yielded slightly stronger item loadings and clearer simple structures for certain items (notably Q05 and Q11). However,due to PCA's intrinsic design, which does not separate unique from common variance,these principal components should be interpreted as descriptive, variance-preserving composites rather than latent constructs. Consequently, researchers primarily interested in reliability, latent factor structures, or subsequent structural equation modeling analyses should prefer the EFA approach, whereas those seeking concise summaries of observed variance might opt for the PCA solution described here.

10.4 Conclusion

This chapter provides a comprehensive guide on the practical implementation of EFA and PCA, particularly highlighting current best practices involving polychoric correlations suitable for handling ordinal Likert-scale data. Through detailed step-by-step coding examples and thorough methodological discussions, the chapter has illustrated how these techniques effectively uncover meaningful insights from survey data collected from bank customers in a large city in India. While EFA highlighted theoretically grounded latent constructs, offering robust interpretative frameworks, PCA delivered clear, descriptive, variance-maximizing summaries ideal for concise data representation. The choice between EFA and PCA, therefore, should closely align with specific research objectives, the nature of the data, and the subsequent analytical processes required.

In our study, both EFA and PCA indicated that the 18-item questionnaire can be dimensionally reduced to three factors or principal components. In the context of customer satisfaction studies, such techniques help banks and financial institutions to focus their efforts on the most impactful aspects of their services, potentially leading to more efficient resource allocation and improved customer experiences. It should also be acknowledged that this study has certain limitations. Firstly, data collection was limited to Kanpur city in Uttar Pradesh, which may not be representative of the entire Indian banking sector. Secondly, the study focused on PSBs, excluding other types of banks from the analysis. As the adoption of digital banking increases rapidly, customer expectations and perceptions of service quality will evolve. Future research could expand the geographical scope and include a wider range of bank types to provide a more comprehensive picture of customer satisfaction in the Indian banking sector.

In conclusion, this study not only provides specific insights into customer satisfaction in Indian public sector banks but also demonstrates the power of dimensionality reduction techniques to uncover lower dimensional structures in complex datasets. As industries continue to expand rapidly, generating substantial volumes of data, data dimensionality reduction techniques such as Exploratory Factor Analysis (EFA) and Principal Component Analysis (PCA) will be essential to understand and address both organizational and customer requirements. Readers who wish to gain a broader yet introductory understanding can start with Yong et al. (2013) for EFA and Rudolph et al. (2023) for PCA. For advanced materials, Holgado-Tello et al. (2010) and Shrestha (2021) are recommended for EFA, while Abdi and Williams (2010) and Greenacre et al. (2022) are suggested for PCA.

Appendix A: Questionnaire Items and Scale

See Tables 10.1 and 10.2.

Table 10.1 Bank Customer Survey Constructs

Code	Constructs
Q01	The hours of operation of the bank are very convenient.
Q02	The bank provides complete information about different terms and conditions, interest rates, charges, offers, etc.
Q03	I am able to complete my transactions quickly.
Q04	Employees of this bank are consistently courteous to me.
Q05	The bank uses new and modern equipment/machines.
Q06	Bank employees always listen attentively to my problems.
Q07	The bank always takes very little time in solving my problems.
Q08	Bank premises have good security provisions.
Q09	Employees of this bank give me personal attention.
Q10	Behavior of employees of this bank instills confidence in me toward the bank.
Q11	The bank's facilities like furniture, etc., are comfortable and put me at ease.
Q12	The bank employees always have neat and attractive appearance.
Q13	Various information boards, pamphlets, advertisements in the bank premises are visually appealing and informative.

(continued)

Table 10.1 (continued)

Code	Constructs
Q14	Forms needed for various services in the bank (like pay-in-slip, withdrawal form, etc.) are readily available.
Q15	There are no hidden charges in the services offered by the bank.
Q16	The bank officials are always available when I need them.
Q17	Location of the bank branch is easy to access.
Q18	Transaction processes are easy, which saves a lot of labor.
C01	Overall I am satisfied with the services provided by the bank.
E01	If you don't use mobile banking are you very likely going to start using mobile banking in next 6 months?
E02	If you don't use e-wallets (like PayTm, SBI Buddy) are you very likely going to start using e-wallets in next 6 months?

Table 10.2 Likert Scale Used for the Questions Q01 - C01

1	Completely Disagree
2	Partly Disagree
3	Neither Agree nor Disagree
4	Partly Agree
5	Completely Agree

Appendix B: Getting Started with Python

Anyone who wishes to run code in Python need not to worry about the cumbersome installation process that it may entail, as they can run Python code directly on Google's servers. The process is detailed in the following steps.

1. One must have or create a Google account.
2. They can go to https://colab.research.google.com/ directly.
3. It will open an welcome window like this -

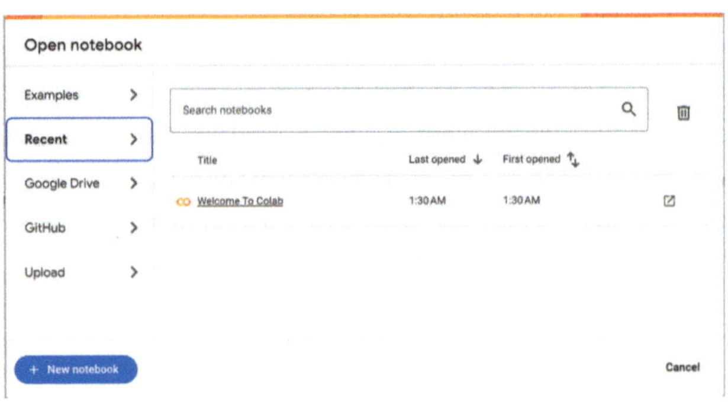

4. One can exit this window and click on "File" in the top-left and select "upload notebook" as shown

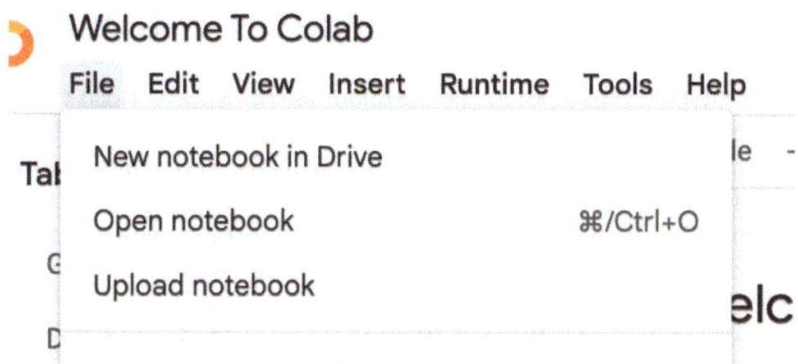

5. One must then upload the file "ch11_code.ipynb" and then the file will be open in a Jupyter Notebook environment along with text, code and pre-run results

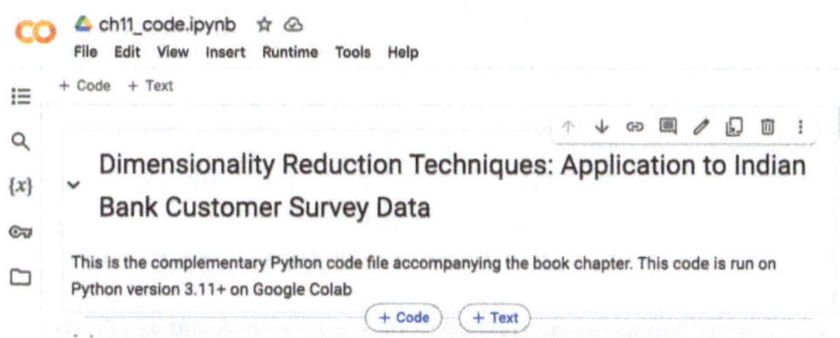

6. The code chunks must be executed sequentially to reproduce the results.

References

Abdi H, Williams LJ (2010) Principal component analysis. Wiley Interdisc Rev: Comput Stat 2(4):433–459

Bandalos DL, Finney SJ (2018) Factor analysis: exploratory and confirmatory. In: The reviewer's guide to quantitative methods in the social sciences, Routledge, pp 98–122

Bartlett MS (1954) A note on the multiplying factors for various χ 2 approximations. J Royal Stat Soc Ser B (Method), 296–298

Bentler PM, Kano Y (1990) On the equivalence of factors and components. Multivar Behav Res 25(1):67–74

Cattell RB (1966) The scree test for the number of factors. Multivar Behav Res 1(2):245–276

Chawla D, Joshi H (2023) Role of mediator in examining the influence of antecedents of mobile wallet adoption on attitude and intention. Glob Bus Rev 24(4):609–625

Costello AB, Osborne J (2019) Best practices in exploratory factor analysis: Four recommendations for getting the most from your analysis. Pract Assess Res Eval 10(1):7

Coughlin KB (2013) An analysis of factor extraction strategies: A comparison of the relative strengths of principal axis, ordinary least squares, and maximum likelihood in research contexts that include both categorical and continuous variables. University of South Florida

Cronbach LJ (1951) Coefficient alpha and the internal structure of tests. psychometrika 16(3):297–334

Dabbagh A, Seens H, Fraser J, MacDermid J (2023) Construct validity and internal consistency of the home and family work roles questionnaires: A cross-sectional study with exploratory factor analysis. BMC Women's Health 23(1):56

DiStefano C, Zhu M, Mindrila D (2019) Understanding and using factor scores: Considerations for the applied researcher. Pract Assess Res Eval 14(1):20

Drasgow F (2006). Polychoric and polyserial correlations. In: Kotz S, Read CB, Balakrishnan N, Vidakovic B, Johnson NL (eds) Encyclopedia of statistical sciences.

Fabrigar LR, Wegener DT (2012) Exploratory factor analysis. Series in understanding statistics. Oxford University Press, Oxford New York.

Fabrigar LR, Wegener DT, MacCallum RC, Strahan EJ (1999) Evaluating the use of exploratory factor analysis in psychological research. Psychol Methods 4(3):272

Field A (2009) SPSS Discovering statistics using SPSS. Porto Alegre

Finch AP, Brazier JE, Mukuria C, Bjorner JB (2017) Using PCA and CFA to identify bolt-on dimensions: An eq-5d case study. Value Health 20(10):1362–1375

Flora DB, Curran PJ (2004) An empirical evaluation of alternative methods of estimation for confirmatory factor analysis with ordinal data. Psychol Methods 9(4):466

Garg R, Rahman Z, Qureshi M (2014) Measuring customer experience in banks: Scale development and validation. J Model Manage 9(1):87–117

Greenacre M, Groenen PJF, Hastie T, D'Enza AI, Markos A, Tuzhilina E (2022) Principal component analysis. Nat Rev Methods Primers 2:100

Habing B (2003) Exploratory factor analysis. Exploratory factor analysis Brian Habing—University of South Carolina

Hasan BMS, Abdulazeez AM (2021) A review of principal component analysis algorithm for dimensionality reduction. J Soft Comput Data Min 2(1):20–30

Henson RK, Roberts JK (2006) Use of exploratory factor analysis in published research. Educ Psychol Measur 66(3):393–416

Holgado-Tello FP, Chacón-Moscoso S, Barbero-García I, Vila-Abad E (2010) Polychoric versus pearson correlations in exploratory and confirmatory factor analysis of ordinal variables. Qual Quant 44:153–166

Horn JL (1965) A rationale and test for the number of factors in factor analysis. Psychometrika 30:179–185

Jolliffe IT (2002) Principal component analysis. Springer Series in Statistics. Springer, New York, Second edition.

Jolliffe IT, Cadima J (2016) Principal component analysis: A review and recent developments. Philos Trans R Soc A Math Phys Eng Sci 374(2065):20150202

Kaiser HF (1958) The varimax criterion for analytic rotation in factor analysis. Psychometrika 23(3):187–200

Kaiser HF (1960) The application of electronic computers to factor analysis. Educ Psychol Measur 20(1):141–151

Kaiser HF (1974) An index of factorial simplicity. Psychometrika 39(1):31–36

Kaura V, Durga Prasad CS, Sharma S (2015) Service quality, service convenience, price and fairness, customer loyalty, and the mediating role of customer satisfaction. Int J bank Mark 33(4):404–422

Khosla N (2004) Dimensionality reduction using factor analysis. Griffith University, Australia

Kline P (2014) An easy guide to factor analysis. Routledge

Kolenikov S, Angeles G (2009) Socioeconomic status measurement with discrete proxy variables: Is principal component analysis a reliable answer? Rev Income Wealth 55(1):128–165

Lenka U, Suar D, Mohapatra PK (2009) Service quality, customer satisfaction, and customer loyalty in Indian commercial banks. J Entrepreneurship 18(1):47–64

Li C (2013) Little's test of missing completely at random. The Stata Journal 13(4):795–809

Ligon E, Malick B, Sheth K, Trachtman C (2019) What explains low adoption of digital payment technologies? Evidence from small-scale merchants in Jaipur, India. PLoS ONE 14(7):e0219450

Mabel OA, Olayemi OS (2020) A comparison of principal component analysis, maximum likelihood and the principal axis in factor analysis. Am J Math Stat 10(2):44–54

Mittal S, Gera R, Batra DK (2015) An evaluation of an integrated perspective of perceived service quality for retail banking services in India. Int J Bank Market 33(3):330–350

Olsson U (1979) Maximum likelihood estimation of the polychoric correlation coefficient. Psychometrika 44(4):443–460.

Parasuraman A, Zeithaml VA, Berry L (1988) Servqual: A multiple-item scale for measuring consumer perceptions of service quality. J Retail 64(1):12–40

Pérez E, Medrano L (2014) Exploratory factor analysis: Conceptual and methodological basis. Revista Argentina de Ciencias del Comportamiento 6(3):71–80

Rencher AC, Christensen W (2002) Méthods of multivariate analysis. Wiley 727:2218–0230

Rudolph A, Krois J, Hartmann K (2023) Principal Component Analysis. Statistics and Geodata Analysis using Python

Ruscio J, Roche B (2012) Determining the number of factors to retain in an exploratory factor analysis using comparison data of known factorial structure. Psychological Assess 24(2):282

Salem N, Hussein S (2019) Data dimensional reduction and principal components analysis. In: Procedia computer science (Proc. 16th International Learning & Technology Conference), Volume 163, 292–299 pp

Sankaran R, Chakraborty S (2021) Factors impacting mobile banking in India: Empirical approach extending UTAUT2 with perceived value and trust. IIM Kozhikode Soc Manag Rev 11(1):7–24

Saxena N, Gera N, Taneja M (2022). An empirical study on facilitators and inhibitors of adoption of mobile banking in india. Electronic Commerce Research, pp 1–32

Shlens J (2014). A tutorial on principal component analysis. https://doi.org/10.48550/ARXIV.1404.1100

Shrestha N (2021) Factor analysis as a tool for survey analysis. Am J Appl Math Stat 9(1):4–11

Sivathanu B (2019) Adoption of digital payment systems in the era of demonetization in India: An empirical study. J Sci Technol Policy Manag 10(1):143–171

Stevens J (2009) Applied multivariate statistics for the social sciences. Routledge, New York, 5th ed edition.

Streiner DL (1994) Figuring out factors: The use and misuse of factor analysis. The Canadian J Psychiatry 39(3):135–140

Taherdoost H, Sahibuddin S, Jalaliyoon N (2014) Exploratory factor analysis; concepts and theory. Adv Appl Pure Math 27:375–382

Van Buuren S, Oudshoorn CG (2000) Multivariate imputation by chained equations, TNO

Watkins MW (2018) Exploratory factor analysis: A guide to best practice. J Black Psychol 44(3):219–246

Yong AG, Pearce S et al (2013) A beginner's guide to factor analysis: Focusing on exploratory factor analysis. Tutorials Quant Methods Psychol 9(2):79–94

Zumbo BD, Gadermann AM, Zeisser C (2007) Ordinal versions of coefficients alpha and theta for likert rating scales. J Modern Appl Statistical Methods 6(1):4

[illegible]

[illegible]

[illegible] (2002) [illegible] multivariate analysis [illegible]

[illegible]

[illegible]

Chapter 11
Classification Using Logistic Regression and Support Vector Machine: Application to Identify Potential Loan Customers

Usha Ananthakumar

Abstract Classification models play a crucial role in predictive analytics by assigning observations to distinct categories based on predictor characteristics. This chapter discusses two widely used classification techniques: logistic regression and support vector machines (SVM). Logistic regression, a classical statistical model, is valued for its interpretability, while SVM provides a more flexible approach by accommodating nonlinear decision boundaries. The chapter explores the theoretical foundations of both methods and demonstrates their application in classifying bank loan customers to identify potential borrowers. Model performance is assessed using accuracy measures and visualization tools to aid in model selection. Additionally, practical implementation in R and Python is provided, offering step-by-step guidance for applying these techniques in real-world loan classification tasks.

11.1 Introduction

Classification is a major area in predictive analytics that pertains to classifying observations into different classes using the predictors that are either categorical or numerical. Some examples of classification problems are:

- Predicting whether a specific product would be bought by customers based on customer and product characteristics.
- Predicting the income bracket (low, medium, high) of individuals for deciding the beneficiaries of a scheme.

Supplementary Information The online version contains supplementary material available at https://doi.org/10.1007/978-981-97-9060-9_11.

U. Ananthakumar (✉)
Shailesh J. Mehta School of Management, Indian Institute of Technology Bombay, Mumbai, India
e-mail: usha@som.iitb.ac.in

F. Hamid and D. Mukherjee (eds.), *Data Science for Modeling Managerial and Socioeconomic Problems*, Contributions to Management Science,
https://doi.org/10.1007/978-981-97-9060-9_11

- Predict the nature of an email as primary, promotional, social or updates on the basis of the content of the email.
- Predict whether a transaction is fraudulent using various features of transactions.

In these applications, the categories of the dependent variables, also referred to as classes, are known apriori and such datasets are called labelled datasets. The presence of the label guides the learning process, and thus classification falls under the larger umbrella of supervised learning problems. Classification models are trained to learn on the given data thus enabling prediction on unseen observations or objects. For good discussion on classification, one can refer to Duda et al. (2000) and Hand (1981).

Classification problems with binary outcomes occur most commonly in practice and hence the focus of this chapter is only on binary outcomes. The next section discusses logistic regression and Sect. 11.3 provides the details of support vector machines. Evaluation metrics are presented in Sect. 11.4 and class imbalance is discussed briefly in the next section. Section 11.6 provides the details of implementation of these techniques on a real-life application using the open-source software R and Python. Section 11.7 concludes the chapter and also provides a brief discussion on the link between logistic regression and support vector machine.

11.2 Logistic Regression

Logistic regression is a widely used model for classification, particularly when the target is binary. The effectiveness and simplicity of logistic regression models have made it to be seen as an analogue to ordinary least squares when outcomes are categorical. Linear regression as well as logistic regression get covered within the class of generalized linear models. An excellent introduction to these models is given in Dobson (2002).

One common thought that may cross the reader's mind is why linear regression is not the natural choice for categorical outcomes. An example can be considered to drive this point. Consider the relationship between level of education, say in number of years, and a categorical outcome denoting the presence or absence of a specific credit card. It is reasonable to believe that a greater number of users of credit card (CC) would have more years of education (YE). A scatterplot in this case results in two parallel lines and does not provide a clear picture about the relationship between the two variables as shown in Fig. 11.1. Instead, if number of years of education is grouped into different intervals and the proportion of people having that credit card in each such interval is obtained, one can expect to see an S-shaped curve as shown in Fig. 11.2. This curve is linear in the middle and curved at the ends as the credit card usage is expected to slowly pick up with an increase in years of education and plateau later. This clearly shows that it requires an approach other than linear regression.

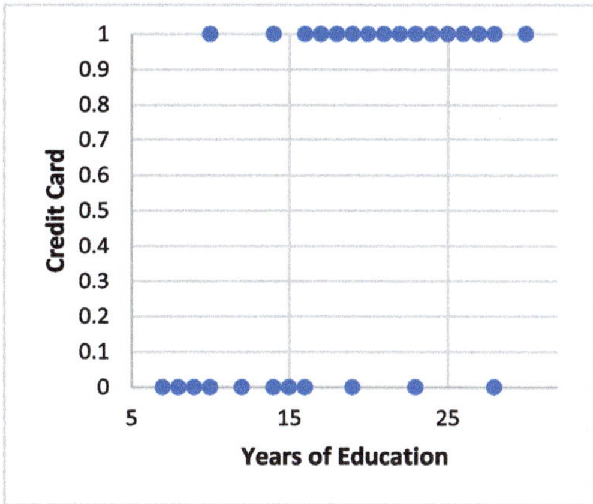

Fig. 11.1 Scatterplot of CC by YE

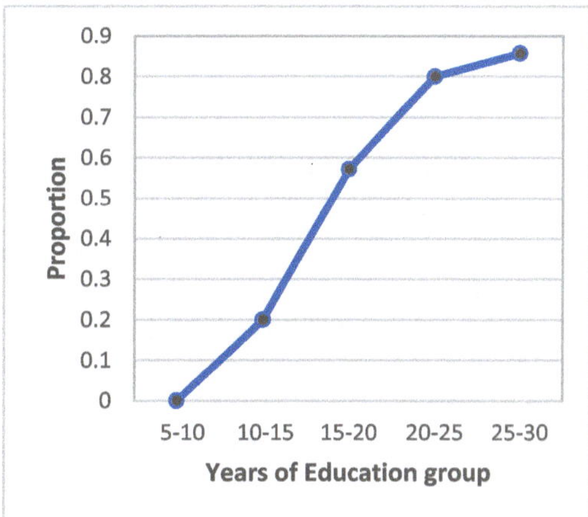

Fig. 11.2 Plot of CC proportion in each YE group

This functional relationship resembles a plot of the cumulative distribution of a continuous random variable. Though various distribution functions have been suggested in literature to capture this relationship, one popular choice is logistic distribution due to its mathematical tractability and interpretability.

In a general set up, let Y be the categorical dependent variable with binary outcomes 0 and 1 (event) and X be the predictor. If $P(Y = 1|X = x) = \pi(x)$, it follows that the expected response is the probability that $Y = 1$. Thus, $E(Y|X) = \pi(x)$.

The model takes the form:

$$\pi(x) = \frac{e^{\beta_0+\beta_1 x}}{1 + e^{\beta_0+\beta_1 x}}, \tag{11.1}$$

where β_0 and β_1 are regression coefficients.

This sigmoidal function provides a closed form expression for probability of success and ensures the model $\pi(x)$ is bounded between 0 and 1. Further, the errors are not normal, but binomially distributed unlike linear regression.

A desirable transformation of $\pi(x)$ that is linear in its parameters is the logit transformation, the natural logarithm of odds of Y given by:

$$\eta(x) = \ln\left(\frac{\pi(x)}{1-\pi(x)}\right) = \beta_0 + \beta_1 x. \tag{11.2}$$

Here, the ratio of probabilities $\frac{\pi(x)}{1-\pi(x)}$ is known as odds. In the example relating level of education and credit card users, if this ratio is 0.2, it means that individuals are overall one-fifth as likely to be possessing the credit card as they are to be without this credit card.

It can be noted from (11.1) that the relationship involving $\pi(x)$ is non-linear in x, whereas the one represented by logit in (11.2) is linear in x, offering flexibility of a linear model.

11.2.1 Estimating the Parameters

The parameters β_0 and β_1 are generally estimated by the maximum likelihood method. This method provides the estimates for the unknown parameters as those values where the probability of obtaining the data that has been observed is maximum. The likelihood function refers to the probability of the observed data as a function of the parameters.

For sample observations $(x_i, y_i), i = 1, 2, ..., n$, each observation follows Bernoulli distribution and hence the likelihood function can be represented as:

$$L(\mathbf{y}, \boldsymbol{\beta}) = \prod_{i=1}^{n} \pi(x_i)^{y_i}(1-\pi(x_i))^{1-y_i}. \tag{11.3}$$

For mathematical simplicity, the natural log of (11.3) is considered and let the log likelihood be denoted by $\ln[L(y, \boldsymbol{\beta})]$. Differentiating this expression with respect to the parameters β_0 and β_1 and equating to zero results in the following likelihood equations:

$$\begin{aligned} &\sum_{i=1}^{n} (y_i - \pi(x_i)) = 0 \\ &\sum_{i=1}^{n} x_i(y_i - \pi(x_i)) = 0. \end{aligned} \tag{11.4}$$

Unlike linear regression, the above equations do not provide closed form solutions as they are non-linear in the parameters β_0 and β_1. Numerical methods that are iterative in nature are being used to obtain the estimates and are available as inbuilt algorithms of any standard statistical software. Let us denote the maximum likelihood estimates as $\hat{\beta}_0$ and $\hat{\beta}_1$.

The estimated log odds are $\hat{\eta} = \hat{\beta}_0 + \hat{\beta}_1 x$ and thus the fitted value of the model (11.1) is:

$$\hat{\pi}(x) = \frac{e^{\hat{\eta}}}{1 + e^{\hat{\eta}}} = \frac{e^{\hat{\beta}_0 + \hat{\beta}_1 x}}{1 + e^{\hat{\beta}_0 + \hat{\beta}_1 x}}. \tag{11.5}$$

11.2.2 Interpreting the Parameters and Odds Ratio

Having fitted the model, it is important to interpret the coefficient corresponding to the slope parameter β_1. The estimated log odds $\hat{\eta}$ provide a familiar approach as it is linear in the coefficients. For a given value of x, the estimated log odds are:

$$\hat{\eta}(x) = \hat{\beta}_0 + \hat{\beta}_1 x. \tag{11.6}$$

For one unit increase in x, it becomes:

$$\hat{\eta}(x + 1) = \hat{\beta}_0 + \hat{\beta}_1 (x + 1). \tag{11.7}$$

The difference between these two equations is:

$$\hat{\eta}(x + 1) - \hat{\eta}(x) = \hat{\beta}_1. \tag{11.8}$$

Using (11.2), the above expression can be written as:

$$\ln\left(\frac{odds_{x+1}}{odds_x}\right) = \hat{\beta}_1, \tag{11.9}$$

where the quantity $\frac{odds_{x+1}}{odds_x}$ is called the odds ratio.

Thus, $\hat{\beta}_1$ denotes the amount by which log odds changes for every unit change in x. Interpreting on the scale of log odds may not be straightforward. Equivalently, one could interpret the slope coefficient using its relationship with the odds ratio given as:

$$\text{Odds Ratio} = e^{\hat{\beta}_1}. \tag{11.10}$$

Thus, an increase of one unit in the predictor value results in the odds of success increasing by a factor of $e^{\hat{\beta}_1}$. At times, it may be more meaningful to see the impact of change of k units in the predictor X. It may be observed that odds ratio in this case becomes $e^{k\hat{\beta}_1}$ implying it multiplies the odds of success by $e^{k\hat{\beta}_1}$.

To better understand the interpretation, let a logistic model for the example that was considered earlier between level of education (X) and the possession of a specific credit card (Y) be given by $Y = -2.89 + 0.19X$. Here, it may be of interest to see the effect of few additional years of education on having the credit card. An increase of 5 years in education results in an odds ratio of 2.586. This implies, for every five years of increase in education, that the odds of a person possessing this specific credit card are 2.586 times as large as they are at that point of time.

If, in the above example, the predictor is categorical, say, user or non-user of internet banking (X), this corresponds to an interesting problem of predicting if the user has a specific credit card on the basis of whether or not a customer uses internet banking. Here, a unit change in X would mean the difference between users and non-users of internet banking. Thus, an odds ratio of 3 can be interpreted to mean that the odds of internet banking users having this credit card is three times more than the corresponding odds among non-users of internet banking.

11.2.3 Is the Model Significant?

The next obvious question is to know if the model is significant. Testing the significance of the model involves testing the hypothesis that all beta coefficients are zero. This involves testing a hypothesis based on likelihood ratio tests using a statistic called deviance. This quantity examines the log likelihood of the fitted model with reference to the perfect model also called the saturated model. Deviance is defined as:

$$D = -2\ln\left(\frac{\text{likelihood of the fitted model}}{\text{likelihood of the saturated model}}\right). \tag{11.11}$$

In the considered setting, where the dependent variable is either 0 or 1, the likelihood is exactly 1 for the saturated model. This is because $\hat{\pi}_i(x_i) = y_i$ for a saturated model and thus substitution in Eq. (11.3) results in the likelihood being 1.

Hence the deviance is:

$$D = -2\ln\,(\text{likelihood of the fitted model}) \tag{11.12}$$

This quantity is analogous to the sum of squares due to residuals considered in linear regression. A small value of this model deviance implies that the fitted model does well as its fit is close to the saturated model. Equivalently, the model is said to perform well if it provides an improvement over the model that contains only the intercept. The difference in their deviances is the test statistic. In case of p predictors, this statistic follows chi square distribution with p degrees of freedom. If this statistic is significant, this implies at least one of the predictors is significant.

11.2.4 Significance of the Predictors

To examine if a specific predictor is significant, it amounts to evaluating the model with and without this predictor. This is obtained by the difference in the deviance values as given by:

$$\begin{aligned} G = & D(\text{for the model without the predictor}) \\ & -D(\text{for the model with the predictor}). \end{aligned} \tag{11.13}$$

This quantity G being large implies the predictor is significant and it follows chi square distribution with 1 degree of freedom. This way, the significance of each of the predictors could be assessed.

11.2.5 Validation of Model Prediction

From the estimated parameters, the logit $\hat{\eta}$ can be obtained, which in turn can be transformed back to predicted probabilities $\hat{\pi}$. The predicted probabilities can be validated with the actual outcome to see if indeed high probabilities are associated with actual occurrence of event and low probabilities with the non-event. One such measure that provides this association is c-statistic. The calculation of this statistic is carried out as follows:

(i) Consider all possible pairs of observations where one among this pair corresponds to the outcome being an event (1) and the other one corresponds to the non-event (0). Let the number of such pairs be T.
(ii) A pair of such observations is said to be concordant if the predicted probability of success of the observation corresponding to outcome being 1 is more than the predicted probability of success of the observation corresponding to outcome being 0.
(iii) c-statistic = No. of concordant pairs/T.

This statistic ranges between 0.5 and 1 where 0.5 corresponds to a model with observations being assigned randomly into categories and a value of 1 corresponds to an ideal model.

Logistic regression is quite flexible in handling both categorical and continuous predictors. The ability of the fitted logistic model in providing the predicted probabilities has been addressed enough in the literature, particularly in the process of classification problems, the topic of interest in this chapter. A good account of logistic regression is provided in DeMaris (1995) and good general books on logistic regression include Agresti (2012) and Hosmer et al. (2013).

11.3 Support Vector Machines

Support Vector Machines (SVM) belong to a class of highly effective and flexible techniques utilized in classification and regression problems. Developed by Vladimir Vapnik in early 1960s, the method became quite popular and considered as one of the 'out of the box' classifiers in early 1990s. This machine learning model is built with the help of a limited number of observations to obtain the best classification results, which is the main area of focus in this section.

11.3.1 Classification with Linear Boundary

We shall begin by delving deeper into a classification problem that is linearly separable with two predictors, say (X_1, X_2), that can be correspondingly extended to a classification problem with p predictors, say $\left(X_1, X_2, \ldots, X_p\right)$. Consider n sample observations $\left\{\boldsymbol{x}_i^T = (x_{i1}, x_{i2}); i = 1, 2, \ldots, n\right\}$ belonging to two distinct classes with binary class labels $y_i\{-1, 1\}$. It is inclined to obtain a clear segregation of the observations into their respective classes. This is carried out by introducing a surface, popularly termed as the hyperplane, which in this case is a linear boundary. Vapnik (1996) discusses in detail the concept of linear separating hyperplane. The hyperplane $f(X) = \beta_0 + \beta_1 X_1 + \beta_2 X_2 = 0$ separates the observations into the two classes based on the sign of $\beta_0 + \beta_1 X_1 + \beta_2 X_2$.

Consider a sample of observations as shown in Fig. 11.3 belonging to two distinct classes—squares and circles. It is clear that there are an infinite number of hyperplanes that could perform the task of separating the observations according to their class. Which hyperplane shall one choose? SVM defines the criterion of selecting the hyperplane that is maximally far away from its closest data point as shown in Fig. 11.3. Maximizing this distance, also known as the margin (denoted further as M), is the prime objective of this classifier. Hence, an optimization problem is formulated, as shown in the next section. Moguerza and Munoz (2006) provide a good introduction to SVM along with its important features.

11.3.1.1 Constructing the Optimization Problem

The hyperplane with the maximum margin is obtained by solving the following optimization problem:

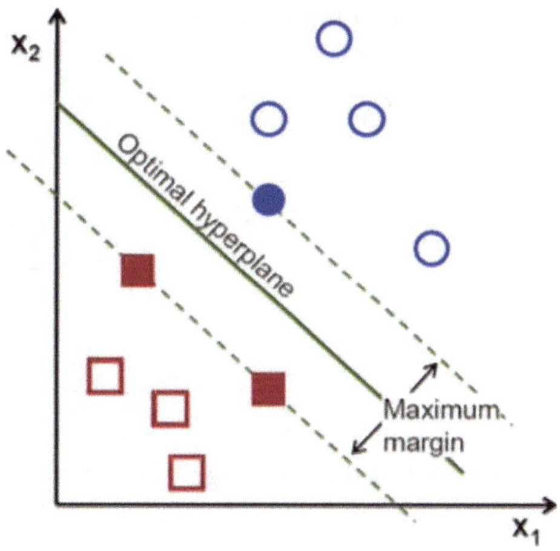

Fig. 11.3 Illustration of support vector machine

$$\underset{\beta_0,\beta_1,\beta_2,\beta_1^2+\beta_2^2=1}{\text{Maximize}} \quad M \tag{11.14}$$

provided that

$$y_i(\beta_0 + \beta_1 x_{i1} + \beta_2 x_{i2}) \geq M, \quad i = 1, 2, \ldots.., n.$$

The given constraint ensures that the distance of the observations from the hyperplane is at least M.

Let $\boldsymbol{\beta} = (\beta_1, \beta_2)$. Considering the $||\boldsymbol{\beta}|| = 1$ condition, the above constraint can be expressed as $y_i(\beta_0 + \beta_1 x_{i1} + \beta_2 x_{i2}) \geq M||\boldsymbol{\beta}||$. As positively scaled multiples of the coefficients β_1, β_2 also satisfy the above equation, it can be arbitrarily set $||\boldsymbol{\beta}|| = 1/M$.

Therefore, the optimization problem (11.14) can be rewritten analogously in the following manner:

$$Minimize \quad \frac{||\boldsymbol{\beta}||^2}{2} \tag{11.15}$$

provided that

$$y_i(\beta_0 + \beta_1 x_{i1} + \beta_2 x_{i2}) \geq 1, i = 1, 2, \ldots, n.$$

The Lagrangian form to the constrained optimization problem is given by:

$$L = \frac{||\boldsymbol{\beta}||^2}{2} - \sum_{i=1}^{n} \alpha_i \left[y_i(\beta_0 + \beta_1 x_{i1} + \beta_2 x_{i2}) - 1 \right]; \alpha_i \geq 0 \tag{11.16}$$

Minimizing (11.16) results in,

$$\beta_1 = \sum_{i=1}^{n} \alpha_i y_i x_{i1}$$
$$\beta_2 = \sum_{i=1}^{n} \alpha_i y_i x_{i2} \quad (11.17)$$
$$\sum_{i=1}^{n} \alpha_i y_i = 0.$$

The dual of the Lagrange function can thus be written as:

$$L = \sum_{i=1}^{n} \alpha_i - \frac{1}{2}\sum_{i=1}^{n}\sum_{j=1}^{n} \alpha_i \alpha_j y_i y_j \boldsymbol{x}_i^T \boldsymbol{x}_j, \quad (11.18)$$

where $\boldsymbol{x}_i^T = (x_{i1}, x_{i2}); i = 1, 2, \ldots, n$.

Maximizing the above function provides estimates of alpha parameters. It can be seen that the Lagrangian multipliers $\alpha's$ are non-zero solely for the training observations that fall on the margin boundary. Such observations are called support vectors and play an important role in determining the hyperplane and thus the classifier.

Though details regarding the estimates of α parameters are not provided here, it can be noted that only the inner product of the training observations is required for obtaining it. The α's can be substituted in (11.17) to obtain $\widehat{\beta_1}, \widehat{\beta_2}$. Further, $\widehat{\beta_0}$ can be obtained from any support vector, say k by solving $y_k\left(\widehat{\beta_0} + \widehat{\beta_1}x_{k1} + \widehat{\beta_2}x_{k2}\right) = 1$.

The corresponding decision rule classifies a test observation $\boldsymbol{x}^* = \left(x_1^*, x_2^*\right)$ based on the sign of $\hat{f}(\boldsymbol{x}*) = \hat{\beta}_0 + \hat{\beta}_1 x_1^* + \hat{\beta}_2 x_2^*$. If $f(\boldsymbol{x}^*) > 0$, the observation is assigned to class 1, and if $\hat{f}(\boldsymbol{x}^*) < 0$, it is assigned to class -1.

It is an interesting fact that the decision rule is solely a function of only the support vectors, making this method unique when compared to other classifiers.

For more details, interested readers can go through Hastie et al. (2009).

11.3.1.2 Extension to 'P' Dimensional Setting

The classification problem explained above in a 2-dimensional setting can be easily extended to any p-dimensional space consisting of the training observations $\left\{\boldsymbol{x}_i^T = \left(x_{i1}, x_{i2}, \ldots, x_{ip}\right); i = 1, 2, \ldots, n\right\}$ separated into two distinct classes by the hyperplane $\beta_0 + \boldsymbol{\beta}^T\boldsymbol{X} = 0$ where $\boldsymbol{\beta}^T = \left(\beta_1, \beta_2, \ldots, \beta_p\right)$ and $\boldsymbol{X}^T = \left(X_1, X_2, \ldots, X_p\right)$. The optimization problem in (11.15) can thus be denoted in the below format:

$$Minimize \ \frac{||\boldsymbol{\beta}||^2}{2} \tag{11.19}$$

provided that

$$y_i(\beta_0 + \boldsymbol{\beta}^T \boldsymbol{x}_i) \geq 1, \quad i = 1, 2, \ldots, n.$$

Here again, the test observation $\boldsymbol{x}^* = \left(x_1^*, x_2^*, \cdots, x_p^*\right)$ is classified on the basis of the sign of $\hat{\beta}_0 + \hat{\beta}_1 x_1^* + \hat{\beta}_2 x_2^* + \ldots \hat{\beta}_p x_p^*$ as before (i.e.)

$$y_{test} = \begin{cases} 1 \ \text{if} \ \hat{\beta}_0 + \hat{\beta}_1 x_1^* + \hat{\beta}_2 x_2^* + \cdots + \hat{\beta}_p x_p^* > 0 \\ -1 \ \text{if} \ \hat{\beta}_0 + \hat{\beta}_1 x_1^* + \hat{\beta}_2 x_2^* + \cdots + \hat{\beta}_p x_p^* < 0 \end{cases} \tag{11.20}$$

Equivalently, the decision rule can be represented as:

$$y_{test} = \begin{cases} 1 \ \text{if} \ \hat{\beta}_0 + \sum_{i=1}^{n} \alpha_i y_i \mathbf{x_i^T} \mathbf{x}^* > 0 \\ -1 \ \text{if} \ \hat{\beta}_0 + \sum_{i=1}^{n} \alpha_i y_i \mathbf{x_i^T} \mathbf{x}^* < 0 \end{cases} \tag{11.21}$$

11.3.1.3 Soft Margin for Overlapping Classes

The occurrence of two classes perfectly separable by a hyperplane is an ideal case. Cortes and Vapnik (1995) provide extension to accommodate the case of classes not being completely separable. This poses the need of imposing a relaxation on the rigidity of the margin created by the hyperplane thus giving some observations the flexibility of being on the other side of the hyperplane or the margin. The soft margin gives rise to an error term $_i$ associated with each misclassified observation $\boldsymbol{x}_i$. Bounding $\sum \varepsilon_i$ at a value C say restricts the total number of misclassifications at C. An optimization problem can be correspondingly constructed as shown below:

$$\text{Minimize} \quad \frac{||\beta||^2}{2} + C \sum_i \epsilon_i \tag{11.22}$$

provided that $y_i(\beta_0 + \beta^{\mathbf{T}} \mathbf{x_i}) \geq 1 - \epsilon_i, \quad i = 1, 2, \ldots, n$

$$\epsilon_i > 0, \quad i = 1, 2, \ldots, n.$$

The objective here is to maximize the distance of the observations from the hyperplane while allowing for some points to lie on the other side of the margin. These two conflicting aspects are controlled by adding a penalty to each misclassified point through the tuning parameter C. This parameter is directly proportional to the distance of the point from the hyperplane. A large C leads to a minimization of the number of misclassified observations due to high penalty thus resulting in a hyperplane with a smaller margin.

11.3.2 Classification with Non-Linear Boundary

Controlling the C parameter to obtain an accurate classification is only useful as long as the data is linearly separable. But what if a non-linear boundary separates the data as shown in the left side of Fig. 11.4 and cannot be partitioned logically using a linear hyperplane?

The right side of Fig. 11.4 displays the mapping function ϕ: $(x_{i1}, x_{i2}) \rightarrow (x_{i1}, x_{i1}^2, x_{i2}^2)$; $i = 1, 2, \ldots, n$ that transforms the input data into a dataset clearly separable by a hyperplane. The original training data $\boldsymbol{x}_i$ is projected onto a higher-dimensional feature space in order to obtain linearly separable classes. This method of enlarging the feature space allows the accommodation of a non-linear boundary between classes.

The optimization problem (11.15) can be expressed as follows:

$$Minimize \ \frac{||\boldsymbol{\beta}||^2}{2} \tag{11.23}$$

provided that $y_i(\beta_0 + \boldsymbol{\beta}^T\phi(\boldsymbol{x}_i)) \geq 1, i = 1, 2, \ldots, n$

whose corresponding Lagrange dual form can be written as:

$$L = \sum_{i=1}^{n} \alpha_i - \frac{1}{2}\sum_{i=1}^{n}\sum_{j=1}^{n} \alpha_i\alpha_j y_i y_j \phi(\boldsymbol{x}_i)^T \phi(\boldsymbol{x}_j). \tag{11.24}$$

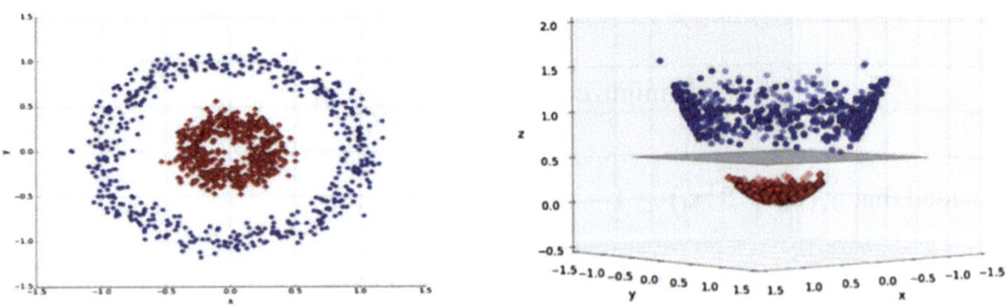

Fig. 11.4 Left to right: Non-linearly separable to linearly separable

Transformation of the training data into a higher-dimensional feature space can thus be done through a mapping function ϕ. Now, a linear SVM model can be used to classify the data in this new feature space. However, the choice of ϕ is quite challenging and the method is also computationally very expensive.

SVM reduces the complexity of the choice of ϕ by utilizing a kernel function represented as the inner product of ϕ in the following way:

$$K(\boldsymbol{x}_i, \boldsymbol{x}_j) = \phi(\boldsymbol{x}_i).\phi(\boldsymbol{x}_j).$$

Popular kernel functions with γ, r, d as kernel parameters are given below:

Linear Kernel	$K(\boldsymbol{x}_i, \boldsymbol{x}_j) = \boldsymbol{x}_i.\boldsymbol{x}_j$
Polynomial Kernel	$K(\boldsymbol{x}_i, \boldsymbol{x}_j) = (\gamma.\boldsymbol{x}_i.\boldsymbol{x}_j + r)^d$
RBF Kernel	$K(\boldsymbol{x}_i, \boldsymbol{x}_j) = \exp(-\gamma.\lVert\boldsymbol{x}_i - \boldsymbol{x}_j\rVert^2)$
Sigmoid Kernel	$K(\boldsymbol{x}_i, \boldsymbol{x}_j) = \tanh(\gamma.\boldsymbol{x}_i.\boldsymbol{x}_j + r)$.

It can be noted that one does not have to calculate $\phi(\boldsymbol{x}_i)$ for solving the optimization problem. Therefore, the Lagrange dual form (11.24) can be solved by simply knowing the kernel function.

$$L = \sum_{i=1}^{n} \alpha_i - \frac{1}{2} \sum_{i=1}^{n} \sum_{j=1}^{n} \alpha_i \alpha_j y_i y_j K(\boldsymbol{x}_i, \boldsymbol{x}_j). \tag{11.25}$$

The utilization of the kernel function $K(\boldsymbol{x}_i, \boldsymbol{x}_j)$ avoids the explicit evaluation of $\phi(\boldsymbol{x}_i)$ since it acts implicitly through the closed form of K. Boser et al. (1992) recommend using the kernel function for accommodating non-linear classification boundaries. This kernel trick allows for easy computations to take place and offers a more efficient, less expensive method to compute the Lagrange multipliers in the dual form in order to obtain the hyperplane.

Thus, the decision rule (11.21) for classifying a test observation takes the form:

$$y_{test} = \begin{cases} 1 \;\; if \;\; \hat{\beta}_0 + \sum_{i=1}^{n} \alpha_i y_i K(\mathbf{x_i}, \mathbf{x}^*) > 0 \\ -1 \;\; if \;\; \hat{\beta}_0 + \sum_{i=1}^{n} \alpha_i y_i K(\mathbf{x_i}, \mathbf{x}^*) < 0 \end{cases} \tag{11.26}$$

A comprehensive treatment of support vector machines is given in Vapnik (2010).

11.4 Evaluation of Model Performance

An important step in building an effective model is in evaluating how the model performs by utilizing appropriate evaluation metrics. These metrics help in understanding how well the model performs on the existing data, thus providing an opportunity to improve its performance by tuning the hyper-parameters and generalizing it on the new unseen data.

For example, consider a model used to classify devices based on the presence of malware in it. Such classification models are very useful for security experts to take appropriate actions for combating it. Conventionally, the labels used for binary classes are 0 and 1 where the class of primary interest is termed as positive (1) and the other class as negative (0). In this example, presence of a malware is positive (1) and its absence is negative (0). Other labels such as -1 and 1 are also used in literature, as seen in the section on SVM. However, for ease of understanding, the labels 0 and 1 shall be used (as seen later in the section on implementation in R).

When a classification model is evaluated for its performance, two elements are of key importance-true outcome and the predicted outcome. Each prediction from the model can be one of the following types:

1. True Positive (TP): A positive class observation being correctly predicted as positive.
2. True Negative (TN): A negative class observation being correctly predicted as negative.
3. False Positive (FP): A negative class observation being falsely predicted as positive.
4. False Negative (FN): A positive class observation being falsely predicted as negative.

This can be represented in a table, as shown below. Such a table is also known as the confusion matrix.

		PREDICTED OUTCOME	
		NEGATIVE (0)	POSITIVE (1)
TRUE OUTCOME	NEGATIVE (0)	TN	FP
	POSITIVE (1)	FN	TP

11.4.1 Measures

One of the most commonly used metrics to evaluate the performance of a classification model is accuracy. It is the fraction of correct predictions obtained by the model:

$$Accuracy = \frac{TP + TN}{TP + TN + FP + FN}.$$

One of the main advantages of using accuracy is that it is easily computable and less complex. Equivalently, the model can be evaluated on the basis of its incorrect predictions, also known as the error rate which is the complement of accuracy. However, these measures work poorly when dealing with an imbalanced dataset by not being able to represent the minority instances well. For example, consider an imbalanced dataset with 90% observations labelled positive and 10% labelled negative. If the classifier were to predict the class of all its observations as positive, the accuracy value would be 90%. This could be depicted as good performance, though the model has been unable to predict any of the negative labelled observations correctly. This justifies the need for a better measure. To deal with this, literature (He et al. (2009), Kuhn et al. (2013)) suggests use of alternative measures, namely precision, sensitivity, specificity and F-measure that offer better understanding of the model.

Precision explains how many cases predicted by the model as positive are actually positive. It is useful in situations where obtaining a false positive is of higher concern than obtaining a false negative. For example, when dealing with recommendation systems on e-commerce sites, false positives could lead to customer churn and prove harmful for business, thus making precision an important metric:

$$Precision = \frac{TP}{TP + FP}.$$

Sensitivity or recall explains the proportion of actual positive cases the model was able to predict correctly. It is a useful metric thereby focusing on identifying more true positive cases. It is important in critical medical cases where raising a false alarm on the presence of a disease is less harmful than letting an actual positive case go undetected:

$$Sensitivity = \frac{TP}{TP + FN}$$

Specificity explains the proportion of actual negative cases the model was able to predict correctly. Usually, a trade off exists between specificity and sensitivity:

$$Specificity = \frac{TN}{TN + FP}$$

The F-Measure is a combined measure of recall and precision where the weight on either recall or precision can be varied using the coefficient β. As both precision and recall are rates, F-Measure considers their reciprocals, as can be seen clearly with the choice of $\beta = 1$ resulting in the harmonic mean, thus balancing both precision and recall:

$$F - measure = \frac{(1 + \beta)^2.Precision.Recall}{\beta^2.Recall + Precision}$$

The above metrics are sensitive to the choice of positive class, predominantly for imbalanced data. Combined performance measure would help in understanding the overall model in terms of the predictive ability of both positive and negative classes.

G-mean is an important metric that balances the true rates of both the classes by considering the geometric mean of sensitivity and specificity. However, it is difficult to compare the precision on each class of different models due to the symmetric nature of G-mean on the true rates.

$$G - Mean = \sqrt{Sensitivity.Specificity}$$

Though F-measure and G-mean are significantly better performance metrics, visualization tools are far superior when it comes to the comparison of the performance of various classifiers.

11.4.2 Visualization Tools

In this section, two useful visualization tools, namely receiver operating characteristic curves and lift charts, are described.

11.4.2.1 ROC Curves

Receiver operating characteristic curves (ROC) are two-dimensional curves that depict the relative trade-off between sensitivity and specificity corresponding to various thresholds of probability values considered for classifying observations to a particular class. In the case of logistic regression, the default threshold considered is 0.5. Lowering or increasing the threshold gives rise to varied confusion matrices. Similarly, in support vector machines, the performance of the model would vary by shifting the original hyperplane $f(x) = 0$ to $f(x) = t$ for a range of values of t, resulting in varied confusion matrices. Thus, ROC curve visually conveys the plot of the model by plotting 1-specificity v/s sensitivity for a range of thresholds as illustrated by a sample curve in Fig. 11.5. Conventionally, 1-specificity is considered for better representation. This curve can be used to select the appropriate threshold that maximizes sensitivity and specificity.

Another use of ROC is to visually compare the performance of various classifiers by superimposing them on the same graph. The model with a high sensitivity and specificity value would be closer to the top left corner of the graph. Such a model can also be observed to have a large area under the curve or AUC, which serves as a useful metric for evaluating the model. It is interesting to note that c-statistic described in Sect. 11.2.5 is the same as this measure. The AUC ranges from 0.5 to 1. A model with an AUC of 0.5 is depicted as a ROC curve coinciding with the 45-degree diagonal passing through the graph indicating a model with no discrimination

Fig. 11.5 A sample ROC curve

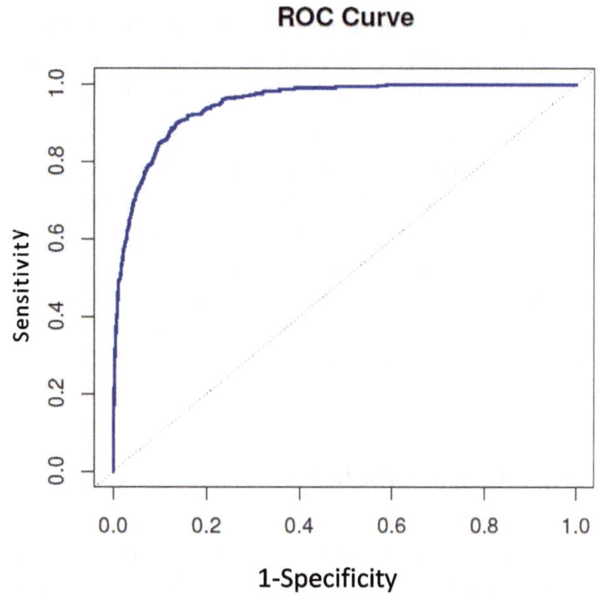

capacity. Larger the AUC, better is the performance of the model in distinguishing between the classes, making the model more favourable for classification.

For better understanding of ROC curves, one can refer to Altman and Bland (1994), Brown and Davis (2006) and Fawcett (2006).

11.4.2.2 Lift Charts

A lift chart graphically represents the improvement that a model provides when compared against a model based on randomly selected records. This change is measured in terms of a lift score that defines the effectiveness of the predictive model (Ling and Li 1998).

Lift curves are very useful in target marketing models where it is of interest to predict whether a certain audience will react to a specific marketing campaign. For example, if one were to target specific customers based on the predictions of a model, it would be resourceful for the organization to target only those with positive predicted response but by a considerable margin, so as to minimize cost and time. It would also be useful for the organization to have a ranking of the customers based on the probability of their predictions.

Lift is calculated as follows:

1. The observations are arranged in descending order of the probability values predicted by the classification model.
2. The dataset is divided into bins.

3. The lift corresponding to a particular bin is then calculated by taking the ratio of the number of positive observations in that bin to the expected number of positives in the same bin based on the predictions of a random model.

A plot of each bin against the lift value is known as the lift chart. Commonly used bin size are the ones containing 10% of the observations and the corresponding plot is known as a decile chart. The lift chart is extremely useful in segmenting the target population on the basis of high probabilities of the observations to provide a positive response. It helps to evaluate which section of the population to focus on to obtain the maximum benefits corresponding to the objective.

The next section gives a brief overview of imbalance in datasets, a challenging area in classification domain.

11.5 Class Imbalance

Imbalance in classes is an important phenomenon where the number of observations of one of the classes is much more than the other class. Class imbalance occurs in many real-life situations including customer churn prediction, financial distress prediction and fraud detection. In such applications, the underrepresented class generally happens to be of prime interest. Standard classifiers fail to classify the important class correctly due to its biasness towards the over represented class.

Literature provides a variety of approaches for dealing with imbalanced data (Lopez et al. 2013). Chawla et al. (2002) proposed a sampling technique SMOTE (Synthetic Minority Over-sampling TEchnique) which is quite popular and this method generates synthetic observations of the minority class. Following this, several such techniques including Borderline SMOTE (Han et al. 2005), ADASYN (He et al., 2008) and MWMOTE (Barua et al. 2014) have been proposed. More details on SMOTE and its applications are discussed in Fernandez et al. (2018a). A more recent method that weighs the observations differently for handling class imbalance combined with class overlap is provided by Shahee and Ananthakumar (2021). Interested readers may go through Fernandez et al. (2018b) for more details on imbalanced learning.

11.6 An Application to Identify Potential Loan Customers

This section pertains to the identification of potential loan customers using the dataset *Universal Bank* from Kaggle.

11.6.1 Universal Bank Case

In the real world, banks have always been interested in growing rapidly more into the loan business. One way of achieving this goal is to convert the existing non-loan taking customers into loan customers. This method is always more profitable and less costly for the banks, instead of acquiring new loan customers. Universal Bank is a hypothetical example of such a bank, which is interested in exploring ways to convert their existing customers to loan takers. The bank has run a campaign on a sample of 5000 of its customers to understand which of them would potentially accept loan offers in the future. The customers' demographic information (age, income, etc.), customers' association with the bank (mortgage, securities account, etc.) and their response to the loan campaign (accepted loan offer or not) are recorded and assessed by the bank. The larger goal is to identify potential customers, among the bank's database, to target for taking up a loan by devising a model based on the available data. The attributes obtained for each customer are described below:

Variable Name	Description
ID	Customer ID
Age	Age (in completed years)
Experience	Number of years of professional experience
Income	Annual income ($000)
ZIPCode	Home ZIP code
Family	Family size
CCAvg	Monthly average spending on credit cards ($000)
Education	Education Level. 1: Undergrad; 2: Graduate; 3: Advanced/Professional
Mortgage	Value of house mortgage, if any ($000)
Personal Loan	Did this customer accept the personal loan offered in the last campaign?
Securities Account	Does the customer have a securities account with the bank?
CD Account	Does the customer have a certificate of deposit (CD) account with the bank?
Online	Does the customer use internet banking facilities?
CreditCard	Does the customer use a credit card issued by the bank?

Variable Description (Shmueli et al. (2017))

Initially, data cleaning is carried out to see the completeness and appropriateness of the data for the analysis. During this process, it is observed that the variable ZIPCode has inconsistent entries and hence dropped from further analysis. Further, the variable ID does not have any role to play in the analysis carried out here and hence not considered further.

It can be noted that the variable Personal_Loan is the response with the binary labels—1: customer availing the loan, and 0: customer not availing the loan. The binary variables Securities Account, CD Account, Online and CreditCard are assigned the labels 1 for the presence and 0 for the absence of the corresponding characteristic of the variables.

The next section makes use of logistic regression and SVM to build the classifiers for predicting customers availing the loan offer.

11.6.2 Implementation in R

R is a well-known statistical software for data analysis and has the advantage of being an open-source language. It is a comprehensive statistical platform providing a wide variety of data analytic techniques. Further, R can be used to create effective and customized graphs due to its powerful graphic capabilities. Good sources of texts in R include Zumel and Mount (2020), Boehmke and Greenwell (2019) and Bilder and Loughin (2014).

The necessary packages are firstly installed in R using the *install.packages(" ")* function and then imported using the *library()* function. The *caret* package (Classification And REgression Training) contains a number of functions for streamlining the training process for classification and regression models. Various tasks like data pre-processing, splitting, model evaluation, tuning parameters, model optimization and feature selection can be efficiently carried out using this package. *e1071* is a package that provides functions for various algorithms like support vector machine, fuzzy classifier, naive Bayes classifier, bagged clustering, etc. It also facilitates classification in SVM by using the kernel trick. The *pROC* package is used to build the roc function.

```
> install.packages("caret")
> listall.packages("e1071")
> listall.packages("pROC")
> library ("caret")
> library("e1071")
> library("pROC")
```

The working directory is set using the *setwd()* function and the Universal Bank dataset in a csv format is imported as a dataframe. The categorical predictors are represented in a factor format. The dataset is partitioned into train and test sets based on a ratio of 60:40, and denoted as *utrain* and *utest,* respectively. The *set.seed()* method is applied to make the answers replicable.

```
> setwd("...")
> ub < - read.csv("UniversalBank.csv")
> ub < - ub[, -C(1, 5)]
> ub$Education < - factor(ub$Education, levels = c(1, 2, 3), labels = c("Undergrad",
"Graduate", "Advanced/Professional"))
```

```
> ub$Securities_Account < - as.factor(ub$Securities_Account)
> ub$CD_Account < - as.factor(ub$CD_Account)
> ub$Online < - as.factor(ub$Online)
> ub$CreditCard < - as.factor(ub$CreditCard)
> set.seed(1)
> train.index < - sample(c(1:dim(ub)[1]), dim(ub)[1]*0.6)
> utrain < - ub[train.index,]
> utest < - ub[-train.index,]
```

11.6.2.1 Logistic Regression Model

The logistic regression model *utrainlogit* is created by applying the *glm()* function with family = "binomial" on the training dataset. This results in the following output:

```
> utrainlogit < - glm(Personal_Loan ~ ., data = utrain, family = "binomial")
> summary(utrainlogit)
Call:
Glm(formula = Personal_Loan ~ ., family = "binomial", data = utrain)
Deviance Residuals:
```

Min	*1Q*	*Median*	*3Q*	*Max*
–3.0741	–0.1898	–0.0701	–0.0232	4.1048

Coefficients:	*Estimate*	*Std.Error*	*z value*	*Pr(>|z|)*
(Intercept)	–12.1171912	2.3152298	–5.234	0.000000166160550 81***
Age	–0.0324946	0.0858273	–0.379	0.704982
Experience	0.0390812	0.0851023	0.459	0.646072
Income	0.0591487	0.0038527	15.353	< 0.0000000000002***
Family	0.5787243	0.0993320	5.826	0.000000056176278***
CCAvg	0.1548502	0.0571994	2.707	0.006785**
EducationGraduate	3.8214901	0.3469562	11.014	< 0.00000000000002***
EducationAdvanced/ Professional	3.9324934	0.3449324	11.401	< 0.00000000000002***
Mortgage	0.0011437	0.0007617	1.501	0.133241
Securities_Account1	–0.7739587	0.3957411	–1.956	0.050498
CD_Account1	3.7923306	0.4801242	7.899	0.00000000000000282***
Online1	–0.66266600	0.2103154	–3.873	0.001628**
CreditCard1	–1.1142969	0.2877270	–3.873	0.000108***

—

*Signif. Codes: 0'***' 0.001 **' 0.01*' 0.05'.' 0.1' " 1.*
(Dispersion parameter for binomial family taken to be 1).
Null Deviance: 1838.26 on 2999 degrees of freedom.
Residual deviance: 704.97 on 2987 degrees of freedom.
AIC: 730.97.
Number of Fisher scoring iterations: 8

To assess if the model is significant, one can recall that the quantity null deviance–residual deviance follows chi square distribution with 12 (2999–2987) degrees of freedom. Thus, it can be seen that the above model is significant. Further, it can be observed that all the variables other than age, experience and mortgage are significant. The continuous and categorical predictors need to be interpreted appropriately. For instance, the continuous predictor CCAvg with a coefficient of 0.155 indicates that for every unit increase in the customer's credit card spending, the increase in log odds of availing the loan offer increases by 0.155 units, when all other variables are held constant. For the categorical predictor CD account, the log odds of availing the loan offer for someone having the CD account with the bank are 3.79 times the log odds of a customer not having the CD account when the other variables are held fixed. Equivalently, the odds ratio of this variable is 44.2564 ($e^{3.79}$), implying the odds of availing the loan offer are roughly 44 times more for customers holding a CD account with the bank in comparison with those not holding this CD account when all other variables are held constant. Further, few categorical predictors Securities_Account, Online and CreditCard have negative coefficients indicating inverse relationship and thus customers having a securities account with the bank, those who use online banking facility or customers having the bank's credit card correspond to less chance of availing the loan.

The *predict()* function with type = "response" is used to predict the probability associated with each observation in the training dataset. The predicted probability value, denoted in the output as *predtrain*, for each observation is merged with the original training dataset *utrain* as a column using *cbind()* to form the dataframe *utrain1*. The corresponding confusion matrix can be constructed on comparing the predicted response v/s the actual response in the training dataset. A predicted probability value of less than 0.5 leads to it being classified with the label 0, otherwise 1. Though the default threshold of 0.5 is being used here, one could iterate upon better choices of the threshold depending on the context of the problem.

```
> predtrain < - predict(utrainlogit, utrain, type = c("response"))
> utrain1 < - cbind(utrain, predtrain)
> table(utrain1$Personal_Loan, utrain1$predtrain > = 0.5)
##confusion matrix
```

	False	True
0	2701	24
1	92	183

The confusion matrix shows an accuracy of 0.961, sensitivity value of 0.665 and specificity of 0.991 for the training data. To know how the model performs in general, *utrainlogit* is then applied on the unseen test dataset *utest*. The predicted probability values of *utest* are stored in the vector *lr_predtest* that is merged with the test dataset as a column to form *lrutest1*. The confusion matrix based on the test dataset is created.

```
> lr_predtest < - predict(utrainlogit, utest,type = c("response"))
> lrutest1 < - cbind(utest, Lr_predtest)
> table(lrutest1$Personal_Loan, lrutest1$lr_predtest > = 0.5)
##confusion matrix
```

	False	True
0	1774	21
1	67	138

For the test set, the model provides 0.956, 0.673 and 0.988 as the values of accuracy, sensitivity and specificity, respectively. The model does not overfit and the results are comparable to the training set results. However, the model performs not so well on positive cases and class imbalance could be a possible reason.

11.6.2.2 SVM Model

An SVM model is then created further on to examine its performance on the dataset in comparison with the logistic regression model. The svm() function performs classification by setting the response variable Personal_Loan to a factor format in R. Initially, an SVM model *ub.svm* is constructed on the training dataset without scaling where a linear kernel is used with cost parameter $C = 0.1$

```
> set.seed(700)
> utrain$Personal_Loan <- as.factor(utrain$Personal_Loan)
> utest$Personal_Loan <- as.factor(utest$Personal_Loan)
> ub.svm = svm(Personal_Loan ~ ., data = utrain, kernel = "linear", cost = 0.1, Scale
= FALSE, probability=TRUE)
```

```
> summary(ub.Svm)
Call:
Svm(formula = Personal_Loan ~ ., data = utrain, kernel = "linear", cost = 0.1, Scale
= FALSE, probability = TRUE).
Parameters:

    SVM-Type:                    C-classification
    SVM-Kernel:                  linear
    Cost:                        0.1
    Number of Support Vectors:   383
                                 (195 188).
    Number of classes:           2
    Levels:                      0 1
```

It is observed that 383 support vectors are utilized by the model, of which 195 observations are of class 0 and 188 are of class 1.

The obtained svm model is applied on the test set and its confusion matrix is created using the function *confusionMatrix()*. The parameter *positive* = *'1'*, in the given case, sets the second level Personal_Loan = 1 as the 'positive' result, the level that holds more importance in this application.

```
> svm_predtest = predict(ub.svm,newdata = utest, probability = TRUE).
> confusionMatrix(Table(Svm_predtest, utest$Personal_Loan), Positive = '1')
Confusion Matrix and Statistics
```

Svm_predtest	0	1
0	1780	67
1	15	138

```
Accuracy: 0.959
95% CI: (0.9494, 0.9673)
No Information Rate: 0.8975
p-value [Acc > NIR]: < 0.0000000000000022
Kappa: 0.749
McNemar's Test P-value: 0.000000001781
Sensitivity: 0.6732
Specificity: 0.9916
Pos Pred Value: 0.9020
Neg Pred Value: 0.9637
Prevalence: 0.1025
Detection Rate: 0.0690
Detection Prevalence: 0.0765
Balanced Accuracy: 0.8324
'Positive' Class: 1
```

The above output displays an accuracy of 0.959. The sensitivity value is 0.6732 and the specificity value is 0.9916. Other measures are also listed in the output. The 'No Information Rate (NIR)' refers to the proportion of correct classifications if all have been assigned to the larger class, namely '0'. The p-value listed next asserts that the overall accuracy is greater than the NIR, thus asserting the usefulness of the model. The Kappa statistic measures the accuracy by considering the accuracy obtained on random chance assignment. McNemar test assesses the performance of the model compared to random classification. Positive predicted value is the same as the precision and negative predicted value is the ratio of true negative to number of elements predicted as negative. Prevalence shows the proportion of positive class elements in the data. Detection rate refers to the ratio of true positive to the total number of elements and detection prevalence shows the ratio of positive elements to the total number of elements. Balanced accuracy refers to the average of sensitivity and specificity, a better measure when compared to overall accuracy for imbalanced data. The performance indicators for the SVM model are found to be marginally higher than the logistic regression model, though these results are based on a single run.

The *tune()* function is used on a range of cost values less and greater than 0.1 to enable a decision of the optimal cost value that reduces the cross-validation error of misclassification. Using the summary command, one obtains the detailed results of average and spread of cross-validation errors of various folds, thus enabling the choice of optimal cost. *bestmod_linear* stores the model corresponding to the optimal cost value. The corresponding confusion matrix is created by applying *bestmod_linear* on *utest*.

```
> set.seed(3)
> tune.out = tune(svm,Personal_Loan ~ .,data = utrain,kernel = "linear",ranges = list(Cost = c(0.01,0.1,1,5,10,50)))
> summary(tune.out)
Parameter tuning of 'svm':
-sampling method: Tenfold cross validation
-best parameters:
cost
10
-best performance: 0.038
-Detailed performance results:
```

	cost	error	dispersion
1	0.01	0.05466667	0.01634504
2	0.10	0.04166667	0.01467761
3	1.00	0.03900000	0.01133660
4	5.00	0.03833333	0.01167989
5	10.00	0.03800000	0.01102186
6	50.00	0.03833333	0.01057017

> bestmod_linear = tune.out$best.model
> summary(bestmod_linear)
Call:
best.tune(method = svm, train.x = Personal_Loan ~ ., data = utrain, ranges = list(cost = c(0.01, 0.1, 1, 5, 10, 50)), Kernel = "linear")
Parameters:

SVM-Type:	*C-classification*
SVM-Kernel:	*Linear*
Cost:	*10*
Number of Support Vectors:	*320*
	(162 158)
Number of Classes:	*2*
Levels:	*0 1*

> confusionMatrix(table(predict(bestmod_linear,newdata = utest),utest$Personal_Loan),positive = '1')
Confusion Matrix and Statistics

Svm_predtest	0	1
0	1785	69
1	10	136

Accuracy: 0.9605
95% CI: (0.951, 0.9686)
No Information Rate: 0.8975
P-value [Acc > NIR]: < 0.0000000000000022
Kappa: 0.7539
McNemar's Test P-value: 0.000000006777

Sensitivity: 0.6634
Specificity: 0.9944
Pos Pred Value: 0.9315
Neg Pred Value: 0.9628
Prevalence: 0.1025
Detection Rate: 0.0680
Detection Prevalence: 0.0730
Balanced Accuracy: 0.8289
'Positive' Class: 1

It is observed that the best model of linear kernel with cost parameter $C = 10$ on the test set results in an increased accuracy of 0.966 and specificity of 0.9944. However, sensitivity value is 0.6634 and has taken a slight hit.

To see whether the performance can be made better, a radial kernel is applied on the training dataset. The radial kernel utilizes the cost parameter C and the γ parameter. A grid search is carried out among various values of C and γ to choose their optimal values. The optimal model is stored in *bestmod_radial.*

```
> set.seed(3)
> tune.out = tune(svm,Personal_Loan ~ .,data = utrain,kernel = "radial",ranges = list(cost = c(0.01,0.1,1,5,10), gamma = c(0.01,0.1,1,5)))
> summary(tune.out)
Parameter tuning of 'svm':
-sampling method tenfold cross validation
-best parameters:
```

Cost	*Gamma*
5	0.1

-best performance: 0.01566667

-Detailed performance results:

	cost	gamma	error	dispersion
1	0.01	0.01	0.09166667	0.017933347
2	0.10	0.01	0.09166667	0.017933347
3	1.00	0.01	0.04333333	0.016101530
4	5.00	0.01	0.03166667	0.013076225
5	10.00	0.01	0.02766667	0.011552350
6	0.01	0.10	0.09166667	0.017933347
7	0.10	0.10	0.06533333	0.018539248
8	1.00	0.10	0.02433333	0.014660773
9	5.00	0.10	0.01566667	0.009692813
10	10.00	0.10	0.01666667	0.010183502
11	0.01	1.00	0.09166667	0.017933347
12	0.10	1.00	0.09166667	0.017933347
13	1.00	1.00	0.06633333	0.017430639
14	5.00	1.00	0.06233333	0.017655832
15	10.00	1.00	0.06233333	0.017655832
16	0.01	5.00	0.09166667	0.017933347
17	0.10	5.00	0.09166667	0.017933347
18	1.00	5.00	0.09100000	0.017430639
19	5.00	5.00	0.09166667	0.017655832
20	10.00	5.00	0.09166667	0.017655832

```
> bestmod_radial = tune.out$best.model
> summary(bstmod_radial)
Call:
best.tune(method = svm, train.x = Personal_Loan ~ ., data = utrain, ranges = list(cost
= c(0.01,0.1,1,5,10), gamma = c(0.01,0.1,1,5)), kernel = "radial")
Parameters:

   SVM-Type:                  C-classification
   SVM-Kernel:                Radial
   Cost:                      5
   Number of support vectors  263
   (144,                      119)

   > tune.out$best.parameters
```

cost	*Gamma*
5	*0.1*

The optimal cost parameter C is found to be 5 and the optimal γ parameter is 0.1 as indicated by *best performance* of *0.01566667,* the average of the cross-validation errors of different runs.

The confusion matrix is constructed by applying the optimal radial kernel model on the test dataset.

> confusionMatrix(table(predict(bestmod_radial, newdata = utest),utest$Personal_Loan),positive = '1')
Confusion Matrix and Statistics

	0	*1*
0	*1789*	*33*
1	*6*	*172*

Accuracy: 0.9805
95% CI: (0.9734,0.9861)
No Information Rate: 0.8975
P-value [Acc > NIR]: < 0.0000000000000022
Kappa: 0.8874
Mcnemar's Test P-value: 0.00003136
Sensitivity: 0.8390
Specificity: 0.9967
Pos Pred Value: 0.9663
Neg Pred Value: 0.9819
Prevalence: 0.1025
Detection Rate: 0.0860
Detection Prevalence: 0.0890
Balanced Accuracy: 0.9178
'Positive' Class: 1

It can be observed that the radial kernel results in an increased value of accuracy 0.9805. The sensitivity value witnesses a considerable increase to 0.839 and the specificity value observes a marginal increase to 0.9967, deeming this to be a better algorithm than the linear kernel for the dataset at hand. The choice of radial kernel captures the decision boundary well for this dataset and thus leads to a far more appropriate decision rule, resulting in better performance measures.

A comparative analysis of the three models—logistic regression, SVM with linear kernel and SVM with radial kernel with respect to various measures—is presented below:

Model	Accuracy	Sensitivity	Specificity	Precision	F-measure	G-mean	AUC
Logistic Regression	0.956	0.673	0.988	0.868	0.758	0.816	0.966
SVM linear (bestmod_linear)	0.966	0.663	0.994	0.931	0.775	0.812	0.960
SVM radial (bestmod_radial)	0.980	0.839	0.997	0.966	0.898	0.914	0.988

This clearly shows the SVM with radial kernel to be the best-performing model for the considered dataset.

11.6.2.3 Visual Analysis

A visual comparative analysis of the performance of the three models—logistic regression, svm with linear kernel and svm with radial kernel—can be carried out using the ROC curves and lift charts, discussed in Sects. 11.4.2.1 and 11.4.2.2, respectively.

The best linear and radial models are stored in *ub_svm_linear* and *ub_svm_radial,* respectively. Setting the argument probability = TRUE ensures that for each prediction, a vector of probabilities belonging to each class of the response variable is returned. This helps in building the ROC curves for the models, denoted by *roc_lr, roc_svm_linear, roc_svm_radial.*

```
> svm linear best model
> ub_svm_linear = svm(Personal_Loan ~ .,data = utrain, Kernel = "linear",cost = 10, probability = TRUE)
> svm_linear_predtest < - predict(ub_svm_linear,newdata = utest, probability = TRUE)
#svm radial best model
> ub_svm_radial = svm(Personal_Loan ~ .,data = utrain, kernel = "radial",cost = 5,gamma = 0.1,probability = TRUE)
> svm_radial_predtest < - predict(ub_svm_radial,newdata = utest, probability = TRUE)
> par(pty = "s")
> roc_lr < - roc(lrutest1$Personal_Loan, lrutest1$lr_predtest, plot = TRUE,print.auc = TRUE, col = "blue",lwd = 2, legacy.axes = TRUE, main = "ROC Curves")
> roc_svm_linear < - roc(utest$Personal_Loan, attr(svm_linear_predtest, "probabilities")[,2], plot = TRUE,print.auc = TRUE, col = "green",lwd = 2, legacy.axes = TRUE, print.auc.y = 0.4,add = TRUE)
> roc_svm_radial < - roc(utest$Personal_Loan, attr(svm_radial_predtest, "probabilities")[,2], plot = TRUE,print.auc = TRUE, col = "red",lwd = 2, legacy.axes = TRUE, print.auc.y = 0.6,add = TRUE)
> legend("bottomright",legend = c("Logistic Regression","Linear SVM","Radial SVM"),col = c("Blue","green","red"),lwd = 4, cex = 0.5)
```

The package *lift* in R is used to compute the top decile lift and also to plot the lift curve. This package also supports plotting cumulative lift curves. The corresponding

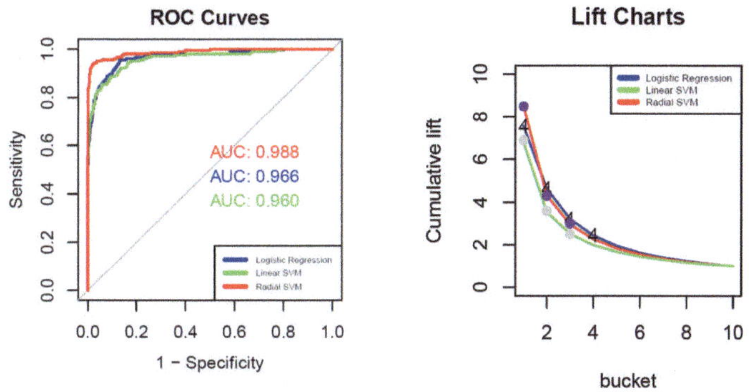

Fig. 11.6 ROC curves and lift charts of various models

lift charts for logistic regression, linear kernel and radial kernel are then plotted (Fig. 11.6).

```
library(lift)
plotLift(lr_predtest, prutest1$Personal_Loan, cumulative = TRUE, col = "blue",lwd
= 2, ylim = c(0,10), main = "Lift Charts")
par(new = TRUE)
plotLift(svm_radial_predtest, utest$Personal_Loan, cumulative = TRUE, col =
"red",lwd = 2,ylim = c(0,10))
par(new = TRUE)
plotLift(svm_linear_predtest, utest$Personal_Loan, cumulative = TRUE, col =
"green",lwd = 2,ylim = c(0,10))
legend("topright",legend = c("Logistic Regression","Linear SVM","Radial
SVM"),col = c("Blue","green","red"),lwd = 4,cex = 0.5)
```

The ROC curve plotted above clearly displays that the performance of SVM with radial kernel model performs much better than the linear kernel SVM model and the logistic regression model. This can also be observed by looking at their AUC, as shown in the graph.

The lift chart provides interesting insights. The cumulative lift corresponding to the 1st decile for the logistic regression, linear kernel and radial kernel model is 7.6, 7 and 8.3, respectively. The selection of the top decile, i.e. the top 10% of customers, as identified by the logistic regression model results in identifying 7.6 times the number of actual loan takers predicted by the random model. It can be observed that this factor decreases as the other deciles are considered. Focusing on particularly the top one or two deciles identifies most of the actual loan takers. It was observed in the earlier sections that the radial SVM model was the best among the three models in classifying the observations. This aspect gets corroborated further by the fact that the radial SVM model identifies 8.3 times the number of actual loan takers as that of the random model, being the maximum of all the three models. Applying this model on the larger database of potential customers of the bank, it would be favourable and profitable for the bank to just target those customers within the 1st decile.

Thus, implementing the right model would result in considerable profit to the bank and also would prevent any wastage of resources by targeting just a fraction of their customers for availing their loan product.

11.6.3 Implementation in Python

Python is a widely used, object-oriented programming language for a variety of applications. Much of its popularity is due to the simple and clear syntax structure it offers along with being highly portable as it runs on different operating systems such as Linux, macOS or Windows.

Python offers a large set of libraries, of which we have utilized pandas, numpy, sklearn, seaborn, scikit-plot and matplotlib in this implementation. NumPy is an open-source python library which is used for performing a wide variety of operations on arrays. It provides multi-dimensional arrays and matrix structures which are widely used for processing numerical data in python. Pandas is yet another python library which has been built on top of NumPy and is typically utilized for data analysis and data processing. It simplifies analysing large datasets by providing operation and data structures which can manipulate numerical and time-series data.

Scikit-learn or sklearn (Pedregosa et al. 2011) provides efficient tools for machine learning and statistical modelling. It provides support for regression, clustering, dimension reduction, selection of models and data pre-processing. It is utilized extensively for both supervised and unsupervised machine learning use cases as it covers many algorithms like support vector machines, random forests, naïve-bayes and k-means clustering to name a few.

Matplotlib is a visualization library which is used extensively for creating plots and interactive figures which can be easily customized to the required visual style and layout. Further, a popular library scikit-plot is used to visualize and plot the scikit-learn objects. It is built on top of the matplotlib library and primarily helps in visualizing the machine learning model evaluation metrics such as generating ROC curves from actual response and predicted probabilities. Yet another visualization library is Seaborn which is used for plotting graphs and visualizing random distributions. This is closely integrated with the pandas data structures and it aids in creating high-quality statistical plots with attractive styles and colour palettes.

The task that was done through R can be carried out using the corresponding Python code and it may be of interest to know that both R and Python would generally lead to similar results. For beginners, good sources of texts in Python include Muller et al. (2016) and Sarkar et al. (2018).

Both R and Python are open-source languages, where R is specifically meant for statistical analysis and Python is a general-purpose language well suited for various machine learning tasks. The user could choose either of these and a wide range of materials and resources are available for beginners.

11.7 Conclusion and Discussion

Classification is one of the key topics in the domain of predictive analytics due to its numerous applications. This chapter provides lucid explanation of two important classification models, logistic regression and support vector machines. An illustration of these methods to an important real-life application of classifying loan customers is presented. Visual analysis of performance of models like ROC curves and lift charts is discussed in addition to various evaluation metrics.

It needs special mention that though the underlying approaches of logistic regression and SVM are seemingly different, deep connections exist between these two methods. SVM appears to be unique in its way of finding a hyperplane to separate the classes when compared to logistic regression that uses a more traditional approach to achieve the same objective. However, it can be observed that the optimization problem corresponding to SVM and the likelihood of logistic regression can be rewritten in a more familiar loss function form along with regularization used in traditional approaches like ridge regression and Lasso regression. Doing this would make one understand and appreciate the similarity in the loss functions of both the techniques. Further, the flexibility of a non-linear boundary in SVM using the kernel trick makes the technique more interesting and novel. Again, this is not unique to just SVM as logistic regression too can be carried out with non-linear kernel, though it is widely applied in SVM. For more discussion on the link between these two techniques, the reader can refer to James et al. (2021).

Logistic regression and SVM considered here for two class problems can be further extended to multiclass problems. Considerable research is happening in the domain of class imbalance and attracts enough attention from industry due to its numerous applications.

References

Agresti A (2012) Categorical data analysis, 3rd edn. Wiley

Altman D, Bland J (1994) Diagnostic tests 3: receiver operating characteristic plots. BMJ 309:188

Barua S, Islam MM, Yao X et al (2014) MWMOTE - Majority weighted minority oversampling technique for imbalanced data set learning. IEEE Trans Knowl Data Eng 26(2):405–425

Bilder CR, Loughin TM (2015) Analysis of categorical data with R. Chapman & Hall/CRC Press

Boehmke B, Greenwell B (2019) Hands-on machine learning with R. Chapman & Hall/CRC Press

Boser B, Guyon I, Vapnik V (1992) A training algorithm for optimal margin classifiers. In: Proceedings of the fifth annual workshop on computational learning theory, pp 144–152

Brown C, Davis H (2006) Receiver operating characteristics curves and related decision measures: a tutorial. Chemom Intell Lab Syst 80(1):24–38

Chawla NV, Bowyer KW, Hall LO et al (2002) Smote: synthetic minority over-sampling. J Artif Intell Res 16:321–357

Cortes C, Vapnik V (1995) Support-vector networks. Mach Learn 20(3):273–297

DeMaris A (1995) A tutorial in logistic regression. J Marriage Fam 57(4):956–968

Dobson A (2002) An introduction to generalized linear models. Chapman & Hall/CRC Press

Duda R, Hart P, Stork D (2000) Pattern classification, 2nd edn. Wiley

Fawcett T (2006) An introduction to ROC analysis. Pattern Recogn Lett 27(8):861–874
Fernández A, García S, Herrera F et al (2018) SMOTE for learning from imbalanced data: progress and challenges, marking the 15-year anniversary. J Artif Intell Res 61(1):863–905
Fernández A, García S, Galar M et al (2018) Learning from imbalanced data sets. Springer
Han, H, Wang WY, Mao BH (2005) Borderline-SMOTE: A new over-sampling method in imbalanced data sets learning. Lect Notes Comput Sci 3644 (PART I), p 878–887
Hand D (1981) Discrimination and classification. Wiley
Hastie T, Tibshirani R, Friedman J (2009) The elements of statistical learning: data mining, inference, and prediction, 2nd edn. Springer
He H, Bai Y, Garcia EA et al (2008) Adasyn: Adaptive synthetic sampling approach for imbalanced learning. In: IEEE international joint conference on neural networks. IJCNN 2008. (IEEE world congress on computational intelligence). IEEE, pp 1322–1328
He H, Garcia EA (2009) Learning from Imbalanced data. IEEE Trans Knowl Data Eng 21(9):1263–1284
Hosmer DW, Lemeshow S, Sturdivant RX (2013) Applied logistic regression, 3rd edn. Wiley
James G, Witten D, Hastie T et al (2021) An introduction to statistical learning: with applications in R, 2nd edn. Springer
Kuhn M, Johnson K (2013) Applied predictive modeling. Springer
Ling C, Li C (1998) Data mining for direct marketing: problems and solutions. In: Proceedings of the fourth international conference on knowledge discovery and data mining, pp 73–79
Lopez V, Fernandez A, Garcıa S et al (2013) An insight into classification with imbalanced data: Empirical results and current trends on using data intrinsic characteristics. Inf Sci 250:113–141
Moguerza JM, Munoz A (2006) Support vector machines with applications. Stat Sci 21(3):322–336
Muller AC, Guido S (2016) Introduction to machine learning with python: A guide for data scientists. O'Reilly
Pedregosa F, Varoquaux G, Gramfort A et al (2011) Scikit-learn: machine learning in Python. J Mach Learn Res 12:2825–2830
Sarkar D, Bali R, Sharma T (2018) Practical machine learning with python: a problem-solver's guide to building real-world intelligent systems. Apress, Berkeley, CA
Shahee SA, Ananthakumar U (2021) An overlap sensitive neural network for class imbalanced data. Data Min Knowl Disc 35:1654–1687
Shmueli G, Bruce PC, Yahav I et al (2017) Data mining for business analytics—concepts, techniques and applications in R, Wiley
Vapnik V (1996) The nature of statistical learning theory. Springer
Vapnik V (2010) The nature of statistical learning theory, 2nd edn. Springer
Zumel N, Mount J (2020) Practical data science with R, 2nd edn. Manning

Chapter 12
Adaptive Learning in Forecasting: Application to European Agricultural Markets

Foteini Kyriazi, Dimitrios D. Thomakos, and Efthymios Xylangouras

Abstract This chapter introduces adaptive learning forecasting, a novel approach that enhances forecasting accuracy by incorporating both forecast error learning and forecast averaging. The univariate and multivariate versions of adaptive learning forecasting are thoroughly explained. It presents the theoretical framework along with the necessary mathematical derivations that establish the desirable properties and practical usefulness of the method. In particular, the chapter demonstrates how *ex ante* MSE-based improvements are available from the use of the method. Additionally, it offers a step-by-step guide on implementing adaptive learning techniques in real-world applications. Using agricultural price data, the chapter illustrates the methodology and highlights the advantages of multivariate adaptive learning over its univariate counterpart, making the method a particularly appealing part of the forecaster's toolbox.

We would like to thank Deep Mukherjee and Faiz Hamid for giving us the opportunity to contribute to this excellent volume.

Supplementary Information The online version contains supplementary material available at https://doi.org/10.1007/978-981-97-9060-9_12.

F. Kyriazi (✉)
Department of Agribusiness and Supply Chain Management, Agricultural University of Athens, Athens, Greece
e-mail: fkyriazi@aua.gr

D. D. Thomakos
Department of Business Administration, School of Economics and Political Science, National and Kapodistrian Utniversity of Athens, 10 559 Athens, Greece

International Centre for Economic Analysis, Toronto, Canada

D. D. Thomakos
e-mail: dthomakos@ba.uoa.gr

E. Xylangouras
Adam Smith Business School, University of Glasgow, G12 8QQ Glasgow, United Kingdom
e-mail: efthymiosxyl@gmail.com

F. Hamid and D. Mukherjee (eds.), *Data Science for Modeling Managerial and Socioeconomic Problems*, Contributions to Management Science,
https://doi.org/10.1007/978-981-97-9060-9_12

12.1 Introduction

Understanding and prognosticating the future is the ultimate purpose of all disciplines which has captured humankind from antiquity. Prognosis, meaning knowing beforehand, is the process of making a prediction, that is decision based on past and present data (and subjective judgment if needed), with some degree of accuracy. In order to consider the issue of predictability, we must work on the assumption of the possible presence of three important conditions: one, a (mostly unknown) set of causal factors that might drive future events; two, an information set containing data on present and past events, and three, an available regular procedure that links past and present events with what might actually happen in the future. Forecasting, another word for prediction and predictability, works on these assumptions and modern methods and models abide by them as well. Our forecasting devices, based on statistics and econometrics rather than the sayings of an oracle, do depend on principles derived from the above assumptions: an underlying mathematical model, a mechanism that tries to explain the evolution of the values which we want to forecast; a history of observations from the past up to the present which make up our data; and, finally, a procedure to generate the forecasts, linking the model and the data. Forecasting is a logical, internally consistent, process that transforms information that we already have (data), through a mathematical model or method (approximation), to what future information might look like (extrapolation). In the realm of data science, forecasting has a significant role in enabling organizations to plan effectively, anticipate changes, and allocate resources efficiently. Thus, forecasting, when accompanied by effective judgment, becomes an essential tool for decision making in a number of fields, ranging from logistics to climate forecasting and economics and finance management.

A key concept in forecasting is the choice of forecasting horizon, which denotes the time period into the future for which predictions are made. The selection of the forecasting horizon depends on the specific application and the time frame of interest to the decision maker. Short-term forecasts may focus on predicting timely immediate trends within hours, days, weeks, or for example up to a month ahead, while long-term forecasts may extend several months or years into the future. Another key concept related to the employed data is the forecasting frequency. This is connected to how frequently observations are gathered and examined. Daily, weekly, monthly, or annual data are examples of typical frequencies. The selection of forecasting frequency depends on data availability and the level of granularity required for accurate predictions. Moreover, a common family of classical forecasting models widely analyzed in the academic setting but also employed in the industry due to their simplicity are the mathematical models that have an autoregressive structure. These autoregressive models operate on the premise that future values in a time series can be estimated using its historical values. They aim to model the relationship between a data series and its delayed versions, enabling the detection of trends and patterns within the process that produces the data. Furthermore, alongside forecasting an expected value of a time series (i.e., the future value), forecasting often includes estimating other statistical properties or moments, such as variance, skewness, and kurtosis. These moments provide insights into the distributional characteristics and volatility of the data, which are important for making robust forecasts and drawing inferences regarding the complexity and trend of the data.

By design, forecasting entails selecting and fitting a specific model to historical data in order to draw conclusions about future outcomes. Whether the selection of the model refers to autoregressive models, exponential smoothing, or machine learning algorithms, each forecasting approach relies on a particular set of assumptions and methodologies. These models are tailored to capture different aspects of the underlying data-generating process and may differ in terms of computational requirements and complexity. On the other hand, several methodologies exist that are not bound to a specific forecasting model. These can be applied across different forecasting models and techniques, aiming to enhance their forecastability. One such framework, which is not model specific and is used for post estimation performance improvement, is the method of adaptive learning, which we will analyze in the following chapter.

For a detailed study on the basics of forecasting, several standard textbooks provide in-depth explanations of fundamental concepts, methodologies, and practical applications in forecasting. A textbook that offers comprehensive coverage and covers both classical and modern approaches, providing real-world examples and case studies, is *Forecasting: Principles and Practice* by Rob J. Hyndman and George Athanasopoulos. Additionally, *Introductory Econometrics for Finance* by Chris Brooks provides a comprehensive introduction to econometrics and forecasting, focusing on finance-related topics such as asset pricing models, risk management, and financial forecasting.

12.2 Adaptive Learning in Forecasting

This section provides a comprehensive introduction to a novel method of forecasting called *adaptive learning*, in its univariate and multivariate versions.[1] Adaptive learning was recently introduced into the literature in the paper of Kyriazi et al. (2019a) for the univariate version and in Kyriazi et al. (2019b) for the multivariate version. This new forecasting method has a significant feature that makes it highly useful for empirical work: it accepts as input any other forecast and, under well-defined theoretical conditions, can improve the performance of that forecast. Furthermore, that this method can provide performance enhancements has been shown to hold *ex ante*, that is theoretically, under said conditions. Adaptive learning is a very useful method for the practitioner's toolbox because it can be either a sort of "cherry-topping" even for the most advanced forecasting method, being a last attempt to improve forecasting performance, or a very (computationally and conceptually) easy way to obtain fast performance enhancements from even the most basic benchmark forecast. We present the whole theoretical overview of adaptive learning, both

[1] For some other forecast methodology, see Kyriazi and Thomakos (2020a, b), Kyriazi et al. (2023) and Tarani and Kyriazi (2024).

univariate and multivariate, and provide the reader with all necessary information to build a successful adaptive learning forecasting system that can assist in real-life decision making.[2] The chapter ends with a complete empirical application that shows how the method can be applied in practice; we use agricultural prices for this application and this is highly relevant in the current global situation. The issue of forecasting agricultural time series holds practical significance, with the characteristics of the goods themselves indicating a range of influencing factors that can be employed to produce these predictions, as demonstrated later in the application.

The reader will find useful literature in the papers of: Adaptive Learning Forecasting with Applications in Forecasting Agricultural Prices by Kyriazi et al. (2019a) and *Forecasting: theory and practice* by Petropoulos et al. (2022). Furthermore, more foundational work on forecasting can be found, among many other excellent volumes, in the three volumes of the *Handbook of Economic Forecasting* (2006, 2013), in the *Oxford Handbook of Economic Forecasting* (Clements and Hendry, 2011), the volume on *Forecasting: Principles and Practice* (Hyndman and Athanasopoulos, 2013), and the volume on *Introduction to Time Series Analysis and Forecasting* (2016).

The remainder of this section is organized in the following manner: in the following section we present the univariate version of the adaptive learning method, while in Sect. 12.4 we present the multivariate version of the adaptive learning method; in Sect. 12.5 we illustrate the necessary steps considering the preparation and implementation of the adaptive learning approach; in Sect. 12.6 we provide an end-to-end empirical application considering a set of forecasting models and Sect. 12.7 provides some final thoughts and potential developments in the ongoing research. Additionally, tables depicting the results of the empirical application are provided.

12.3 Univariate Adaptive Learning

12.3.1 Adaption as (Forecast) Error Learning

We start-off our discussion with an example, which will show to the reader what we interpret as "learning" in our forecasting context. We explore the simple exponential smoothing forecast, given as:

$$\widehat{X}^{ES}_{n+1|n} \stackrel{\text{def}}{=} \alpha X_n + (1-\alpha)\widehat{X}^{ES}_{n|n-1} \tag{12.1}$$

[2] Our presentation here depends on and adapts the results from Kyriazi et al. (2019a; b). The reader can consult these references for additional details.

where $\widehat{X}^{ES}_{n+1|n}$ and $\widehat{X}^{ES}_{n|n-1}$ are the forecasts corresponding to two consecutive time periods and $n+1|n$ refers to the out-of-sample prediction of the unknown additional data point in the future based on the available observations.[3] Re-writing the above expression we obtain:

$$\widehat{X}^{ES}_{n+1|n} \stackrel{\text{def}}{=} \widehat{X}^{ES}_{n|n-1} + \alpha \widehat{e}^{ES}_{n|n-1} \tag{12.2}$$

where $\widehat{e}^{ES}_{n|n-1} \stackrel{\text{def}}{=} X_n - \widehat{X}^{ES}_{n|n-1}$ is the prediction error. The forecast relies on a straightforward approach: it adjusts the current forecast $\widehat{X}^{ES}_{n|n-1}$ by adding a proportion of the existing forecast error, essentially "learning" from past errors. This approach represents a "prototype" of forecast error learning, and although it is quite simple, it is not the only standard forecast that incorporates this type of learning from previous forecast errors. To illustrate this, consider the first-order autoregressive model forecast, provided that the model is accurately specified (meaning the model aligns with the data-generating process):

$$\widehat{X}^{AR}_{n+1|n} \stackrel{\text{def}}{=} \alpha X_n = \alpha \left(\widehat{X}^{AR}_{n|n-1} + e_n \right) \tag{12.3}$$

since we have that:

$$\widehat{X}^{AR}_{n+1|n} \stackrel{\text{def}}{=} \alpha(\alpha X_{n-1} + e_n) = \alpha^2 X_{n-1} + \alpha e_n \tag{12.4}$$

and $\widehat{X}^{AR}_{n|n-1} \stackrel{\text{def}}{=} \alpha X_{n-1}$. Note that we have $\widehat{e}^{AR}_{n|n-1} \equiv e_n$, because the model is correctly specified. It is evident that the forecast is once again built on a straightforward structure that incorporates past forecast errors and a "learning" parameter.[4]

These two examples, along with others that can be easily identified, can be incorporated into a broader framework with the following structure:

$$\widehat{X}^{m}_{n+1|n} \stackrel{\text{def}}{=} \gamma_1 \widehat{X}^{m}_{n|n-1} + \gamma_2 \widehat{e}^{m}_{n|n-1} \tag{12.5}$$

for a given method or model m and certain constants (γ_1, γ_2). We utilize a recursive structure where the information is derived from a single method or model. While multiple methods or models can be incorporated into the equation mentioned above, this equation imposes specific, and potentially overly strict, conditions by relying on just one method or model and learning solely from it. Adaptive learning generalizes

[3] In this equation, and also in what follows, the parameter α is assumed to be in the unit interval, $\alpha \in [0, 1]$.

[4] The reader will immediately see here that Eq. (12.4) is recursive and upon successive substitutions we arrive at the well-known form of an autoregression being a moving average of infinite order, i.e., $\widehat{X}^{AR}_{n+1|n} = \sum_{j=0}^{\infty} \alpha^{j+1} e_{n-j}$. In that case, "learning" is simply a discounted sum of all previous forecast errors.

in two ways: first, it eliminates the recursive nature of the prediction itself, but retains it within the forecast error (which is essential for learning from it); and, second, it allows any forecast to serve as the input for the new, adaptive forecast. In the next section, we outline how these two generalizations lead to univariate adaptive learning.

12.3.2 The Types and Properties of Adaptive Learning

Let $X^{adt}_{n+1|n}$ denote the (final) adaptive learning forecast, $X^{avg}_{n+1|n}$ denote the input forecast (we use the *avg* notation to indicate that it might even be a forecast average from many input models), and let $e^{adt}_{n+1|n}$ denote the adaptive learning forecast error. Should we want to be very general about the impact of learning via the forecast error we could write two formulas for adaptive learning as follows:

$$\widehat{X}^{adt}_{n+1|n} \stackrel{\text{def}}{=} (1-\gamma)\widehat{X}^{avg}_{n+1|n} + \gamma\mathcal{S}\left(\widehat{e}^{adt}_{n|n-1}\right) \tag{12.6}$$

which is a formula that involves forecast smoothing (of the input forecast) and the simpler formula:

$$\widehat{X}^{adt}_{n+1|n} \stackrel{\text{def}}{=} \widehat{X}^{avg}_{n+1|n} + \gamma\mathcal{S}\left(\widehat{e}^{adt}_{n|n-1}\right) \tag{12.7}$$

which does not involve forecast smoothing. Here $\mathcal{S}(\cdot)$ is a function of the previous adaptive forecast errors, but is only written to indicate the generality of the method; in our analysis we will use $\mathcal{S}(x) = x$ as the identity function. This allows us to obtain analytical results, although one is not constrained to use another form such as an average of past forecast errors. It is important to note, and more details can be found in Kyriazi et al. (2019b), that we are initiating a method and not a model. Thus, we need not link the method with a data-generating process and we can proceed with unconditional moments in uncovering the properties of the method.[5]

If we examine the forecast error that results from the adaptive forecasting method, without smoothing, i.e., subtracting X_{n+1} from both sides of Eq. (12.7) and set $\mathcal{S}(x) = x$, we define:

$$\widehat{e}^{adt}_{n+1|n} \stackrel{\text{def}}{=} \widehat{e}^{avg}_{n+1|n} - \gamma\widehat{e}^{adt}_{n|n-1}. \tag{12.8}$$

This seems to resemble a first-order autoregression in $\widehat{e}^{adt}_{n|n-1}$, with the learning rate γ serving as the autoregressive parameter; however, this interpretation is not entirely

[5] In the paper of Kyriazi et al. (2019b), the connection between conditional and unconditional analysis is explained in detail. It is also interesting to note that the expressions for adaptive learning can be considered within the framework of the "intercept correction" approach of Clements and Hendry (1996).

accurate, as the series of forecast errors cannot be instantly observed. Therefore, the equation is actually structured as a filter that is constructed recursively.[6]

In order to progress, we calculate the forecast error variance of the adaptive forecast based on its components and derive:

$$\sigma_{adt}^2 \overset{\text{def}}{=} \sigma_{avg}^2 + \gamma^2\sigma_{adt}^2 - 2\gamma \mathsf{E}\left[\hat{e}_{n+1|n}^{avg}\hat{e}_{n|n-1}^{adt}\right], \tag{12.9}$$

which illustrates the two components that constitute the variance: the variance of the average forecast and the covariance between the input/average forecast and the output/adaptive forecast. By utilizing the recursive nature of the adaptive forecast error, we can demonstrate that:

$$\mathsf{E}\left[\hat{e}_{n+1|n}^{avg}\hat{e}_{n|n-1}^{adt}\right] = \mathsf{E}\left[\hat{e}_{n+1|n}^{avg}\left(\hat{e}_{n|n-1}^{avg} - \gamma\hat{e}_{n-1|n-2}^{adt}\right)\right] = \mathsf{E}\left[\hat{e}_{n+1|n}^{avg}\hat{e}_{n|n-1}^{avg}\right] = \sigma_{avg}(1) \tag{12.10}$$

thus, the covariance between the average forecast and the adaptive forecast represents the first-order autocovariance of the average forecast error.[7] Plugging the cross-covariance and the autocovariance terms back into our prior expression for the forecast error variance, we arrive at:

$$\sigma_{adt}^2 \overset{\text{def}}{=} \sigma_{avg}^2 + \gamma^2\sigma_{adt}^2 - 2\gamma\sigma_{avg}(1); \tag{12.11}$$

next, by dividing both sides by σ_{avg}^2 and defining the variance ratio $\nu^2 \overset{\text{def}}{=} \sigma_{adt}^2/\sigma_{avg}^2$, we can derive

$$\nu^2 \overset{\text{def}}{=} \frac{1 - 2\gamma\rho_{avg}(1)}{1-\gamma^2}, \tag{12.12}$$

which is smaller than one when $\gamma \leq 2\rho_{avg}(1)$, with $\rho_{avg}(1) > 0$, where $\rho_{avg}(1) \overset{\text{def}}{=} \sigma_{avg}(1)/\sigma_{avg}^2$ represents the first-order autocorrelation of the average forecast error. This outcome is quite valuable in practical terms, as we will now elaborate.

To improve the efficiency of the adaptive forecast (reducing the variance of forecast errors), we find that the input forecast (the average forecast) needs to exhibit

---

[6] Note that the sign of the past forecast error will enter the computation of the new forecast and the forecast error directly. Thus, there is no need to consider an approach that treats positive and negative forecast errors differently, unless the costs of forecast errors with different signs are themselves different.

[7] Note that the covariance between the average forecast error and the past adaptive forecast error is zero by design, as the average forecast is computed independently of past adaptive forecast errors; it is only the adaptive forecast that feeds on past adaptive forecast errors.

some degree of positive forecast error autocorrelation, even if it is minimal.[8] If autocorrelation is observed, an adjustment to the average forecast can be made, which, in theory, will improve performance. Additionally, it's important to recognize that the factor of two in front of the autocorrelation coefficient suggests that even minor values of autocorrelation can enable the adaptive forecast to perform effectively. When the average forecast achieves full efficiency (as $\rho_{avg}(1) \rightarrow 0$), the learning rate approaches zero and the adaptive forecast converges to the average forecast.

The result above is clear and practical; however, it is reasonable to question its utility. We are merely provided with a limit for the learning rate, and there is no additional direction on how to determine the value of γ. For instance, we might choose $\gamma = \rho_{avg}(1)$, where the first-order autocorrelation of the average forecast error is estimated in a recursive manner as the forecast errors build up. This approach would be acceptable, but it would result in the learning rate being virtually constant and the variance ratio ν^2 remaining nearly constant as well, assuming that the estimated value of $\rho_{avg}(1)$ does not change significantly over time. The learning rate determines the rate that the model adapts to new information. An increased learning rate facilitates quicker modifications but might result in instability, whereas a decreased learning rate promotes stability but may result in slower convergence to optimal solutions. We can approach the problem in two distinct ways: firstly, by identifying the learning rate that reduces the variance of the adaptive forecast; secondly, by utilizing heuristics that are estimated recursively, which aim to take advantage of the changes in the performance of the forecasts that contribute to the average forecast. The issue of maintaining a nearly constant learning rate is not entirely eliminated in the first approach mentioned, but it does serve as a theoretically robust reference point in the quest for alternatives. Consequently, we will take Eq. (12.11) and minimize it with respect to γ to derive the first-order conditions:

$$2\gamma\left(1-\gamma^2\right)^{-2}\left[\sigma_{avg}^2 - 2\gamma\sigma_{avg}(1)\right] = \left(1-\gamma^2\right)^{-1}\sigma_{avg}(1), \tag{12.13}$$

which reduces to the subsequent second-degree polynomial in the learning rate:

$$\gamma^2\sigma_{avg}(1) - \gamma\sigma_{avg}^2 + \sigma_{avg}(1) = \gamma^2\rho_{avg}(1) - \gamma + \rho_{avg}(1) = 0, \tag{12.14}$$

the two solutions of which, γ^*, are given by:

$$\gamma^* \stackrel{\text{def}}{=} \frac{1 \pm \sqrt{1 - 4\rho_{avg}(1)^2}}{2\rho_{avg}(1)}. \tag{12.15}$$

It is necessary to ensure that $1 - 4\rho_{avg}(1)^2 \geq 0$, which lead us to the first necessary condition that $\rho_{avg}(1) \leq 0.5$. Additionally, the solution must also meet the requirement that the variance ratio ν^2 is smaller than one, i.e., $\gamma^* \leq 2\rho_{avg}(1)$. This implies

[8] This is a natural result: if the input forecast is "perfect", insofar as forecasts go, one cannot hope of improving it!

that we require that:

$$1 \pm \sqrt{1 - 4\rho_{avg}(1)^2} \leq 4\rho_{avg}(1)^2, \tag{12.16}$$

which holds for one of the two solutions when we have that:

$$1 - 4\rho_{avg}(1)^2 \leq \sqrt{1 - 4\rho_{avg}(1)^2}. \tag{12.17}$$

As a result, the optimal value of the learning rate, which minimizes the adaptive forecast variance, is observed to be:

$$\gamma^* = \frac{1 - \sqrt{1 - 4\rho_{avg}(1)^2}}{2\rho_{avg}(1)}. \tag{12.18}$$

This "optimal", variance-minimizing, learning rate exhibits several noteworthy characteristics, which can be readily observed in Fig. 12.1. It is important to note that the learning rate is a non-linear function of the autocorrelation, displaying two distinct segments: a highly non-linear segment for $\rho_{avg}(1)$ values ranging approximately from 0.3 to 0.5, and a predominantly linear segment for values below roughly 0.3. This observation is quite intuitive: the learning rate increases linearly (indicating slower learning) for smaller autocorrelation values, but exhibits non-linear behavior (signifying faster learning) as the input forecast becomes increasingly inefficient.

While the previous discussion effectively addresses the selection of γ from a variance-minimizing perspective, heuristics can serve as a different method for determining the learning rate. More details can be found in Kyriazi et al. (2019b), where additional improvements on the learning rate are presented and their empirical application is also discussed.

12.4 Multivariate Adaptive Learning

The idea of adaptive learning is compelling as it provides us with a method that, under conditions, will give us *a priori* MSE-reductions. One might assume that transitioning from the univariate case to the multivariate case would provide even better performance. By incorporating related variables in our multivariate input forecasts and adjusting for more than one forecast error, we anticipate improved accuracy. Consequently, in this section we present the mechanics of multivariate adaptive learning. Obtaining theoretical results is considerably more difficult in the multivariate case but, as we will illustrate later on, we can find empirical performance enhancements over the univariate approach with the use of multivariate adaptive learning.

Consider now M forecasts in the $(M \times 1)$ vector $\widehat{\boldsymbol{X}}_{n+1|n}$. These forecasts might refer to the same variable across different cross-sectional units, i.e.,

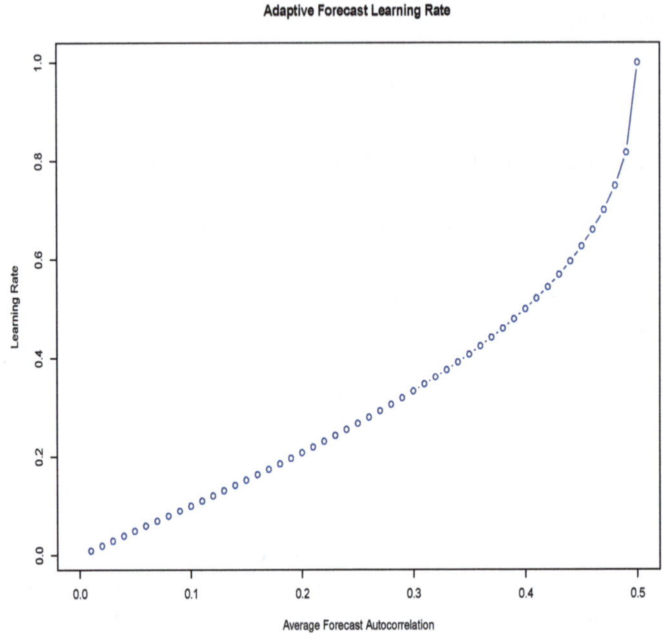

Fig. 12.1 Variance-minimizing learning rate γ^* of the adaptive forecast. Figure adapted from Kyriazi et al. (2019a)

$\widehat{X}_{n+1|n} \stackrel{\text{def}}{=} \left[\widehat{X}^1_{n+1|n,}, \widehat{X}^2_{n+1|n,}, \ldots, \widehat{X}^M_{n+1|n,}\right]^\top$ and this is a panel data structure, or might refer to M different variables, i.e., $\widehat{\boldsymbol{X}}_{n+1|n} \stackrel{\text{def}}{=} \left[\widehat{X}_{n+1|n,}, \widehat{Y}_{n+1|n,}, \ldots, \widehat{Z}_{n+1|n,}\right]^\top$ and this is a multiple equations data structure.[9] Each forecast is obtained by any method desired and for compatibility with the univariate case we will denote the input forecasts as $\widehat{\boldsymbol{X}}^{avg}_{n+1|n}$, with avg denoting an average forecast and with the implicit assumption that each of the components of the forecast vector may or may not come from forecast averaging.

Let $\widehat{\boldsymbol{X}}^{adt}_{n+1|n}$ indicate the adaptive, or product, forecast vector and let $\boldsymbol{\Gamma}$ denote the $(M \times M)$matrix of learning coefficients. Then, we define the adaptive learning forecast via the following equation, assuming no smoothing:

$$\widehat{\boldsymbol{X}}^{adt}_{n+1|n} \stackrel{\text{def}}{=} \widehat{\boldsymbol{X}}^{avg}_{n+1|n} + \boldsymbol{\Gamma}\widehat{\boldsymbol{e}}^{adt}_{n|n-1} \tag{12.19}$$

with the identifying assumption that $\boldsymbol{\Gamma} \in \mathcal{G} \in [0, 1]^{M^2}$, a subset of the unit hypercube. Subtracting the forecast from the actual value $\boldsymbol{X}_{n+1}$ results in the forecast error updating equation:

[9] This is a distinct advantage of the multivariate adaptive learning, in that it can handle panel and simultaneous equations data structures at the same time.

$$\hat{\boldsymbol{e}}^{adt}_{n+1|n} \stackrel{\text{def}}{=} \hat{\boldsymbol{e}}^{avg}_{n+1|n} - \boldsymbol{\Gamma}\hat{\boldsymbol{e}}^{adt}_{n|n-1} \tag{12.20}$$

which has the same format as the univariate case but now can account for all the cross-dependencies among the forecasts.

In the multivariate case, we can compute the adaptive/output forecast error variance-covariance matrix $\boldsymbol{\Sigma}_{adt}$ as follows:

$$\boldsymbol{\Sigma}_{adt} \stackrel{\text{def}}{=} \boldsymbol{\Sigma}_{avg} - \boldsymbol{\Gamma}\boldsymbol{\Sigma}_{avg}(1) - \boldsymbol{\Sigma}(1)^{\top}_{avg}\boldsymbol{\Gamma}^{\top} + \boldsymbol{\Gamma}\boldsymbol{\Sigma}_{adt}\boldsymbol{\Gamma}^{\top} \tag{12.21}$$

where $\boldsymbol{\Sigma}_{avg}$ is the average/input forecast error variance-covariance matrix and where $\boldsymbol{\Sigma}_{avg}(1) \stackrel{\text{def}}{=} \mathsf{E}\big[\hat{\boldsymbol{e}}^{avg}_{n|n-1}\hat{\boldsymbol{e}}^{avg}_{n+1|n}\big] \equiv \mathsf{E}\Big[\hat{\boldsymbol{e}}^{adt}_{n|n-1}\hat{\boldsymbol{e}}^{avg}_{n+1|n}\Big]$ is the first-order autocovariance matrix of the input forecast error. Please be aware that we adhere to the same assumptions per the structure of these covariance matrices as in the univariate case.

To solve for the explicit form of $\boldsymbol{\Sigma}_{adt}$, we apply the $\mathsf{vec}[\boldsymbol{A}]$ operator, that stacks the columns of matrix $\boldsymbol{A}$ into a single vector, to both sides of the variance-covariance equation above to obtain:

$$\begin{aligned}[\boldsymbol{I} - (\boldsymbol{\Gamma} \otimes \boldsymbol{\Gamma})]\boldsymbol{\sigma}_{adt} &= \boldsymbol{\sigma}_{avg} - \big[\big(\boldsymbol{\Sigma}_{avg}(1)^{\top} \otimes \boldsymbol{I}\big) + \big(\boldsymbol{I} \otimes \boldsymbol{\Sigma}_{avg}(1)^{\top}\big)\boldsymbol{K}\big]\boldsymbol{\gamma} \\ &= \boldsymbol{\sigma}_{avg} - \boldsymbol{S}_{avg}\boldsymbol{\gamma}\end{aligned} \tag{12.22}$$

where we defined $\boldsymbol{\sigma}_{adt} \stackrel{\text{def}}{=} \mathsf{vec}[\boldsymbol{\Sigma}_{adt}]$, $\boldsymbol{\sigma}_{avg} \stackrel{\text{def}}{=} \mathsf{vec}\big[\boldsymbol{\Sigma}_{avg}\big]$, $\boldsymbol{\gamma} \stackrel{\text{def}}{=} \mathsf{vec}[\boldsymbol{\Gamma}]$, in the usual fashion, and then $\boldsymbol{S}_{avg} \stackrel{\text{def}}{=} \big[\big(\boldsymbol{\Sigma}_{avg}(1)^{\top} \otimes \boldsymbol{I}\big) + \big(\boldsymbol{I} \otimes \boldsymbol{\Sigma}_{avg}(1)^{\top}\big)\boldsymbol{K}\big]$ as the composite matrix that depends on the first-order autocovariance matrix $\boldsymbol{\Sigma}_{avg}(1)$ and finally $\boldsymbol{K}$ as the commutation matrix.

Because of the identifying conditions for $\boldsymbol{\Gamma}$, we have that the $[\boldsymbol{I} - (\boldsymbol{\Gamma} \otimes \boldsymbol{\Gamma})]$ matrix is invertible and thus we can find the final expression (in vector form) of the adaptive forecast error variance-covariance as in:

$$\boldsymbol{\sigma}_{adt} = [\boldsymbol{I} - (\boldsymbol{\Gamma} \otimes \boldsymbol{\Gamma})]^{-1}\big(\boldsymbol{\sigma}_{avg} - \boldsymbol{S}_{avg}\boldsymbol{\gamma}\big) \tag{12.23}$$

which takes a very similar form as in the corresponding equation of the univariate case.

As we cannot obtain an explicit expression for the re-matricized version of $\boldsymbol{\sigma}_{adt}$ of Eq. (12.23) and then solve these optimization problems analytically—the main difficulty lying in the presence of the term $[\boldsymbol{I} - (\boldsymbol{\Gamma} \otimes \boldsymbol{\Gamma})]^{-1}$—we proceed with numerical methods and optimize under the constraints on the elements of $\boldsymbol{\Gamma}$ lying within $\mathcal{G}$.

Let $\boldsymbol{\Sigma}_{avg}(\boldsymbol{\gamma})$ be the re-matricized version of Eq. (12.23), denote by $Q(\boldsymbol{\gamma})$ the chosen objective function (taken as the Frobenius matrix norm), and consider the two optimization problems note above in generic form as:

$$\boldsymbol{\gamma}^* \stackrel{\text{def}}{=} \underset{\boldsymbol{\gamma}}{\operatorname{argmin}}\, Q(\boldsymbol{\gamma}) \;\; \text{s.t.} \;\; \boldsymbol{\gamma} \in \mathcal{G} \in [0, 1]^{M^2} \tag{12.24}$$

where we have that $Q(\boldsymbol{\gamma}) = ||\Sigma_{avg}(\boldsymbol{\gamma})||_F^2$. The numerical optimization of the above problem gives us the optimal learning matrix $\boldsymbol{\Gamma}^*$ and, tentatively, solves the multivariate adaptive learning problem.

However, we have two more issues to consider. First, in the multivariate case we cannot explicitly show *a priori* MSE-improvements; one would assume that more information in terms of more cross-related time series will lead to such improvements in practice. Second, we have to address the (usual) "curse of dimensionality": as M increases the complexity increases by M^2 if we do not impose any additional assumptions regarding the structure of $\boldsymbol{\Gamma}$; for example, in the bivariate case of $M = 2$ we only have to optimize with respect to $M^2 = 4$ parameters; for $M = 4$ (a very reasonable dimension of a multivariate problem) we would have to optimize with respect to $M^2 = 16$ parameters—still feasible but considerably more demanding of any optimizer. For even larger problems, the number of parameters to estimate is increasing exponentially and therefore additional restrictions have to be placed.

One potential solution is to introduce some sparsity into the elements of the $\boldsymbol{\Gamma}$ matrix; for example, the simplifying assumption that $\boldsymbol{\Gamma} = \mathsf{diag}[\gamma_1, \ldots, \gamma_M]$ is a diagonal matrix and then the complexity of the optimization is reduced for larger M—this assumption is not overly restrictive for it implies that the cross-sectional heterogeneity is considered in the estimation of the γ's but there is no cross-sectional correlation among the adaptive learning correction that is being implemented. However, the appropriateness of this solution depends on its suitability on the data being analyzed. The subject of how to implement additional restrictions in the multivariate adaptive learning case is a subject of current research.

12.5 Summary of How to Use Adaptive Learning

For a practitioner, the theoretical results of the previous sections are instructive. However, even more useful would be a guide as to how to apply adaptive learning with real-world data. We provide such a summary below that will aid the preparation and implementation of the adaptive learning approach.

Step 1. Prepare all the time series to be forecasted with appropriate transformations (e.g., differencing, logarithmic transformations, and seasonal adjustments).

Step 2. Consider the number of benchmark forecasts that will be computed individually, compute them and store their values. A common approach to computing benchmark forecasts is to use historical averages or naive forecasting methods. For example, the mean or median of past observations can be used as a benchmark for forecasting future values. Similarly, naive methods such as the last observation or a simple moving average can serve as benchmark forecasts.

Step 3. Either check for the best individual forecast, based on some performance measure, or compute an average forecast or keep both the best individual forecast and the average forecast (recommendation: it is worthwhile keeping also the forecast from a plain benchmark, e.g., the naíve forecast or a simple sample mean forecast depending on the data).

Step 4. For the set of forecasts that you kept in the previous step, apply univariate adaptive learning, with and without smoothing. Examine your results and see where improvements are being made. Consider also the cost of estimation of your chosen models: adaptive learning might actually perform equally well with a plain benchmark as with a more advanced benchmark. You may want to apply adaptive learning directly to the plain benchmark.

Step 5. If more series are available and suitable, consider multivariate adaptive learning. Here the options are even more: you can apply multivariate adaptive learning in the univariate adaptive learning forecasts by pooling them together or, and most usually, you consider multivariate forecasting models to generate your forecasts and then pass them on to the multivariate adaptive learning algorithm.

This is a simple sequence of steps that one can consider in preparation for an application of adaptive learning with real-world data. In the next section, we will see such an application and the associated results.

12.6 An Illustrative Empirical Application

12.6.1 Data

In our analysis, we examine the time series data of agricultural product price indices across ten European countries. Additionally, we incorporate real Gross Domestic Product and a productivity index (measured as real labor productivity per hours worked) as explanatory variables within bivariate models and methods. The countries in our dataset comprise the Czech Republic, Greece, Latvia, Lithuania, Luxembourg, Malta, the Netherlands, Romania, Slovenia, and Spain. We analyze seven key categories of agricultural commodities, namely olive oil, milk, eggs, cereals, fresh vegetables, tomatoes, and potatoes. Given that data frequency significantly influences price transmission dynamics (Nazlioglou 2011) and economic variables are reported on a quarterly basis, we limited our analysis to quarterly observations covering the period from the first quarter of 2000 (2000Q1) to the second quarter of 2017 (2017Q2). The dataset is sourced from the official Eurostat website and we employ two rolling windows of 36 and 46 quarters for estimating the model parameters.[10] Our results encompass analyses at different transformations, including levels,

[10] Ashley (2003) provides forecasting insights regarding the number of observations that are capable of achieving statistically significant outcomes. Due to the limited length of our time series, our primary focus is on optimizing forecast efficiency rather than merely reducing forecast error variance. This topic is further explored in the subsequent section.

log-levels, as well as first and seasonal differences for all series. Moreover, in the discussion section, we present a selection of findings drawn from the available results.

12.6.2 Forecasting Models

There is a multitude of approaches that can be taken here and we stick to something simple: we use the naíve and naíve with drift benchmarks with bivariate VAR(1) and VAR(AIC) models on pairs of products or pairs of countries and combine the forecasts in different ways. We also consider the application of the univariate adaptive learning approach as a natural benchmark as well. The naive (last observation) benchmark forecasts future values based solely on the most recent observation, while the naive with drift extends this by incorporating a constant trend. In bivariate VAR models, these benchmarks forecast each variable simultaneously based on the lagged observations of all variables within the system, with VAR(1) using only the previous observations and VAR(p) using the previous p observations.

For the multivariate adaptive learning, we consider three approaches for making the vector of forecasts: using the forecasts of (a) the VAR(1) model, (b) the VAR(AIC) model, as well as utilizing the forecasts of the (c) VAR(1) and VAR(AIC) model for each of the two variables; combining them with a diagonal restriction for $\mathbf{\Gamma}$ and a full-sized $\mathbf{\Gamma}$ we have a total of 8 multivariate combinations.

Naive	naíve forecast
Naive-D	naíve forecast with drift
VAR(1)	*VAR*(1) forecast
VAR(p)	*VAR*(*p*) forecast, *p* by the AIC
ADL-NS	univariate adt. learning forecast, smoothing excluded
ADL	univariate adt. learning forecast, smoothing included
MALF-D-1	multivariate adt. learning, VAR(1) diagonal
MALF-D-p	multivariate adt. learning, VAR(p), diagonal
MALF-1	multivariate adt. learning, VAR(1) non-diagonal
MALF-p	multivariate adt. learning, VAR(p), non-diagonal
MALF-D-1x	multivariate adt. learning, VAR(1)+VAR(p), diagonal, VAR(1) forecast
MALF-D-2x	multivariate adt. learning, VAR(1)+VAR(p), diagonal, VAR(p) forecast
MALF-1x	multivariate adt. learning, VAR(1)+VAR(p), non-diagonal, VAR(1) forecast
MALF-2x	multivariate adt. learning, VAR(1)+VAR(p), non-diagonal, VAR(p) forecast

12.6.3 Forecast Evaluation

To evaluate our forecasts, we rely on standard descriptive metrics. That is, a properly adjusted regression, the Mincer-Zarnowitz (MZ) regression (Mincer and Zarnowitz 1969), and fundamental significance tests. We particularly favor the Mincer-Zarnowitz (MZ) regression approach as it offers a straightforward yet effective means of comparing forecasting performance. Additionally, it has been refined to function appropriately within the type of experimental framework we consider in our case (see West 2006). Since our primary objective extends beyond identifying the optimal forecasting model to examining the performance enhancements achievable through adaptive learning, the Mincer-Zarnowitz (MZ) regression serves as an ideal tool for this purpose. Nonetheless, we also present robust forecast comparisons.[11]

Our investigation adheres to the conventional structure of the majority of forecasting studies. We generate one-step-ahead forecasts $h = 1$ using a dataset comprising n observations, partitioned as $n = n_0 + n_1$. Here, n_0 represents the rolling window utilized for estimating model parameters and generating forecasts (set at 36 and 46 quarters, as mentioned earlier), while forecast evaluation is conducted on the remaining n_1 observations. For performance evaluation, we employ the mean absolute error (MAE) and (root) mean squared error (RMSE), both expressed relative to the corresponding values of a naC/ve forecast. Additionally, the MZ-regression measures are derived from the estimates of the associated regression model. Consequently, the standard equations for these descriptive statistics are as follows:

$$\begin{aligned} RMSE(m, n_1) &\stackrel{\text{def}}{=} \sqrt{n_1^{-1} \sum\nolimits_{t=n_0+1}^{n} \hat{e}_t^{m,2}} \\ MAE(m, n_1) &\stackrel{\text{def}}{=} \sum\nolimits_{t=n_0+1}^{n} |\hat{e}_t^{m}|, \end{aligned} \tag{12.25}$$

And the corresponding equations for the MZ-regression are as follows:

$$\begin{aligned} X_{t+1} &= \delta_0 + \delta_1 \widehat{X}_{t+1}^{m} + \eta_{t+1}^{m} \\ \hat{e}_{t+1}^{m} &= \delta_0 + (\delta_1 - 1)\widehat{X}_{t+1} + \eta_{t+1}^{m}; \end{aligned} \tag{12.26}$$

In this approach, the forecast error is regressed on the forecast, and if efficiency holds, the joint hypothesis $H_0^{E1} : \delta_0 = 0, \delta_1 = 1$ should be satisfied. For an unbiased forecast, it is sufficient to examine whether $H_0^{E2} : \delta_1 = 1$ holds. We estimate $(\widehat{\delta}_0, \widehat{\delta}_1)$ using the least squares method, calculate standard errors based on a HAC-based covariance matrix, and then apply a standard F-type test for H_0^{E1} or t-test for H_0^{E2}. In the latter case, the test statistic is adjusted appropriately based on the sample split between n_0 and n_1 observations. For these tests, conventional F and normal distribution critical values are applicable. In the tables, t^* represents the p-value associated with the test for H_0^{E2}, while F^* denotes the p-value for the test of H_0^{E1}.

[11] Makridakis and Hibon (1979) provide an early reference on the challenge of measuring and assessing forecasting accuracy.

When both forecasts are efficient (i.e., $\delta_1 = 1$) , the regressions mentioned above reduce to their respective regression errors. Consequently, we can compare the forecasts' RMSEs. Let $g_t \stackrel{\text{def}}{=} g(\widehat{e}_t^{m1}) - g(\widehat{e}_t^{m2})$, where in our application, we set $g(x) = x^2$. This leads to the following test regression:

$$g_t = \delta + \eta_t, \tag{12.27}$$

To assess whether the RMSE performances are equal, we test the null hypothesis $H_0^V : \delta = 0$, relying solely on the least squares estimate $\widehat{\delta}$ and its corresponding HAC-based t-statistic. In cases where forecasts originate from nested methods or models, and when $g(x) = x^2$, standard normal critical values can also be applied within the modified test regression.

$$g_t + (\widehat{X}_t^{m1} - \widehat{X}_t^{m2})^2 = \delta + \eta_t, \tag{12.28}$$

This leads to a modified estimator, $\widehat{\delta}$, which accounts for the mean squared difference between the forecasts generated by the two methods or models. The tables present two sets of test results for evaluating RMSE performance: first, a comparison between the best-performing individual method/model and the naC/ve benchmark, and second, a comparison between the same top-performing method/model and the adaptive learning forecast. We use τ_0^* to indicate the p-value of the first test and τ_1^* to represent the p-value of the second test.[12]

A final note on how we implement the adaptive learning forecasting approach in the empirical results presented. Initially, we generate an average forecast by combining the two top-performing individual benchmark models, selected based on their RMSEs, to assess the marginal improvements achievable through adaptive learning. The process begins with an equally weighted forecast, which then transitions to an adaptively weighted forecast, where we compute the relevant quantities for the variance-minimizing value of γ^*.

12.6.4 Representative Results

To enhance readability and facilitate discussion, we maintain a consistent format across all tables when presenting our results. There is a multitude of approaches that can be taken here and we stick to something simple: we employ both the naíve benchmark and the naíve benchmark with drift with bivariate VAR(1) and VAR(AIC) models on pairs of products or pairs of countries and combine the forecasts in different ways as described below: we also consider the application of the univariate adaptive learning approach as a natural benchmark as well.

---

[12] As all our method/model forecasts outperform the naïve benchmark with ease, the reported τ_0^* p-values are effectively zero and are therefore not discussed in greater detail.

Table 12.1 RMSE evaluation, Netherlands, $n_0 = 36$, ΔX_t

Forecasting schemes	Tomatoes	Fresh vegetables
Naive	1.000	1.000
Naive-D	1.024	1.006
VAR(1)	0.622	0.810
VAR(p)	0.368	0.736
ADL-NS	0.393	0.782
MALF-D-1	0.620	0.823
MALF-D-p	0.345	0.731
MALF-1	0.596	0.805
MALF-p	0.350	0.733
MALF-D-1x	0.615	0.810
MALF-D-2x	0.343	0.731
MALF-1x	0.613	0.801
MALF-2x	0.340	0.733

Table 12.1 presents results for the Netherlands, specifically for the products Tomatoes and Fresh Vegetables. A rolling window of 36 quarters was applied, and the analysis follows a system format. The variables are expressed in their first differences. Among the models assessed, the VAR(p) model emerges as the best-performing individual model based on RMSE. This top-performing VAR(p) model is statistically significant when compared to the naíve benchmark, as indicated by the τ_0^* p-value. However, while the adaptive learning approach (both with and without smoothing) achieves a slight reduction in RMSE, its parameter estimates remain comparable to those of other models. The corresponding τ_1^* p-value , which is not displayed in the table, suggests that the multivariate adaptive learning forecast is statistically indistinguishable from the best individual forecast.

Table 12.2 presents results for the same country, the Netherlands, and the same products, Tomatoes and Fresh Vegetables. However, in this case, the accuracy metric used is MAE. The series are expressed in first differences, and the analysis is conducted using a rolling window of 36 quarters. Once again, the best-performing model overall is the vector autoregressive model of order *p*, which remains statistically significant compared to the naíve benchmark, as indicated by the τ_0^* p-value. Notably, the multivariate adaptive learning approach without smoothing yields statistically significant improvements and substantially enhances forecast efficiency in terms of MAE.

Table 12.3 presents results based on seasonal differences, specifically annual growth rates, and follows a panel format. This table introduces a change in the analyzed products, now including eggs alongside Tomatoes and Fresh Vegetables, while also expanding the country set to include both the Netherlands and Malta. The rolling window remains at 36 quarters. The results indicate greater marginal reductions and notable improvements in RMSE values. We can also see that the adaptive learning approach outperforms all other benchmarks. Specifically, univariate

Table 12.2 MAE evaluation, Netherlands, $n_0 = 36$, ΔX_t

Forecasting schemes	Tomatoes	Fresh vegetables
Naive	1.000	1.000
Naive-D	1.024	1.006
VAR(1)	0.607	0.771
VAR(p)	0.381	0.706
ADL-NS	0.403	0.748
MALF-D-1	0.603	0.760
MALF-D-p	0.360	0.698
MALF-1	0.579	0.765
MALF-p	0.371	0.700
MALF-D-1x	0.611	0.771
MALF-D-2x	0.355	0.689
MALF-1x	0.605	0.754
MALF-2x	0.351	0.701

adaptive learning provides numerical improvements over all models, including the bivariate VAR. Even more, the bivariate ADL offers additional performance enhancements over the univariate one having the smallest values of the RMSE accuracy metric.

Similar to the findings in the previous table, Table 12.4 presents results for the product eggs, with Malta and the Netherlands as the analyzed countries. The rolling window remains at 36 quarters, and the variables are expressed in seasonal differences. Performance improvements are clearly evident when evaluated using the MAE metric. The univariate adaptive learning approach serves as the baseline case, exhibiting the lowest MSE values. Additionally, the bivariate adaptive learning method further enhances performance beyond what is achieved by the univariate approach.

Table 12.5 presents qualitatively similar results, following a system format and analyzing eggs for Malta and the Netherlands with the same rolling window of 36 quarters. However, in this case, the variables are expressed in first differences. The best-performing model, VAR(1), is statistically significant compared to the naíve benchmark, as indicated by the τ_0^* p-value. When transitioning to the multivariate approach, we observe a statistically significant reduction in RMSE for the adaptive learning forecasts relative to the top-performing individual forecast.

Table 12.6 examines the same product and countries as in Table 12.5, but with MAE as the accuracy metric. Similar to previous cases, the bivariate adaptive learning approach yields the most significant improvements in the MAE statistic, surpassing the performance of the best individual models. Notably, the top-performing model in this case is the vector autoregressive model of order 1 (VAR(1)). When incorporated into the adaptive learning framework, the MAE is further reduced, reinforcing the pattern observed in prior analyses.

Table 12.3 RMSE evaluation, EGGS, $n_0 = 36$, $\Delta_s X_t$

Forecasting schemes	Malta	Netherlands
Naive	1.000	1.000
Naive-D	1.020	1.015
VAR(1)	0.984	0.881
VAR(p)	1.363	1.137
ADL	1.051	0.756
MALF-D-1	0.984	0.704
MALF-D-p	1.363	1.151
MALF-1	0.984	0.703
MALF-p	1.363	1.154
MALF-D-1x	0.951	0.677
MALF-D-2x	1.363	1.200
MALF-1x	0.955	0.704
MALF-2x	1.330	1.227

Table 12.4 MAE evaluation, EGGS, $n_0 = 36$, $\Delta_s X_t$

Forecasting schemes	Malta	Netherlands
Naive	1.000	1.000
Naive-D	1.020	1.012
VAR(1)	0.988	0.880
VAR(p)	1.240	1.220
ADL	0.990	0.768
MALF-D-1	0.988	0.684
MALF-D-p	1.240	1.217
MALF-1	0.988	0.689
MALF-p	1.240	1.219
MALF-D-1x	0.964	0.641
MALF-D-2x	1.240	1.258
MALF-1x	0.959	0.684
MALF-2x	1.208	1.249

To summarize the findings presented in this section, the bivariate adaptive learning methodology effectively aligns with the theoretical expectations and demonstrates significant potential for enhancing forecasting performance, even surpassing the improvements achieved by univariate adaptive learning. Our approach to presenting results has been deliberately conservative, as we compare the proposed method against the *a posteriori* best forecasting models, those that have demonstrated the strongest performance when evaluated based on historical data or actual outcomes. One might reasonably question the necessity of adaptive learning once the optimal model or method has already been identified. While this perspective is valid, we

Table 12.5 RMSE evaluation, EGGS, $n_0 = 36$, ΔX_t

Forecasting schemes	Malta	Netherlands
Naive	1.000	1.000
Naive	1.017	1.033
VAR(1)	0.601	0.442
VAR(p)	0.959	1.332
ADL	0.776	0.731
MALF-D-1	0.600	0.442
MALF-D-p	0.951	1.322
MALF-1	0.601	0.441
MALF-p	0.959	2.225
MALF-D-1x	0.590	0.436
MALF-D-2x	0.929	1.327
MALF-1x	0.587	0.700
MALF-2x	0.928	1.344

Table 12.6 MAE evaluation, EGGS, $n_0 = 36$, ΔX_t

Forecasting schemes	Malta	Netherlands
Naive	1.000	1.000
Naive	1.021	1.034
VAR(1)	0.581	0.461
VAR(p)	0.981	1.315
ADL	0.791	0.749
MALF-D-11	0.580	0.461
MALF-D-1p	0.972	1.307
MALF-1	0.580	0.461
MALF-p	0.981	2.252
MALF-D-1x	0.576	0.456
MALF-D-2x	0.960	1.311
MALF-1x	0.573	0.653
MALF-2x	0.960	1.365

counter that the methodology is designed to integrate seamlessly with any a priori input, including the naíve model or other benchmarks. More importantly, our results clearly highlight the value and potential of the method, as it consistently enhances forecasting accuracy even when applied to top-performing models. Adaptive learning naturally adjusts to different conditions; if the input forecast is fully efficient and free of forecast error autocorrelation, it will not introduce any inefficiencies. However, in all other cases, it offers improvements ranging from marginal to substantial, with most of these gains being statistically significant.

12.7 Concluding Remarks

The primary contribution of this chapter is to demonstrate both the theoretical and empirical advantages of a novel forecasting methodology, which we refer to as adaptive learning, in both its univariate and bivariate forms. This approach introduces new theoretical insights and results, illustrating how past forecast errors can be leveraged to enhance the performance of any given forecast. Our findings indicate that, under specific conditions, adaptive forecasts can achieve inherent performance improvements by reducing forecast error variance relative to the input forecast. Furthermore, the rate of learning or adaptation is a non-linear function of the input forecast error autocorrelation: a fully efficient forecast requires no adjustments and, consequently, no learning, whereas for all other forecasts that deviate from this condition, adaptive learning provides a mechanism for performance enhancement.

In addition, we present an empirical application using agricultural price indices, marking our second key contribution, an exploration of the forecastability of agricultural prices. Our analysis reveals that agricultural price forecasting is feasible across a range of models. Moreover, the empirical results confirm that adaptive learning holds substantial practical value in forecasting applications, consistently yielding improved forecasts across most agricultural products and countries examined. These improvements are evident in key performance measures such as RMSE and MAE.

The use of univariate and multivariate learning can be easily added in the forecaster's toolbox and allow the expansion of the available, and powerful, methods that offer the researcher and practitioner the potential to provide significant forecasting performance enhancements in any context that time series forecasts are being used.

Sample R Code

```
# (c) Foteini Kyriazi, Dimitrios D. Thomakos \& Efthymios Xylangouras

# Load required packages
library(dplyr) # For data manipulation
library(quantmod) # For fetching financial and economic data

# Fetch historical stock prices (e.g., AAPL for Apple Inc.)
stock_symbol <- "AAPL"
getSymbols(stock_symbol)
stock <- as.matrix(AAPL[, c("AAPL.Close")])

# Retrieve the available dates
dates_raw <- rownames(stock)
# Convert the AAPL stock column to appropriate numerical format,
    excluding the dates
stock <- apply(stock, 2, as.numeric)
# Add back the dates as index in the table
rownames(stock) <- dates_raw

# Take log-prices?
do_logs <- F
if (do_logs)
{
  stock <- log(stock)
}

# Retrieve the available dates of the pre-processed data set
dates <- rownames(stock)

# Select how many periods to forecast - 1 step ahead forecasting
periods_fc <- 63
nobs <- NROW(dates)
stock <- stock[(nobs-periods_fc+1):nobs,1]

# Forecasting with the Naive Benchmark Model

# One step ahead forecasting for the naive model is just the last
    observation
forecast <- stock[1:((periods_fc)-1)]
actual <- stock[2:periods_fc]
errors <- actual - forecast

# Get the Mean Squared and Mean Absolute Error for each estimated
    observation
MSE_series <- (errors ^ 2)
```

```
MAE_series <- abs(errors)
neval <- NROW(MAE_series)

# Application of the Adaptive Learning Methodology

# Define the adaptive learning parameter
adl_param <- seq(0.01, 0.25, 0.01)

# Pre define the matrix to save the re-evaluated error terms
num_cols <- length(adl_param)
adl_output <- matrix(NA, nrow = neval, ncol = num_cols)

# Loop over each value of the adaptive learning parameter
for (a in seq_along(adl_param)) {

  ADLM <- matrix(NA, nrow=neval, ncol=1)

  # Loop over the rows of produced forecast errors starting from the
      second row
  for (i in 1:neval) {
    if (i == 1) {
      ADLM[i] <- forecast[i]
    }
    if (i > 1) {
      ADLM[i] <- forecast[i] + adl_param[a] * (actual[i - 1] - ADLM[i -
          1])
    }
  }

  # Store the re-evaluated MAE terms based on the Adaptive learning
      methodology
  adl_output[, a] <- ADLM
}

# Calculate the final MAE of the Naive
MSE_naive <- mean(MSE_series)
MAE_naive <- mean(MAE_series)

# Calculate the MAE for each application of the adaptive learning
    parameter
MSE_adl <- NA
MAE_adl <- NA
for (i in 1:ncol(adl_output)){
  MSE_adl[i] <- mean((actual - adl_output[,i])^2)
  MAE_adl[i] <- mean(abs(actual - adl_output[,i]))
```

```
print(paste(''The MSE for the Adaptive learning parameter",
    adl_param[i],''is:",
                round(MSE_adl[i],5),''compared to the Naive
                    benchmark MSE of:",round(MSE_naive,4)))
  print(paste(''The MAE for the Adaptive learning parameter",
    adl_param[i],''is:",
              round(MAE_adl[i],5),''compared to the Naive
    benchmark MAE of:",round(MAE_naive,4)))
  print("
    ________________________________________________
    ")
}
```

References

Ashley RA (2003) Statistically significant forecasting improvements: how much out-of-sample data is likely necessary? International Journal of Forecasting 19:229–240

Brockwell PJ, Davis RA (2016) Introduction to time series and forecasting. Springer, New York

Brooks C (2019) Introductory Econometrics for Finance, 4th edn. Cambridge University Press, Cambridge

Clements MP, Hendry DF (2011) The Oxford handbook of economic forecasting. Oxford University Press, New York

Clements MP, Hendry DF (1996) Intercept corrections and structural change. Journal of Applied Econometrics 11:475–494

Elliott G,Timmermann A (2013a) Handbook of economic forecasting:, vol 2A. Elsevier, Amsterdam

Elliott G, Timmermann A (2013b) Handbook of economic forecasting:, vol 2B. Elsevier, Amsterdam

Elliott G, Granger Clive WJ, Timmermann A (2006) Handbook of economic forecasting. Elsevier, Amsterdam

Hyndman, R.J., & Athanasopoulos, G. (2013) *Forecasting: Principles and Practice.* OTexts

Kyriazi F, Thomakos D, Guerard J (2019a) Adaptive learning forecasting, with applications in forecasting agricultural prices. International Journal of Forecasting. 35:1356–1369

Kyriazi, F., Thomakos, D., & Guerard, J. (2019b). Multivariate Adaptive Learning. *Presented at the 39th International Symposium on Forecasting (ISF) in Thessaloniki, 16-19 June*

Kyriazi F, Thomakos D (2020a) Foreign Exchange Rate Predictability: Seek and Ye Shall Find It. Handbook of Applied Investment Research, World Scientific Publishers, Chapter 19:511–556

Kyriazi F, Thomakos D (2020b) Distance-based nearest neighbour forecasting with application to exchange rate predictability. IMA Journal of Management Mathematics 31:469–490

Kyriazi, F., Tarani, S., & Thomakos, D., (2023). Median-adaptive portfolios: a minimum criteria approach to asset allocation. *Annals of Operations Research*

Mincer J, Zarnowitz V (1969) The evaluation of economic forecasts. In: Mincer J (ed) Economic forecasts and expectations. Columbia University Press, New York

Montgomery DC, Jennings CL, & Kulahci M (2015) *Introduction to time series analysis and forecasting.* John Wiley & Sons.

Muth JF (1960) Optimal properties of exponentially weighted forecasts of time series with permanent and transitory components. Journal of the American Statistical Association 55:299–306

Makridakis S, Hibon M (1979) Accuracy of forecasting: an empirical investigation. Journal of the Royal Statistical Society, Series A 142:97–145

Nazlioglou S (2011) World oil and agricultural commodity prices: Evidence from nonlinear causality. Energy Policy 39:2935–2943

Petropoulos F, Apiletti D, Assimakopoulos V, Babai M, Barrow DK, Ben Teb S, Bergmeir C, Bessa RJ, Bijak J, Boylan J, Browell J, Carnevale C, Castle JL, Cirillo P, Clements M, Cordeiro C, Oliveira FL, De Baets S, Dokumentov A, Ziel F (2022) Forecasting: theory and practice. International Journal of Forecasting 38(3):705–871

Tarani S, Kyriazi F (2024) Market Timing & Predictable Profitability. International Journal of Portfolio Analysis and Management 2:379–393

West, K. D. (2006). Forecast evaluation. In G. Elliott, C. W. J. Granger, & A. Timmermann (eds.), *Handbook of economic forecasting*, vol. 1 (ch. 3, pp. 99-134). Elsevier

Chapter 13
Decision Trees and Random Forest to Predict Rare Events: Case of Child Health in India

Dweepobotee Brahma and Debasri Mukherjee

Abstract This chapter examines the use of Random Forest, a nonparametric machine learning algorithm based on decision trees, for predicting rare events. A common challenge in binary classification is data imbalance, where one outcome occurs far less frequently than the other, such as loan defaults. Accurately predicting these rare events and evaluating model performance under such conditions poses special challenges. The chapter explores various resampling techniques that enhance prediction accuracy in imbalanced datasets. Additionally, it introduces the Precision-Recall curve, a key metric for assessing rare-event prediction. These methodologies are demonstrated using real-world survey data on child health in India, with all analyses conducted using the R programming language.

13.1 Introduction

In many prediction exercises, the response variable/dependent variable is often a discrete variable (taking non-negative integer values). The goal in these problems is to assign/classify observations into different categories or classes. Thus, these types of problems are called classification problems. The most common case is when there are only two such categories: these are called binary classification problems. Banks may be interested to know whether borrowers are likely to default on their loans or not. They will try to use historical data on previous loans and borrowers'

Supplementary Information The online version contains supplementary material available at https://doi.org/10.1007/978-981-97-9060-9_13.

D. Brahma (✉)
Centre for Mathematical and Computational Economics, School of AI and Data Science, Indian Institute of Technology Jodhpur, Jodhpur, Rajasthan, India
e-mail: dweepobotee@iitj.ac.in

D. Mukherjee
Western Michigan University, Kalamazoo, MI, USA
e-mail: debasri.mukherjee@wmich.edu

F. Hamid and D. Mukherjee (eds.), *Data Science for Modeling Managerial and Socioeconomic Problems*, Contributions to Management Science,
https://doi.org/10.1007/978-981-97-9060-9_13

characteristics to try to categorize new borrowers into two classes: likely to default and not likely to default. Similarly, school administrators/education policymakers might be interested in preventing high school students from dropping out. They might use data on previous students to classify students in an incoming cohort as likely to drop out, or not likely to drop out. A clinician might be interested in using genetic data to predict whether a patient is likely to develop a rare disease or not.

In many such binary classification exercises, very often one class has a low proportion of instances in the data. This imbalance can significantly affect the predictive power of the model. For example, the instances of developing a rare disease ('events') might be much less compared to its opposite ('non-events') happening. Therefore, the data on the binary response variable (that a researcher wants to predict) will be imbalanced against one category, usually the events (such as defaulting on loans) which are often much rarer (less than even 5% in the data) than the non-events. Examples of class imbalance are ubiquitous in applied problems. The issue of imbalanced data can also arise in case of multi-class classification[1] problems. For the sake of brevity, this chapter focuses on binary classification and demonstrates how Machine Learning (ML) techniques can be employed in binary classification problems (involving imbalanced data) for predictive modelling. It further discusses appropriate evaluation metrics for assessing predictive performance of such models, and some resampling strategies to subdue the class imbalance problem.

Section 13.2 elaborates the process of growing decision trees using a hypothetical example inspired by the data that is used. Section 13.3 demonstrates the process of deriving predictions from decision trees. Section 13.4 demonstrates the process of combining trees into forest. Section 13.5 discusses evaluation metrics for prediction accuracy for imbalanced data. Section 13.6 illustrates commonly used resampling strategies for imbalanced data. Section 13.7 illustrates the applications of these algorithms and concepts using an example from development economics. Section 13.8 concludes with a discussion on future research directions.

13.2 Decision Trees

The theories of ML for predictive modelling have been developed across several chapters in textbooks such as James et al. (2021) and Hastie et al. (2009). This section focuses on one such predictive modelling technique and presents it in an intuitive fashion along with some necessary algorithms that can help clarify the underlying intuition as well as provide an outline for applied researchers. For detailed and step-by-step analysis of the underlying *theories* behind the algorithms that are illustrated here, readers are advised to refer to the above-mentioned textbooks.

Following James et al. (2021), the decision tree algorithm is demonstrated below. A decision tree involves repetitively partitioning the predictor space following some

[1] The issue becomes more complicated in multi-class classification problems and is beyond the scope of this chapter.

splitting rule into a number of simple non-overlapping regions. These splitting rules can be represented graphically as a tree giving the algorithm its name. Each splitting point is called a node (referred to using index m) and the terminal nodes are called leaves (referred to using R_j). The final predictions are generated in the R_j s. To generate a prediction for a new test observation, the users can drop it down the tree following the splitting rules. The most frequently appearing class in the region where the observation lands (based on its characteristics) is simply the predicted class for this observation.

The process of building a decision tree can be summarized in the following steps.

Step 1: Let us divide the predictor space into J number of mutually exclusive and exhaustive regions: $R_1, R_2, \ldots, R_J$ following a splitting rule.

Step 2: For any test observation that lands in region R_j following the splitting rule, the predicted *class* for R_j is the most frequently appearing class in that region R_j.

One can start at the beginning of the tree where all the observations are initially in the same region and successively split that region into two regions (hence binary split), and then again split those resulting two regions each into two further regions and continue growing the tree. This is known as the process of *recursive binary partitioning*. 'Binary' implies that each node is split into two regions only. 'Recursive' refers to the fact that this is a repetitive process where binary partitioning is done repetitively until a stopping rule is reached. It is also called a greedy algorithm. This is because at each node the best split is chosen such that prediction error is minimized at that stage instead of selecting a split that could produce further prediction error minimization in future steps.

The process of building the tree requires making two choices at each step—which predictor to use as the splitting variable at each node X_j, and what value of the predictor (s) to split at (when the predictor is numeric/continuous). All the predictors and all possible cut points for each of the predictors are considered, and finally, the pair X_j, s that minimizes the prediction error[2] is chosen. For a binary classification problem, a natural measure of prediction error is the classification error rate. This is the proportion of observations in a given region not belonging to the most commonly occurring class.

$$E = 1 - max_k\left(\widehat{p_{mk}}\right)$$

where $\widehat{p_{mk}}$ is the fraction of training data points in the m-th region that are in the kth class (K = {1,2} here). This splitting process is repeated at every node searching for the best predictor and best cut point to divide the data further minimizing the prediction error in each of the resulting regions. This splitting process continues until a stopping condition is fulfilled. Once the tree is fully grown, the prediction

[2] Note that there can be other measures of building a tree such as entropy which are not covered here for the sake of brevity.

for the test observations for each of the regions $R_1, R_2, \ldots, R_J$ is generated as the commonly occurring class in that region.

Note that in practice however, the misclassification error is often not appropriate as it is not sensitive enough to small changes in splitting. Instead, a different measure of prediction accuracy, Gini Index is used while building the tree. Gini Index is defined as

$$G = \sum_{k=1}^{K} \widehat{p_{mk}}\left(1 - \widehat{p_{mk}}\right)$$

The value of Gini Index will be small value when $\widehat{p_{mk}}$ is either close to 0 or 1 implying that most observations in that region predominantly belong to one class (also known as node purity). For more details see James et al. (2021).

Decision trees can be presented graphically for a general audience. They are a nonparametric technique, i.e. they do not make strong parametric assumptions regarding the functional form in the data and generally predict better than traditional parametric techniques. Note that unless the underlying functional relations are correct, parametric models suffer from misspecification error and substantially inaccurate predictions. Nonparametric techniques avoid such biases and thereby are more likely to generate higher prediction accuracy.

A hypothetical decision tree using a subset of the predictors in the dataset is illustrated in Fig. 13.1. The variable names at the nodes correspond to the predictors used as the splitting variable and the values along the branches refer to the values for observations that lie along that branch. Minority class observations or events are denoted as '1' in the data while majority class observations or non-events are denoted as '0' in the data. For example, the first internal node corresponds to 'first-born'. The left branch labelled 'Yes' corresponds to observations who are first-borns, coded as '1' in the data. This branch leads to the internal node where 'income' is used as the splitting variable. The left branch from this node corresponds to observations who are first-borns and have household income per month less than Rs 1220. This leads to the terminal node (also known as leaf) for which the predicted class generated by the majority class of all observations that land in this region is 'dead'. The right branch from the 'income' node corresponds to observations who are first-borns but have household income per month of Rs 1220 or more. This branch leads to the terminal node where the predicted class generated by the majority class of all observations that land in this region is 'alive'. The right branch from the 'first-born' node corresponds to infants who are not first-borns and coded as '0' in the data. This branch leads to the internal node where 'prior deaths' is used as the splitting variable. The left branch is made up of observations who are not first-borns and whose mothers have had infant deaths from prior births. This branch leads to a terminal node with a predicted class of 'dead'. The right branch from the 'prior deaths' terminal node refers to observations who are not first-borns and their mothers did not have any infant deaths before. The observations along this branch are split further based on 'total births'. The left branch from the 'total births' node carries observations of infants who are not first-borns,

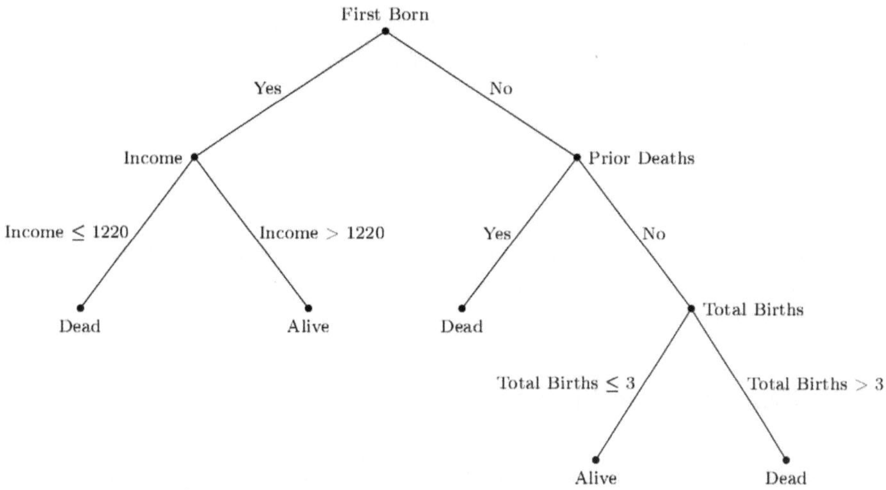

Fig. 13.1 Hypothetical decision tree

whose mothers did not have prior infant deaths and whose mothers have 3 or less total births. This leads to a terminal node which generates a prediction of 'alive'. The right branch from the 'total births' node similarly carries observations of infants who are not first-borns, whose mothers did not have prior infant deaths and had more than 3 births. This leads to a terminal node with a predicted class of 'dead'.

13.3 Generating Predictions from Decision Trees

Once a decision tree following splitting rules has been built, it can be used to generate predictions for previously unseen/unused data at hand, i.e. 'out-of-sample prediction'. This process is demonstrated using the following hypothetical examples using the decision tree in Fig. 13.1. Each of these previously unused observations listed in Table 13.1 is taken and dropped down the tree following the splitting rules used to grow the tree. Suppose the first such observation (in the test data) is a first-born (1 = yes) and so it follows the left branch from the topmost node. At the node 'income', it takes the right branch because income has a value of 1500 for this observation. It eventually lands in the right terminal node with a predicted class of 'alive'. This prediction can be compared to the observed value of the response in the last column of Table 13.1 and one can conclude that this prediction is correct. Similarly, suppose the second observation is not a first-born (0 = no) and so it follows the right branch from the top terminal node. At the node 'prior deaths', this observation takes the right branch (0 = no) and reaches the internal node 'total births'. Here, it follows the right branch (total births > 3) to reach the terminal node with the predicted class

Table 13.1 Hypothetical test observations from the empirical example

Obs	Income	First born	Prior death	Total birth	Infant death (Observed)	Infant death (Predicted)
#1	1500	1	1	2	0	0
#2	1250	0	0	4	0	1

'dead'. Comparing this prediction with the observed value, one can notice that this is a prediction error produced by the algorithm.

13.4 From Trees to Forests

As was noted earlier, decision trees are nonparametric structures as opposed to a simple traditional parametric model (e.g. linear regression). When a simple parametric modelling restriction is imposed, it often suffers from misspecification error, leading to bias or inaccuracy in the results. Due to this in many ML settings (including in decision tree), no such modelling assumption is superimposed, and the algorithm learns only from the training data. However, due to lack of general model and any other exposure, the algorithm overfocuses on the training data leading to overfitting of the same; thus, it performs poorly for new data (in terms of out-of-sample predictions) though it produces good in-sample predictions. For any new data, the results can *vary* substantially causing poor out-of-sample predictions. Thus, a decision tree has less bias compared to a simple general model but comes with its own cost of high variance. This common bias-variance (accuracy from complexity vs. low variability from simplicity or generalization) trade-off in econometric analysis calls for a hard-to-achieve 'Goldilocks-zone' solution. One such solution comes in the form of Random Forest technique where the decision tree algorithm is applied many times to build many trees based on many resampled (bootstrap resamples with replacements) data thereby building a 'forest'. This exposure (to forest instead of one tree) brings the effect of generalization without compromising accuracy. This essentially makes out-of-sample prediction more reliable. In practice, predictions from each tree in the forest are taken and then the average of the predictions (or the majority decision in case of a binary situation) from all trees is constructed as the final prediction. Thus, the prediction exercise becomes an outcome of all ("aggregate") resampled ("bootstrapped") trees—calling it bagging ("bootstrap aggregating").

To clarify further, bagging or bootstrap aggregation is a method by which decision trees are applied to several bootstrapped resamples of the data. Bootstrapping is a data and computer-based simulation technique where observations can be repeatedly sampled from the original data by resampling *with* replacements to generate multiple versions of the original dataset. This is helpful for out-of-sample prediction by exposing the analysis to various possible resamples and hence mimicking the unknown real life (which one is trying to predict) better than using a single sample.

See Efron and Tibshirani (1994) for details on basic bootstrap methods. Each of the decision tree algorithms applied to each bootstrap resample will generate a prediction of a class for each observation in the test data. To generate the final prediction, one can simply take the class predicted by the majority of the trees for each observation in the test data. Such aggregation is used to reduce the variance in an estimated prediction function. See Hastie et al. (2009).

However, the bagging technique still suffers from some drawbacks. Suppose the dataset has few predictors that are highly correlated with the response among the set of predictors. In building decision trees, these variables will usually be chosen by the algorithm to make a split towards the top portions of most of the decision trees. Because of this, the trees from each bootstrapped sample can be very similar to each other. Naturally, in such settings, bagging will lose its advantage.

To overcome this drawback, a researcher can use a small tweak in the process of building decision trees. At each node, while choosing which predictor to split by, the user can consider only a randomly selected subset of predictors. Among that subset of predictors, one is selected as the splitting variable following the splitting rules explained earlier. At every node, a fresh randomly selected subset of predictors is considered. Decision trees are grown on each bootstrapped resamples and the predictions are aggregated using a majority rule in the same manner as previously described. This process is called Random Forest. *Thus, in a Random Forest the predictors are also sampled at each node along with taking bootstrapped resamples of the data.* Thus, Random Forest technique introduces much more variations/exposures to the information set (observations and predictors) across trees thereby leading to improved robustness and accuracies in the out-of-sample prediction compared to a single decision tree analysis. Taking a subset of candidate predictors for a split at each node implies that at nodes, the predictors that are highly correlated with the response will not always be picked and that the other predictors will get a chance to be evaluated. This will reduce correlations among the trees across different bootstrapped resamples and will make the prediction more reliable and robust.

In the case of Random Forest, the previously described Gini Index can be used to construct a measure of predictor importance which can be used to rank order predictors. For every instance where predictor X_j makes a split, the algorithm calculates the difference in the Gini Index in the resulting child nodes and the mother node of a tree. For a predictor X_j this reduction in Gini Index is totalled across the forest for each instance in which X_j was used to construct a split. To obtain the '*mean decrease in Gini Index*' for X_j, this sum is normalized by dividing by the number of trees in the forest. This *mean decrease in Gini Index* can be used as a measure of predictor importance and can be used to rank predictors in decision trees and Random Forests. Predictors with a higher 'mean decrease in Gini Index' are better than those with a low 'mean decrease in Gini Index'. The variable importance ranking of predictors in the dataset using mean decrease in Gini Index is presented later in Fig. 13.6 in the Appendix.

To generate predictions from Random Forests, the present analysis takes each of the test observations and drop it down each of the decision trees grown on each bootstrapped resampled data. Suppose $b\,(= 1, 2, \ldots, B)$ is the index for the trees grown

Table 13.2 Generating prediction for the hypothetical observations from Random Forest

Obs	Income	First born	Prior death	Total birth	Infant death (response)	Prediction from Tree 1	Prediction from Tree 2	...	Prediction from Tree B	Final prediction (majority vote rule)
#1	1500	1	1	2	0	0	0	...	1	0
#2	1250	0	0	4	0	1	0	...	0	0

on bootstrapped resamples. These trees will look somewhat different from each other due to the variability/exposure (in the information set) that has been purposefully introduced through the bootstrapped resampling process as well as through subsampling of predictors. Irrespective of what they look like, each test observation is assigned to one of the terminal regions in each of these B trees (following the process explained in Section 13.3), thereby yielding the corresponding prediction produced by each tree for that observation. To generate the final class prediction, one can take the majority vote across all the B trees for each observation in the test data. Table 13.2 illustrates this process.

However, often one is interested not just in a final class prediction but also a probability of belonging to the minority class. One can use the vote share of the minority class across all the B trees as a measure of probability of belonging to the minority class. Suppose one runs a Random Forest with 100 trees of which 62 trees predict the class of a particular observation to be the minority class. The predicted probability of belonging to the minority class generated for that particular observation is $\frac{62}{100} = 0.62$. Hence for that particular observation the probability of belonging to the majority class is $1 - 0.62 = 0.38$. Reverting to class prediction from probability is quite simple. A particular cut-off value is chosen for these probabilities which is used as the rule to assign test observations into different classes. For instance, the most commonly used cut-off for binary classification is 0.5 where observations in the test data with predicted probabilities ≥ 0.5 are classified into the minority class (events) and observations with predicted probabilities < 0.5 are classified into the majority class (non-events).

13.5 Evaluation of Prediction Accuracy

For any given cut-off point, a simple confusion matrix (Table 13.3) demonstrates the possibilities that arise from a typical binary classification exercise.

Recall that in the present study, minority class instances which are few in number are the 'events' or 'positives' while the majority class instances which are more in number are the 'non-events' or 'negatives'. From Table 13.3, true positive (TP) refers to observations in the test data that actually belong to the minority class and are predicted to belong to the minority class by the algorithm. Similarly, true negative (TN) refers to observations in the test data that belong to the majority class and are

Table 13.3 Confusion matrix for a binary classification problem

	Predicted Positive Class	Predicted Negative Class
Observed Positive Class	True Positive (TP)	False Negative (FN)
Observed Negative Class	False Positive (FP)	True Negative (TN)

predicted by the algorithm to belong to the majority class. Both true positives and true negatives are correct predictions made by the algorithm and the goal is to have as many observations that fall into these categories as possible. False positives (FP) are observations in the test data that belong to the majority class but are predicted to be in the minority class by the algorithm used. False negatives (FN) are observations in the minority class but are predicted to be in the majority class by the algorithm. False positives and false negatives are essentially prediction errors made by the algorithm and the goal is to have as few of these as possible. There lies an inherent trade-off between these two types of errors—false positive and false negative which correspond to Type-I and Type II errors (James et al. 2021, p. 151) readers might be familiar with.

It is a common practice in ML to look at the prediction accuracy over a full range of possible cut-off values (to produce a continuous ROC or PRC curve, to be discussed below). This is because one might be interested in understanding the nature of the trade-off between these two errors. Also, instead of looking at the absolute numbers, the literature often looks at two relative measures—true positive rate (TPR) or sensitivity and false positive rate (FPR) or (1-specificity). $\text{TPR} = \frac{\text{TP}}{\text{TP+FN}}$ is the probability that an actual positive instance is classified as positive. $\text{FPR} = \frac{\text{FP}}{\text{FP+TN}}$ is the probability that an actual negative instance will be misclassified as positive.

The true positive rate and false positive rate vary as one changes the probability cut-off used to convert probabilities into class predictions. This relation between the true positive rate and false positive rate across varying probability cut-offs can be plotted using a receiver operator curve or ROC. For details on building an ROC curve, see Aggarwal and Ranganathan (2018). Here, the perfect algorithm is the one that correctly classifies all positive instances as positive and all negative instances as negative and has a value of TPR = 1 and FPR = 0 at any given cut-off value. On the other hand, the worst algorithm would be the one with no prediction capability and randomly assigns instances across the two classes—a random guess. This random guess classifier is used as a benchmark in most binary classification[3] problems. The benchmark serves as the baseline prediction performance which machine learning algorithms are supposed to beat and can also be plotted in an ROC. ROC can be used to compare performance of multiple ML algorithms.

While the ROC curve itself is useful in visualizing the trade-off between TPR and FPR at each probability cut-off, one may sometimes want to boil it down to a single metric—area under curve (AUC). For the perfect classifier, this AUC is 1.0 while for

[3] Note that this paper focuses on the binary classification problem. Multi-classification problems can be of many types (ordered, unordered, sequential) and the problems need to be revisited separately for each situation. This is well beyond the scope of this chapter.

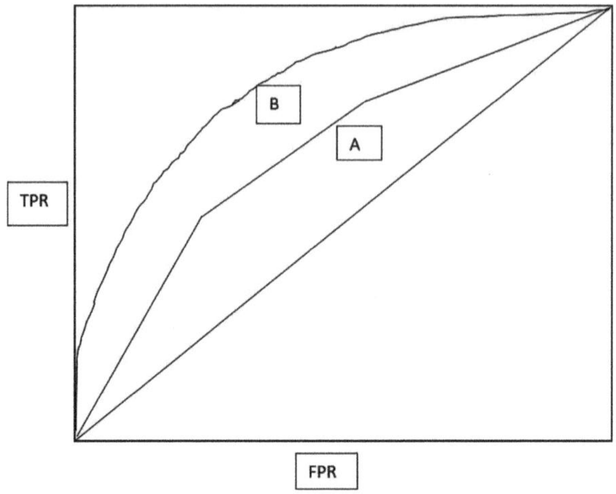

Fig. 13.2 Hypothetical ROC curves. Model B has a higher AUROC than Model A

the baseline random classifier the AUC is 0.5. Most algorithms will have a value of AUC between 0.5 and 1. One can compare across algorithms by comparing the value of AUC with larger values implying higher prediction accuracy. A hypothetical ROC is demonstrated in Fig. 13.2.

In case of imbalanced data, the choice of the *evaluation metric* for prediction accuracy becomes critical. This is usually because the consequences of the misclassification are usually not the same for the two different types of errors (which can be the case even in balanced data). The consequences of misclassifying the minority class as majority class are graver than the consequences of misclassifying the majority class as minority class. For instance, in the present example misclassifying an infant likely to die as likely to remain alive will usually imply that the infant will miss out on any healthcare interventions derived from the results of the algorithm whereas misclassifying an infant who would have remained alive, as likely to die will mean some wastage of intervention. Here, the consequences of a false negative are worse than that of a false positive.

ROC curve and area under it, AUROC, are inappropriate for imbalanced data where misclassification cost of one type of error is substantially higher than the other type, as AUROC can take a high value by just classifying the majority classes alone and missing the minority class instances thereby defeating the purpose of the exercise (see Davis and Goadrich [2006]). In this case, one needs a measure that focuses only on how well the algorithm classifies the minority observations without sacrificing the majority observations too much. Here, the present study uses two specific measures of prediction accuracy.

Recall $= \frac{\text{TP}}{\text{TP+FN}}$ is the ratio of true positives to the total predicted positives in the test data. Put differently, it is the probability that a randomly selected minority class observation is predicted as belonging to the minority class by the algorithm. As demonstrated earlier, this is also known by the name true positive rate or Sensitivity. Precision $= \frac{\text{TP}}{\text{TP+FP}}$ is the ratio of true positives to total observed positives in the test data. Put differently, it is the probability that a randomly selected positive prediction is indeed true. Evidently, both higher precision and recall are desirable. Comparing to ROC, one can replace the measure of false positive rate or specificity (which measures the rate at which the algorithm misclassifies negative instances) with precision that measures how accurate the positive predictions are. This moves the focus away from balancing prediction accuracy across both classes to that of the positive class. Notice that true negatives (TN) do not feature in the formula for either *recall* or *precision* implying that the user does not care about how many negative instances the algorithm correctly predicts. In the binary classification setting, a single metric F1 score combining the measures of precision and recall can also be used. F1 is calculated as $F1 = \frac{2\times\text{Precision}\times\text{Recall}}{\text{Precision}+\text{Recall}}$.

However, there is a trade-off between precision and recall. This stems from the trade-off between correctly making positive predictions and being conservative about making positive predictions. Similar to an ROC, a particular probability cut-off value will generate a set of precision and recall values. By varying the cut-off, one can get a precision-recall curve (PRC). One can also boil down this curve to a single metric called area under the precision-recall curve (AUPRC) and is used to compare the prediction accuracy across algorithms.

In binary classification with imbalanced data, since the goal is to correctly classify the minority class observation, the baseline is no longer random guessing. To correctly classify all the minority class observations, the easiest and most conservative approach would be to adopt a rule of always classifying as positive (called an always positive classifier). An always positive classifier will definitely classify all minority class observations correctly but obviously classify all majority observations incorrectly. This serves as the baseline for this analysis (see, Flach and Kull 2015, p. 5) which produces a recall of 1 (since all minority class observations are classified correctly) and a precision equal to the existing proportion of minority observations in the data (since that is the proportion of minority class predictions that are actually correct). A hypothetical precision-recall curve is shown in Fig. 13.3.

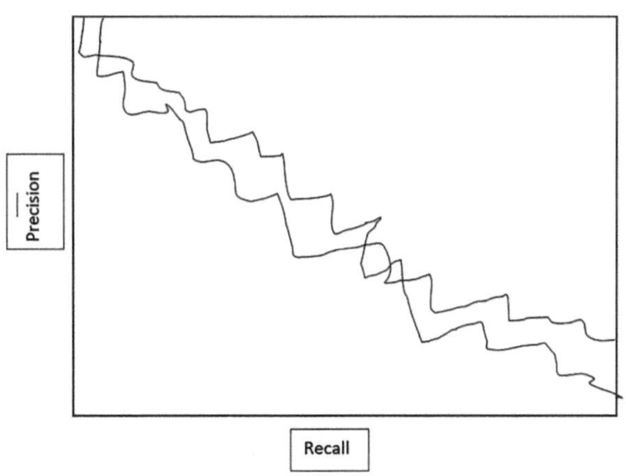

Fig. 13.3 Hypothetical precision-recall curve. Algorithms are compared based on area under precision-recall curves

13.6 Resampling Strategies

The commonly employed strategy for imbalanced data involves using various resampling techniques to change the class distribution in the training data. The idea behind employing such techniques is to counter the paucity of information on the minority class relative to the majority class presented in the training data. Resampling techniques are of three major types—upsampling, downsampling, and generating synthetic data.

Upsampling involves generating simply identical copies of observations of the minority class data points already present in the training data. This increases the proportion of the minority class in the training dataset as well as the total number of observations in the training data. Upsampling techniques provide more information about the minority class observations so that the algorithm is better able to learn about the minority class.

Downsampling involves dropping randomly selected majority class observations from the training data. This decreases the proportion of majority class observations as well as the total number of observations in the training data. Downsampling techniques remove some of the excess information about the majority class so that the algorithm is able to focus better on the minority class observations. This process is illustrated in Fig. 13.4.

The most commonly used technique for generating synthetic examples is Synthetic Minority Oversampling Technique (SMOTE) introduced by Chawla et al.

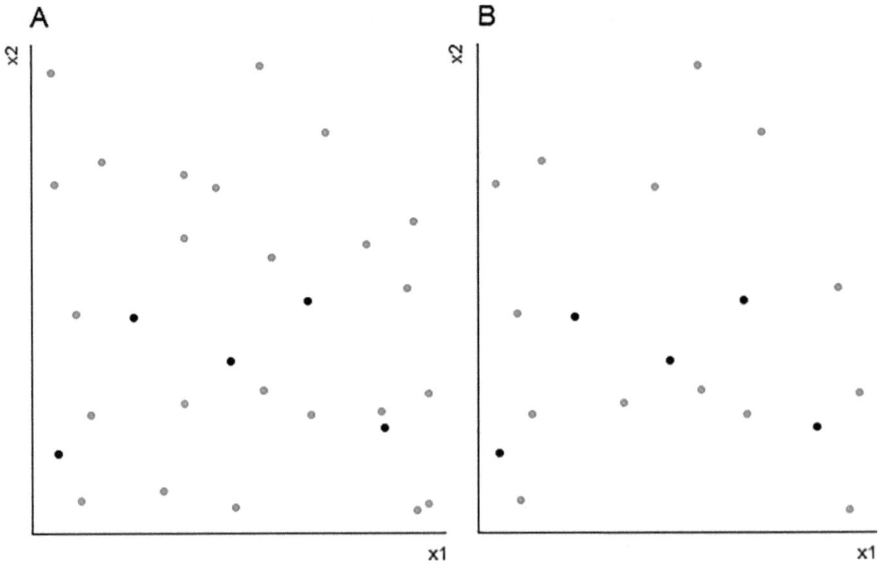

Fig. 13.4 Downsampling. *Note* (A)—black dots represent events and grey dots represent non-events, (B)—a randomly selected subset of non-events is dropped to produce better balancing between the events and non-events

(2002, 2003). In SMOTE, one can generate synthetic observations on 'events' by using information on the already existing events in the data.

Such synthetic observations can be constructed using nearest neighbours of the minority class observations that are present in the data. Suppose d_1 is a minority class observation already present in the data and d_2 is one of its nearest neighbour minority observations. A new synthetic observation can be generated using a weighting scheme such that $d_3 = w(d_1 - d_2) + d_1$. Here, w is a random number generated from a uniform continuous distribution in the interval zero to one. In other words, SMOTE creates new artificial instances of minority observations using a convex combination of existing observed minority class instances already present in the data. Thus, the algorithm has more instances of the minority class observations to learn from.

SMOTE is different from random upsampling in the sense that while random upsampling creates repeated identical instances of the same data, SMOTE creates new instances that are similar to existing minority observations but not identical. Thus, SMOTE introduces more variation into the minority class instances. The process is presented in Fig. 13.5. Random Forest technique as described before can be used after employing upsampling, or downsampling or synthetic minority oversampling technique in the data.

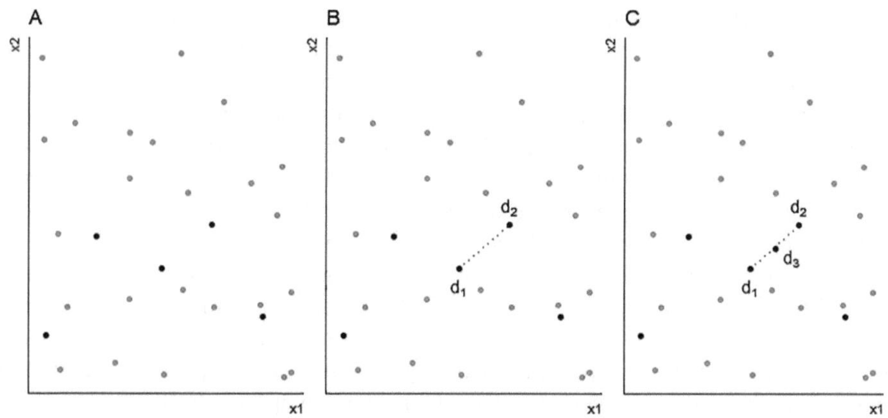

Fig. 13.5 SMOTE. *Note* (A)—black dots represent events and grey dots represent non-events, (B)—it selects an event (say d_1) and identifies its nearest neighbour (d_2) among the events, (C)—a new synthetic event (d_3) is generated along the straight line connecting the two points as a synthetic event

13.7 Empirical Illustration[4]

A commonly used indicator for economic development in any country is its infant mortality rate. Infant mortality rate is defined as the number of deaths within the first year of birth per 1000 live births. A common policy problem is to build a predictive model for early identification of infant deaths. The challenge here is that in large-scale nationally representative household survey data, the instances of infant deaths are few, i.e. the number of births that remain alive (non-events) are far higher in number (majority class) compared to the infant deaths (events) which are much fewer in number (minority class). This makes predicting infant mortality an imbalanced response binary classification problem.

Such problems are common in other areas of applications. Khalilia et al. (2011) predict disease risk from medical data where the incidence of disease is as low as 0.01%. In finance, Bhattacharyya et al. (2011) predict credit card fraud risk where the incidence was as low as 2%. Muchlinski et al. (2015) predict civil war risk where the incidence of civil war in countries is 1%. To demonstrate the application of machine learning algorithms on imbalanced data (that have been discussed in the earlier sections of this chapter), the present analysis uses an example of a commonly used dataset in development economics—a nationally representative household survey data—IHDS-II (Desai et al. 2015). Brahma and Mukherjee (2022) used this dataset[5] to predict neonatal and infant death. Based on crucial insight from that paper, the authors now consider a subset of the sample and focus only on infant deaths in this chapter. Brahma and Mukherjee (2022) show that delivery complication is an important predictor of infant death. Using that *insight*, the present work focuses *only* on the high-risk group of mothers (a subset of the original data that is used in most papers including Brahma and Mukherjee [2022]) who suffered delivery complications during the birth of the infants that are of interest. This gives 2584 observations and an incidence of infant mortality of approximately 2%. This chapter now illustrates how ML methods can be used for prediction in this particular setting.

Following the conventional literature in machine learning algorithms, the dataset is randomly divided into two parts—training and test. This helps one evaluate the out-of-sample prediction accuracy. The algorithms are applied to the training data and this trained algorithm is then applied to test data to generate predictions. The predictions are compared to the actual observed values in the test data to assess the algorithm's out-of-sample prediction accuracy. It is important to hold out a part of the data as test data while training the algorithms. This is because the ultimate purpose of machine learning algorithms is to generate predictions on unseen data and the test data serves as a proxy for that purpose.

However, note that if a resampling-based ML technique is used, the resampling techniques are applied only after one partitions the data into training and test sets. That means on an imbalanced data, resampling is applied only on the training data to train the algorithm. The test data still remains imbalanced to represent the true

[4] The contents in the first part of this section draw heavily from Brahma and Mukherjee (2022).

[5] The full list of predictors and their description can be found in Brahma and Mukherjee (2022).

Table 13.4 Prediction accuracy across different machine learning algorithms

	AUROC	AUPRC	F1
LPM	0.986	0.449	0
Random Forest	0.990	0.539	0.21
Upsampling then Random Forest	0.990	0.521	0.53
Downsampling then Random Forest	0.985	0.520	0.40
SMOTE then Random Forest	0.990	0.576	0.48

state of the world. In this empirical example, data is randomly divided into a 60–40 split. This implies that randomly (without replacement) selected 60% of the sample is in the training data whereas the remaining 40% are in the test data. Each of the algorithms discussed in this chapter is applied—Random Forest, Random Forest with upsampling, Random Forest with downsampling and Random Forest with SMOTE (SMOTE-RF) and then the area under precision-recall curve is presented for each case. We also include AUROC and F1 for each case (Table 13.4).

As a baseline, authors first run a standard econometric technique of linear regression (called linear probability model) to predict infant mortality. For more details on LPM, see Wooldridge (2016) pg. 238. Several things are evident from Tables 13.4 and 13.5. First the metric AUROC barely varies across the different algorithms. Even in the case of the standard econometric model (LPM) where all the minority class observations are miss-classified as majority class, the AUROC is above 0.98. LPM classifies all observations into the majority class and thus has a F1 score of 0. This is also illustrated better through Table 13.6. Thus, LPM achieves a high AUROC by correctly classifying only the majority class observations (even after missing minority classes completely). This illustrates the point that using AUROC can give a misleadingly optimistic view of the performance of any algorithm while having imbalanced classes and justifies our use of AUPRC as the main evaluation metric instead.

While comparing the performance of the algorithms using AUPRC, the present work finds that all the nonparametric machine learning algorithms perform better than the standard econometric model. Their percentage improvements in accuracy over the AUPRC score of LPM are presented in Table 13.5. All the resampling techniques show improved prediction accuracy over standard econometric algorithms. However, comparing within the three types of resampling techniques shows interesting patterns.

Table 13.5 Improvement in AUPRC from standard econometric model (LPM)

Algorithm	Percentage improvement in AUPRC over LPM
LPM	–
Random Forest	20.0%
Upsampling then Random Forest	16.0%
Downsampling then Random Forest	15.8%
SMOTE then Random Forest	28.28%

Table 13.6 Illustration of prediction accuracy from LPM and SMOTE

Obs ID	Observed class	Predicted prob from LPM	Predicted prob from SMOTE-RF	Predicted class at 0.5 threshold for LPM	Predicted class at 0.5 threshold for SMOTE-RF	LPM	SMOTE-RF
1	1	0.330	0.7442	0	1	FN	TP
2	0	0.0070	0	0	0	TN	TN
3	1	0.312	0.7244	0	1	FN	TP
4	1	0.238	0.5184	0	1	FN	TP
5	0	0.005	0	0	0	TN	TN
6	1	0.275	0.636	0	1	FN	TP
7	0	0.003	0	0	0	TN	TN
8	0	0.009	0	0	0	TN	TN
9	1	0.239	0.566	0	1	FN	TP
10	0	0.00549	0	0	0	FN	FN

Both simple upsampling and downsampling perform similarly which are actually slightly lower in prediction accuracy than doing Random Forest on imbalanced data. Recall that upsampling is just including identical copies of the existing minority class observations. In a dataset with extreme imbalance such as the one being used, this implies including many copies of the same data point until the class ratios are equal. This likely leads to overfitting as the algorithm is focusing on the same minority class examples over and over again and is thus trained to recognize only those as minority class instances. This lacks exposure to the variations in the event characteristics in a real-world minority class setting (as in test data). Thus, not surprisingly, it does not show an improvement over Random Forest (in identifying minority class instances in the test data). On the other hand, in case of downsampling, where majority class observations are randomly dropped to create a dataset with balanced class ratios, a researcher is left with a much smaller dataset to work with. Thus, again it fails to improve upon Random Forest (due to having less information and also becomes less efficient) as expected.

The best performance is shown by SMOTE which combines the best of both worlds—increased sample size as well as higher variability and proportions of minority class in the training data to learn from. New *synthetic* observations are generated which are similar to the existing minority class observations but are not identical. This gives the algorithm more information about the minority class to learn from. In addition, the number of observations increases rather than decreases. Hence it performs better than all other algorithms as expected.

The superiority of the SMOTE-RF compared to LPM using 10 observations/instances[6] from the dataset used in this chapter is illustrated in Table 13.6.

[6] Note that the ML algorithms do make mistakes in classifying some of the other minority class observations and some majority class observations too just like any other algorithm. No algorithm is

In Table 13.6, 10 data points from the dataset are used. The second column reproduces the actual observed class for each of the 10 observations. Columns 3 and 4 present the predicted probabilities of each of these observations belonging to the minority class derived from the two algorithms—linear probability model (LPM) and the best-performing algorithm SMOTE-Random Forest. Using the popular threshold of 0.5, one can convert these predicted probabilities into class predictions. This is presented in columns 5 and 6. Finally, columns 7–8 compare the predicted class with the observed class and note the type of classification/misclassification.

For all the observations of negatives, all three algorithms correctly predict them to be negatives. These are the true negatives. However, notice that the best-performing algorithm SMOTE-RF gives a predicted probability of 0 while LPM gives a predicted probability greater than 0 but albeit less than 0.5.

For the observations on events, with a threshold of 0.5 LPM misclassifies all the observations. However, at the threshold of 0.5, the best-performing model SMOTE-RF is able to correctly classify all five observations of events. The superiority of SMOTE-RF in prediction can be illustrated through these observations. SMOTE-RF gives a high predicted probability in the minority cases thereby classifying them correctly as events.

13.8 Concluding Discussion

The contribution of this work is myriad. It focuses on several ML methods from the literature and presents them for applied researchers using illustrative data and example. The chapter focuses only on binary classification problems which occur most frequently. Readers should note that the issue of class imbalance can also occur in multi-class classification setting but it can be much more complex depending on the nature of the response variable, i.e. ordered versus unordered, or the number of imbalanced classes (two or three or more). There is no one-size-fits-all solution, and each problem will require solutions specific to that setting.

This chapter demonstrates how data-driven machine learning tools such as decision trees and the aggregation of different decision trees in the form of Random Forest can be used in binary classification problems with imbalanced data ('rare-event' problem). In such situations, the appropriate evaluation metric for assessing prediction accuracy should be used. PRC which is a more accurate evaluation criterion than the standard ROC criterion is used. Resampling techniques can be used along with these algorithms to boost the prediction performance, especially with data imbalance. However, there is no one-size-fits-all approach to resampling as it is a function of the dataset and hence the appropriate resampling technique should be chosen with care.

---

perfect in a probabilistic world. The ML algorithms do an overall better job in correctly classifying observations as is evident from Tables 13.4, 13.5 and 13.6.

Although machine learning tools are being applied considerably for the purpose of predictive modelling, use of machine learning for causal inference is limited. How machine learning can improve causal inference with imbalanced data is still an uncharted territory. Although there are a few works on causal trees and causal forests, it is not clear in the theoretical literature how they will perform when events are rarer than the non-events. Note that nonparametric machine learning techniques such as Random Forest do not superimpose any restrictions on the underlying distributions or conditional moment functions (as is done in any parametric model) and hence are naturally well suited to produce more accurate predictions as well as causal inference. Thus, future research can potentially use such techniques to produce unbiased and efficient causal inference in the face of imbalanced data.[7]

Appendix: Code Implementation

The data and code used in the illustration are attached with the chapter. The entire data contains 2584 observations. Only 1.8% of these observations, i.e. 47 observations, are infant deaths. Hence, this is a case of highly imbalanced data. R version 4.1.1 is used, and the algorithms are run on R Studio.

The authors divide the data randomly into training and test sets using a 60:40 split. This implies that a randomly selected 60% of the samples are in the training set whereas the remaining 40% are in the test set. All the algorithms are fitted on the training data and their prediction accuracy is evaluated on the test data. R package 'randomforest' is used to fit a Random Forest algorithm on the training data. The Random Forest algorithm has 5000 trees in this study. Since the work is focusing on an imbalanced binary response variable, the stratified resampling option (by specifying the stratifying variable to be the response) is used while building bootstrapped resamples of the data. This prevents any bootstrapped resampled dataset to be formed which has only majority class observations while building the Random Forest. Authors set the 'importance' option as true in the code to generate the variable importance metric using mean decrease in Gini Index. Authors also set the 'norm.votes' option as true to normalize the votes across all the trees in the forest and generate the final predicted class probability.

The 'caret' package in R is used to employ the basic types of upsampling and downsampling the training data. The 'upsample' command will generate repeated versions of the minority class observations in the training data and balance the class distribution of the response variable in the training data. The 'downsample' command will drop a random subsample of the majority class observations and balance the class distribution of the response variable in the training data.

The 'smotefamily' package in R is used to construct synthetic minority class observations using SMOTE. The 'dup_size' argument is used to choose how many

---

[7] The authors are highly indebted to an anonymous referee and the editors for many insightful comments which have enriched the work.

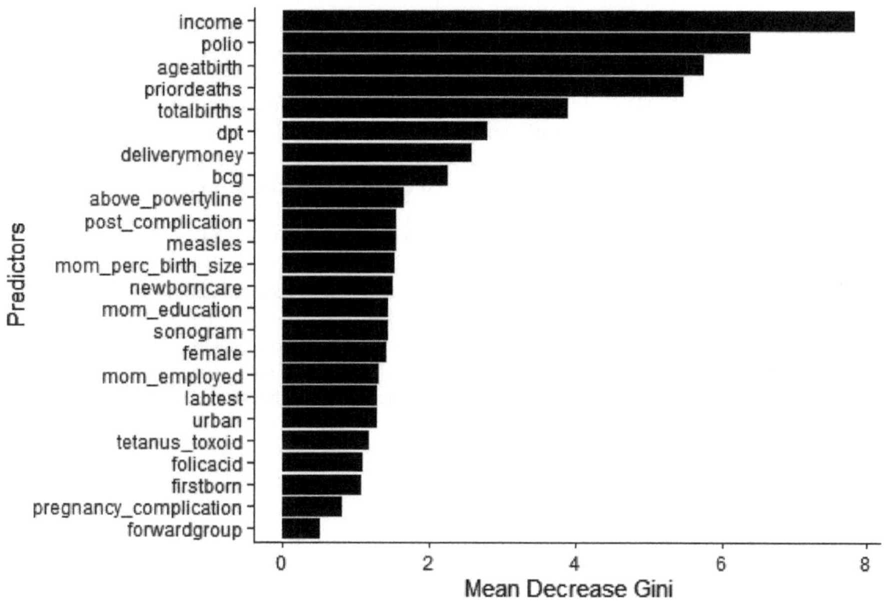

Fig. 13.6 Variable importance ranking using mean decrease in Gini Index

times synthetic minority class observations will be generated. The reported results here are with 'dup_size' = 4, i.e. 4 times new synthetic observations were generated over the original number of minority observations in the data. The 'precrec' package is used to calculate the AUROC and AUPRC values for the algorithms. Also, the 'confusionmatrix' command is used in the 'caret' package to generate the F1 score and variable importance (Fig. 13.6).

References

Aggarwal R, Ranganathan P (2018) Understanding diagnostic tests—Part 3: receiver operating characteristic curves. Perspect Clin Res 9(3):145–148

Bhattacharyya S, Jha S, Tharakunnel K, Westland JC (2011) Data mining for credit card fraud: a comparative study. Decis Support Syst 50(3):602–613

Brahma D, Mukherjee D (2022) Early warning signs: targeting neonatal and infant mortality using machine learning. Appl Econ 54(1):57–74

Chawla NV, Bowyer KW, Hall LO, Kegelmeyer WP (2002) SMOTE: synthetic minority over-sampling technique. J Artif Intell Res 16:321–357

Chawla NV, Lazarevic A, Hall LO, Bowyer KW (2003) SMOTEBoost: improving prediction of the minority class in boosting. In European Conference on Principles of Data Mining and Knowledge Discovery PKDD, pp 107–119

Davis J, Goadrich M (2006) The relationship between Precision-Recall and ROC curves. Proceedings of the 23rd international conference on Machine learning—ICML, 233–240.

Desai S, Vanneman R, New Delhi NCAER (2015) India Human Development Survey-II (IHDS-II), 2011–12.: Inter-university Consortium for Political and Social Research, Ann Arbor, MI. doi.org/https://doi.org/10.3886/ICPSR36151.v2.
Efron B, Tibshirani R (1994) An introduction to the bootstrap. Chapman and Hall/CRC
Flach P, Kull M (2015) Precision-recall-gain curves: PR analysis done right. Advances in Neural Information Processing Systems 28
Hastie T, Tibshirani R, Friedman J (2009) The elements of statistical learning, 2nd edn. Springer, Heidelberg
James G, Witten D, Hastie T, Tibshirani R (2021) An introduction to statistical learning, 2nd edn. Springer, Heidelberg
Khalilia M, Chakraborty S, Popescu M (2011) Predicting disease risks from highly imbalanced data using random forest. BMC Med Inf Decis Making 11(1):51
Muchlinski D, Siroky D, He J, Kocher M (2015) Comparing random forest with logistic regression for predicting class-imbalanced civil war onset data. Polit Anal 24(1):1–17
Wooldridge J (2016) Introductory econometrics: a modern approach, 5th edn. Cengage, Boston

Chapter 14
Anomaly Detection Methods: Application to Automated Vehicle Health Monitoring

Khyati Thakkar, N. Hemachandra, and Tilak Raj Singh

Abstract Anomaly detection is essential for identifying irregular patterns that may indicate system deterioration, component failure, or defects. By leveraging machine learning, it enables continuous condition monitoring, integrating historical data with real-time trends to enhance reliability, reduce downtime, and support proactive maintenance. This chapter explores vehicle health monitoring and early fault diagnosis using anomaly detection methods. Three methods are proposed for failure prediction: two employ Mahalanobis distance after pre-processing and feature engineering, while the third utilizes an Autoencoder neural network with inbuilt feature learning capabilities. Each approach establishes thresholds based on healthy vehicle data, detecting anomalies when deviations occur. The implementation has been carried out using Python programming language.

14.1 Introduction

With the advent of IoT and Industry 4.0, there is a shift in trends in industries with machines becoming more and more smart, integrated, and data driven. Machine defects and failures which are evident in the form of anomalies are unfortunate and lead to severe consequences. These include loss of productivity due to unscheduled

---

Supplementary Information The online version contains supplementary material available at https://doi.org/10.1007/978-981-97-9060-9_14.

K. Thakkar · N. Hemachandra (✉) · T. R. Singh
Department of Industrial Engineering and Operations Research, Indian Institute of Technology Bombay, Mumbai, Maharashtra, India
e-mail: nh@iitb.ac.in

K. Thakkar
e-mail: 173194001@iitb.ac.in

T. R. Singh
e-mail: tilakrajsingh@iitb.ac.in

F. Hamid and D. Mukherjee (eds.), *Data Science for Modeling Managerial and Socioeconomic Problems*, Contributions to Management Science,
https://doi.org/10.1007/978-981-97-9060-9_14

downtime, heavy maintenance costs, and potential injury to workers. Condition-based monitoring with predictive analytics helps in drastic changes in the maintenance schedules of equipment. The goal is to prevent unplanned reactive maintenance, without incurring costs associated with doing too much preventive maintenance. Due to high operating costs and risk exposure, condition monitoring is viewed as a life-saver in asset-heavy industries such as manufacturing, automotive, and aerospace because of its potential to deliver significant cost reductions.

14.1.1 Condition Monitoring

Condition monitoring of machines involves studying sensory data such as temperature, vibration, electric current, and flow rate for the detection and diagnosis of defects. It intends to identify significant pattern changes in underlying sensor data referred to as input signals. Following are a few common examples of condition monitoring technologies and their applications:

- **Vibration Analysis**: Rotating equipment such as compressors, electric motors, and pumps exhibit a certain degree of vibration while in operation. As engineering components incur some fault or degrade, the amount of vibration may change due to the underlying fault. The railway's predictive maintenance team uses sensors such as a vertical accelerometer to capture vibration frequency response signals. Machine learning-based classification techniques such as the nearest neighbour classifier and support vector machine are used to predict potential faults in the railway car body, bogie frame, and axles (Karlsson et al. 2020). High-speed train (e.g., Japan's Shinkansen) uses vibration-based conditional monitoring to increase the total system reliability and reduce downtime due to unplanned maintenance (Tsunashima 2019).
- **Infrared Thermography**: It is a highly popular technology applied for mass screening of febrile individuals during pandemic outbreaks, such as COVID-19, Swine Flu, Ebola, and Dengue (Khaksari et al. 2021). Infrared thermography finds its application in a wide range of system health monitoring such as induction motor fault, monitoring of high-voltage transmissions, and non-invasive health screening (Mahami et al. 2021; Jagadev and Giri 2020). One challenge though is to find a reliable classifier for noisy data captured by infrared thermography. Machine learning techniques ranging from logistic regression to pixel-by-pixel classification using deep neural networks have shown sizable improvement in the underlying prediction quality (Lile and Yiqun 2017).
- **Ultrasonic Diagnosis**: Ultrasound technology, which uses sound waves at frequencies above 20kHz that are inaudible to humans, has long been employed as a safe, non-destructive method in medical applications. In electric vehicles, it is essential to measure the battery's State of Charge (SoC) and State of Health (SoH), which requires technology capable of detecting internal changes within the battery. Similar to how ultrasound is used for medical diagnostics, these methods can

now be applied to batteries. By using capacitive micro-machined ultrasonic transducers (CMUTs), an ultrasonic signal is either reflected or transmitted through the battery cell. This signal is then analyzed and characterized using machine learning algorithms that compile data from thousands of hours of battery cycling at various temperatures and degradation rates to create a computational model for each battery (Hu et al.2015). Additionally, ultrasound sensors combined with spatial clustering are utilized for parking assistance in modern vehicles (Yang et al. 2019).

- **Acoustic**: This method is mainly used to detect leaks in gas, liquid, or vacuum. The use of this method is also observed when it is difficult to accurately position multiple sensors as acoustic sensors can be implemented without physical contact and can help in deriving valuable insights from the data. The development of a data mining model for identifying faults in reciprocating air compressors using acoustic signals is explained in Verma et al. (2015).
- **Lubricant Analysis**: The condition of lubrication oil and its circulation system reflect the health status of the machinery. A vehicle's engine oil under high temperatures and varied moisture conditions can result in the development of the oil sludge, which can create severe malfunction of the vehicle. Predictive maintenance techniques use temperature and particle flow rate along with data-driven methods based on statistical and machine learning algorithms to improve vehicle downtime (Wakiru et al. 2019).
- **Electrical**: Any wear on the machine may lead to increased friction which in turn increases the current drawn by the moving parts. These motor currents are monitored using ammeters. Faults in induction motors like imbalance, misalignment, and bearing tears can be detected using motor current spectrum analysis (Popaleny and Antonino-Daviu 2018).
- **Operational Performance**: Sensors are mounted across the system to measure important parameters like temperature, pressure, flow, vibrations, etc. Mechanical processes including systems like pumps, boilers, and mixers are monitored and optimized using field sensors to improve the overall equipment operating efficiency (Reges et al. 2021).

Methods for condition monitoring of machine involve observing the component of interest to identify changes in operation that can be indicative of a developing fault. The practical application of condition monitoring requires the analysis of multiple data to predict and diagnose faults. In practice, data related to developing fault or faulty conditions is sparse compared to normal operating conditions as any equipment or system is expected to operate in a healthy condition for majority of its lifetime. Because of this data imbalance and scarcity of labeled datasets, supervised learning algorithms cannot be considered. Hence, unsupervised learning methods such as clustering and anomaly detection are fundamental for modeling condition monitoring-based decisions.

14.1.2 Anomaly Detection

Anomalies are defined as observations or instances that deviate from the normal behaviour of the majority of the remaining data. The different types of anomalies defined are as follows (Pawar and Mahindrakar 2015):

- **Point Anomalies**: If any individual data instance is inconsistent concerning the remaining dataset, then that instance or data point is labeled as a point anomaly. For example, data points a1 and a2 in Fig. 14.1 can be defined as point anomalies.
- **Contextual Anomalies**: If a certain data point is unusual or unconventional in a particular situation, then that data instance is termed as a contextual anomaly.
- **Collective Anomalies**: If a group of data instances that are related to each other are inconsistent when compared to the entire dataset, those data points are referred to as collective anomalies. The presence of these data points together as a group is collectively anomalous even when the individual data point may not be anomalous by itself. In Fig. 14.1, data points in the a3 region can be termed as collective anomalies if they are inconsistent with respect to the complete dataset.

There are various techniques for anomaly detection which are defined based on their principle of operation. Some of these are grouped as follows (Riazi et al. 2019):

- **Statistical-Based Methods**: In this case, the anomaly detection model is built using a specific probability distribution as it holds a very strong assumption that the data inherently follows a certain statistical distribution.
- **Clustering-Based Methods**: In this method, the assumption is made that data instances that show similarity among themselves belong to one group with respect to data points present in another group. A threshold is set based on different parameters like number of clusters, size of a cluster, distance to the cluster centroid, etc. Based on this set threshold, the anomaly score is calculated and accordingly, points are classified as normal or anomalous.
- **Nearest Neighbour-Based Methods**: This method assumes that the normal (non-anomalous) data instances usually appear nearby and hence have dense neighborhoods whereas the anomalous data instances are far from the rest of the data points. Nearest neighbour-based methods are further grouped into two types: distance-based nearest neighbour and density-based nearest neighbour methods. A distance

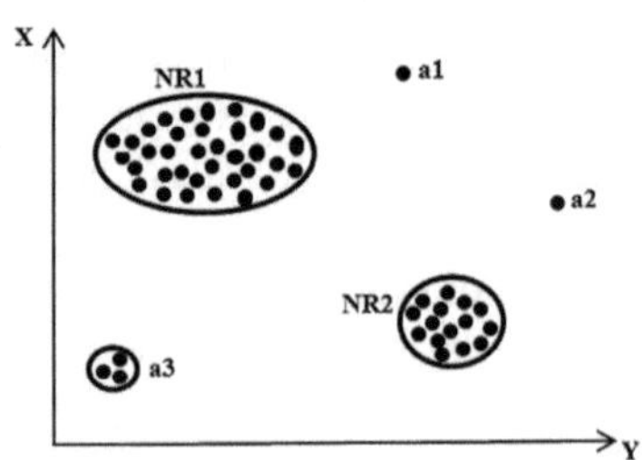

Fig. 14.1 Anomalies in a X-Y plane (Pawar and Mahindrakar 2015)

or similarity measure is used by both of these methods to determine the magnitude of abnormality.

- **Classification-Based Methods**: This method comprises two steps: The first step is the training phase, where the classifier is trained on the labeled training data. The second step is the testing phase, where the learning of the trained model is utilized to classify the test instances as normal or anomalous.
- **Spectral (Subspace)-Based Methods**: This method tries extracting features that best explain the variability present in the training data. With these models, there is an assumption that a lower dimension subspace can be utilized to represent the normal data where they can be differentiated from the abnormal or anomalous data.
- **Deep Learning-Based Methods**: These anomaly detection methods model high-level representations and predict complex patterns of the input data using multiple stacked layers. This eliminates expert knowledge for feature engineering, and thus, the models can be applied directly to raw data.

Some of the factors that are considered while choosing an algorithm are (Souiden et al. 2017):

- Nature of data
- Parametric or Non-Parametric
- Online or Batch setup
- Arrival rate of data: Fixed or Variable
- Supervised, Unsupervised, or Semi-supervised
- Univariate or Multivariate
- Low or High dimension
- The trade-off between Type 1 (false detections) errors, Type2 (missed detections) errors, and Time lag
- Computational requirements: Time and Cost
- Anomaly type: Local or Global
- Output type: Anomaly label or Anomaly score

14.1.3 Automated Vehicle Health Monitoring

Vehicle health monitoring can be considered as a special case of condition monitoring. Test vehicles need to undergo a rough road durability test which requires the vehicle to run on the specifically engineered track with off-road, pave roads, potholes, dips, and other road surface defects precisely placed along the track. These tracks leave a bone-jarring experience to the vehicle drivers and limit their total driving time. Autonomous driving gives flexibility in the vehicle testing schedule and reduces the completion time of long-lasting vehicle tests as the robots can drive 24×7. Adverse effect on human drivers and long vehicle testing time has motivated manufacturers to deploy robot-driven (automated) vehicle for rough road testing (Isaac Bober 2018). However, humans have mastered the art of driving over the years so any onset of fault in the vehicle could be sensed and diagnosed by a human driver even in the absence of sensor systems as compared to a robot-driven vehicle. Also, since human drivers share the same test track, safety measures need to be assured to avoid any collision or mishappenings. This needs to be ensured by taking necessary measures to detect expected mechanical parts failure (e.g., suspension, wheel, and chassis) causing uncontrolled vehicle movement.

An attempt is made to develop a data-driven fault detection system for the automated vehicle that can detect faults with the potential of causing a serious accident. This work aims to create a condition-based health monitoring model which can aid in the early and automatic diagnosis of any deterioration in the vehicle condition by detecting anomalies or unusual behaviour in the vibration pattern of vertical acceleration sensors installed across the length of the test vehicle and trigger an alarm signal for each abnormal behaviour.

14.1.4 Vehicle Health Monitoring via Anomaly Detection

Anomalies are indicative of deterioration of vehicle health, the potential failure of components, or sudden changes in external factors (environmental changes, operating condition changes, etc.). Therefore, the detection of anomalies can aid in the condition-based monitoring of vehicles under test. There are two anomaly detection algorithms considered in this experiment, Mahalanobis distance-based anomaly detection (single-sensor and multiple-sensor approach) and Autoencoder neural network. When a vehicle experiences a certain fault, the vibration signals will deviate from its normal patterns exhibited while the vehicle is in a healthy state. These deviations can be captured using some useful features and used as the basis for unsupervised anomaly detection. It is often difficult to simulate fault conditions or avail faulty data from any setup due to experimental cost and safety issues. Therefore, in such cases, one tries to explore and learn from the sensor data coming from a healthy vehicle, and any deviations from such learned behaviour are marked as anomalies to label a vehicle as healthy or unhealthy (Jin et al. 2012).

14.1.5 Business Problem Formulation

The test vehicle is equipped with multiple vertical acceleration sensors and a GPS location sensor to get continuous data during vehicle runs. The data is captured continuously as the vehicle is made to run across the test track consisting of four segments (alternating smooth and rough) forming a lap. This problem is approached by segmenting the continuous data into individual laps using geospatial data. These laps are further segmented into individual surfaces by detecting changes in road surface from smooth to rough and vice versa using Clustering along with geospatial data as the data available is unlabeled. Since the vehicle is made to run through the same test track, statistical analysis is done for comparison between segments across laps as a part of exploratory data analysis to find any significant trends or patterns.

For anomaly detection, Mahalanobis distance and Autoencoder-based models are used over these segmented surfaces. The sensor data from the known healthy vehicles is used to train and learn useful patterns from this healthy data which models the normal behaviour of vehicles over that surface. Thereafter, unknown vehicle data over the same surface is tested against the learned normal behaviour. Any data point deviating from this normal behaviour beyond the selected threshold is marked as an anomaly in the test data assuming these deviating patterns in the data are a result of a fault condition. It is very difficult to learn from raw vibration signals because of the presence of noise and transients. Therefore, time domain features which are known to explain various statistical properties of the data are extracted from these raw signals. After feature extraction, feature selection is done using principal component analysis (PCA) to derive the most significant features influencing the output. This transformed feature space is then used as an input to our Mahalanobis distance-based anomaly detection algorithm. However, for the Autoencoder network, feature extraction and feature selection are not needed as it is itself said to learn useful representation from vibration data. Therefore, the Autoencoder is trained on smoothed vibration data from a healthy vehicle and is used to determine whether the test data from an unknown vehicle is normal or anomalous. The details of these techniques along with their framework are explained in subsequent sections.

This chapter is organized such that Sect. 14.1 provides an introductory view of condition monitoring, anomaly detection, the business problem of vehicle health monitoring, and the approach followed toward it. Section 14.2 explains machine learning techniques for clustering, dimensionality reduction, and anomaly detection used across the chapter. Section 14.3 defines the experimental setup used for data collection, data pre-processing, and data segmentation. Section 14.4 details how anomaly detection is used for condition-based vehicle health monitoring. Section 14.5 states the experimental results of all three anomaly detection approaches used. This chapter is concluded with the final remarks given in Sect. 14.6 followed by references.

14.2 Machine Learning Techniques

Machine learning plays a vital role in deriving useful patterns from data. This section provides details of some of the techniques that are used in this experiment for analyzing data. Unsupervised machine learning techniques like clustering, PCA, and anomaly detection are defined with their operating principle, types, advantages of each of these types, and their relevance in this experimental study.

14.2.1 *Clustering*

Clustering is a machine-learning technique that is used to form distinct groups from unlabeled data. The clustering algorithm works by analyzing the similarity between instances and putting similar instances in the same cluster and "dissimilar" instances in the different cluster to have minimum intra-cluster distances and inter-cluster distances as maximum as possible (Flach 2012). Some examples of clustering include customer segmentation for marketing campaigns, image segmentation to identify similar species, segmenting geographies based on climatic conditions, etc.

Out of many methods of clustering, two methods, namely Hierarchical clustering and K-means clustering are experimented with in this chapter due to their capabilities in terms of performance, time complexity, space complexity, and ease of implementation.

14.2.1.1 Hierarchical Clustering

Depending upon whether the Hierarchical representation is top-down or bottom-up, the Hierarchical clustering algorithm has two types (Reddy 2018):

1. Agglomerative Hierarchical Clustering Algorithm (Agglomerative Nesting)
2. Divisive Hierarchical Clustering Algorithm (Divisive Analysis)

The operating principles of these two algorithms are exactly opposite of each other.

Agglomerative Hierarchical Clustering

The Hierarchical tree is constructed from the data points upwards, which is why it is called a bottom-up or agglomerative algorithm. In each iteration, the algorithm forms a new group by merging the two closest clusters. The distance between the two clusters (measure of "nearest") can be defined via variation criteria; in a generic sense, *linkage function* is used to define pairwise point distances into pairwise cluster distances.

Following is a list of some distance metrics used in Hierarchical clustering and their pictorial representation in Fig. 14.2 (Rhys 2020):

- Single Linkage: Defines the distance between two clusters as the minimum pairwise distance between elements from each cluster.
- Complete Linkage: Defines the distance between two clusters as the maximum pairwise distance between elements from each cluster.
- Average Linkage: Defines the distance between two clusters as the average pairwise distance between elements from each cluster.
- Centroid Linkage: Defines the distance between two clusters as the distance between the centroids (means) of the clusters.

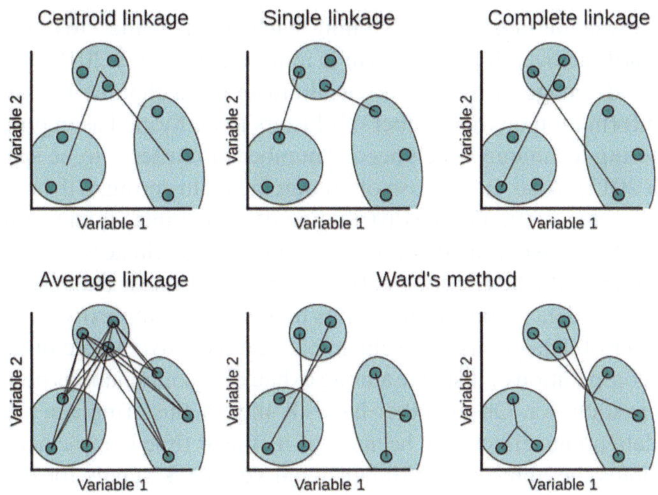

Fig. 14.2 Different linkage methods used in Hierarchical clustering (Rhys 2020)

- Ward Method: Unlike the other methods, Ward's method for hierarchical clustering does not use the distance between two clusters. Instead, it minimizes the total within-cluster variance, defined by the squared Euclidean distance between points.

Using one of the above measures, Hierarchical clustering creates a dendrogram over a given data, which is a tree-like structure connecting the clusters at different levels of hierarchy. It helps us identify how many clusters should be considered in actuality. An example of a dendrogram using the ward method for agglomerative Hierarchical clustering is presented in Fig. 14.7.

Divisive Hierarchical Clustering

Divisive Hierarchical clustering algorithm follows exactly the opposite way by starting with one big cluster and successively partitioning the data points into different clusters until all data points form their separate groups.

14.2.1.2 K-Means Clustering

K-means is a clustering algorithm that relies on distance metrics, where the distance metric indirectly represents the learning objective. The goal is to identify clusters that are compact based on the chosen distance metric. Given n observations say $(x_1, x2, ...x_n)$, the K-means algorithm partition n data to k sets where $(k << n)$, i.e., $S_1, S_2, ..S_k$ by minimizing within the cluster variance. The algorithm keeps on re-iterating, recalculating, and reassigning each point to the closest centroids of $S_1, S_2, ..S_k$ until the overall penalty term is minimized (Reddy 2018).

The main advantage of the Hierarchical clustering technique is that it does not need prior information on the number of clusters to be formed as opposed to the K-means clustering technique which needs "number of clusters" to be specified as an input (Flach 2012). Also, in the case of Hierarchical clustering, a dendrogram gives us a pictorial representation of the distribution of data points among various clusters which makes the decision of selection of the number of clusters more explainable. The only problem with the Hierarchical clustering technique is that it can handle only a small number of data points because of its higher time complexity of $\mathcal{O}(n^3)$, where n is the number of data points to be clustered. The recursive nature of the algorithm requires the computation of distance among exhaustive combinations before deciding to combine in a cluster. On the other hand, K-means clustering can handle a huge number of data points effectively because of its linear time complexity of $\mathcal{O}(nkdi)$, where n denotes the number of d-dimensional vectors, k stands for the number of clusters and i is the number of iterations needed until convergence.

14.2.2 Principal Component Analysis

The extraction of multiple features from raw data leads to a large number of variables. To reduce the effect of the "curse of dimensionality," feature selection is required which helps to reduce the dimensionality of the dataset along with multi-collinearity in the data features for effective model building. One of the widely used techniques for dimensionality reduction is principal component analysis. The main concept of PCA is to reduce the dimension of a dataset that consists of multiple variables or features without losing much information by retaining the maximum variation within the data. This is achievable because of the correlation present within variables in the original space. The way PCA works is it transforms the original feature variables into a new set of feature variables orthogonal to each other, which are known as the principal components. These principal components are arranged such that the variation present within the original variables in the dataset goes on decreasing with the order. Therefore, the first principal component explains the maximum variation present in the original data, the second principal component retains the next highest variation of the original data, and so on. These principal components are orthogonal to each other as they are the eigenvectors of a covariance matrix. The PCA outputs these principal components which can be defined as a linear combination of optimally-weighted variables. The number of these principal components is chosen such that they explain the majority of the variation present in the data as shown in Fig. 14.3. Therefore, it is always less than or equal to the number of original variables. This results in a reduction of feature dimension in the dataset without much loss of information (Tharwat 2016).

The steps usually followed for PCA are as given below:

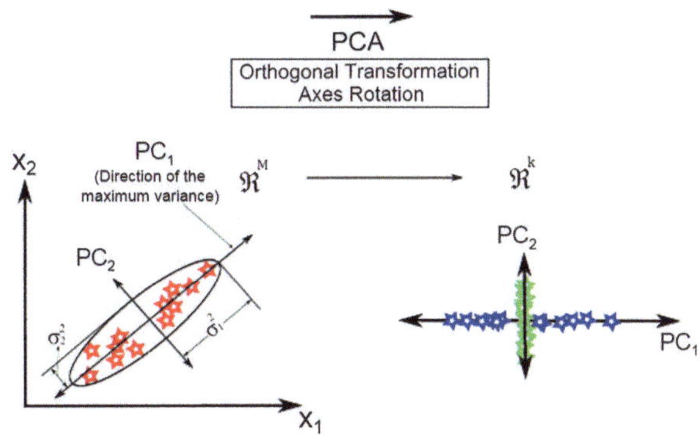

Fig. 14.3 Example of the two-dimensional data where the variance of each variable is graphically represented in the direction of the maximum variance (Tharwat 2016)

1. Normalize the data.
2. Compute the covariance matrix between variables.
3. Determine the eigenvalues of the covariance matrix.
4. Calculate the eigenvectors corresponding to the calculated eigenvalues.
5. Select the number of principal components to retain maximum variation in the data.
6. Transform the original feature space to form a new feature space.
7. Generate the new dataset using transformed feature space.

14.2.3 Anomaly Detection Techniques

The dataset considered in this use case is unlabeled, multivariate has a fixed arrival rate, and anomalies need to be identified with minimum computational requirements giving near real-time output. Based on these factors, two methods are experimented with for anomaly detection, namely Mahalanobis distance which is based on nearest neighbour, and Autoencoder belonging to a deep learning group of techniques. Mahalanobis distance-based anomaly detection is experimented with two settings, one with all sensors and another with only one sensor considered at a time. The details of these techniques are explained in subsequent subsections.

14.2.3.1 Mahalanobis Distance-Based Anomaly Detection

Mahalanobis distance is a generalized distance metric that uses system features to measure the system under test. It represents a multivariate system using a scalar value, i.e., it can be used to convert the multi-dimensional data into one dimension and then apply univariate techniques to it for outlier or anomaly detection. It is very useful in determining the similarities and differences between training and testing (unknown) datasets. Mahalanobis distance-based methods have been successfully applied to health monitoring and anomaly detection of mechanical components like gearboxes, bearings, compressors, cooling fans, pumps, induction motors, etc. Out of various sensor signals that help in revealing health information about the system under test, vibration signals are widely used for mechanical systems sensitive to vibrations (Jin et al. 2017). Useful time-domain, frequency-domain, or time-frequency-domain features are extracted from these raw signals (Caesarendra and Tjahjowidodo 2017) depending upon the system mechanics or domain knowledge. Data collected from the healthy products constitutes the training dataset. The Mahalanobis distances of these training data instances form a reference or a benchmark set that is referred to as the Mahalanobis space which is used to determine a suitable anomaly threshold. The Mahalanobis distances of the test data resemble its closeness with the Mahalanobis space which reflects the similarities and differences between them. These Mahalanobis distances are used to determine the health condition and identify whether the system is anomalous or healthy. Mahalanobis distance-based approach has great potential for online system health monitoring because of very less computational complexity.

Mahalanobis Distance

Mahalanobis distance is a multivariate equivalent of the Euclidean distance with the difference that Mahalanobis distance considers covariance among feature vectors for computing the distance (Narasimhan 2019). Mahalanobis distance is effectively the distance between a data point and a distribution.

The formula used to calculate the Mahalanobis distance is as follows:

$$D^2 = (x - m)^T \cdot C^{-1} \cdot (x - m)$$

where,

D^2 is the squared Mahalanobis distance.

x is the observation vector which is a row of a dataset.

m is the mean values vector of independent variables which is the mean of each column.

C^{-1} is an inverse covariance matrix of independent variables.

The first term $(x - m)$ is the distance of the observation vector from the mean. The second term is the inverse covariance matrix C^{-1} which is multiplied with the first term. Whenever the data has a high correlation with the mean, the value of covariance will be high which in turn will lead to low values for the inverse covariance matrix resulting in a smaller Mahalanobis distance value. However, data with a low correlation with the mean will have a high Mahalanobis distance value. A large value is a stronger indication of the data point being an anomaly or an outlier. This forms the basis for calculating distances of distinct unknown data points from the known distribution of data points. After the calculation of Mahalanobis distance, the threshold is selected beyond which an instance is tagged as an anomaly or outlier.

Anomaly Threshold

Some ways of choosing an appropriate anomaly threshold value are as explained below (Jin et al. 2017):

- **Chi-Square Distribution**: Squared Mahalanobis distances are known to follow chi-square distribution where degree of freedom is equal to the number of features or independent variables if the given data is multivariate normal. Depending on the risk associated with anomalies in the system, a suitable critical value can be considered. For example, the 97.5% or 99.5% quantile (Q) of the chi-square distribution with the required degree of freedom can be computed and considered as a threshold to distinguish anomalous instances from normal instances. The Mahalanobis distance calculated is compared with this critical value. If the distance is greater than the critical value (considered as threshold), that data point is marked as an anomaly.
- **Mahalanobis Distance for Train Set**: If the input data is not multivariate normal or the distribution of squared distances does not follow any standard distribution, one needs to calculate the Mahalanobis distances for healthy (training) data. The maximum value for this calculated distance from the training data can be considered a threshold. While testing, if the distance calculated crosses this threshold then that data point is tagged as an anomaly.
- **Box-Cox Transformation**: Mahalanobis distance is a metric that does not adhere to a Gaussian distribution, with values ranging from zero to infinity. Therefore, a Box-Cox transformation can be applied to convert it into Gaussian data, allowing the properties of the Gaussian distribution to be used for determining the threshold value.

14.2.3.2 Autoencoder Network-Based Anomaly Detection

In the context of condition monitoring using anomaly detection, the idea behind using an Autoencoder network is to represent the sensor readings in a compressed format capturing correlations between these sensor readings. The Autoencoder network is trained on data coming from normal equipment representing sensor behaviour under normal operating conditions. The goal here is to learn the sensor interactions and reconstruct the input coming from these sensor readings again at the output with minimum reconstruction error. Now any normal test equipment readings when given to this pre-trained model will be able to predict the output with the least reconstruction error. If the monitoring equipment is degraded or is unhealthy, the interactions between sensor readings will be changed in the form of changes in pressure, vibration, temperature, etc. If the input of such degraded equipment is given to a trained model, it will show an increased error while reconstructing the input sensor readings at the output. If this reconstruction error increases beyond the pre-defined threshold as determined while training the model, that data point can be marked as an anomaly. Therefore, by monitoring the reconstruction error, one can

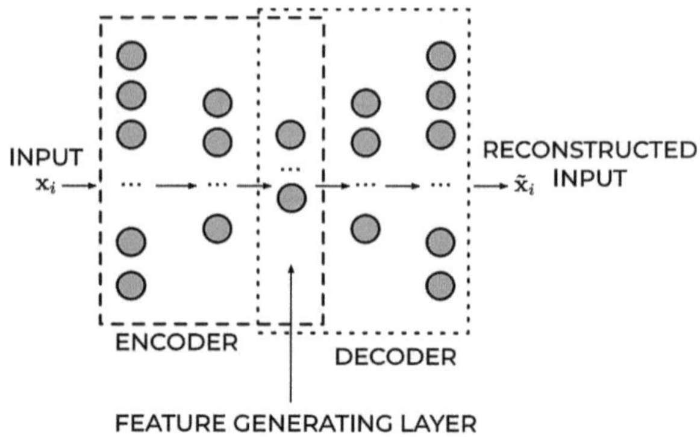

Fig. 14.4 Autoencoder Architecture (Michelucci 2022)

determine the health of the equipment as degraded equipment is expected to yield high reconstruction error.

Autoencoder Architecture

An Autoencoder neural network is considered to be an unsupervised learning algorithm as it does not use labeled data but generates its labels using the training data. An Autoencoder tries to learn an identity function by setting the output the same as the input. It is a feed-forward neural network that applies back propagation and tries to learn important patterns from input data. It then stores these learned patterns in a compressed format in the form of feature representation or code which is then used to reconstruct the input with minimum loss. An Autoencoder network is also known as a replicator neural network as it replicates the data from input to output. The sum-of-squared difference is calculated between the input and the output and it is used as its loss function so that the output and input can be brought as close as possible (Michelucci 2022).

The network consists of three parts, an encoder layer which is a neural network used to convert the input data into a compressed meaningful code representation, a latent feature representation or code layer which is typically lower dimensional layer storing encoder patterns, and a decoder layer which is usually mirror image of encoder layer that uses these stored compressed code patterns to reconstruct the input at the output layer. The detailed architecture is shown in Fig. 14.4.

The following hyper-parameters are mostly used while training the Autoencoder model:

- Code size
- Number of network layers
- Number of nodes per layer
- Activation function for each layer
- Loss function

Autoencoders are not allowed to learn the input completely, i.e., care must be taken such that the Autoencoder does not merely copy the input data and try to mimic the identity function. Otherwise, it will be able to reconstruct the input data perfectly on the training set; however, it will overfit and fail to generalize and reconstruct output effectively for the testing set. That is the reason most of the time the code layer in an Autoencoder structure is kept small as compared to the input layer so that it is forced to learn meaningful representation capturing interactions and correlations between the input variables.

14.3 Data Preparation Framework

This section explains details of the data collection setup, nature of data available, appropriate pre-processing applied, and overall methodology followed on the data to make it suitable for anomaly detection schemes proposed in the subsequent sections.

14.3.1 Data Collection

The test vehicle is equipped with five vertical acceleration sensors as shown in Fig. 14.5. These sensors are mounted at the front left (FR_LH) side, front right (FR_RH) side, rear left (RR_LH) side, rear right (RR_RH) side, and cabin (Cabin) of the vehicle. Along with these, vehicle speed and GPS coordinates are also captured continuously as this robot-driven test vehicle is made to run in the predefined test track with an approximate length of around 1.58km consisting of alternating rough (stone paved) road surfaces of lengths 340 m and 674 m each and smooth road surfaces of lengths 402 m and 166 m each as shown in Fig. 14.6.

The sensors mounted on the vehicle are connected to different channels of the data acquisition system which collects these sensor values and stores them in the form of different files for a pre-defined time, the file name bearing details of the date, time,

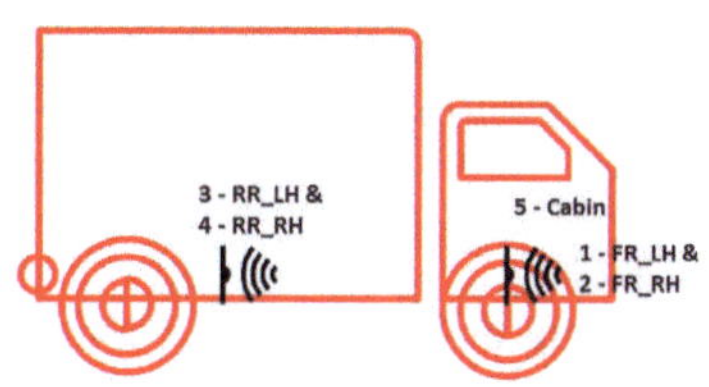

Fig. 14.5 Test vehicle with mounting locations of vertical acceleration sensors

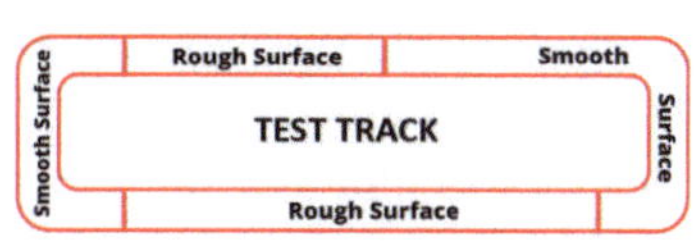

Fig. 14.6 Test track with alternating rough and smooth surfaces

and type of vehicle for which data is collected. The features provided are captured at different sampling rates. These files consist of an arbitrary number of instances depending upon the time for which data is stored. The vehicle speed is maintained at an average of 25km/hr with ramp-up and ramp-down to 35km/hr and 15km/hr respectively.

14.3.2 Data Pre-Processing

The data available for this project has characteristics like different sampling rates, noisy sensor measurements, different road surfaces, etc. that need to be dealt with some pre-processing techniques for converting it into a usable format to draw useful insights and apply machine learning techniques.

Date-Time Indexing
The data capturing and storing is done in time intervals of a few minutes, therefore every file has a different name indicating the date and time of data capturing. Within the file, there is a sampling rate mentioned for all features. This information on the name of the file and sampling rate is used to convert the data into "Date Time Index" type data to maintain continuity and consistency among all files.

Resampling
Some of the features are captured at a sampling rate of 5 msec and some at a sampling rate of 500 msec. For better analysis of data, all the features have to be brought to the same sampling frequency. One can either upsample all the data to 5 msec or downsample to 500 msec. Since downsampling all the features to 500 msec would have resulted in a substantial loss of information, in this experiment upsampling of all the features to 5 msec is chosen.

Interpolation
After the resampling is applied, the features captured at a slow sampling rate have in-between values as "NA" which needs to be interpolated with some suitable measure to make it more usable. The various methods that can be used for interpolation are linear, quadratic, cubic, nearest, polynomial, etc. As in this case, vehicle speed and GPS coordinates need to be interpolated so linear interpolation is chosen as both these features tend to follow a linear trend while the vehicle is running on the test track.

Filtering
Filtering or Smoothing of a time series is done when there are noisy measurements or some irrelevant observations that one wants to remove to analyze and visualize the underlying trends in a better way. In this case, filtering or smoothing is done in two steps: In the first step, the observations with vehicle speed less than 15km/hr, which is termed as an "idle" period of the vehicle, are filtered out. This is done because when the vehicle speed is less than the threshold, the vehicle is stationary at some location due to some reason like robot configuration, vehicle downtime, etc.,

or has just begun to move in the desired direction. During this period the vertical acceleration values are very low (close to zero) which is not indicative of any useful information regarding road surface or vehicle condition. In the second step, moving average and moving standard deviation values are calculated considering the window length of 500 samples which calculates to 2.5sec of data which is enough to average out the fluctuations present due to sensor noise.

The formulae used to calculate simple moving average and moving standard deviation are as follows:

$$SMA = \frac{P_1 + ... + P_n}{n} \tag{14.1}$$

where,

P_i is the datapoint in the ith interval.

n is the number of intervals considered.

$$MSD = \sqrt{\frac{(P_1 - SMA)^2 + ... + (P_n - SMA)^2}{n}} \tag{14.2}$$

where,

P_i is the datapoint in the ith interval.

SMA is the simple moving average calculated for nth interval.

n is the number of intervals considered.

14.3.3 Lap Segmentation

After the pre-processing of data, the next step followed is to segment the data into multiple laps. Here, lap means completion of an entire round of elliptical test track which consists of a total of four road surfaces with alternating rough and smooth segments. The test vehicle is traveling across the test track continuously so the data is periodic with respect to a lap. Hence, lap segmentation is done to study trends and other patterns in the time series data effectively. Lap segmentation is achieved using available geospatial data by taking a reference location as the start of the track (lap 1) and detecting when the same location is repeated indicating the start of the track again (lap 2) and so on. The accuracy of the GPS sensor is taken into account while developing the segmentation algorithm.

14.3.4 Surface Segmentation

After lap segmentation is performed, each lap needs to be segmented into individual surfaces, i.e., Rough and Smooth road surfaces. For surface segmentation, geodesic distance calculation is performed along with Agglomerative Hierarchical and K-Means clustering techniques, the details of which are explained in Sect. 14.2.1.

The aim of performing clustering on the data is to look for certain structures within the sensor data. The goal is to get some insights into how many clusters are being formed, the distance measures between drawn clusters, and whether the clusters formed based on vertical acceleration sensor values are in sync with test track road surfaces on which the vehicle is made to run. This is done because the behaviour of sensors on a rough surface is very different as compared to a smooth surface, so potential anomalies on the smooth surface could be the normal behaviour on the rough road surface. If the anomaly detection algorithm is modeled without surface segmentation, the anomalies on the smooth surface may get overlooked because of similar normal instances on rough surfaces and never be taken into consideration. So change in road surface can be understood as a change in the operating condition of the vehicle and it becomes important to account for this change to apply anomaly detection techniques for effective results. The results obtained by both these clustering algorithms are also cross-validated using Geospatial coordinates.

The details of each of these setups are explained in subsequent sections.

14.3.4.1 Hierarchical Clustering

Agglomerative Hierarchical clustering is applied to the filtered vertical acceleration sensor values using "ward" linkage.

The dendrogram plotted after clustering is as shown in the Fig. 14.7.

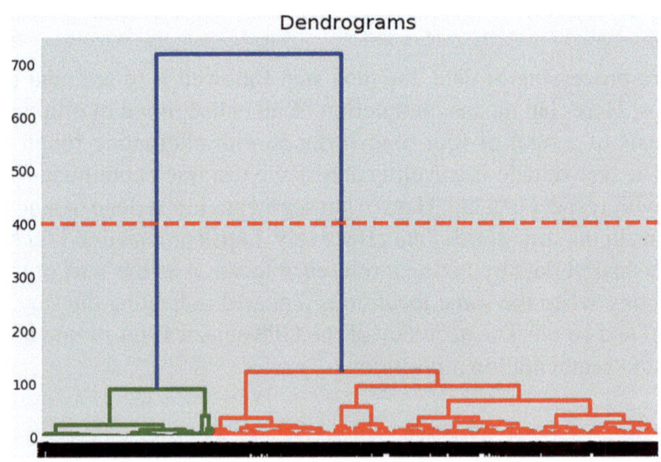

Fig. 14.7 Dendrogram showing two distinct clusters based on filtered vertical acceleration sensor values

14.3.4.2 K-Means Clustering

K-means clustering is applied on the filtered vertical acceleration sensor values tuning following three hyper-parameters:

1. **n_clusters**: The n_clusters allow us to input the desired number of clusters for the given data. This parameter is selected as two as it is evident from the dendrogram and gives consistent results similar to Hierarchical clustering.
2. **max_iteration**: It signifies the maximum number of iterations that the K-means algorithm goes through in a single run. The value of this parameter is tuned to be 1000 ensuring that the algorithm comfortably converges giving consistent results.
3. **algorithm**: It specifies which K-means algorithm to use. There are predominantly two algorithms: the first being the classical EM-style algorithm named "full," and the second one named "elkan" which is more efficient as it uses the triangle inequality, but it does not work well with sparse data. Here, the algorithm chosen is "auto" which automatically selects "elkan" for dense data and "full" for sparse data (Reddy 2018).

In this experiment, the Hierarchical clustering algorithm is applied to statistically derive the number of clusters using a subset of the data which is then provided as an input to the K-means algorithm to cluster the remaining subset of data. This is done to leverage the advantages of both these clustering techniques. Application of both these methods gave consistent and data-driven results with two distinct clusters, each corresponding to rough and smooth road surfaces respectively. Also, this shows that the statistical properties of both surfaces of the same type (smooth surfaces and rough surfaces) are similar. The applied clustering algorithms can distinguish between these road surfaces effectively; however, there is some time lag that is seen between the time the surface change occurs and the time when it is detected by the clustering algorithm. Owing to the context of this problem, this time lag is acceptable and does not affect the results significantly.

14.3.4.3 Geodesic Distance Calculation

Geographical distance, or geodesic distance, is the measurement of distance along the Earth's surface, calculated between points defined by their geographical coordinates, such as latitude and longitude. Since the test vehicle is equipped with a GPS sensor and the vehicle is known to use the same test track, geodesic distance calculation along with clustering can help to segment surfaces with more accuracy as it purely segments based on the distance traveled by the vehicle. The details of this approach are as explained:

- **Step 1**: Geodesic distance is calculated using the 2D Vincenty distance formula between each pair of data points with available GPS latitude and longitude data corresponding to each data point. The cumulative distance traveled by vehicle is calculated using the previously calculated distances between data points across the dataset.

- **Step 2**: Clustering results are used to obtain the indices of the data points at which there is surface change, i.e., surface changes from smooth to rough or rough to smooth. These indices mark the beginning of each of these surfaces. These indices are saved and the cumulative geodesic distance covered corresponding to individual indices is extracted. This gives us the stretch (length) of each of these road surfaces which can be used to validate the algorithm's accuracy by comparing it with known information.
- **Step 3**: Distances calculated from the previous step will remain constant across laps for each surface as the vehicle is traveling across the same test track. The length of each surface derived from the previous step is hard-coded and used for the segmentation of surfaces across all other laps. For any new lap dataset, the distances between consecutive points and cumulative distances are calculated. As soon as the cumulative distance reaches the value derived from the previous step the segmentation is carried out. This procedure is repeated for the entire lap dataset and it results in four datasets corresponding to individual road surfaces within the lap.

The results obtained by geodesic distance calculation along with clustering algorithms validate ground truth concerning test track surfaces. After these data preparation steps, anomaly detection is carried out to determine the health of the test vehicle.

14.4 Anomaly Detection for Vehicle Health Monitoring

In this experiment, anomaly detection is performed on two datasets, "Feb" and "March," belonging to the same vehicle which are segmented into rough surface data and smooth surface data. One of them, the "Feb" dataset is confirmed to be from the healthy vehicle. The "March" dataset is known to have anomalies as a breakdown occurred in the front part of the vehicle while this dataset was recorded. So it is expected that unusual patterns reflecting breakdown might have been captured in the raw vibration signals. Here, a dataset known to be from a healthy vehicle is used to train (Train Dataset) the model and the "Feb" dataset (healthy termed as Test1 Dataset) and the "March" dataset (unhealthy termed as Test2 Dataset) is used as test datasets. The details of these datasets are given in Table 14.1. The features considered are the five vertical acceleration sensor values captured along the length of the vehicle. An example of the dataset considered in this experiment is shown in below Fig. 14.8.

Inference is to be drawn if the experimented algorithms can distinguish between these two datasets and if they can help in analyzing the differences reflected between them. Also, an attempt is made to understand if this breakdown is reflected in advance

DateTime	FR_LH [m/s2]	FR_RH [m/s2]	RR_LH [m/s2]	RR_RH [m/s2]	Cabin [m/s2]
2020-02-25 16:44:05.545	0.100473	0.292175	-1.257048	-2.306725	0.057444
2020-02-25 16:44:05.550	0.751109	0.805315	-1.192182	-2.509827	-0.368753
2020-02-25 16:44:05.555	-0.395714	-0.780379	-1.384547	-2.121707	-1.137764
2020-02-25 16:44:05.560	-1.351551	-0.489236	-1.609597	-2.620094	-0.982462
2020-02-25 16:44:05.565	-0.074189	-0.203695	-1.431848	-2.502993	0.066187
...	...	...	...	...	...
2020-02-25 15:22:49.850	0.165165	-0.074006	-0.535016	1.069233	0.866391
2020-02-25 15:22:49.855	-0.223721	0.647041	-0.746756	0.860075	0.863193
2020-02-25 15:22:49.860	0.817461	0.130095	-0.769459	0.078495	0.577989
2020-02-25 15:22:49.865	1.196808	1.338246	-0.712049	0.262588	-0.092126
2020-02-25 15:22:49.870	0.696986	0.894795	-0.485566	-0.163781	0.232745

Fig. 14.8 Example of the input dataset

Table 14.1 Dataset details

Dataset	Smooth raw dataset	Rough raw dataset
Train	105928 rows × 5 columns	164932 rows × 5 columns
Test1 (Feb)	120509 rows × 5 columns	187517 rows × 5 columns
Test2 (March)	122041 rows × 5 columns	186734 rows × 5 columns

before its onset on the vehicle. If it could be detected beforehand by raising an alert or an alarm, an accident/catastrophe can be prevented.

14.4.1 Mahalanobis Distance Based Approach

The Mahalanobis distance based approach used is shown in Fig. 14.9 and summarized below.

14.4.1.1 Feature Extraction

The presence of random fluctuations and noise in the sensor data makes it very important to derive important features from raw signals which can help in determining useful statistical properties of data. In this work, seven critical time domain features are derived for each sensor signal (Riaz et al. 2017). The details of these features calculated for each sensor value (FR_LH, FR_RH, RR_LH, RR_RH, and Cabin) are given in Table 14.2. These values are calculated with a sliding window of 2000

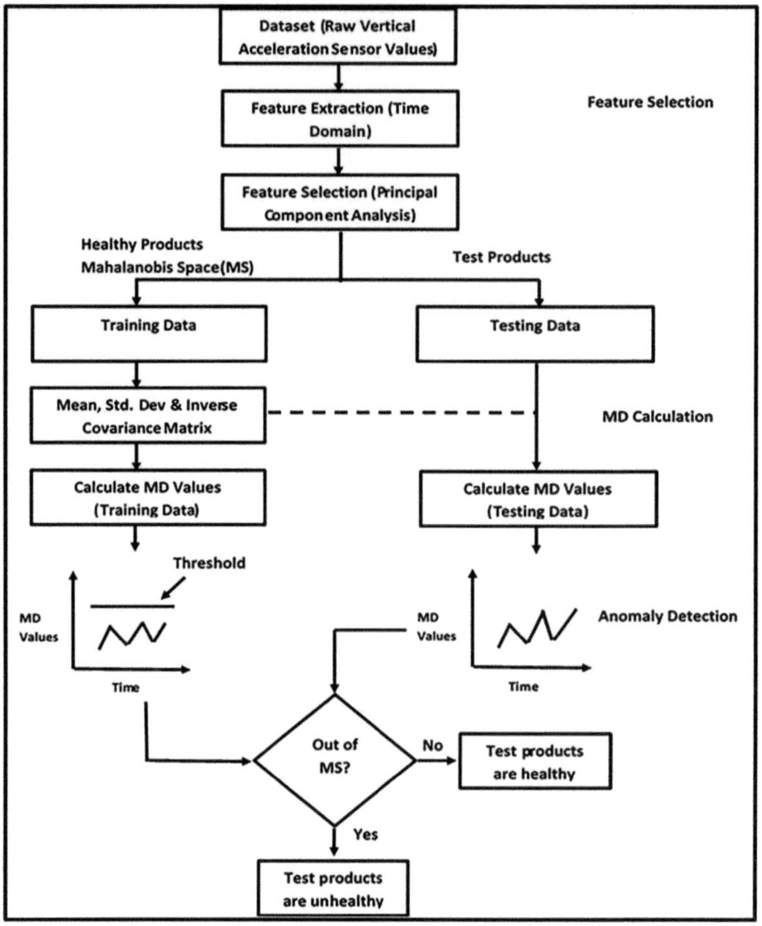

Fig. 14.9 Flow chart explaining the steps of Mahalanobis distance based approach for anomaly detection

data points and a jump of 200 data points. For example, if a total of 20000 data points are considered, these seven features are calculated by considering the first 2000 values from 1 to 2000 (sliding window of 2000), and the next set of features are calculated by considering values from 200 to 2200 (a jump of 200) and so on till all the data points are considered. Different values for the sliding window and jump are tested and the best values for optimized results are chosen. This sliding window also helps in reducing the noise by smoothing transients in the raw signal. So a total of thirty-five features are extracted corresponding to five sensors.

Table 14.2 List of time-domain features extracted for each sensor signal

Feature	Description	Formula
Mean	It is the first statistical moment which describes the central tendency of data	$Mean = \frac{1}{N}\sum_{i=1}^{N} x_i$
Variance	It is the second statistical moment and gives information about the spread of the distribution	$Var = \frac{\sum_{i=1}^{N}(x_i-m)^2}{(N-1)\sigma^2}$
Skewness	It is the third statistical moment and tells about the symmetry of the distribution. For symmetric distributions, the value is zero	$Skew = \frac{\sum_{i=1}^{N}(x_i-m)^3}{(N-1)\sigma^3}$
Kurtosis	It is the fourth statistical moment and measures peakedness and provides a measure of the tailedness of the distribution	$Kurt = \frac{\sum_{i=1}^{N}(x_i-m)^4}{(N-1)\sigma^4}$
Peak Value	It measures the maximum value encountered in the raw signal	$Peak = max\lvert x_i\rvert$
Root Mean Square	It measures the power content in the raw vibration signals	$RMS = \sqrt{\frac{1}{N}\sum_{i=1}^{N} x_i^2}$
Crest Factor	It is defined as the ratio of the Peak Value to the RMS value and is used to detect changes in the signal pattern due to impulsive vibration sources	$CF = \frac{PeakValue}{RMS}$

14.4.1.2 Feature Selection

In this experiment, PCA is used for dimensionality reduction which transforms the original data into lower dimensional space without much loss of information. Here, the number of dimensions for the new feature set is determined based on the graph plotted of a number of principal components versus explained variance. As seen in Fig. 14.10, the elbow of the graph came to be around ten which explains around 96% of the total variance in the original data. After this point, it saturates such that any increase in dimension of output feature set will only marginally increase explained variance in the input feature set. This transformed feature space with ten dimensions is then used for further Mahalanobis distance calculations.

14.4.1.3 Mahalanobis Space Determination

The training dataset consists of data collected from healthy vehicles. The mean and inverse covariance matrix of the feature set of this healthy vehicle is calculated and their Mahalanobis distances form a reference set that is also called the Mahalanobis Space. In this case, the input data is not multivariate normal, therefore, the anomaly threshold here is determined by the distribution of the calculated Mahalanobis distance values forming Mahalanobis space, and any test data point beyond this threshold is termed as an anomaly (Jin et al. 2012). For instance, Mahalanobis

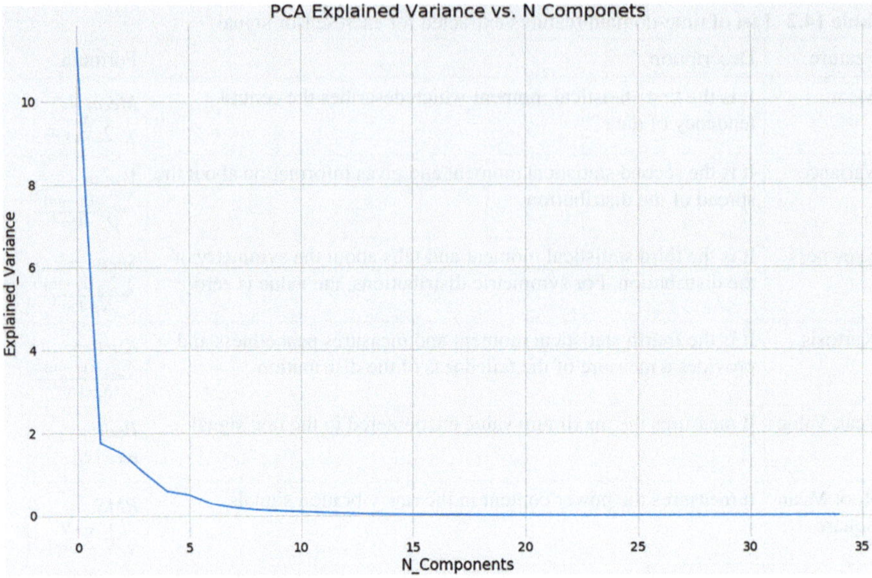

Fig. 14.10 Elbow graph of the number of principal components versus explained variance

space for the smooth dataset after PCA is derived as zero to forty. So the threshold is set as forty and any data point crossing this threshold is marked as an anomaly.

14.4.1.4 Mahalanobis Calculation

The mean and inverse covariance matrix of the feature set of healthy vehicles (training dataset) is used to calculate Mahalanobis distance values for all data points. The farther the test points are from the multivariate train dataset, the larger the Mahalanobis distance values will be which helps in flagging the anomalies. In conclusion, Mahalanobis space is used to discriminate whether the vehicle under test is healthy or unhealthy.

14.4.2 Autoencoder Network Based Approach

In Autoencoder based anomaly detection, a similar training and testing procedure with the same five sensor values (FR_LH, FR_RH, RR_LH, RR_RH, and Cabin) is followed as it is for the Mahalanobis approach. The Autoencoder network based approach used is shown in Fig. 14.11 and summarized below:

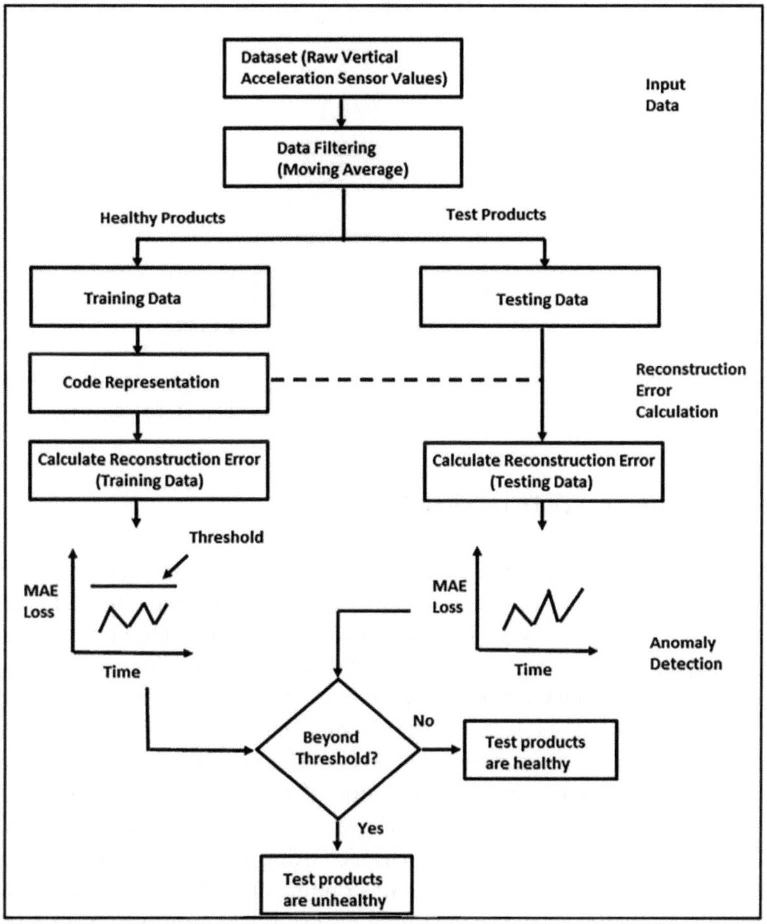

Fig. 14.11 Flow chart explaining the steps of Autoencoder based approach for anomaly detection

14.4.2.1 Raw Data Smoothing

Feature engineering is not needed here as Autoencoder is known to learn useful patterns or representations from input data. The raw vibration data collected from sensors is very noisy so the training and testing datasets are smoothed using a moving average with a window size of 500 samples, which looked good for effective learning after experimenting with multiple window size values.

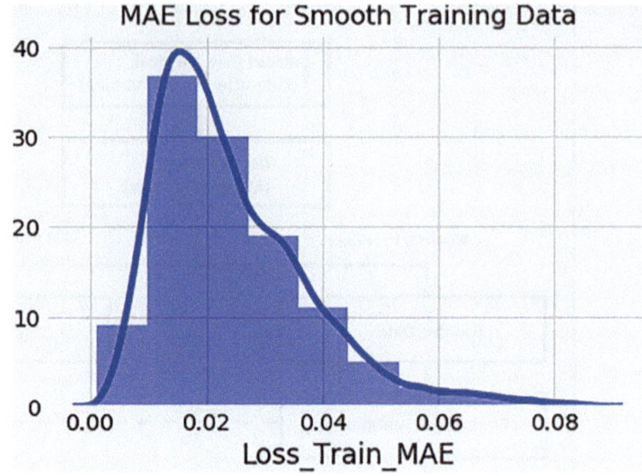

Fig. 14.12 Distribution plot of training reconstruction error over smooth surface

14.4.2.2 Healthy Space Determination

For the creation of healthy space and selection of threshold, the data available from known healthy vehicles is considered as in the case of Mahalanobis distance based approach. Different sets of hyper-parameters like the number of hidden units, number of layers, activation function, number of epochs, batch size, optimizer, and validation loss are experimented and the hyper-parameter set that result in minimum reconstruction error while training is selected. The Autoencoder used here consists of one layer each in the encoder, code, and decoder layers. The encoder layer consists of four hidden units with a "tanh" activation function, the code layer with two units and a "relu" activation function and the decoder layer consists of four hidden units with the "tanh" activation function. The model is trained using a hundred epochs with a batch size chosen as sixty-four. It uses an "Adam" optimizer with mean squared error as a validation loss function. The distribution of the error values derived from this model is used to determine the threshold such that any value beyond this threshold can be tagged as an anomaly. This set of reconstruction error values for which data point will be considered normal constitutes the healthy space. For instance, consider the distribution plot of reconstruction error for a healthy dataset as shown in Fig. 14.12. This distribution shows healthy vehicle's training loss values in the range of 0 to 0.1. Therefore, with some tolerance, the healthy space for the considered dataset is derived as 0 to 0.12. So any data point with a value beyond 0.12 is treated as an anomaly.

14.4.2.3 Reconstruction Error Calculation

The representation learned in the code layer is expected to save useful patterns of healthy vehicles. These patterns are then used by the decoding layer to reconstruct the input at the output layer. While testing, any data point whose reconstruction error goes beyond the threshold value derived in Sect. 14.4.2.2 is marked as an anomalous data point. If the vehicle is faulty, it will have significant differences in its input patterns. When the representation learned from the healthy training dataset is used to reconstruct the input for a faulty vehicle there be will significant differences leading to high reconstruction errors. Therefore, by monitoring the reconstruction error, one can determine the health of the equipment.

14.5 Anomaly Detection Results

The results from the setup defined in Sect. 14.4 are explained in the following subsections. Here, the results for each of these anomaly detection methods are shown over the smooth surface for reference. Similar results are also obtained over rough road surfaces using the same setup. As an example, the Autoencoder model result over rough surfaces is also shown to draw comparisons between different road surfaces.

14.5.1 Multiple-Sensor Mahalanobis Distance Based Model

In the multiple-sensor model, all thirty-five feature values derived for five sensors are considered for modeling. Initially, the training and testing were performed using all the calculated features for all the sensor values. However, since the number of features is large, the model failed to distinguish between the two test datasets, one of which is anomalous. Therefore feature selection is performed using PCA and transformed data into ten dimensional space is used for model building. The graph of the calculated Mahalanobis distances for both the test datasets over smooth surfaces is shown in Fig. 14.13.

As can be seen, the Test2 dataset (March) shows higher Mahalanobis distance values as compared to the Test1 dataset (Feb); and this is expected as the vehicle is known to be unhealthy in March. It also shows many observations of the Test2 dataset crossing the threshold indicating the unhealthy status of the vehicle whereas the Test1 dataset has all the values below the threshold. So it can be inferred that multi-sensor Mahalanobis distance approach can detect the difference between healthy and unhealthy vehicles. However, the difference between the two datasets in terms of their Mahalanobis distances is not very significant. Also, it is difficult to determine which section of the vehicle is leading to this anomalous behaviour without analyzing individual sensor data separately.

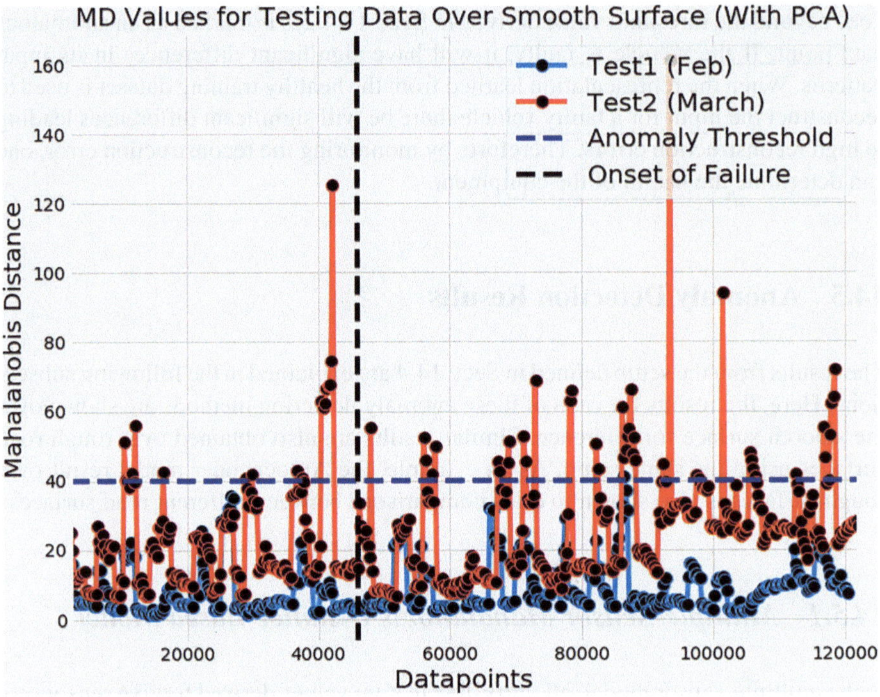

Fig. 14.13 Mahalanobis distance calculation for test datasets using multiple-sensor model over a smooth surface with X-axis as number of observations and Y-axis as Mahalanobis distance values

14.5.2 Single-Sensor Mahalanobis Distance Based Model

To overcome the shortcomings of the multiple-sensor approach, one sensor at a time is taken into account and the same analysis is carried out. So for each sensor, seven features are extracted taking one sensor at a time for getting Mahalanobis space (training) and Mahalanobis distance calculation (testing). Since there are only seven features, there is no need for dimensional reduction and all the features are considered for the analysis. The Mahalanobis calculation plot of one front sensor (FR_LH) is shown in Fig. 14.14 and one rear sensor (RR_LH) is shown in Fig. 14.15.

It is observed that the Test2 dataset (March) shows very high Mahalanobis distance values as compared to the Test1 dataset (Feb) for front sensors whereas the two rear sensors show normal behaviour for the majority of the data. Most of these values from the Test2 set for front sensors are crossing the threshold especially the ones that are after the vertical black dotted line which indicates the actual onset of failure on the vehicle. These data points beyond the threshold are therefore considered anomalous. The Test1 dataset has all the values below the threshold level indicating the healthy state of the vehicle. The anomalous behaviour in the front three sensors as compared to two rear sensors indicates chances of a breakdown in the front part of the vehicle which is in line with the knowledge of physical breakdown observed in the front

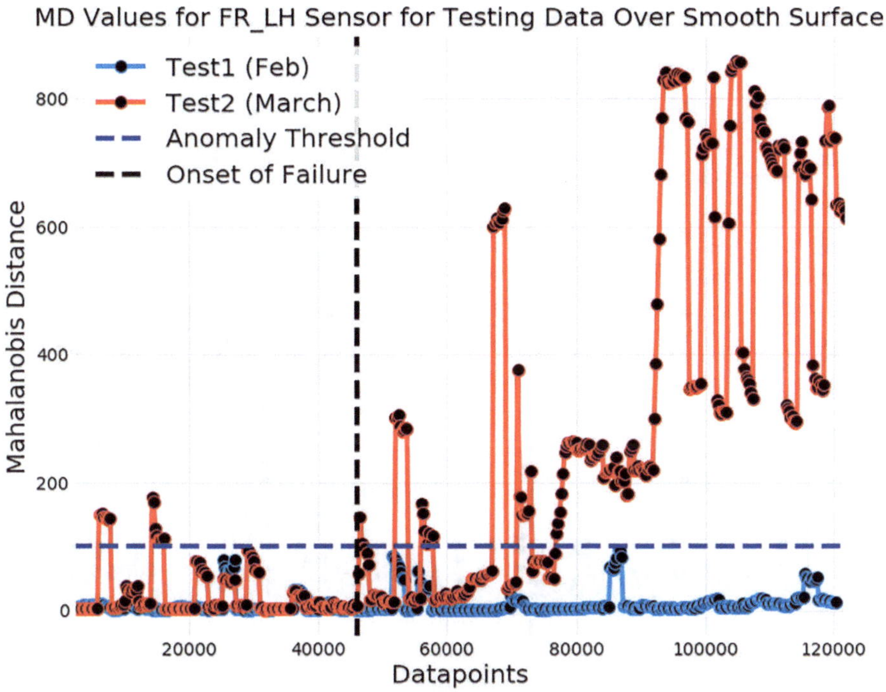

Fig. 14.14 Mahalanobis distance calculation for test datasets using single-sensor model for front left sensor over a smooth surface with X-axis as number of observations and Y-axis as Mahalanobis distance values

vehicle. So it can be inferred that this approach can detect the difference between healthy and unhealthy vehicles and it also helps in the localization of faults for troubleshooting and effective decision-making as some faults are not reflected in all sensor values. For instance, anomalous behaviour of any individual sensor indicates a problem with a specific location of the vehicle where that sensor is mounted, sensor failure, or connection issues. Similarly, all front sensors indicate a problem with the front part of the vehicle, all rear sensors indicate a problem with the rear part of the vehicle, and all sensors indicate a problem with either the entire vehicle or the influence of external factors.

14.5.3 Autoencoder Model

The Autoencoder model trained as explained in Sect. 14.4.2 is then used to predict the two test sets: Test1 (Feb) and Test2 (March). The reconstruction errors for both these test sets on a smooth surface are shown in Fig. 14.16.

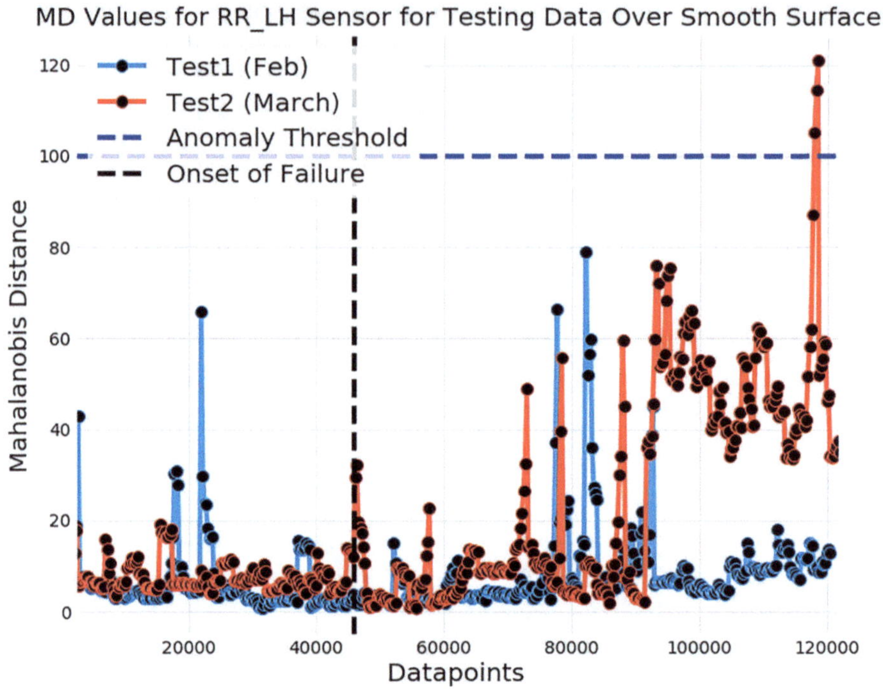

Fig. 14.15 Mahalanobis distance calculation for test datasets using single-sensor model for rear left sensor over a smooth surface with X-axis as number of observations and Y-axis as Mahalanobis distance values

The reconstruction errors for both these test sets on a rough surface are shown in Fig. 14.17.

The threshold values selected for both road surfaces, smooth and rough, are indicated by the blue horizontal dotted line in Figs. 14.16 and 14.17 respectively. The reconstruction error is very high for the Test2 (March) dataset as compared to the Test1 (Feb) dataset for both these road surfaces. This happens because the vibration data present in the Test2 dataset shows deviation from normal vibration patterns learned from the healthy vehicle dataset. These deviations do not allow the decoder section of the Autoencoder to reconstruct the input data well at the output layer as compared to the healthy dataset. Thus, it can be inferred that the Test2 dataset is anomalous. This shows that the Autoencoder network can successfully identify the vehicle as healthy or unhealthy based on the detected anomalies. Also, the onset of failure in the vehicle is indicated by the black vertical dotted line in Figs. 14.16 and 14.17. However, anomalies are marked in the data before the actual occurrence of the failure on the vehicle (left side of black dotted line) showing that the mentioned algorithm is capable of capturing the patterns related to the onset of failure before the actual occurrence which can help in preventing any major breakdowns or adverse scenarios.

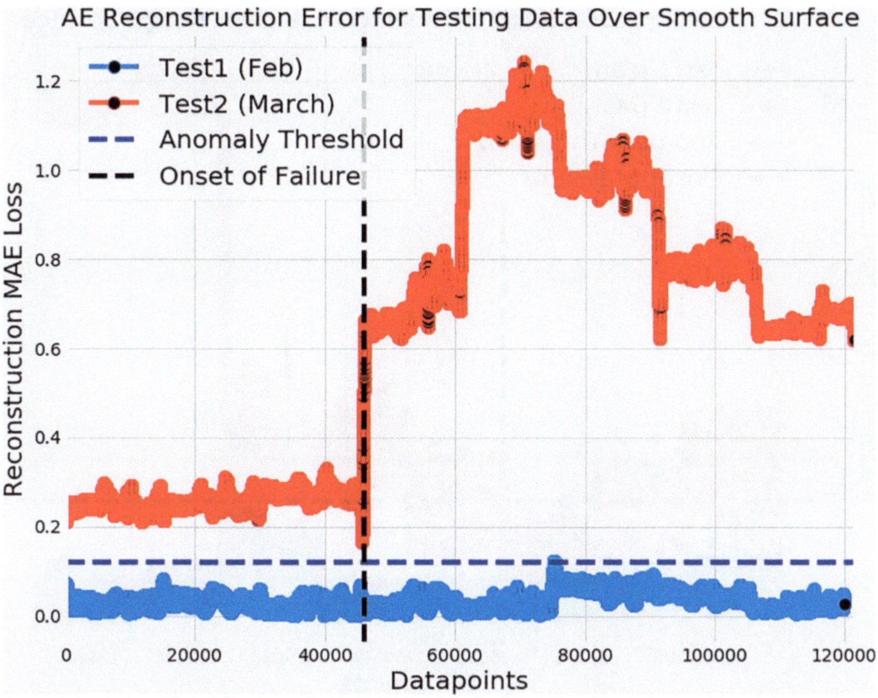

Fig. 14.16 Autoencoder reconstruction error on testing datasets over a smooth surface with X-axis as number of observations and Y-axis as reconstruction MAE loss values

These experiments for anomaly detection are performed on a server having 20-core Intel(R) Xeon(R) CPU E5-2670 v2 2.50GHz processing power and 128 GB RAM and Table 14.3 gives the computational time required for each of these methods.

The details of the data and code files used for anomaly detection algorithms explained in this chapter are available at git repository (Khyati Thakkar 2024).

14.6 Final Remarks

The following section summarizes the conclusions drawn from the experimental analysis conducted so far and the scope for future developments in this area.

14.6.1 Conclusion

The following conclusions are drawn from the above analysis:

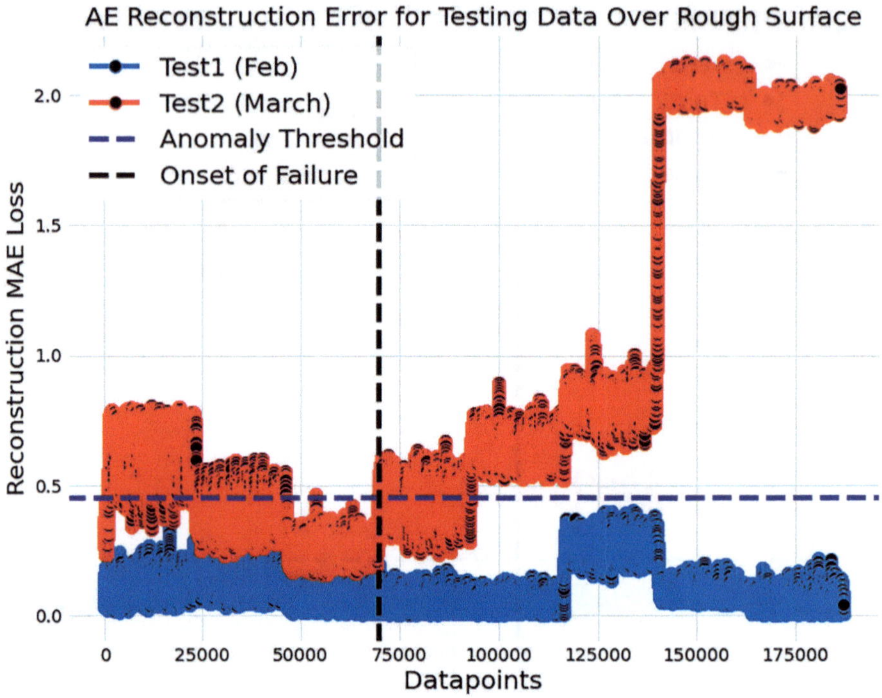

Fig. 14.17 Autoencoder reconstruction error on testing datasets over a rough surface with X-axis as number of observations and Y-axis as reconstruction MAE loss values

Table 14.3 Computational time for anomaly detection approaches

Approach	Feature extraction	Feature selection	Modeling
Single-Sensor Mahalanobis Distance	15sec	0.8sec	1.5sec
Multiple-Sensor Mahalanobis Distance	15sec	Not required	1.5sec
Autoencoder Network	Not required	Not required	2.2sec

- Out of the two clustering approaches, K-means is more suitable as it can handle large data and gives accurate results in very little time as compared to Hierarchical clustering.
- All three anomaly detection approaches: the multiple-sensor Mahalanobis distance model, the single-sensor Mahalanobis distance model, and the Autoencoder model can distinguish faulty vehicles by marking anomalies either in the form of high Mahalanobis distance values or high Autoencoder reconstruction error as compared to a normal vehicle. The training for all three approaches happens offline using historical healthy vehicle data and testing is done on unknown vehicle data.

- In the case of a multiple-sensor approach, feature extraction and feature selection are required and then Mahalanobis distance calculation is used over the transformed space for effective learning. Mahalanobis distance calculation has very little computational complexity, but feature engineering takes substantial time and in particular domain knowledge.
- In the case of a single-sensor approach, feature extraction is required but feature selection is not needed. Apart from distinguishing between the healthy and unhealthy status of the vehicle, the single-sensor approach helps us with the localization of faults, unlike the multiple-sensor approach.
- In the case of the Autoencoder model, feature engineering is not required as it learns useful representation from input data using non-linear activation functions. This saves the time and effort needed to manually select features from raw data. This model is capable of distinguishing vehicle status as faulty or normal based on the flagged anomalies before the onset of the fault on the vehicle.
- Even though all the approaches are successfully able to distinguish anomalies indicating deteriorated health of the vehicle, the Autoencoder model proves to be most effective in terms of the computational time required with inherent feature engineering capability.

14.6.2 Future Scope

Following are some ideas for improvement on the existing work:

- The same methods are to be applied to an existing benchmark dataset and a similar analysis is to be conducted. This will help in validating this study and quantifying the evaluation metrics with higher confidence.
- Explore other unsupervised anomaly detection techniques and draw comparative studies between different approaches to understand which method proves the best for such data.
- Frequency domain features using Fourier transform and time-frequency domain features using Wavelet transform can be explored to reveal some useful patterns with respect to anomalies or outliers.
- Consider different fault types and failure data to understand sensor behaviour to these failure modes and test the robustness of all the anomaly detection algorithms against these faults.

References

Bowler AL, Pound MP, Watson NJ (2022) A review of ultrasonic sensing and machine learning methods to monitor industrial processes. Ultrasonics 124:106776

Caesarendra W, Tjahjowidodo T (2017) A review of feature extraction methods in vibration-based condition monitoring and its application for degradation trend estimation of low-speed slew bearing. Machines 5(4):21

Flach P (2012) Machine learning: the art and science of algorithms that make sense of data. Cambridge University Press

Hu X, Li SE, Yang Y (2015) Advanced machine learning approach for lithium-ion battery state estimation in electric vehicles. IEEE Trans Transp Electrification 2(2):140–149

Isaac Bober: Toyota Uses Robot For Rough-Road Durability Testing (2018) https://practicalmotoring.com.au/car-news/toyota-uses-robot-rough-road-durability-testing/

Jagadev P, Giri LI (2020) Infrared thermography and machine learning based patient health monitoring. In: Infrared sensors, devices, and applications X, August, vol. 11503. SPIE, pp. 59–65

Jin X, Ma EW, Cheng LL, Pecht M (2012) Health monitoring of cooling fans based on Mahalanobis distance with mRMR feature selection. IEEE Trans Instrum Meas 61(8):2222–2229

Jin X, Wang Y, Chow TW, Sun Y (2017) MD-based approaches for system health monitoring: a review. IET Sci Meas Tech 11(4):371–379

Karlsson H, Qazizadeh A, Stichel S, Berg M (2020) Condition monitoring of rail vehicle suspension elements: a machine learning approach. In: Advances in dynamics of vehicles on roads and tracks: Proceedings of the 26th Symposium of the International Association of Vehicle System Dynamics, IAVSD 2019, August 12–16, 2019, Gothenburg, Sweden. Springer International Publishing, pp 119–127

Khaksari K, Nguyen T, Hill B, Quang T, Perreault J, Gorti V, Gandjbakhche AH (2021) Review of the efficacy of infrared thermography for screening infectious diseases with applications to COVID-19. J Med Imaging 8(S1):010901–010901

Khyati Thakkar KT (2024) Anomaly-Detection-Using-Mahalanobis-Distance-Autoencoder-Automated-Vehicle-Health-Monitoring (Version 1.0.0) [Computer software]. https://github.com/thakkarkhyati/Anomaly-Detection-Using-Mahalanobis-Distance-Autoencoder-Automated-Vehicle-Health-Monitoring

Lile C, Yiqun L (2017) Anomaly detection in thermal images using deep neural networks. In: 2017 IEEE International conference on image processing (ICIP), September. IEEE, pp. 2299–2303

Mahami A, Rahmoune C, Bettahar T, Benazzouz D (2021) Induction motor condition monitoring using infrared thermography imaging and ensemble learning techniques. Adv Mech Eng 13(11). https://doi.org/10.1177/16878140211060956

Michelucci U (2022) An introduction to autoencoders. arXiv preprint arXiv:2201.03898

Narasimhan S (2019) Application of data mining techniques for sensor drift analysis to optimize nuclear power plant performance. Int J Innovative Tech Exploring Eng 9(1):3087–3095

Pawar AM, Mahindrakar MS (2015) A comprehensive survey on online anomaly detection. Int J Comput Appl 119(17)

Popaleny P, Antonino-Daviu J (2018) Electric motors condition monitoring using currents and vibrations analyses. In: 2018 XIII International Conference on Electrical Machines (ICEM), September. IEEE, pp 1834–1840

Reddy CK (2018) Data clustering: algorithms and applications. Chapman and Hall/CRC

Reges G, Fontana M, Ribeiro M, Silva T, Abreu O, Reis R, Schnitman L (2021) Electric submersible pump vibration analysis under several operational conditions for vibration fault differential diagnosis. Ocean Eng 219:108249

Rhys H (2020) Machine Learning with R, the tidyverse, and mlr. Simon and Schuster

Riazi M, Zaiane O, Takeuchi T, Maltais A, Günther J, Lipsett M (2019) Detecting the onset of machine failure using anomaly detection methods. In: Big Data Analytics and Knowledge Discovery: 21st International Conference, DaWaK 2019, Linz, Austria, August 26–29, 2019, Proceedings 21. Springer International Publishing, pp 3–12

Riaz S, Elahi H, Javaid K, Shahzad T (2017) Vibration feature extraction and analysis for fault diagnosis of rotating machinery-a literature survey. Asia Pacific J Multi Res 5(1):103–110

Souiden I, Brahmi Z, Toumi H (2017) A survey on outlier detection in the context of stream mining: review of existing approaches and recommadations. In: Intelligent Systems Design and Applications: 16th International Conference on Intelligent Systems Design and Applications (ISDA 2016) held in Porto, Portugal, December 16–18, 2016. Springer International Publishing, pp 372–383

Tharwat A (2016) Principal component analysis-a tutorial. Int J Appl Pattern Recogn 3(3):197–240

Tsunashima H (2019) Condition monitoring of railway tracks from car-body vibration using a machine learning technique. Appl Sci 9(13):2734

Verma NK, Sevakula RK, Dixit S, Salour A (2015) Intelligent condition based monitoring using acoustic signals for air compressors. IEEE Trans Reliab 65(1):291–309

Wakiru JM, Pintelon L, Muchiri PN, Chemweno PK (2019) A review on lubricant condition monitoring information analysis for maintenance decision support. Mech Syst Signal Process 118:108–132

Yang Q, Chen H, Su J, Li J (2019) Towards high accuracy parking slot detection for automated valet parking system (No. 2019-01-5061). SAE Technical Paper

Chapter 15
Fuzzy Linear Regression: Analyzing Employment-Poverty Relationship in MENA Countries

Belhadj Besma and Kaabi Firas

Abstract Fuzzy logic, particularly through Gaussian membership functions, provides a flexible approach to modeling uncertainty by representing varying degrees of truth rather than relying on binary classifications. Unlike probability, fuzzy logic quantifies the degree of membership within a set, making it well-suited for complex, real-world relationships. This chapter introduces a fuzzy linear regression model that overcomes the limitations of traditional regression techniques. The model is employed in R platform to analyze the employment-poverty relationship in selected countries of the Middle East and North Africa (MENA) region. By leveraging Gaussian membership functions, the model captures subtle variations, non-linearities, and enhances predictive accuracy. The estimated model offers valuable guidance for policymakers, providing a deeper understanding of the complex dynamics between employment and poverty.

15.1 Introduction

Linear regression is a commonly applied technique for analyzing the relationship between a dependent variable and one or more independent variables. This method models how the dependent variable changes in response to variations in the independent variables. In linear regression, one of the most common estimation techniques is the ordinary least-squares (OLS) method (Bourbonnais 2018). The OLS estimator

Supplementary Information The online version contains supplementary material available at https://doi.org/10.1007/978-981-97-9060-9_15.

B. Besma (✉)
University of ElManar, Tunis, Tunisia
e-mail: besma.belhadj@fsegt.utm.tn

K. Firas
University of Bordeaux, Nouvelle-Aquitaine, France

F. Hamid and D. Mukherjee (eds.), *Data Science for Modeling Managerial and Socioeconomic Problems*, Contributions to Management Science,
https://doi.org/10.1007/978-981-97-9060-9_15

seeks to reduce the total squared differences between the actual values of the dependent variable and the predicted values generated by the linear regression model. It is considered a desirable estimator because it possesses several desirable properties under certain assumptions.

The consistency of the OLS estimator implies that as the sample size increases, the estimated coefficients converge to the true population values. This property holds when the regressors (independent variables) are exogenous, meaning that they are not correlated with the error term in the regression equation. Another important assumption for OLS is that the errors are homoscedastic, which means that the variances of the errors are constant across all levels of the independent variables. Additionally, the errors should be serially uncorrelated, meaning that there is no autocorrelation among the errors.

Under these assumptions, OLS provides the minimum-variance mean-unbiased estimation, meaning that among all unbiased estimators, it has the smallest variance. This property makes OLS efficient in terms of estimating the true coefficients of the regression model. Furthermore, if the errors in the regression model are assumed to be normally distributed, OLS becomes the maximum likelihood estimator. This assumption allows for various statistical inference procedures, such as hypothesis testing, confidence interval estimation, and prediction interval construction.

By assuming normality of the errors, one can perform hypothesis tests on the estimated coefficients to determine their statistical significance. Confidence intervals indicate a range of possible values for the true population coefficients, whereas prediction intervals estimate the range in which future observations are expected to lie. It is important to note that the assumptions of OLS should be carefully evaluated in practice. Violations of these assumptions, such as heteroscedasticity or autocorrelation, may lead to biased or inefficient estimates. In such cases, alternative regression methods or adjustments, like robust standard errors or autoregressive models, may be employed.

In this chapter, the focus is on examining the hypothesis of normality in the distribution of variables and translating it into a more realistic representation. Instead of assuming a strict normal distribution, a fuzzy version of the normal distribution is adopted, referred to as the Gaussian membership function. This approach allows for a more flexible and nuanced modeling of uncertainty.

The concept of fuzzy mathematical modeling is then introduced to accommodate the fuzzy predictor and fuzzy parameters in a linear regression framework. Fuzzy predictors refer to variables or inputs that possess a degree of fuzziness or uncertainty, which is represented using fuzzy sets. Similarly, fuzzy parameters refer to the coefficients or parameters of the linear regression model that are described with fuzzy sets.

By incorporating fuzzy sets into the linear regression model, it becomes possible to capture and represent uncertainty in both the predictors and the parameters. This fuzzy mathematical modeling approach enables a more realistic representation of real-world scenarios where variables and parameters may not be precisely defined.

The construction of a fuzzy mathematical model involves defining the fuzzy membership functions for the fuzzy predictor and fuzzy parameters. These membership functions assign degrees of membership or likelihood to each value within their respective fuzzy sets.

Once the fuzzy mathematical model is established, various operations and techniques from fuzzy logic and fuzzy set theory can be applied. This includes fuzzy inference systems, fuzzy rule-based systems, and fuzzy reasoning to make predictions and draw conclusions based on the fuzzy inputs and fuzzy parameters. By adopting this fuzzy mathematical modeling approach, the linear regression model can handle situations where the data or variables exhibit a certain level of imprecision or uncertainty. It provides a framework for capturing and representing the inherent vagueness and ambiguity that may be present in real-world data and relationships.

It's worth noting that this fuzzy approach to linear regression is just one of the many ways to account for uncertainty and imprecision in modeling. Other techniques, such as Bayesian inference or robust regression, may also be used depending on the specific context and objectives of the analysis. The Gaussian membership function can be considered as a deduction of real behavior and can indeed be viewed as a generalization of the normal distribution for classical sets. By using the Gaussian membership function, we can capture the degree of truth or membership to a fuzzy set, which is often associated with the concept of probability. However, it's important to note that the degree of truth or membership is conceptually distinct from probability, although they can sometimes be confused.

The Gaussian membership function quantifies the extent to which an element belongs to a fuzzy set, assigning values between 0 and 1. A value of 1 signifies full membership or truth, while 0 represents no membership or falsity. The shape of the Gaussian membership function resembles the bell curve of the normal distribution, which is why it is often used to model fuzzy sets. In the context of fuzzy modeling, the approach described here aims to capture natural behaviors observed in nature. By employing fuzzy sets and fuzzy logic, we can model and analyze complex and uncertain systems, such as human decision-making, environmental processes, or economic behaviors.

Fuzzy modeling allows for a more nuanced representation of real-world phenomena, where uncertainties, imprecisions, and vagueness are inherent. It provides a means to describe and analyze systems that cannot be easily captured by crisp or binary representations. Fuzzy modeling is widely applied in various domains, such as control systems, pattern recognition, artificial intelligence, and expert systems, to capture natural behaviors observed in nature. This approach enables the integration of expert knowledge and linguistic variables for representing and reasoning about complex systems.

Nevertheless, it is essential to recognize that fuzzy modeling is just one approach among many in the broader field of modeling and analysis. Depending on the specific context and objectives, other modeling techniques, such as statistical models, machine learning algorithms, or other forms of mathematical modeling, may also be employed. The choice of modeling approach depends on the nature of the problem, the available data, and the desired level of abstraction and interpretability. In order to provide an empirical illustration of the proposed approach, let's assume a normal behavior for the poverty rate variable. We will estimate a simple fuzzy regression model that links the poverty rate to the unemployment rate using data from a group of MENA (Middle East and North Africa) countries.

The first step in the analysis is to gather the necessary data on the poverty rate and unemployment rate for the selected MENA countries. The data should be collected for the same time period to ensure consistency. Once the data is collected, we can proceed with estimating the fuzzy regression model. The fuzzy regression model allows us to capture the uncertainty and imprecision in the relationship between the poverty rate and the unemployment rate.

In this case, we assume a normal behavior for the poverty rate variable. This means that we can use the Gaussian membership function to represent the fuzzy sets associated with the poverty rate. The membership function will assign degrees of truth or membership to different poverty rate values based on their similarity to the assumed normal behavior. Next, we estimate the fuzzy regression model by linking the poverty rate, represented using the Gaussian membership function, to the unemployment rate. This can be done using techniques such as fuzzy least squares or fuzzy optimization.

The estimated fuzzy regression model will provide insights into the relationship between the poverty rate and the unemployment rate, considering the fuzzy nature of the variables and the uncertainty associated with them. It's important to interpret the results of the fuzzy regression model in the context of fuzzy logic and fuzzy sets. The estimated fuzzy parameters can provide information about the strength and direction of the relationship between the poverty rate and the unemployment rate, considering the inherent uncertainty and imprecision in the data. By conducting this empirical illustration, we can gain a deeper understanding of the relationship between the poverty rate and the unemployment rate in the selected MENA countries, while accounting for the fuzziness and uncertainty in the data. This approach allows for a more realistic and nuanced analysis of the relationship, considering the complex dynamics of economic and social variables.

This chapter is structured as follows: Sect. 15.2 presents numerical illustrations on fuzzy number generation. Section 15.3 highlights the distinction between the normal distribution and the Gaussian membership function. Section 15.4 introduces an alternative estimation method for a linear regression model with fuzzy predictors and fuzzy parameters. In this section, the construction of estimators for a linear fuzzy regression model is presented, and the consistency of these estimators is established. Section 15.5 discusses relationship between unemployment and poverty from economics point of view. Section 15.6 presents an application of the proposed method using data from a group of MENA countries in 2015 and 2020–2024 to estimate the relationship between employment and poverty. This application serves to justify the effectiveness of the model. Finally, Sect. 15.7 provides concluding remarks.

15.2 Numerical Illustrations on Fuzzy Number Generation

15.2.1 Fuzzy Number

A fuzzy number is a special case of a fuzzy set that represents imprecise numerical values. It is characterized by a membership function, which maps each number in the domain to a value between 0 and 1. This value (called the membership grade) indicates the degree of membership of the element in the fuzzy number. Key properties of fuzzy numbers are:

Convexity: The membership function must be convex.
Normality: The maximum membership value should be 1.
Piecewise Continuity: The membership function is typically continuous or piecewise continuous.
Support: The set of elements with non-zero membership.

15.2.2 Numerical Illustrations on Fuzzy Number Generation

Example 1: Generating a Triangular Fuzzy Number (TFN)
A triangular fuzzy number (TFN) is represented as $\{A\} = (a, b, c)$, where the parameters *a, b,* and *c* represent the lower limit, the peak point (most likely value), and the upper limit, respectively (Fig. 15.1).

Generate a TFN where the lower limit is 3, the peak point is 5, and the upper limit is 8 (Fig. 15.1). The membership function $\mu_A(x)$ is defined as:

$$\mu_A(x_i) = \begin{cases} 0 & x_i \le a \\ \dfrac{1}{b-a}x_i - \dfrac{a}{b-a} & a \le x_i < b \\ \dfrac{-1}{c-b}x_i + \dfrac{c}{c-b} & b \le x_i < c \\ 0 & x_i \ge c \end{cases}.$$

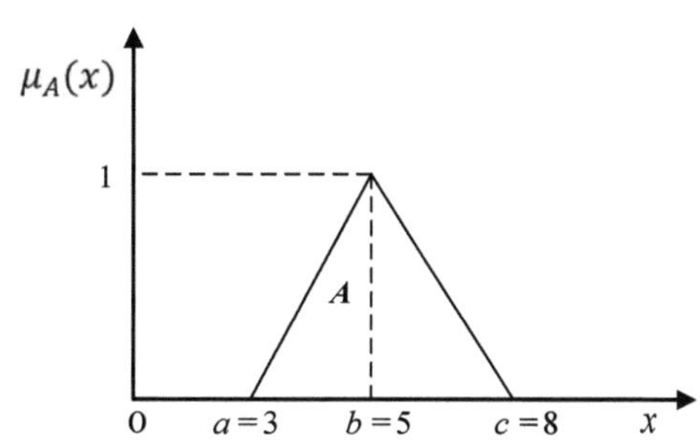

Fig. 15.1 Triangular fuzzy number. Source: Belhadj (2022)

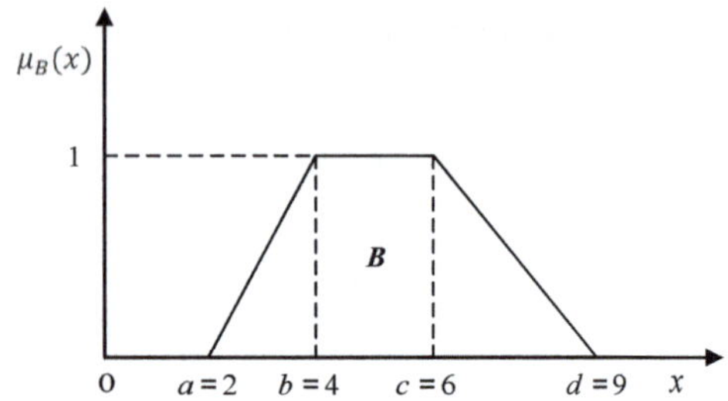

Fig. 15.2 Trapezoidal fuzzy number. Source: Belhadj (2022)

Example Calculation: For $x = 4$, $\mu_A(4) = 0, 5$ and for $x = 6$, $\mu_A(6) = 0, 67$.

Example 2: Generating a Trapezoidal Fuzzy Number (TrFN)

A trapezoidal fuzzy number (TrFN) is represented as $\{B\} = (a, b, c, d)$, where the parameters *a, b, c,* and *d* represent the lower limit, left shoulder, right shoulder, and upper limit, respectively (Fig. 15.2).

Generate a TrFN where the lower limit is 2, the left shoulder is 4, the right shoulder is 6, and the upper limit is 9 (Fig. 15.2). The membership function is $\mu_B(x)$ defined as:

$$\mu_B(x_i) = \begin{cases} 0 & x_i \leq a \\ \dfrac{1}{b-a}x_i - \dfrac{a}{b-a} & a \leq x_i < b \\ 1\, b \leq x_i < c & \\ \dfrac{-1}{d-c}x_i + \dfrac{d}{d-c} & c \leq x_i < d \\ 0 & x_i \geq d \end{cases}.$$

Example Calculation: For $x = 3$, $\mu_A(3) = 0, 5$ and for $x = 8$, $\mu_A(8) = 0, 33$.

Example 3: Generating a Gaussian Fuzzy Number

A Gaussian fuzzy number is characterized by its mean m and standard deviation σ. The membership function is defined as in Fig. 15.3.

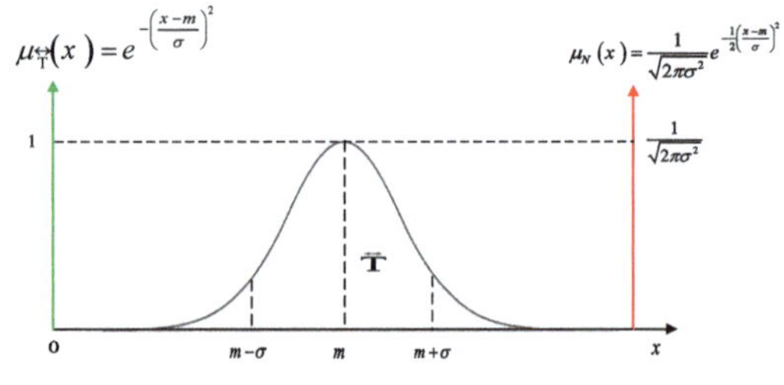

Fig. 15.3 Gaussian membership function. Source: Belhadj (2022)

Generate a Gaussian fuzzy number $\{\overleftrightarrow{T}\}$ with a mean of *0* and a standard deviation of *1*.

Example Calculation: For $x = 0$, $\mu_{\overleftrightarrow{T}}(0) = 1$; for $x = 1$, $\mu_{\overleftrightarrow{T}}(1) \cong 0,607$ and for $x = -1$, $\mu_{\overleftrightarrow{T}}(1) \cong 0,607$. This Gaussian fuzzy number is symmetric about the mean.

15.3 Gaussian Membership Function

The normal distribution, commonly referred to as the Gaussian distribution, plays a crucial role in various analytic studies and is widely utilized across different fields. Its significance stems from several key properties that make it a fundamental tool for statistical analysis. One important characteristic of the normal distribution is that it has the maximum entropy among all continuous distributions with a specified mean and variance. This means that when the mean and variance are fixed, the normal distribution represents the distribution with the greatest uncertainty or randomness. It provides a natural choice when the exact form of the distribution is unknown but the mean and variance are known or estimated.

Furthermore, the normal distribution is distinguished by the fact that all cumulants beyond the mean and variance are equal to zero. Cumulants are statistical measures that describe the shape and behavior of a distribution. The normal distribution's cumulants beyond the second order vanish, which sets it apart from other distributions. This property simplifies calculations and allows for convenient analytical solutions in various statistical problems.

An important feature of the normal distribution is that the mean and variance derived from a set of independent samples are statistically independent. This property remains valid regardless of the sample size. In practical applications, this means that the mean and variance can be estimated separately without influencing each other, simplifying statistical inference and estimation processes.

The normal distribution belongs to the class of elliptical distributions, which encompasses a family of distributions that share certain similarities in their shapes and properties. The distribution is symmetric about its mean, meaning that it is evenly balanced around the central value. Additionally, the normal distribution has non-zero probability density over the entire real line, making it suitable for modeling continuous variables. Analytical tractability is another advantage of the normal distribution. It has a well-defined probability density function (PDF), allowing for explicit mathematical expressions. This analytical form enables the derivation of various statistical methods and facilitates the development of closed-form solutions for statistical inference. For example, least-squares parameter fitting, a widely used technique for estimating model parameters, can be derived analytically when the underlying variables are assumed to follow a normal distribution.

The normal distribution's importance in analytic studies arises therefore from its maximum entropy property, its zero cumulants beyond the mean and variance, the independence between the mean and variance, its membership in the class of elliptical distributions, its symmetry, non-zero probability density over the entire real line, and its analytical tractability. These characteristics make the normal distribution a powerful and versatile tool for statistical analysis and modeling in a wide range of applications.

Let $\mu_N(x)$ the probability density function of the normal distribution Fig. 15.1. For an element x of X, $\mu_N(x) = \frac{1}{\sqrt{2\pi\sigma^2}} e^{-\frac{1}{2}\left(\frac{x-m}{\sigma}\right)^2}$ quantifies the probability of the element x to the series N, where m is the mean and also its median and mode, while the parameter σ is its standard deviation. The probability density function is such as $0 < \mu_N(x) \leq \frac{1}{\sqrt{2\pi\sigma^2}} < 1$ (Fig. 15.3, red axis).

In fuzzy logic, the probability density function becomes $\mu_{\overset{\leftrightarrow}{\mathrm{T}}}(x) = e^{-\left(\frac{x-m}{\sigma}\right)^2}$. The parameter m is always the mean, the median, and the mode, and σ is the standard deviation. $\mu_{\overset{\leftrightarrow}{\mathrm{T}}}(x)$ is called Gaussian membership function. It represents membership in vaguely defined sets, not likelihood of some event or condition. For an element x of X, $0 < \mu_{\overset{\leftrightarrow}{\mathrm{T}}}(x) \leq 1$ quantifies the grade of membership of the element x to the fuzzy set $\overset{\leftrightarrow}{\mathrm{T}}$ (Fig. 15.3, green axis). A value that tends to 0 means that x gets out of the fuzzy set; the value 1 means that x is fully a member of the fuzzy set. The values between 0 and 1 characterize fuzzy members, which belong to the fuzzy set only partially.

15.4 Linear Regression with Fuzzy Predictor and Fuzzy Parameters

The fuzzy linear regression model, first proposed by Tanaka et al. (1982), has played a key role in fuzzy regression analysis. Their groundbreaking research framed linear regression models as mathematical programming problems. Over time, this approach has evolved, giving rise to numerous studies that generally fall into two main categories: the possibilistic approach, which relies on linear programming techniques, and the fuzzy least-squares method.

The possibilistic approach, building upon Tanaka's initial work, has been further developed by various researchers. Tanaka (1987) and in collaboration with Ishibuchi (1991) continued to refine and expand upon their original ideas. Diamond and Tanaka (1999), contributed to the field by exploring the application of fuzzy regression models within the framework of possibility theory. Additionally, Hong and Yi (2003), and Hong et al. (2004) further extended the possibilistic approach to fuzzy linear regression modeling.

On the other hand, the fuzzy least-squares methods provide an alternative approach within the fuzzy regression framework. Diamond (1988) proposed a fuzzy least-squares regression model that aimed to minimize the sum of squared residuals. D'Urso and Gastaldi (2000) further enhanced this approach by incorporating fuzzy clustering techniques into the modeling process. D'Urso (2003) introduced a modified version of fuzzy least-squares regression based on the idea of nonadditive fuzzy integrals. Yang and Liu (2003) continued this line of research by considering different fuzzy membership functions in the least-squares framework. Shapiro (2005) provides simple introduction to fuzzy regression models.

More recently, Belhadj and Kaabi (2021), and Bouanani and Belhadj (2022) have contributed to the field by proposing novel fuzzy regression models that incorporate elements of fuzzy logic and fuzzy set theory. Their studies aimed to improve the accuracy and robustness of fuzzy regression modeling by introducing new concepts and techniques.

Overall, the field of fuzzy linear regression modeling has witnessed significant advancements since Tanaka et al.'s seminal work in 1982. The possibilistic approach, based on linear programming methods, and the fuzzy least-squares methods have both played important roles in the development of this field. These approaches, along with subsequent studies by various researchers, have expanded the theoretical foundations and practical applications of fuzzy regression analysis, making it a valuable tool for handling uncertainty and imprecision in regression modeling. What follows next is a description of fuzzy linear regression model.

Fuzzy regression is not a statistical method (Kim and Chen 1997). When the model is indefinite, the relationships between model parameters are unclear, the sample size is small, or the data is hierarchically structured, fuzzy regression offers an alternative to statistical regression. When there is some degree of vagueness (e.g., input data could be imprecise), fuzzy regression analysis provides a fuzzy functional relationship between the independent and dependent variables. In some cases, we may need to consider that the relationship expressed as $y_i = ax_i + b + \varepsilon_i$ may be fuzzy.

Indeed, three cases are possible:

The case where the predictor variable is fuzzy, but the parameters are crisp,

$$\tilde{y}_i = a + b\tilde{x}_i + \tilde{\varepsilon}_i, \tag{15.1}$$

the case of a crisp predictor and fuzzy parameters,

$$\tilde{y}_i = \tilde{a} + \tilde{b}x_i + \tilde{\varepsilon}_i, \tag{15.2}$$

and finally, the case of a fuzzy predictor and fuzzy parameters.

$$\tilde{y}_i = \tilde{a} + \tilde{b}\tilde{x}_i + \tilde{\varepsilon}_i \tag{15.3}$$

where $\tilde{y}_i$'s are the *ith*, $i = 1, ..., n$, fuzzy responses.

In this section, we estimate the parameters of model (15.3) when the predictor and unknown parameters are Gaussian fuzzy numbers. The Gaussian fuzzy numbers curve represents the modification of an interval of confidence from 0 to 1 having a deflection in its slope as shown in Fig. 15.2). This curve is the Gaussian membership function.

Let $\tilde{S} = (m - \sigma, m + \sigma)$ a Gaussian fuzzy number where $m - \sigma$ and $m + \sigma$ are the left and right "end" points of the corresponding bell (Fig. 15.4).

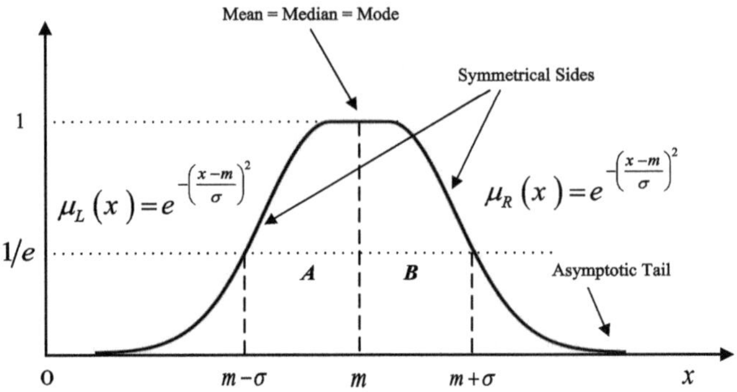

Fig. 15.4 Gaussian fuzzy number. Source: This study

The membership of fuzzy response $\tilde{y}_i$, $\forall i = 1, \ldots, n$ of model (15.3) is as follows:

$$\mu_{\tilde{y}_i}(u_i)=\begin{cases} e^{-\left(\frac{u_i-m}{\sigma}\right)^2} & -\infty < u_i < m \\ e^{-\left(\frac{u_i-m}{\sigma}\right)^2} & m \le u_i < +\infty \end{cases}, \tag{15.4}$$

We suppose that parameters $\tilde{a}$ and $\tilde{b}$, as well as the predictor $\tilde{x}$ of model (15.3), are Gaussian fuzzy numbers, where

$\tilde{a} = (a - \sigma_a, a + \sigma_a)$, $\tilde{b} = (b - \sigma_b, b + \sigma_b)$, and $\tilde{x} = (x - \sigma_x, x + \sigma_x)$.

Then (Kaufmann and Gupta 1991):

$$\tilde{a}(+)\tilde{b}\tilde{x} = ((a - \sigma_a) + (b - \sigma_b)(x - \sigma_x), (a + \sigma_a) + (b + \sigma_b)(x + \sigma_x)).$$

From which, we have:

$\forall u \in R$:

$$\begin{aligned} \mu_{\tilde{a}(+)\tilde{b}x_i}(u_i) &= e^{-\left(\frac{u_i-((a-\sigma_a)+(b-\sigma_b)(x_i-\sigma_x))}{\sigma_a+\sigma_b+\sigma_x}\right)^2} \quad -\infty < u_i < (a - \sigma_a) + (b - \sigma_b)(x_i - \sigma_x) \\ &= e^{-\left(\frac{u_i-((a+\sigma_a)+(b+\sigma_b)(x_i+\sigma_x))}{\sigma_a+\sigma_b+\sigma_x}\right)^2} \quad (a + \sigma_a) + (b + \sigma_b)(x_i + \sigma_x) \le u_i < +\infty \end{aligned} \tag{15.5}$$

To estimate the parameters of model (15.3), we propose two fuzzy methods.

15.4.1 Error Minimization Fuzzy Method

Starting from the model (15.3), we can write the membership function of the error $\tilde{\varepsilon}_i$, $\forall i = 1, \ldots, n$ as follows:

$$\mu_{\tilde{\varepsilon}_i}(\varepsilon_i) = \mu_{\tilde{y}_i(-)\,\tilde{a}+\tilde{b}\tilde{x}_i}(\varepsilon_i), \tag{15.6}$$

The term to the right of Eq. (15.6) is a subtraction of fuzzy numbers which is written as follows:

$$\mu_{\tilde{y}_i(-)\,\tilde{a}+\tilde{b}\tilde{x}_i}(\varepsilon_i) = \bigvee_{\varepsilon_i}\left(\mu_{\tilde{a}+\tilde{b}\tilde{x}_i}(a+bx_i) \wedge \mu_{\tilde{y}_i}(y_i)\right), \tag{15.7}$$

which corresponds to

$$\begin{aligned}&\tilde{y}_i(-)\tilde{a}+\tilde{b}\tilde{x}_i\\ &\quad = (m-\sigma-(a+\sigma_a)-(b+\sigma_b)(x_i+\sigma_x), m+\sigma-(a-\sigma_a)-(b-\sigma_b)(x_i-\sigma_x)).\end{aligned}$$

Now using Eq. (15.4), let $\forall\alpha \in [0, 1]$ $\alpha = \left(e^{-\left(\frac{u_i-m}{\sigma}\right)^2}\right)^{(\alpha)}$

$$\alpha = \left(e^{-\left(\frac{u_i-m}{\sigma}\right)^2}\right)^{(\alpha)}$$

From which

$$\begin{aligned}\underset{i\alpha}{\breve{y}} &= \left(y_{1i}^{(\alpha)}, y_{2i}^{(\alpha)}\right)\\ &= \left[\alpha - e^{-\left(\frac{u_i-m}{\sigma}\right)^2}, \alpha - e^{-\left(\frac{u_i-m}{\sigma}\right)^2}\right]\end{aligned} \tag{15.8}$$

Now using Eq. (15.5), we obtain $\alpha = \left(e^{-\left(\frac{u_i-((a-\sigma_a)+(b-\sigma_b)(x_i-\sigma_x))}{\sigma_a+\sigma_b+\sigma_x}\right)^2}\right)^{(\alpha)}$

$$\alpha = \left(e^{-\left(\frac{u_i-((a+\sigma_a)+(b+\sigma_b)(x_i+\sigma_x))}{\sigma_a+\sigma_b+\sigma_x}\right)^2}\right)^{(\alpha)}.$$

From which

$$\begin{aligned}\left(\tilde{a}+\tilde{b}x_i\right)_\alpha &= \left(\nu_{1i}^{(\alpha)}, \nu_{2i}^{(\alpha)}\right)\\ &= \left[-\alpha + e^{-\left(\frac{u_i-((a-\sigma_a)+(b-\sigma_b)(x_i-\sigma_x))}{\sigma_a+\sigma_b+\sigma_x}\right)^2}, -\alpha + e^{-\left(\frac{u_i-((a+\sigma_a)+(b+\sigma_b)(x_i+\sigma_x))}{\sigma_a+\sigma_b+\sigma_x}\right)^2}\right].\end{aligned} \tag{15.9}$$

We must then have

$$\breve{y}_{i\alpha}(-)\left(\breve{a}+\breve{b}\,\breve{x}_i\right)_\alpha = \left(y_{1i}^{(\alpha)}, y_{2i}^{(\alpha)}\right)(-)\left(\nu_{1i}^{(\alpha)}, \nu_{2i}^{(\alpha)}\right)$$

$$= \left[y_{1i}^{(\alpha)} - \nu_{2i}^{(\alpha)}, y_{2i}^{(\alpha)} - \nu_{1i}^{(\alpha)}\right].$$

Thus, if we define $y_{1i}^{(\alpha)} - \nu_{2i}^{(\alpha)} = \alpha - e^{-\left(\frac{u_i - m}{\sigma}\right)^2} + \alpha - e^{-\left(\frac{u_i - ((a+\sigma_a)+(b+\sigma_b)(x_i+\sigma_x))}{\sigma_a+\sigma_b+\sigma_x}\right)^2}$

$$y_{2i}^{(\alpha)} - \nu_{1i}^{(\alpha)} = \alpha - e^{-\left(\frac{u_i - m}{\sigma}\right)^2} + \alpha - e^{-\left(\frac{u_i - ((a-\sigma_a)+(b-\sigma_b)(x_i-\sigma_x))}{\sigma_a+\sigma_b+\sigma_x}\right)^2}.$$

From which, we have
$\forall u \in R$:

$$\mu_{\tilde{\varepsilon}}(u_i) = e^{-\left(\frac{u_i - (m-\sigma-(a-\sigma_a)-(b-\sigma_b)(x_i-\sigma_x))}{\sigma+\sigma_a+\sigma_b+\sigma_x}\right)^2} \quad -\infty < u_i < m - \sigma - (a - \sigma_a) - (b - \sigma_b)(x_i - \sigma_x)$$

$$= e^{-\left(\frac{u_i - (m+\sigma-(a+\sigma_a)-(b+\sigma_b)(x_i+\sigma_x))}{\sigma+\sigma_a+\sigma_b+\sigma_x}\right)^2} \quad m + \sigma - (a + \sigma_a) - (b + \sigma_b)(x_i + \sigma_x) \le u_i < +\infty. \tag{15.10}$$

The estimate of a and b of model (15.3) consists in solving the following minimization problem:

$$\min \mu_{\tilde{\varepsilon}_i}\left(\tilde{a}, \tilde{b}\right). \tag{15.11}$$

Using (15.7), (15.11) becomes:

$$\min \mu_{\tilde{\varepsilon}_i}\left(\tilde{a}, \tilde{b}\right) = \min_{\varepsilon_i} \vee\left(\mu_{\tilde{a}+\tilde{b}x_i}(u_i) \wedge \mu_{\tilde{y}_i}(u_i)\right) \tag{15.12}$$

We show that to solve (15.12) is to solve the Eq. (15.13)

$$\partial/\partial a, \partial b \frac{\sum_{i=1}^{n} (u_i - (m + \sigma - (a + \sigma_a) - (b + \sigma_b)(x_i + \sigma_x)))^2}{(\sigma + \sigma_a + \sigma_b + \sigma_x)^2} = 0. \tag{15.13}$$

We prove that error's membership function $\left(\mu_{\tilde{\varepsilon}_i}\left(\tilde{a}, \tilde{b}\right)\right)$ of (15.11) is also the common areas of Fig. 15.3 given by the difference between the membership function of the observed value $\left(\mu_{\tilde{y}_i}(u_i)\right)$ and the membership function of estimated value of $y_i\left(\mu_{\tilde{a}+\tilde{b}x_i}(u_i)\right)$. Therefore, solving (15.12) is equivalent, amounts, to minimizing the common areas of Fig. 15.3; hence the following method.

15.4.2 Dissemblance Method

Starting from Eqs. (15.8) and (15.9), we define $\forall i = 1, \ldots, n$

$$A = \left[y_{1i}, y_{2i}\right] \text{and}\, B = [\nu_{1i}, \nu_{2i}].$$

Or, from (15.4), $\forall \alpha \in [0, 1]$ $\alpha = e^{-\left(\frac{u_i - m + \sigma}{\sigma}\right)^2} \Rightarrow u_i = \sigma\sqrt{-\log\alpha} + m - \sigma$

$$\alpha = e^{-\left(\frac{u_i - m - \sigma}{\sigma}\right)^2} \Rightarrow u_i = \sigma\sqrt{-\log\alpha} + m + \sigma$$

$$A_\alpha = \left[\sigma\sqrt{-\log\alpha} + m, \sigma\sqrt{-\log\alpha} + m + \sigma\right],$$

And, from (15.5),

$$B_\alpha = \begin{bmatrix} ((a - \sigma_a) + (b - \sigma_b)(x_i - \sigma_x)) + (\sigma_a + \sigma_b + \sigma_x)\sqrt{-\log\alpha}, \\ ((a + \sigma_a) + (b + \sigma_b)(x_i + \sigma_x)) + (\sigma_a + \sigma_b + \sigma_x)\sqrt{-\log\alpha} \end{bmatrix}$$

Note that, in the right of Fig. 15.4, A and B intersect at the two points given by (15.14) and (15.15):

$$\sigma\sqrt{-\log\alpha} + m + \sigma = ((a + \sigma_a) + (b + \sigma_b)(x_i + \sigma_x)) + (\sigma_a + \sigma_b + \sigma_x)\sqrt{-\log\alpha}$$

gives

$$\alpha_1 = e^{-\left(\frac{((a+\sigma_a)+(b+\sigma_b)(x_i+\sigma_x))-m-\sigma}{(\sigma-(\sigma_a+\sigma_b+\sigma_x))}\right)^2} \tag{15.14}$$

$$m + \sigma = (a + \sigma_a) + (b + \sigma_b)(x_i + \sigma_x)$$

Gives

$$\alpha_2 = 0. \tag{15.15}$$

We then have:

$$\begin{aligned} y_{2i}^{(\alpha)} - v_{2i}^{(\alpha)} &= \sigma\sqrt{-\log\alpha} + m + \sigma - ((a + \sigma_a) + (b + \sigma_b)(x_i + \sigma_x)) \\ &\quad - (\sigma_a + \sigma_b + \sigma_x)\sqrt{-\log\alpha} \\ &= m + \sigma - ((a + \sigma_a) + (b + \sigma_b)(x_i + \sigma_x)) \\ &\quad + (\sigma - (\sigma_a + \sigma_b + \sigma_x))\sqrt{-\log\alpha}. \end{aligned}$$

If we proceed to integrate from $\alpha = 0$ to $\alpha = 1$, we obtain a distance $\aleph(A, B) = \int_{\alpha=0}^{1} \aleph(A_\alpha, B_\alpha)d\alpha$.

$\aleph(A, B)$ is called the dissemblance index of A and B. It is a distance between two fuzzy numbers A and B.

$$\aleph(A, B) = \int_{\alpha=0}^{\alpha_1} \begin{bmatrix} m + \sigma - ((a + \sigma_a) + (b + \sigma_b)(x_i + \sigma_x)) \\ +(\sigma - (\sigma_a + \sigma_b + \sigma_x))\sqrt{-\log\alpha} \end{bmatrix} d\alpha$$

$$+\int_{\alpha_1}^{1}\begin{bmatrix} m+\sigma-((a+\sigma_a)+(b+\sigma_b)(x_i+\sigma_x)) \\ +(\sigma-(\sigma_a+\sigma_b+\sigma_x))\sqrt{-\log\alpha} \end{bmatrix}d\alpha. \tag{15.16}$$

The calculation of (15.16) shows:

$$\aleph(A,B)=[m+\sigma-((a+\sigma_a)+(b+\sigma_b)(x_i+\sigma_x))]\left([\alpha]_{\alpha=0}^{\alpha_1}+[\alpha]_{\alpha_1}^{1}\right)$$
$$+(\sigma-(\sigma_a+\sigma_b+\sigma_x))\int_0^1\left[\sqrt{-\log\alpha}\right]d\alpha.$$

And

$$\aleph(A,B)=[m+\sigma-((a+\sigma_a)+(b+\sigma_b)(x_i+\sigma_x))]\left([\alpha]_{\alpha=0}^{\alpha_1}+[\alpha]_{\alpha_1}^{1}\right)$$
$$+\frac{\sqrt{\pi}}{2}(\sigma-(\sigma_a+\sigma_b+\sigma_x)). \tag{15.17}$$

The estimate of a and b of model (15.3) are obtained by solving the following minimization problem:

$$\min\sum_i\aleph^2(A,B)=\min\sum_i\aleph^2(\tilde{a},\tilde{b}). \tag{15.18}$$

The problem (15.18) is equivalent to problem (15.12). The resolution consists in solving the system (15.19).

$$\begin{cases} \dfrac{\partial\aleph^2(\tilde{a},\tilde{b})}{\partial a}=0 \\ \dfrac{\partial\aleph^2(\tilde{a},\tilde{b})}{\partial b}=0 \end{cases}. \tag{15.19}$$

The solution of (15.19) is:

$$\begin{cases} a+b(\bar{X}+\sigma_x)=\dfrac{\sqrt{\pi}}{2}(\sigma-\sigma_a-\sigma_b-\sigma_x)+m+\sigma+n\sigma_a+(\bar{X}+\sigma_x)\sigma_b \\ a(\bar{X}+\sigma_x)+b\overline{(X+\sigma_x)^2}=\left(\dfrac{\sqrt{\pi}}{2}(\sigma-\sigma_a-\sigma_b-\sigma_x)-(m+\sigma)+\sigma_a\right)^{,} \\ (\bar{X}+\sigma_x)+\sigma_b\overline{(X+\sigma_x)^2} \end{cases} \tag{15.20}$$

which gives:

$$\widehat{b}=\frac{(\overline{X}+\sigma_x)(\sigma_b-n\sigma_a)}{\sigma^2_{X+\sigma_x}}+\sigma_b, \tag{15.21}$$

and

$$\widehat{a} = \frac{\sqrt{\pi}}{2}(\sigma - \sigma_a - \sigma_b - \sigma_x) + m + \sigma + n\sigma_a - \frac{\left(\overline{X} + \sigma_x\right)^2(\sigma_b - n\sigma_a)}{\sigma^2_{X+\sigma_x}}, \tag{15.22}$$

where $\overline{X} = \frac{1}{n}\sum_i x_i$, $\overline{(X + \sigma_x)^2} = \frac{1}{n}\sum_i (x_i + \sigma_x)^2$ and $\sigma^2_{X+\sigma_x} = \overline{(X + \sigma_x)^2} - \left(\overline{X} + \sigma_x\right)^2$.

The fuzzy modeling of linear regression, as presented in the chapter, holds potential for addressing the inherent complexity of systems, particularly in domains like social economy, where various uncertain factors exist within the boundary environment of the systems. Given the intricate nature of these systems and the presence of multiple influencing factors, the proposed fuzzy modeling approach can offer valuable insights.

To demonstrate the applicability and benefits of the proposed fuzzy modeling, the authors conducted estimation in the subsequent section of the paper. Specifically, they examined the relationship between employment and poverty in a group of MENA (Middle East and North Africa) countries for the years 2015 and 2020–2024. This empirical analysis aimed to shed light on the often ambiguous relationship between poverty and unemployment, which may vary over time.

15.5 Relationship Between Unemployment and Poverty

Understanding the relationship between unemployment and poverty is crucial in economics as it provides valuable insights into the dynamics of labor markets and the well-being of individuals and societies. We highlight the importance of their relationship through the following discussion.

1. Unemployment and poverty are key indicators of the overall economic health of a nation (Brian et al. 2001). High levels of unemployment often correlate with increased poverty rates, signaling economic challenges such as recession or structural issues within the labor market. Indeed, unemployment and poverty are often regarded as leading indicators, providing early signals of economic distress. When unemployment rates rise significantly, it typically implies a reduction in income levels for a substantial portion of the population, leading to an increase in poverty rates. This scenario is indicative of broader economic challenges. High levels of unemployment coupled with a rise in poverty rates are commonly associated with economic recessions. During a recession, businesses may cut jobs due to reduced consumer spending and investment. The resulting surge in unemployment contributes to an uptick in poverty as households experience a decline

in income. Persistent high levels of unemployment and poverty can signal structural issues within the labor market. These structural problems may include a mismatch between the skills demanded by employers and those possessed by the workforce, inhibiting the efficient functioning of the job market. Analyzing the relationship between unemployment and poverty allows economists to distinguish between cyclical and structural unemployment. Cyclical unemployment is related to economic downturns, while structural unemployment stems from fundamental changes in the economy. Both forms can contribute to increased poverty, but targeted policy responses are required for each. Unemployment directly impacts household income, as individuals lose their primary source of earnings. The resulting reduction in income levels often pushes families below the poverty threshold. The extent to which unemployment influences poverty rates depends on factors such as the duration of joblessness and the availability of social safety nets. Governments and policymakers closely monitor the relationship between unemployment and poverty to formulate effective policy responses. For instance, during times of economic downturn, policymakers may implement stimulus measures, job creation initiatives, or social welfare programs to mitigate the adverse effects of unemployment on poverty. Persistent high levels of unemployment and poverty can pose long-term challenges to economic growth and development. A large pool of unemployed or underemployed individuals hampers human capital development, potentially leading to a less skilled and less competitive workforce, which can have lasting effects on the economy. Nations experiencing high unemployment and poverty rates may face challenges in maintaining global competitiveness. Additionally, the social stability of a nation can be compromised, as high levels of poverty often correlate with increased social unrest and dissatisfaction.

2. Unemployment negatively affects human capital development (Michael and Markus 2022). Prolonged periods of joblessness can lead to skill depreciation and a decline in productivity among the workforce. This, in turn, perpetuates poverty as individuals find it difficult to secure well-paying jobs. Indeed, prolonged unemployment can lead to skill depreciation as individuals may not have the opportunity to apply and enhance their skills in a work environment. Skills that are not regularly used can become outdated or less relevant to evolving industry needs. This can result in a decline in the overall skill level of the workforce. In rapidly evolving industries, extended periods of unemployment can render individuals technologically obsolete. Advancements in technology and changes in industry practices may outpace the skills of those who have been out of the workforce, making it challenging for them to compete for jobs in sectors requiring updated skills. Being out of work for a prolonged period can erode work-related habits and routines. Individuals may lose the discipline, time-management skills, and professional habits developed through regular employment. This erosion of work-related habits can hinder their ability to adapt quickly when re-entering the workforce. Unemployment can lead to a decline in overall productivity among the workforce. When individuals are not engaged in productive activities, their work-related skills and efficiency may diminish. This reduced productivity can have

broader economic implications, impacting the competitiveness and efficiency of the labor market. Prolonged unemployment can lead to stigmatization, making it more difficult for individuals to secure well-paying jobs once they re-enter the job market. Employers may perceive gaps in employment history negatively, and individuals may face challenges in convincing potential employers of their continued relevance and value. Extended periods of joblessness can interrupt career progression. Individuals may miss out on opportunities for promotions, skill advancement, and networking that are essential for climbing the career ladder. This interruption can have lasting effects on earning potential and overall career trajectories. The decline in human capital development due to unemployment contributes to a cycle of poverty. Individuals facing difficulty securing well-paying jobs may find themselves trapped in low-wage or unstable employment, perpetuating a cycle where limited income and opportunities hinder their ability to escape poverty. Prolonged unemployment can result in gaps in education and training. Changes in industry requirements may occur during periods of joblessness, leaving individuals with outdated or inadequate qualifications. This can further impede their ability to access well-paying jobs in sectors demanding contemporary skills and knowledge.

3. Understanding the intricate connection between unemployment and poverty is essential for policymakers to effectively address income inequality (Emilie et al. 2018). During periods of economic downturn, certain segments of the population, particularly those already vulnerable, are often disproportionately impacted, leading to a widening income gap. By delving into this relationship, policymakers can gain valuable insights that inform the development of targeted policies aimed at mitigating inequality. In times of high unemployment, individuals and families may face challenges in meeting their basic needs, such as housing, health care, and education. This economic strain can push people further into poverty, creating a cycle that is difficult to break. As unemployment rates rise, the risk of poverty increases, and the burden falls disproportionately on those who are already marginalized or lack access to resources. Analyzing the link between unemployment and poverty involves examining various factors, including the socioeconomic background of affected individuals, the availability of social safety nets, and the overall economic conditions. Policymakers can utilize this information to design interventions that specifically address the needs of vulnerable groups, ensuring that support reaches those who need it the most. Moreover, understanding this relationship allows policymakers to anticipate and respond to the potential long-term consequences of economic downturns. Persistent unemployment can lead to a generational impact on income and wealth disparities, as individuals who experience prolonged joblessness may face challenges in building financial stability and accessing opportunities for advancement.

In crafting targeted policies, policymakers can consider measures such as investing in education and skills training programs to enhance employability, expanding social welfare programs to provide a safety net for those facing unemployment, and implementing economic stimulus measures to spur job creation. Additionally, addressing systemic issues that contribute to inequality, such as discrimination in the workplace or lack of access to equal opportunities, can be crucial in breaking the cycle of poverty linked to unemployment.

4. High levels of unemployment and poverty not only have profound economic implications but can also be significant contributors to social unrest and instability (Atkinson et al. 1998). Analyzing the intricate relationship between these factors is imperative for policymakers seeking to design effective social safety nets and programs. Such initiatives can play a pivotal role in alleviating economic hardships, fostering social cohesion, and ultimately contributing to overall stability in a society. When large segments of the population experience unemployment and poverty, feelings of frustration, disenchantment, and hopelessness can permeate communities. This can give rise to social unrest as individuals, particularly those facing economic hardship, may become more prone to participating in protests, demonstrations, or other forms of civil disobedience. Understanding the social implications of high unemployment and poverty rates is crucial for policymakers aiming to address not only the economic fallout but also the potential social ramifications. In analyzing this relationship, policymakers should consider the impact of economic disparities on social cohesion. Communities facing widespread unemployment and poverty may witness strained social bonds, increased crime rates, and a heightened sense of inequality. These factors can create a volatile environment that poses challenges to the overall stability of a society. To counteract these potential consequences, policymakers can develop and implement robust social safety nets and targeted programs. Social safety nets, including unemployment benefits, welfare programs, and access to affordable health care, can provide a buffer for individuals and families facing economic challenges. By ensuring that basic needs are met, these safety nets contribute not only to poverty alleviation but also to social stability. Moreover, policymakers can invest in programs that address the root causes of unemployment and poverty. This may involve initiatives focused on education and skills development, job creation, and economic diversification. By empowering individuals with the tools and opportunities to escape poverty, policymakers can contribute to a more stable and harmonious society. In the context of global interconnectedness, addressing high levels of unemployment and poverty is not only a matter of domestic concern but also a crucial element of international stability. Economic hardships can contribute to migration flows, exacerbate geopolitical tensions, and impact global security. Therefore, a comprehensive understanding of the relationship between unemployment, poverty, and social stability is essential for crafting policies that promote resilience and well-being on a broader scale.
5. Governments play a crucial role in shaping economic policies aimed at tackling the twin challenges of unemployment and poverty (Stiglitz 2012). Understanding the intricate relationship between these two factors is essential for policymakers to craft targeted interventions that address the root causes and dynamics at play. By employing an evidence-based approach, policymakers can develop a nuanced understanding of the factors contributing to unemployment and poverty, allowing for the creation of more effective and tailored solutions. Analyzing the interplay between unemployment and poverty is a multifaceted endeavor. Unemployment

often serves as a key driver of poverty, as individuals without stable employment face challenges in meeting their basic needs and achieving economic security. At the same time, poverty can perpetuate cycles of unemployment, as individuals trapped in poverty may lack the resources and opportunities necessary to acquire the skills and education needed to secure gainful employment. Policymakers must consider various factors influencing unemployment, such as technological advancements, shifts in market demand, and global economic trends. Additionally, understanding the demographic and regional variations in unemployment rates is crucial for devising policies that address specific needs and challenges faced by different segments of the population. Targeted interventions can take various forms, including education and skills training programs, job placement services, and initiatives to promote entrepreneurship. By tailoring these interventions to the specific needs of vulnerable populations, policymakers can maximize their impact and address the underlying causes of unemployment and poverty. The effectiveness of economic policies designed to alleviate unemployment and poverty must be regularly assessed through rigorous evaluation mechanisms. This involves monitoring key indicators, such as employment rates, income distribution, and access to education and health care. By continuously evaluating the impact of policies, governments can make informed adjustments, ensuring that resources are allocated efficiently and that interventions remain relevant and responsive to evolving economic conditions. An evidence-based approach not only provides a solid foundation for policy formulation but also enhances transparency and accountability in governance. Stakeholders, including citizens and international organizations, can assess the rationale behind policy decisions and hold governments accountable for their outcomes. This transparency fosters public trust and confidence in the government's ability to address pressing economic challenges.

6. Unemployed individuals typically have lower purchasing power, impacting consumer spending (Blanchard and Johnson 2013). Analyzing the relationship between unemployment and poverty is essential for understanding the potential drag on economic growth due to reduced consumer demand. Consumer spending is a key driver of economic growth in many countries, constituting a substantial portion of the Gross Domestic Product (GDP). When a significant portion of the population is unemployed, it directly translates into a reduction in consumer spending. This decline in spending can create a negative feedback loop, as decreased demand for goods and services may lead to further job losses in various sectors, exacerbating the initial problem of unemployment. Analyzing the relationship between unemployment and poverty becomes crucial in this context, as it helps policymakers grasp the potential drag on economic growth resulting from reduced consumer demand. In times of high unemployment, businesses across industries experience a decrease in sales, prompting them to cut costs, lay off workers, or even shut down operations. This, in turn, contributes to a vicious cycle where unemployment begets more unemployment. The consequences of reduced consumer spending extend beyond the immediate impact on businesses. As businesses struggle, investments and expansion plans are often put

on hold, hindering overall economic development. Additionally, the government may experience a decline in tax revenues due to lower corporate profits and individual incomes, limiting its capacity to fund social programs and public services that are crucial for poverty reduction. Moreover, the interplay between unemployment, reduced consumer spending, and poverty can have lasting effects on social and economic inequality. Individuals and families facing financial hardship may find it challenging to break out of the cycle of poverty, hindering social mobility and perpetuating disparities in society. To address these challenges, policymakers can implement targeted measures to stimulate consumer spending and mitigate the negative impact of unemployment on the economy. This may include fiscal policies such as tax incentives, social safety nets, and direct financial assistance to those most affected by unemployment. Additionally, these efforts to boost job creation through infrastructure projects, workforce development programs, and support for small businesses can contribute to breaking the cycle of reduced consumer spending and unemployment.
7. Nations with high levels of unemployment and poverty may struggle to compete globally. A skilled and employed workforce is essential for economic competitiveness (World Economic Forum 2020). Understanding the relationship between unemployment and poverty enables countries to invest in education and training programs to enhance their human capital.

So, measuring and analyzing the relationship between unemployment and poverty is integral to a comprehensive understanding of economic dynamics. It facilitates the development of targeted policies, helps in the assessment of economic health, and contributes to the creation of a more inclusive and stable society.

By employing the fuzzy modeling approach, the authors aimed to capture the inherent uncertainty and imprecision in the employment-poverty relationship, which can arise due to diverse contextual factors and regional specificities. The use of fuzzy logic and fuzzy set theory allows for the representation and analysis of vague or indeterminate relationships, which traditional deterministic approaches may struggle to capture adequately.

Through their estimation, the authors sought to provide empirical evidence that supports the importance of considering fuzzy modeling in understanding and interpreting the complex dynamics between poverty and unemployment. By accounting for various uncertain factors and the intricate nature of the social economy, the fuzzy modeling approach can potentially offer more nuanced insights and contribute to a deeper understanding of the relationship under investigation.

Overall, the application of the proposed fuzzy modeling approach to examining the employment-poverty relationship in MENA countries in pre-COVID-19 and post-COVID-19 years serves as a justification for the relevance and usefulness of the approach in addressing complex systems and incorporating the influence of uncertain factors. It highlights the potential of fuzzy modeling to provide valuable insights into intricate social and economic phenomena where traditional models may fall short in capturing the full complexity of the relationships involved.

15.6 Analyzing the Employment-Poverty Relationship Using the Dissemblance Method

15.6.1 Key Concepts and Challenges

Unemployment and poverty are two key socioeconomic concepts that are often measured to assess the well-being of a population. Various indicators and methods are used to quantify and analyze these phenomena. Definitions and thresholds for both unemployment and poverty may vary between countries. The quality and accuracy of data collection are crucial for reliable measurements. Social and economic policies can influence these indicators, making it essential to consider contextual factors. Measuring unemployment and poverty involves therefore a combination of quantitative indicators and qualitative considerations to provide a comprehensive understanding of the economic well-being of a population. Here's an overview of the measurement of unemployment and poverty.

15.6.1.1 Measurement of Unemployment

The most commonly used indicator is the unemployment rate. It is calculated as the percentage of the labor force that is unemployed and actively seeking employment. Labor force participation rate measures the proportion of the working-age population that is either employed or actively seeking employment. Structural, frictional, cyclical, and seasonal unemployments are different types that can be assessed separately to understand the underlying causes. Some measures also include underemployment, accounting for individuals working part-time who desire full-time employment. Long-term unemployment indicator looks at the percentage of the unemployed who have been without a job for an extended period.

15.6.1.2 Measurement of Poverty

The poverty line is a threshold income level below which individuals or families are considered to be in poverty. It is often determined based on the cost of a basic basket of goods and services needed for a minimum standard of living. The poverty rate indicates the percentage of the population living below the poverty line. Some measures distinguish between regular poverty and deep poverty, which represents individuals living well below the poverty line. Gini coefficient or other measures of income distribution can be used to assess the degree of income inequality within a society. Multidimensional poverty index (MPI) considers factors beyond income, such as health and education, providing a more comprehensive picture of poverty. Relative poverty measures poverty in relation to the overall distribution of income and standards of living in a society.

15.6.2 Empirical Literature Review on the Relationship Between Unemployment and Poverty

The interconnection between unemployment and poverty has been a subject of extensive empirical research, shedding light on the intricate dynamics that link these two socioeconomic phenomena. Scholars have employed diverse methodologies and examined various aspects to gain a comprehensive understanding of the relationship. This literature review synthesizes key findings from empirical studies conducted in this domain.

Unemployment as a Catalyst for Poverty

Numerous studies have provided robust evidence supporting the idea that unemployment serves as a significant catalyst for poverty. Individuals and households experiencing joblessness often face income loss, leading to a decline in their standard of living. The duration and severity of unemployment are crucial factors influencing the extent of poverty experienced (Grosso and Smith 2012).

Cyclical Nature of the Relationship

The cyclical nature of the relationship between unemployment and poverty has been explored in-depth. Economic downturns and recessions tend to exacerbate both unemployment rates and poverty levels. Conversely, periods of economic growth may alleviate these challenges, emphasizing the importance of macroeconomic factors (Jones and Brown 2005).

Labor Market Dynamics and Skill Mismatch

Researchers have delved into labor market dynamics to understand how structural factors contribute to the unemployment-poverty nexus. Skill mismatches, inadequate training, and shifts in demand for certain skill sets play a pivotal role in determining the vulnerability of individuals to both unemployment and poverty (Garcia and Gomez 2012).

Policy Interventions and Mitigation Strategies

The literature also examines the effectiveness of policy interventions in mitigating the impact of unemployment on poverty. Social safety nets, educational programs, and targeted employment initiatives are among the strategies explored to break the cycle of unemployment-induced poverty (Babatunde et al. 2017).

Geographical Disparities and Urban-Rural Divide

Geographical disparities and the urban-rural divide are additional dimensions considered in the literature. Studies highlight how regional variations in job opportunities and living costs contribute to differential experiences of unemployment and poverty across different locales (Giannakis and Bruggeman 2020).

Gender and Demographic Perspectives

Gender and demographic perspectives have been integrated into the analysis, revealing nuanced insights into how unemployment affects various demographic groups differently. Women, youth, and marginalized populations may face distinct challenges in navigating the intersection of unemployment and poverty (Chen and Williams 2018).

Long-Term Effects and Inter-Generational Transmission

Some studies explore the long-term effects of unemployment on poverty, including its potential for inter-generational transmission. Understanding how early-life exposure to parental unemployment influences future socioeconomic outcomes adds depth to the literature (Johnson et al. 2011).

In conclusion, the empirical literature underscores the multifaceted nature of the relationship between unemployment and poverty. Recognizing the complex interplay between individual, structural, and policy factors is essential for developing targeted interventions and policies aimed at alleviating the adverse consequences of unemployment on poverty.

15.6.3 Estimating the Connection Between Employment and Poverty

15.6.3.1 By OLS Method

The observation that employment and access to decent earning opportunities are essential factors in poverty reduction serves as the foundation for estimating the relationship between the poverty headcount ratio at $1.90 per day and the unemployment rate in 2015 across 20 Middle East and North Africa (MENA) countries. The MENA region is particularly vulnerable to fragility and ranks third among developing regions in terms of extreme poverty. In 2015, 5% of the population lived on less than $1.90 per day, amounting to 18.6 million people living in poverty (World Bank 2015).

The dataset is constructed as a cross-sectional collection of country-level observations obtained from the World Bank's database (Unemployment, total (% of total labor force) (modeled ILO estimate); MENA Economic Monitor Report—October 2015). The year 2015 was selected based on the availability and consistency of data across countries. The estimation of this relationship is carried out using the ordinary least-squares (OLS) method, which provides a standard approach for estimating linear regression models and yields the following results:

1. The OLS method provides estimates for the coefficients of the regression model. In this case, the coefficients indicate the relationship between the poverty headcount ratio at $1.90 and the unemployment rate. The coefficient estimates quantify the expected change in the poverty headcount ratio associated with a one-unit change in the unemployment rate.
2. OLS estimation also allows for assessing the statistical significance of the coefficient estimates. This involves determining whether the estimated relationships are unlikely to have occurred by chance. Typically, statistical significance is assessed using p-values or confidence intervals.
3. OLS estimation provides a measure of how well the estimated model fits the observed data. This is typically evaluated using metrics such as the R-squared value, which indicates the proportion of variation in the dependent variable (poverty headcount ratio) that can be explained by the independent variable (unemployment rate). A higher R-squared value suggests a better fit of the model to the data.

The specific results of the OLS estimation, including the coefficient estimates, their statistical significance, and the goodness-of-fit measures, need to be provided in order to discuss the findings in more detail and draw conclusions about the relationship between the poverty headcount ratio (P) at $1.90 and the unemployment rate (U) in the selected group of MENA countries (the list of countries included in the sample is provided in Table 15.3 of the Appendix) in 2015.

$$\begin{aligned} P_i &= 0.20 - 0.33U_i \\ &\quad (8.03)(-3.10) \\ &\left(R^2 = 0.348;\ i = 1, \ldots, 20\,\text{countries};\ \text{t values in parentheses}\right). \end{aligned} \tag{15.23}$$

Based on the findings of the OLS estimation, a negative linear relationship is observed between the poverty headcount ratio at $1.90 and the unemployment rate in the selected group. This relationship is also illustrated in Fig. 15.5. To further investigate this relationship, we conduct another regression on the same group of MENA countries as in Eq. (15.23), examining the link between the poverty headcount ratio at $2.15 and the unemployment rate. This analysis is based on general estimates for recent years (2020–2024), with the results reported in Eq. (15.24).

Between 2020 and 2024, the MENA region faced severe poverty and food insecurity, with over half the population in nine countries experiencing acute food shortages in 2023. This crisis was driven by conflicts, economic instability, and rising food prices. Analyzing this entire period rather than a single year provides a more accurate picture of unemployment and poverty trends, which fluctuate due to economic crises, government policies, and global events like COVID-19, inflation, and conflicts. A single-year analysis may not accurately capture long-term trends. By examining multiple years, we gain a deeper understanding of structural changes and economic stability. A broader timeframe helps smooth out short-term distortions in data, providing a more balanced perspective. Moreover, a multi-year dataset improves the accuracy of regression models by capturing consistent patterns over

Fig. 15.5 Negative relationship between unemployment rate and poverty rate (2015). Source: This study

time rather than temporary fluctuations. Most global institutions, such as the World Bank, IMF, and OECD, publish economic indicators over multi-year periods. Using a broader timeframe ensures consistency, facilitates data collection, and enhances comparative analysis.

$$P_i = 0.28 - 0.43U_i$$
$$(10.62)(-3.72)$$
$$\left(R^2 = 0.435;\ i = 1, \ldots, 20\,\text{countries};\ \text{t values in parentheses}\right). \tag{15.24}$$

According to this second regression, the relationship between the poverty rate and the unemployment rate still appears to be linear, but it is not strong due to the dispersion of data points (Fig. 15.6). The negative relationship between the poverty headcount ratio and the unemployment rate can be explained by the following factors:

1. Unemployment rate in developing middle-income countries: It is noted that the unemployment rate tends to be higher in developing middle-income countries. These countries often experience challenges in creating sufficient employment opportunities to accommodate their growing workforce. The higher unemployment rates in these countries can contribute to a lower poverty headcount ratio, as individuals who are unemployed may not be included in the calculation of the poverty rate.
2. Poverty rate in developing low-income countries: On the other hand, developing low-income countries typically face higher poverty rates. These countries may have limited resources and face significant socioeconomic challenges, which can

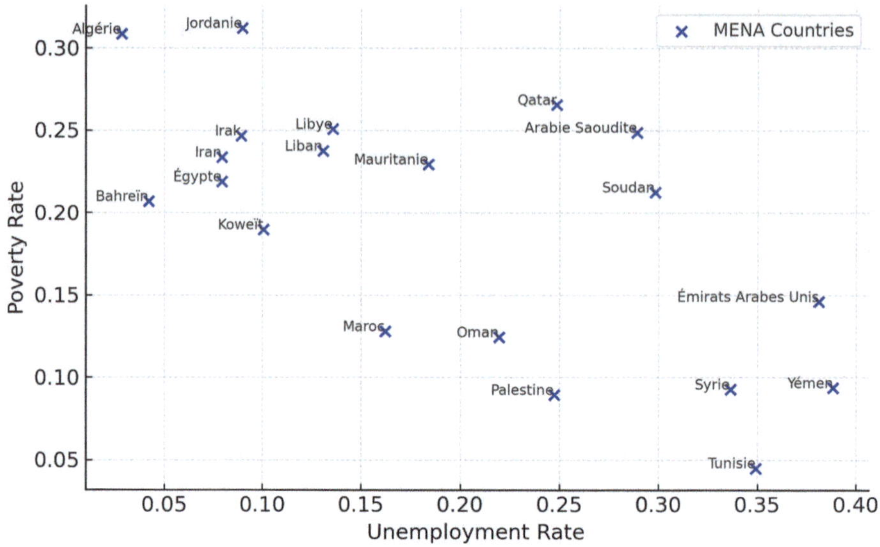

Fig. 15.6 Negative relationship between unemployment rate and poverty rate (2020–2024). Source: This study

result in a larger population living below the poverty line. Despite potentially lower unemployment rates in these countries, the higher poverty rates contribute to an overall negative relationship between the poverty headcount ratio and the unemployment rate.

15.6.3.2 By Dissemblance Method

To further analyze the relationship, the dissemblance method is applied to the same sample as mentioned in equations (15.23) and (15.24). Additionally, solutions (15.21) and (15.22) are calculated, which presumably provide further insights into the relationship between the poverty headcount ratio and the unemployment rate. However, in order to provide more specific details and interpretation of the results obtained through the dissemblance method and the calculated solutions (15.21) and (15.22), it is necessary to provide the actual results and their implications. These results would shed light on any additional findings or patterns observed, enabling a deeper understanding of the relationship between the poverty headcount ratio at $1.90 and the unemployment rate in the selected group of developing countries.

Guide to Software Tools for Calculating Solutions

For computed solutions, various software tools can be employed, such as R. One can use the R package fuzzyreg (Škrabánek and Martínková 2021). Figure 15.7 presents the R codes used for the calculations and graphs presented in this chapter.

```
# Load necessary libraries
library(stats)

# Definition of variables
X_bar <- mean(X) # Calculate the mean of X
sigma_x <- sd(X) # Calculate the standard deviation of X
sigma_a <- sd(a) # Calculate the standard deviation of a
sigma_b <- sd(b) # Calculate the standard deviation of b
sigma_5 <- sd(c) # Another parameter, sigma_5
n <- length(X) # Sample size
m <- mean(c) # Mean of c

# Calculation of b (Equation 21)
b_hat <- (X_bar + sigma_x) * (sigma_5 - n * sigma_a) + sigma_5
b_hat <- b_hat / sigma_x

# Calculation of a (Equation 22)
a_hat <- sqrt(pi / 2) * (sigma_a - sigma_b * sigma_x) + m + sigma +
      n * sigma_a + ((X_bar + sigma_x)^2 * (sigma_5 - n * sigma_a)) / sigma_x
```

Fig. 15.7 R code for solving equations. Source: This study

Results

Table 15.1 indicates that a fuzzy determination coefficient $R^2 = 0.901$, suggesting a good fit of the model to the data. The Fuzzy R^2 (Munish and Kuljit 2016) is a measure of goodness-of-fit in fuzzy regression models, particularly useful when dealing with uncertain or imprecise data. It assesses how well a fuzzy regression model explains the variability in a fuzzy-dependent variable. The general formula for Fuzzy R^2 is $R^2 = 1 - D_E \big/ D_Y$, where D_E represents the total fuzziness of the residuals (errors) in the fuzzy regression model and D_Y represents the total fuzziness of the dependent variable. To determine the fuzziness of output D_Y, it must compute the total fuzziness (spread) in the observed dependent variable Y. This is typically done using a fuzziness measure such as: $D_Y = \sum_{i=1}^{n} \text{spread}(Y_i)$. The spread of a fuzzy number can be computed as the difference between its upper and lower bounds. Residuals in fuzzy regression are the difference between the observed fuzzy outputs Y_i and the predicted fuzzy values $\hat{Y}_i$. The spread of these residuals is calculated similarly: $D_E = \sum_{i=1}^{n} \text{spread}\left(Y_i - \hat{Y}_i\right)$.

We observe that with fuzzy estimation, the R^2 value improves under fuzzy regression compared to OLS, indicating that the fuzzy regression model provides a better fit to the data. This improvement suggests that fuzzy regression captures uncertainties and variations in the dataset more effectively than traditional ordinary least-squares (OLS) regression, which assumes precise relationships between variables. As a result, fuzzy regression may offer more reliable insights in cases where the data contain vagueness or imprecision.

Table 15.1 indicates too that the constant term in the regression model represents the incidence of poverty when the unemployment rate is assumed constant. The presence of a non-zero constant term implies that even when the unemployment rate

```
# Display the results
print(paste("Value of b_hat:", b_hat))
print(paste("Value of a_hat:", a_hat))

# Calculate fuzzy R2 for Equations 21 and 22
residuals_21 <- sigma_5 - n * sigma_a
residuals_22 <- sigma_a - sigma_b * sigma_x

fuzzy_R2_21 <- 1 - var(residuals_21) / var(sigma_5)
fuzzy_R2_22 <- 1 - var(residuals_22) / var(sigma_a)

print(paste("Fuzzy R2 for Equation 21:", fuzzy_R2_21))
print(paste("Fuzzy R2 for Equation 22:", fuzzy_R2_22))

# Ordinary Least Squares (OLS) regression for estimating Equation 23
# Poverty rate as a function of unemployment rate

# Generate example data
data <- data.frame(
  unemployment_rate = rnorm(100, mean = 7, sd = 2),  # Example unemployment rate
  poverty_rate = rnorm(100, mean = 15, sd = 5)  # Example poverty rate
)

# Apply OLS regression
ols_model <- lm(poverty_rate ~ unemployment_rate, data = data)

# Summary of the model
summary(ols_model)

# Calculate R2 for Equation 23
R2_23 <- summary(ols_model)$r.squared
print(paste("R2 for Equation 23:", R2_23))

# Visualization of the regression
plot(data$unemployment_rate, data$poverty_rate, main = "OLS Regression: Poverty Rate vs.
Unemployment Rate",
     xlab = "Unemployment Rate (%)", ylab = "Poverty Rate (%)", pch = 19)
abline(ols_model, col = "blue", lwd = 2)
```

Fig. 15.7 (continued)

Table 15.1 Employment-poverty relationship using the dissemblance index method 2015

	a	b	Fuzzy R^2
The left-Gaussian-fuzzy-number-estimates	23.45 (3.06)	−0.58 (−2.03)	0.879
The right-Gaussian-fuzzy-number-estimates	14.06 (3.97)	0.43 (3.77)	0.901

Fuzzy R^2 tries to capture the proportion of variance explained by the model. This is calculated based on the fuzzy membership values, typically involving the use of fuzzy sets to represent the model's predicted values and the actual observed data. Source: This study remains constant, there is still a visible incidence of poverty. This suggests that factors other than the unemployment rate alone contribute to the occurrence of poverty. The constant term captures the baseline level of poverty that exists irrespective of the unemployment rate.

Furthermore, the parameter measuring the effect of unemployment on poverty is not consistently negative, as indicated in Eqs. (15.23) and (15.24) of this chapter. The sign of this parameter depends on the type of fuzzy number used to represent the relationship between poverty and unemployment. In the case of a right-Gaussian fuzzy number, the relationship between poverty and unemployment is positive. This is evident from coefficient $b = 0.43$ in Table 15.1 for the year 2015, which indicates a slightly positive relationship between employment and poverty, with an uncertainty of ± 3.77. Similarly, coefficient $b = 0.49$ in Table 15.2 for the years 2020–2024 also suggests a slightly positive relationship, with an uncertainty of $\pm$ 2.97. These findings imply that an increase in unemployment is associated with a higher incidence of poverty.

However, when a left-Gaussian fuzzy number is used, the relationship between poverty and unemployment becomes negative, suggesting that higher unemployment is associated with a lower incidence of poverty. This is evident from coefficient $b = -0.58$ in Table 15.1 for the year 2015, which indicates a slightly negative relationship between employment and poverty, with an uncertainty of ± 2.03. Similarly, coefficient $b = -0.61$ in Table 15.2 for the years 2020–2024 also suggests a slightly negative relationship, with an uncertainty of ± 3.01.

These findings support the notion that the relationship between poverty and unemployment is not straightforward and can vary depending on different factors and contexts. The complexity of this relationship justifies the ambiguity observed over the years in understanding the exact nature of the association between poverty and unemployment. It is important to note that while being unemployed often leads to a decline in one's standard of living due to the lack of income, it is also possible for

Table 15.2 Employment-poverty relationship using the dissemblance index method 2020–2024

	a	b	Fuzzy R^2
The left-Gaussian-fuzzy-number-estimates	20.92 (3.52)	−0.61 (−3.01)	0.803
The right-Gaussian-fuzzy-number-estimates	12.77 (3.22)	0.49 (2.97)	0.852

The values in parentheses are the fuzzy Student's t-statistics. For symmetric fuzzy numbers (e.g., Gaussian or triangular), a fuzzy coefficient is given as (b, s) where b_i is the central value and s_i is the dispersion (fuzziness); the t-statistic can be computed as: $t = b \div s$. If the t-statistic interval excludes 0, the coefficient remains significant despite fuzziness. If it includes 0, the effect is uncertain. To refine the interpretation, we can use: defuzzification (e.g., taking the central value or a weighted average) or alpha-cut analysis (testing significance at different confidence levels). The computed t-statistic can then be compared to a critical value from the t-distribution for hypothesis testing. Source: This study individuals to be employed and still experience poverty. This highlights the multi-dimensional nature of poverty, where factors such as low wages, inadequate access to resources, and inequality can contribute to poverty even among the employed population.

Overall, the findings underscore the need for a comprehensive understanding of the factors influencing poverty, beyond the sole consideration of the unemployment rate. They emphasize the importance of considering multiple dimensions and contextual factors when analyzing the relationship between poverty and unemployment. The relationship between the incidence of poverty and unemployment is indeed complex and subject to various factors related to labor force dynamics and income statistics. Interpretation of this relationship becomes challenging, especially when considering workers who face employment issues but whose family income does not fall below the poverty level. This highlights the intricacies involved in understanding the link between poverty and unemployment and the need for comprehensive analysis.

In order to accurately depict and analyze such phenomena, it is crucial to employ realistic and reliable methodologies. For instance, when dealing with natural effects like fog, steam, smoke, and fire, it is essential to use appropriate methods for importance sampling. A survey conducted by Raab et al. (2006) provides an overview of the most well-known methods for importance sampling in these scenarios.

In line with this, the approach presented in the current study offers an alternative strategy for importance sampling that can lead to more robust estimators. This approach specifically focuses on the distribution of variables involved, allowing for valid calculations of the relationship between these variables. By considering the underlying distribution of the variables, the proposed approach aims to provide a more accurate representation of the relationship and yield more reliable estimations.

Employing variables distribution in importance sampling can help overcome potential biases or limitations that may arise from other sampling strategies. This approach takes into account the probabilistic nature of the variables, enabling a more comprehensive understanding of their relationship and reducing the potential for errors or misinterpretations. By emphasizing the importance of realistic depictions of phenomena and adopting appropriate sampling strategies, the proposed approach contributes to enhancing the accuracy and reliability of estimations in analyzing the relationship between variables, such as poverty and unemployment.

Overall, the study recognizes the challenges in interpreting the relationship between poverty and unemployment and acknowledges the need for robust methodologies. By focusing on variables distribution and employing an alternative importance sampling strategy, the approach presented in the study aims to provide a more accurate and comprehensive analysis of this relationship, thereby contributing to a better understanding of these complex phenomena.

Indeed, the relationship between unemployment and poverty can exhibit both positive and negative aspects, which can be further understood by considering the distribution of each variable. One possible scenario is the existence of employed individuals who still experience poverty, often referred to as the "working poor." This situation arises when individuals have employment but their income is insufficient to lift them above the poverty line. On the other hand, there can also be instances where individuals are unemployed but not classified as poor. This can occur when other sources of income, such as the earnings of other family members or benefits provided to the unemployed, prevent them from falling below the poverty threshold.

This illustrates that the distribution of each variable, as well as the specific circumstances and dynamics involved, play a crucial role in determining the relationship between unemployment and poverty. The interplay between various factors, such as income sources and household composition, contributes to the complex nature of this relationship.

It is important to consider the measurement and unit of analysis for each variable when examining their relationship. Poverty is commonly measured using income, typically assessed on an annual basis in poverty studies. This provides a broader perspective on the individual or household's overall economic situation. On the other hand, unemployment is a variable that captures labor force status at a specific point in time. It represents an individual's employment status at the time of measurement, which may not fully reflect their employment experience over an entire year.

The temporal mismatch between the measurement of unemployment and income introduces additional complexity to the relationship. For example, an individual may be unemployed at the time of the interview but may have been employed for a significant portion of the year. This highlights the importance of considering the specific circumstances, duration of unemployment, and fluctuations in employment status when analyzing the relationship between unemployment and poverty.

So, the distribution of each variable, the existence of the "working poor," and the complexities introduced by the measurement and unit of analysis highlight the multifaceted nature of the relationship between unemployment and poverty. A comprehensive understanding of these dynamics requires considering various factors, such as income sources, household composition, and the temporal aspects of employment status and income measurement.

Scholars have brought up an important aspect related to the unit of analysis in the context of labor force status and poverty research. In determination of labor force status, the focus is typically on the individual level, considering factors such as employment or unemployment status for each person. On the other hand, poverty research often takes into account the incomes of all members within a family or household.

This distinction in unit of analysis highlights the potential disparity between individual incomes and the poverty status of the families they belong to. This becomes particularly significant in situations where there are dual-earner families, where both partners are employed. In such cases, the incomes of individuals within the family may not accurately reflect the overall economic well-being of the household.

The growth of dual-earner families has further emphasized this gap between individual incomes and the potential poverty status of the families they reside in. While the income earned by each individual may be above the poverty line, the combined income of the family may still fall below the threshold, resulting in a higher risk of poverty for the household as a whole.

This widening gap between individual incomes and family poverty status highlights the need to consider the household as the unit of analysis in poverty research. It acknowledges that poverty is not solely an individual-level phenomenon but is influenced by the economic circumstances of the entire household. By taking into account the income dynamics within households, researchers can gain a more comprehensive understanding of the potential poverty risk faced by families, even when individual incomes may seem sufficient. This recognition is crucial for accurately assessing and addressing poverty, as it provides insights into the economic conditions of households and the resources available to them as a collective unit.

So, the distinction in unit of analysis between labor force status at the individual level and poverty research at the household level underscores the importance of considering the incomes of all members within a family or household. This recognition is especially significant in the context of dual-earner families and helps bridge the gap between individual incomes and the potential poverty status of the families in which they reside.

15.7 Conclusion

In this chapter, a novel method is proposed to address the challenges posed by fuzziness in regression modeling. The method focuses on estimating the parameters of a fuzzy regression model when both the predictor variable and the parameters themselves are fuzzy, utilizing unimodal Gaussian membership functions. The proposed method is designed to capture the inherent fuzziness in the data while maintaining the essence of the classical linear regression process. It employs an integral abstraction of reality, allowing for a comprehensive representation of fuzzy relationships within the regression framework. This abstraction ensures that the proposed method retains the fundamental principles and interpretability associated with linear regression.

One key advantage of the proposed method is that it preserves a Gaussian distribution within a real-world context. Gaussian distributions are widely used in statistical analysis due to their well-understood properties and the availability of analytical techniques. By maintaining a Gaussian distribution, the proposed method facilitates calculations and enhances the interpretability of the estimated parameters. By utilizing unimodal Gaussian membership functions, the method provides a flexible and intuitive way to model the fuzziness inherent in both the predictor variable and the parameters. Unimodal Gaussian membership functions are a common choice due to their simplicity and ability to capture a wide range of fuzzy relationships.

Overall, the proposed method presents an innovative approach to address fuzziness in regression modeling. It offers a way to estimate the parameters of a fuzzy regression model, accommodating fuzzy predictors and fuzzy parameters. By leveraging the integral abstraction of reality and maintaining a Gaussian distribution, the method combines the benefits of interpretability and computational tractability, making it an attractive choice for handling fuzziness in a real-world context.

Fuzzy regression proposed in this chapter can be a better alternative to ordinary least-squares (OLS) or maximum likelihood estimation (MLE) techniques under certain conditions due to its ability to address limitations inherent in those methods.

Both OLS and MLE rely on the assumption that the relationship between variables can be modeled precisely with deterministic coefficients, along with specific probabilistic error structures (e.g., normally distributed residuals in the case of OLS). However, if the data contains ambiguous, imprecise, or vague information that cannot be adequately represented within a probabilistic framework, these methods may fail to provide accurate or interpretable results.

We have shown in this chapter that fuzzy regression is specifically designed to handle imprecision and vagueness in data by representing coefficients and residuals as fuzzy numbers or intervals. This approach allows for modeling relationships where exact numerical values may not be precisely known or where there is inherent ambiguity in the system being studied. Unlike OLS and MLE, fuzzy regression does not require stringent assumptions such as linearity, homoscedasticity (constant variance of errors), independence, or normally distributed errors. It can accommodate non-linear relationships and does not depend on probabilistic distributions, making it particularly robust in cases where statistical assumptions are difficult to meet. Additionally, fuzzy regression is effective with small datasets or incomplete information by relying on qualitative or interval-based inputs rather than precise numerical values. This makes it especially useful in fields like the social sciences, where data collection is often challenging. Another key advantage of fuzzy regression is its ability to integrate subjective and linguistic information into the analysis. By using fuzzy numbers to express imprecise variables, the model can incorporate human expertise directly into the modeling process. Fuzzy regression also excels in handling non-linear and complex systems by leveraging fuzzy rules and membership functions, providing a natural way to model systems with inherent ambiguity or uncertainty. Moreover, fuzzy regression is less sensitive to outliers because it focuses on capturing the range of possible values rather than precise point estimates, making it a robust choice for datasets with outliers or noisy data.

Appendix

Table 15.3 Unemployment and poverty rates in MENA Countries 2015

Country	Unemployment rate	Poverty rate
Algérie	0.17	0.16
Bahreïn	0.38	0.06
Égypte	0.30	0.09
Iran	0.25	0.11
Irak	0.09	0.13
Jordanie	0.09	0.20
Koweït	0.05	0.07
Liban	0.35	0.14

(continued)

Table 15.3 (continued)

Country	Unemployment rate	Poverty rate
Libye	0.25	0.16
Maroc	0.29	0.04
Mauritanie	0.04	0.16
Oman	0.39	0.07
Palestine	0.34	0.04
Qatar	0.11	0.24
Arabie Saoudite	0.10	0.24
Soudan	0.10	0.21
Syrie	0.14	0.10
Tunisie	0.22	0.15
Émirats Arabes Unis	0.19	0.18
Yémen	0.14	0.13

Source: World Bank, World Development Indicators. Available at: https://data.worldbank.org

Table 15.4 Unemployment and poverty rates in MENA countries (2020–2024)

Country	Unemployment rate	Poverty rate
Algérie	0.03	0.31
Bahreïn	0.04	0.21
Égypte	0.08	0.22
Iran	0.08	0.23
Irak	0.09	0.25
Jordanie	0.09	0.31
Koweït	0.10	0.19
Liban	0.13	0.24
Libye	0.14	0.25
Maroc	0.16	0.13
Mauritanie	0.18	0.23
Oman	0.22	0.12
Palestine	0.25	0.09
Qatar	0.25	0.27
Arabie Saoudite	0.29	0.25
Soudan	0.30	0.21
Syrie	0.34	0.09
Tunisie	0.35	0.04
Émirats Arabes Unis	0.38	0.15
Yémen	0.39	0.09

Source: World Bank, World Development Indicators. Available at: https://data.worldbank.org

References

Atkinson T, Freyssinet J, Glaude M, Seibel C (1998) Pauvreté et exclusion. La Documentation française 2-11-004046-7

Babatunde KA, Begum RA, Said FF (2017) Application of computable general equilibrium (CGE) to climate change mitigation policy: a systematic review. Renew Sustain Energy Rev 78:61–71

Belhadj (2022) Fuzzy simple linear regression using Gaussian membership functions minimization problem. J Fuzzy Ext Appl 3(4):279–289

Belhadj B, Kaabi F (2021) The relationship between employment poverty using fuzzy regression. In: Betti G, Lemmi A (eds) Analysis of socio-economic conditions: insights from a fuzzy multidimensional approach. Routledge, London and New York. https://doi.org/10.4324/9781003053712-17

Blanchard O, Johnson DR (2013) Macroeconomics, 6th edn. Pearson

Bouanani M, Belhadj B (2022) The dissemblance index method case of the employment poverty relationship in the MENA. Fuzzy Sets Syst 447:137–146

Bourbonnais R (2018) Économétrie. Dunod, Paris

Brian A, Ward B, Shanta D, Alejandro I (2001) Macroeconomic policy and poverty reduction. The International Monetary Fund and the World Bank

Chen Y, Williams M (2018) Subjective social status in transitioning China: trends and determinants. Soc Sci Q 99(1):406–422

Diamond P (1988) Fuzzy least squares. Inf Sci Int J 46(3):141–157

Diamond P, Tanaka H (1999) Fuzzy regression analysis. In: Fuzzy sets in decision analysis, operations research and statistics. Kluwer Academic Publishers, page 387

D'Urso P (2003) Linear regression analysis for fuzzy/crisp input and fuzzy/crisp output data. Comput Stat Data Anal 42(1–2):47–72

D'Urso P, Gastaldi T (2000) A least-squares approach to fuzzy linear regression analysis. Comput Stat Data Anal 34(4):427–440

Emilie R, Christiane M, Agnes M, Carles M, Edwin Ng, Farihah A, Patricia O (2018) Connections between unemployment insurance, poverty and health: a systematic review. Eur J Pub Health 28(2):269–275

Garcia I, Gomez TL (2012) Poverty and unemployment: a cultural approach. Thunderbird Int Bus Rev 54(1):79–90

Giannakis E, Bruggeman A (2020) Regional disparities in economic resilience in the European Union across the urban–rural divide. Reg Stud 54(9):1200–1213

Grosso JL, Smith TL (2012) Poverty and unemployment: a cultural approach. Thunderbird Int Bus Rev 54(1):79–90

Hong D, Yi H (2003) A note on fuzzy regression model with fuzzy input and output data for manpower forecasting. Fuzzy Sets Syst 138(2):301–305

Hong D, Hwang C, Ahn C (2004) Ridge estimation for regression models with crisp inputs and Gaussian fuzzy output. Fuzzy Sets Syst 142(2):307–319

Johnson C, Formby J, Kim KJ (2011) Economic growth and poverty: a tale of two decades. Appl Econ 43(28):4277–4288

Jones JM, Brown WT (2005) Any time is Trinidad Time! Cultural variations in the value and function of time, understanding behavior in the context of time. Psychology Press, p 19

Kaufmann A, Gupta MM (1991) Introduction to fuzzy arithmetic. International Thomson Computer Press

Kim KJ, Chen H (1997) A comparison of fuzzy and nonparametric linear regression. Comput Oper Res 44:505–519

Michael B, Markus P (2022) Unemployment and productivity growth: an empirical analysis within the augmented Solow model. Econ Model 19:105–120

Munish S, Kuljit K (2016) Fuzzy analysis and prediction of commit activity in open source software projects. IET Softw 10(5):136–146

Raab M, Seibert D, & Keller A (2006) Unbiased global illumination with participating media. In Keller A, Heinrich S, Niederreiter H (eds) Monte Carlo and Quasi-Monte Carlo Methods 2006 (pp 591–605), Springer
Shapiro AF (2005) Fuzzy regression models. Art Penn State Univ 102(2):373–383
Škrabánek P, Martínková N (2021) Algorithm 1017: fuzzyreg: an r package for fitting fuzzy regression models. ACM Trans Math Soft (TOMS) 47(3):1–18
Stiglitz JE (2012) The price of inequality: how today's divided society endangers our future. W.W. Norton & Company
Tanaka H (1987) Fuzzy data analysis by possibilistic linear models. Fuzzy Sets Syst 24:363–375
Tanaka H, Ishibuchi H (1991) Identification of possibilistic linear systems by quadratic membership functions of fuzzy parameters. Fuzzy Sets Syst 41(2):145–160
Tanaka H, Uejima S, Asai K (1982) Linear regression analysis with fuzzy model. IEEE Trans Syst Man Cybern 12(6):903–907
World Bank (2015) A measured approach to ending poverty and boosting shared prosperity. World Bank, Washington, DC
World Economic Forum (2020) Global Competitiveness Report 2020
Yang M, Liu H (2003) Fuzzy least-squares algorithms for interactive fuzzy linear regression models. Fuzzy Sets Syst 135:305–316